Ecological Studies, Vol. 198

Analysis and Synthesis

Edited by

M.M. Caldwell, Washington, USA
G. Heldmaier, Marburg, Germany
R.B. Jackson, Durham, USA
O.L. Lange, Würzburg, Germany
H.A. Mooney, Stanford, USA
E.-D. Schulze, Jena, Germany
U. Sommer, Kiel, Germany

Ecological Studies
Further volumes can be found at springer.com

Volume 182
Human Ecology: Biocultural Adaptations in Human Cummunities (2006)
H. Schutkowski

Volume 183
Growth Dynamics of Conifer Tree Rings: Images of Past and Future Environments (2006)
E.A. Vaganov, M.K. Hughes, and A.V. Shashkin

Volume 184
Reindeer Management in Northernmost Europe: Linking Practical and Scientific Knowledge in Social-Ecological Systems (2006)
B.C. Forbes, M. Bölter, L. Müller-Wille, J. Hukkinen, F. Müller, N. Gunslay, and Y. Konstantinov (Eds.)

Volume 185
Ecology and Conservation of Neotropical Montane Oak Forests (2006)
M. Kappelle (Ed.)

Volume 186
Biological Invasions in New Zealand (2006)
R.B. Allen and W.G. Lee (Eds.)

Volume 187
Managed Ecosystems and CO_2: Case Studies, Processes, and Perspectives (2006)
J. Nösberger, S.P. Long, R.J. Norby, M. Stitt, G.R. Hendrey, and H. Blum (Eds.)

Volume 188
Boreal Peatland Ecosystem (2006)
R.K. Wieder and D.H. Vitt (Eds.)

Volume 189
Ecology of Harmful Algae (2006)
E. Granéli and J.T. Turner (Eds.)

Volume 190
Wetlands and Natural Resource Management (2006)
J.T.A. Verhoeven, B. Beltman, R. Bobbink, and D.F. Whigham (Eds.)

Volume 191
Wetlands: Functioning, Biodiversity Conservation, and Restoration (2006)
R. Bobbink, B. Beltman, J.T.A. Verhoeven, and D.F. Whigham (Eds.)

Volume 192
Geological Approaches to Coral Reef Ecology (2007)
R.B. Aronson (Ed.)

Volume 193
Biological Invasions (2007)
W. Nentwig (Ed.)

Volume 194
***Clusia*: A Woody Neotropical Genus of Remarkable Plasticity and Diversity** (2007)
U. Lüttge (Ed.)

Volume 195
The Ecology of Browsing and Grazing (2008)
I.J. Gordon and H.H.T. Prins (Eds.)

Volume 196
Western North American *Juniperus* Communites: A Dynamic Vegetation Type (2008)
O. Van Auken (Ed.)

Volume 197
Ecology of Baltic Coastal Waters (2008)
U. Schiewer (Ed.)

Volume 198
Gradients in a Tropical Mountain Ecosystem of Ecuador (2008)
E. Beck, J. Bendix, I. Kottke, F. Makeschin, R. Mosandl (Eds.)

Erwin Beck • Jörg Bendix • Ingrid Kottke
Franz Makeschin • Reinhard Mosandl
Editors

Gradients in a Tropical Mountain Ecosystem of Ecuador

 Springer

Prof. em. Dr. rer. nat. Dr. h.c. Erwin Beck
Department of Plant Physiology
Bayreuth Centre for Ecology and Ecosystem
 Research, BayCEER
University of Bayreuth
95440 Bayreuth, Germany

Prof. Dr. rer. nat. Jörg Bendix
LCRS, Faculty of Geography
University of Marburg
Deutschhausstr. 10
35032 Marburg, Germany

Prof. ret. Dr. rer. nat. Ingrid Kottke
Special Botany and Mycology
 and Botanical Garden
University of Tübingen
Auf der Morgenstelle 1
72076 Tübingen, Germany

Prof. Dr. agr. Dr. rer. silv. Franz Makeschin
Institute of Soil Science
Faculty of Forest, Geo and Hydro Sciences
Dresden University of Technology
Piennerstr. 19
01737 Tharandt, Germany

Prof. Dr. rer. silv. Reinhard Mosandl
Institute of Silviculture
Technische Universität München
Am Hochanger 13
85354 Freising, Germany

Cover illustration: Change of land use from pristine forest to grassland, abandoned pastures and secondary forest as characterized by the ^{13}C signatures of plant and soil matter. The different types and intensities of land use represent one of the gradients analyzed by the research teams. Another ecological gradient presented here is the altitudinal gradient on the slopes of the Andes of southern Ecuador: see Logo of the Research Group above. TMF: Tropical Mountain Forests.

ISBN 978-3-540-73525-0   e-ISBN 978-3-540-73526-7

Ecological Studies ISSN 0070-8356

Library of Congress Control Number: 2007937095

© 2008 Springer-Verlag Berlin Heidelberg

This work is subject to copyright. All rights are reserved, whether the whole or part of the material is concerned, specifically the rights of translation, reprinting, reuse of illustrations, recitation, broadcasting, reproduction on microfilm or in any other way, and storage in data banks. Duplication of this publication or parts thereof is permitted only under the provisions of the German Copyright Law of September 9, 1965, in its current version, and permissions for use must always be obtained from Springer-Verlag. Violations are liable for prosecution under the German Copyright Law.

The use of general descriptive names, registered names, trademarks, etc. in this publication does not imply, even in the absence of a specific statement, that such names are exempt from the relevant protective laws and regulations and therefore free for general use.

Cover design: WMXDesign GmbH, Heidelberg, Germany

Printed on acid-free paper

9 8 7 6 5 4 3 2 1

springer.com

Preface

This book reports on a comprehensive study of a neotropical mountain rain forest, a type of an ecosystem that has received much less scientific attention than the rain forests of the tropical lowlands. Since the local pastoral population is crowded together in the limited accessible regions of the mountains, and because of the fragility of these ecosystems, tropical mountain forests are more endangered by human activities than most of the lowland rain forests. This holds in particular for the evergreen mountain rain forests of Ecuador, the smallest of the Andean countries which now, according to the 2006 FAO report, suffers the highest annual rate (1.7%) of deforestation in the whole of South America.

In spite of human impact the Ecuadorian Andes still represent one of the "hottest" biodiversity hotspots worldwide. There are many reasons for the outstanding biodiversity of that area, and those applying to the eastern range of the South-Ecuadorian Andes are discussed in this book: the steep altitudinal gradient over more than 1500 m, the upwind and lee effects, the extraordinary edaphic and microclimatic heterogeneity, the outstanding vegetation dynamics due to an enormous frequency of landslides, the limitation of nutrients and last but not least the eventful landscape history since the Pleistocene. This biodiversity is fostered and maintained by an incredible multitude of organismic interactions which significantly contribute to the stabilization of an otherwise fragile ecosystem.

A hotspot of biodiversity on the one hand and the highest deforestation rate on the other, this conflict is quite obvious in many valleys in the eastern Cordillera of southern Ecuador. One of these is the valley of the Rio San Francisco in the provinces of Loja and Zamora, where a widely undisturbed natural forest covers the orographically right slopes whereas on the left side the forest has been and still is – illegally – cleared by slash and burn for grazing livestock. The replacement of the natural ecosystem "tropical mountain rain forest" by a completely different anthropogenic system (pastures) within the same altitudinal range and geographic situation, separated by a horizontal distance of barely more than one kilometre provides one of the rare opportunities for a comprehensive comparison of two historically related ecosystems. Such a comparison not only helps to unravel functional interrelations of ecosystem compartments but is also extremely useful in examining the suspected loss of ecosystem services following human impact.

The autochthonous ecosystem "neotropical mountain rain forest" and its anthropogenic derivatives "tropical pastures" and "abandoned tropical pastures" have been studied in an interdisciplinary endeavour by temporarily up to 30 German/Ecuadorian research groups. The research station "Estación Científica San Francisco" (ECSF), situated above the banks of the Rio San Francisco at 1850 m a.s.l. and close to the communicating road between the two provincial capitals Loja and Zamora, was and still is the centre of the ecological studies reported here. It is owned and operated by the foundation "Nature and Culture International" (Del Mar, California) through its Ecuadorian branch "Naturaleza y Cultura Internacional" (Loja). The project started in the late 1990s with an inventory of the biotic and abiotic compartments of the mountain rain forest. From the very beginning the investigations of soils, hydrology, climate, vegetation and fauna of the area were staged along an altitudinal gradient of almost 1500 m as the major guideline. In addition, all subprojects were carried out on the same core area of about 1000 ha, the so-called Reserva Biológica San Francisco (RBSF). In good time the results of the inventories could be used to address also processual relations between specific elements and compartments of the ecosystem, subsumed under the term "functionality". Challenged by the non-sustainable practices of land use by the settlers, applied research projects were incorporated in the study programme. A gradient of land use intensity was identified which could also be considered as a gradient of human impact or disturbance: starting with a soft use of the natural forest and ending with home garden agriculture. However, due to the use of fire as an agricultural tool, the gained areas, mainly pastures, cannot be used sustainably as they are overgrown by persistent weeds like the bracken fern and amply propagating bushes. These form a new type of climax vegetation which forces abandonment of the areas but also prevents a natural recovery of the indigenous forest. This book therefore also reports on socially compatible measures of forest management and reforestation experiments with indigenous tree species at locations of different land use intensities, especially on already abandoned areas.

Ecuador is tied in a particular way to the name of Alexander von Humboldt whose fundamental description of the land, its people and especially its fauna and flora still merits our highest admiration. Even today, 205 years after Humboldt's expedition to Ecuador, the majority of its biota may still await scientific description; nevertheless, a study of the tropical ecosystem, like that presented in this book, appears as a consequent further development of Humboldt's idea of an ecological landscape portrait.

A book written by 104 authors requires a lot of endeavour and a sense of solidarity from the authors, the editors and last but not least the publisher. All of them are mutually grateful to one another, but special acknowledgement merits our assistant editor, Dr. Esther Schwarz-Weig, Mistelgau, for her fruitful suggestions and untiring efforts for the completion of the book. The authors would also like to thank the sponsor of the research, the German Research Foundation (Deutsche Forschungsgemeinschaft), the foundation Nature and Culture International for providing the facilities, in particular the famous station ECSF with the research area, and our counterparts from the Ecuadorian Universities, above all from the Universidad Técnica Particular de Loja and the Universidad Nacional de Loja. In addition to the authors, numerous colleagues and other highly esteemed persons have contributed to the achievements reported in this book, but as the space of a preface is limited, they hopefully can forgive me for not mentioning them here by name.

Bayreuth, October 2007   Erwin Beck

Contents

Part I   Introduction

1 **The Ecosystem (Reserva Biológica San Francisco)** 1
   E. Beck, F. Makeschin, F. Haubrich, M. Richter, J. Bendix,
   and C. Valerezo

2 **Mountain Rain Forests in Southern Ecuador
   as a Hotspot of Biodiversity – Limited Knowledge
   and Diverging Patterns** 15
   G. Brehm, J. Homeier, K. Fiedler, I. Kottke, J. Illig, N.M. Nöske,
   F.A. Werner, and S.-W. Breckle

3 **The People Settled Around Podocarpus National Park** 25
   P. Pohle

4 **Ecuador Suffers the Highest Deforestation Rate in South America** 37
   R. Mosandl, S. Günter, B. Stimm, and M. Weber

5 **Methodological Challenges of a Megadiverse Ecosystem** 41
   G. Brehm, K. Fiedler, C.L. Häuser, and H. Dalitz

Part II   Gradients in Ecosystem Analysis

6 **Investigating Gradients in Ecosystem Analysis** 49
   K. Fiedler and E. Beck

7 **The Investigated Gradients** 55
   E. Beck, R. Mosandl, M. Richter, and I. Kottke

Part III   The Altitudinal Gradient

Part III.1   Gradual Changes Along the Altitudinal Gradient

8 **Climate** 63
   J. Bendix, R. Rollenbeck, M. Richter, P. Fabian, and P. Emck

9	Soils Along the Altitudinal Transect and in Catchments 75
	W. Wilcke, S. Yasin, A. Schmitt, C. Valarezo, and W. Zech

10 Flora and Fungi: Composition and Function

10.1	Potential Vegetation and Floristic Composition of Andean Forests in South Ecuador, with a Focus on the RBSF 87
	J. Homeier, F.A. Werner, S.R. Gradstein, S.-W. Breckle, and M. Richter

10.2	Past Vegetation and Fire Dynamics......................... 101
	H. Niemann and H. Behling

10.3	Forest Vegetation Structure Along an Altitudinal Gradient in Southern Ecuador....................................... 113
	A. Paulsch, D. Piechowski, and K. Müller-Hohenstein

10.4	Vegetation Structures and Ecological Features of the Upper Timberline Ecotone................................ 123
	M. Richter, K.-H. Diertl, T. Peters, and R.W. Bussman

10.5	Mycorrhizal State and New and Special Features of Mycorrhizae of Trees, Ericads, Orchids, Ferns, and Liverworts 137
	I. Kottke, A. Beck, I. Haug, S. Setaro, V. Jeske, J.P. Suárez, L. Pazmiño, M. Preußing, M, Nebel, and F. Oberwinkler

11 Fauna: Composition and Function

11.1	Bird Species Distribution Along an Altitudinal Gradient in Southern Ecuador and its Functional Relationships with Vegetation Structure................................. 149
	D. Paulsch and K. Müller-Hohenstein

11.2	Seed Dispersal by Birds, Bats and Wind 157
	F. Matt, K. Almeida, A. Arguero, and C. Reudenbach

11.3	Variation of Diversity Patterns Across Moth Families Along a Tropical Altitudinal Gradient 167
	K. Fiedler, G. Brehm, N. Hilt, D. Süßenbach, and C.L. Häuser

11.4	Soil Fauna 181
	M. Maraun, J. Illig, D. Sandman, V. Krashevska, R.A. Norton, and S. Scheu

Part III.2 Processes Along and Within the Gradient

12 Water Relations 193
    W. Wilcke, S. Yasin, K. Fleischbein, R. Goller, J. Boy, J. Knuth,
    C. Valarezo, and W. Zech

13 Nutrient Status and Fluxes at the Field and
    Catchment Scale... 203
    W. Wilcke, S.Yasin, K. Fleischbein, R. Goller, J. Boy, J. Knuth,
    C. Valarezo, and W. Zech

14 Biotic Soil Activities....................................... 217
    S. Iost, F. Makeschin, M. Abiy, and F. Haubrich

15 Altitudinal Changes in Stand Structure and Biomass
    Allocation of Tropical Mountain Forests in Relation
    to Microclimate and Soil Chemistry......................... 229
    G. Moser, M. Röderstein, N. Soethe, D. Hertel, and C. Leuschner

16 Stand Structure, Transpiration Responses in Trees and
    Vines and Stand Transpiration of Different Forest Types
    Within the Mountain Rainforest............................. 243
    M. Küppers, T. Motzer, D. Schmitt, C. Ohlemacher, R. Zimmermann,
    V. Horna, B.I.L. Küppers, and T. Mette

17 Plant Growth Along the Altitudinal Gradient – Role of
    Plant Nutritional Status, Fine Root Activity, and
    Soil Properties 259
    N. Soethe, W. Wilcke, J. Homeier, J. Lehmann, and C. Engels

Part III.3 Gradient Heterogeneities

Part III.3.A Spatial Heterogeneities

18 Spatial Heterogeneity Patterns – a Comparison Between
    Gorges and Ridges in the Upper Part of an Evergreen
    Lower Montane Forest 267
    M. Oesker, H. Dalitz, S. Günter, J. Homeier, and S. Matezki

19 The Unique *Purdiaea nutans* Forest of Southern Ecuador –
    Abiotic Characteristics and Cryptogamic Diversity............. 275
    N. Mandl, M. Lehnert, S.R. Gradstein, M. Kessler, M. Abiy,
    and M. Richter

Part III.3.B Temporal Heterogeneities

20 Climate Variability 281
J. Bendix, R. Rollenbeck, P. Fabian, P. Emck, M. Richter, and E. Beck

21 Growth Dynamics of Trees in Tropical Mountain Ecosystems 291
A. Bräuning, J. Homeier, E. Cueva, E. Beck, and S. Günter

**22 Temporal Heterogeneities – Matter Deposition from
Remote Areas** 303
R. Rollenbeck, P. Fabian, and J. Bendix

Part IV Gradients of Disturbance

Part IV.1 Natural Disturbance

**23 Gap Dynamics in a Tropical Lower Montane Forest
in South Ecuador** 311
J. Homeier and S.-W. Breckle

**24 Landslides as Important Disturbance Regimes – Causes
and Regeneration** 319
R.W. Bussmann, W. Wilcke, and M. Richter

Part IV.2 Disturbance by Human Activities

Part IV.2.A Planned Disturbance as Strategy for a Sustainable Use

**25 Sustainable and Non-Sustainable Use of Natural Resources
by Indigenous and Local Communities** 331
P. Pohle and A. Gerique

**26 Natural Forest Management in Neotropical Mountain Rain
Forests – An Ecological Experiment** 347
S. Günter, O. Cabrera, M. Weber, B. Stimm, M. Zimmermann,
K. Fiedler, J. Knuth, J. Boy, W. Wilcke, S. Iost, F. Makeschin,
F. Werner, R. Gradstein, and R. Mosandl

Part IV.2.B Disturbance by Clearing the Forest

**27 Permanent Removal of the Forest: Construction of Roads
and Power Supply Lines** 361
E. Beck, K. Hartig, K. Roos, M. Preußing, and M. Nebel

28 Forest Clearing by Slash and Burn 371
E. Beck, K. Hartig, and K. Roos

Part V Gradients of Regeneration

29 **Gradients and Patterns of Soil Physical Parameters at Local, Field and Catchment Scales** 375
B. Huwe, B. Zimmermann, J. Zeilinger, M. Quizhpe, and H. Elsenbeer

30 **Visualization and Analysis of Flow Patterns and Water Flow Simulations in Disturbed and Undisturbed Tropical Soils**......... 387
C. Bogner, S. Engelhardt, J. Zeilinger, and B. Huwe

31 **Pasture Management and Natural Soil Regeneration** 397
F. Makeschin, F. Haubrich, M. Abiy, J.I. Burneo, and T. Klinger

32 **Succession Stages of Vegetation Regeneration: Secondary Tropical Mountain Forests** 409
A. Martinez, M.D. Mahecha, G. Lischeid, and E. Beck

33 **Reforestation of Abandoned Pastures: Seed Ecology of Native Species and Production of Indigenous Plant Material** 417
B. Stimm, E. Beck, S. Günter, N. Aguirre, E. Cueva, R. Mosandl, and M. Weber

34 **Reforestation of Abandoned Pastures: Silvicultural Means to Accelerate Forest Recovery and Biodiversity** 431
M. Weber, S. Günter, N. Aguirre, B. Stimm, and R. Mosandl

35 **Successional Stages of Faunal Regeneration – A Case Study on Megadiverse Moths** 443
N. Hilt and K. Fiedler

Part VI Synopsis

36 **Gradients in a Tropical Mountain Ecosystem – a Synthesis** 451
E. Beck, I. Kottke, J. Bendix, F. Makeschin, and R. Mosandl

References 465

Subject Index 511

Taxonomic Index 523

Contributors

Abiy, M.
Institute of Soil Science and Site Ecology, Dresden University of Technology, Pienner Strasse 19, 01737 Tharandt, Germany

Aguirre, N.
Department of Forest Ecology, Universidad Nacional de Loja, Ecuador

Almeida, K.
Instituto de Ciencias Biológicas, Escuela Politécnica Nacional, Ladrón de Guevara s/n, Quito, Ecuador

Arguero, A.
Instituto de Ciencias Biológicas, Escuela Politécnica Nacional, Ladrón de Guevara s/n, Quito, Ecuador

Beck, A.
Spezielle Botanik, Mykologie und Botanischer Garten, Eberhard-Karls-Universität Tübingen, Auf der Morgenstelle 1, 72076 Tübingen, Germany

Beck, E.
Department of Plant Physiology, Bayreuth Centre for Ecology and Ecosystem Research (BayCEER), University of Bayreuth, 95440 Bayreuth, Germany, e-mail: erwin.beck@uni-bayreuth.de

Behling, H.
Department of Palynology and Climate Dynamics, Albrecht-von-Haller-Institute for Plant Sciences, University of Göttingen, Untere Karspüle 2, 37073 Göttingen, Germany, e-mail: Hermann.Behling@bio.uni-goettingen.de

Bendix, J.
Laboratory for Climatology and Remote Sensing (LCRS), Faculty of Geography, University of Marburg, Deutschhausstrasse 10, 35032 Marburg, Germany, e-mail: bendix@staff.uni-marburg.de

Bogner, C.
Soil Physics Group, University of Bayreuth, 95440 Bayreuth, Germany, e-mail: christina.bogner@uni-bayreuth.de

Boy, J.
Department of Soil Geography/Soil Science, Geographic Institute, Johannes
Gutenberg University, Johann-Joachim-Becher Weg 21, 55128 Mainz, Germany

Bräuning, A.
Institute of Geography, University of Erlangen–Nuremberg, Kochstraße 4/4,
91054 Erlangen, Germany, e-mail: abraeuning@geographie.uni-erlangen.de

Breckle, S.-W.
Department of Ecology, University of Bielefeld, Wasserfuhr 24–26, 33619
Bielefeld, Germany

Brehm, G.
Institut für Spezielle Zoologie und Evolutionsbiologie mit Phyletischen Museum,
Friedrich Schiller Universität Jena, Erbertstraße 1, 07743 Jena, Germany,
e-mail: gunnar.brehm@uni-jena.de

Burneo, J.I.
Centro de Transferencia de Technología e Investigación Agroindustrial,
Universidad Técnica Particular de Loja, Loja, Ecuador

Bussmann, R.W.
Nature and Culture International, 1613 Bouldin Avenue, Austin, TX 78704, USA,
e-mail: rbussmann@natureandculture.org

Cabrera, O.
Institute of Silviculture, TU-München, Am Hochanger 13, 85354 Freising,
Germany

Cueva, E.
Department of Plant Physiology, Bayreuth Centre for Ecology and Ecosystem
Research (BayCEER), University of Bayreuth, 95440 Bayreuth, Germany and
Fundacion Naturaleza y Cultura Internacional, Av. Pío Jaramillo Alvarado y
Venezuela, Loja, Ecuador

Dalitz, H.
Institute of Botany and Botanical Gardens, University of Hohenheim (210),
Garbenstraße 30, 70599 Stuttgart, Germany

Diertl, K.-H.
Institute of Geography, FA University of Erlangen, Kochstrasse 4/4, 91054
Erlangen, Germany

Elsenbeer, H.
Institute of Geoecology, University of Potsdam, P.O. Box 601553, 14415
Potsdam, Germany

Emck, P.
Institute of Geography, University of Erlangen, Kochstrasse 4/4, 91054 Erlangen,
Germany

Engelhardt, S.
Soil Physics Group, University of Bayreuth, 95440 Bayreuth, Germany

Engels, C.
Department of Plant Nutrition and Fertilization, Humboldt University of Berlin, Albrecht Thaer Weg 4, 14195 Berlin, Germany

Fabian, P.
Department of Ecoclimate, Institute for Bioclimatology and Immission Research, TU-München, Science Center Weihenstephan, Am Hochanger 13, 85354 Freising, Germany

Fiedler, K.
Department of Population Ecology, Faculty of Life Sciences, University of Vienna, Althanstraße 14, 1090 Vienna, Austria,
e-mail: konrad.fiedler@univie.ac.at

Fleischbein, K.
GeoForschungsZentrum Potsdam (GFZ), Telegrafenberg, 14473 Potsdam, Germany

Gerique, A.
Institute of Geography, University of Erlangen–Nuremberg, Kochstrasse 4/4, 91054 Erlangen, Germany

Goller, R.
Landesamt für Umwelt, Bayrisches Landesamt für Umwelt (Referat 104), Hans-Högn-Str. 12, 95030 Hof/Saale, Germany

Gradstein, S.R.
Department of Systematic Botany, Albrecht von Haller Institute of Plant Sciences, University of Göttingen, Untere Karspüle 2, 37073 Göttingen, Germany

Günter, S.
Institute of Silviculture, TU-München, Am Hochanger 13, 85354 Freising, Germany

Hartig, K.
Department of Plant Physiology, Bayreuth Centre for Ecology and Ecosystem Research (BayCEER), University of Bayreuth, 95440 Bayreuth, Germany

Haubrich, F.
Institute of Soil Science and Site Ecology, Dresden University of Technology, Pienner Strasse 19, 01737 Tharandt, Germany

Haug, I.
Spezielle Botanik, Mykologie und Botanischer Garten, Eberhard-Karls-Universität Tübingen, Auf der Morgenstelle 1, 72076 Tübingen, Germany

Häuser, C.L.
State Museum of Natural History Stuttgart (SMNS), Rosenstein 1, 70191
Stuttgart, Germany

Hertel, D.
Plant Ecology, Albrecht von Haller Institute for Plant Sciences, University of
Göttingen, Untere Karspüle 2, 37075 Göttingen, Germany

Hilt, N.
Department of Animal Ecology I, University of Bayreuth, Spitalgasse 2, 95444
Bayreuth, Germany, e-mail: n-hilt@gmx.de

Homeier, J.
Plant Ecology, Albrecht-von-Haller Institute for Plant Science, University of
Göttingen, Untere Karspüle 2, 37073 Göttingen, Germany, e-mail: jhomeie@
gwdg.de

Horna, V.
Albrecht von Haller Institute for Plant Science, University of Göttingen, Untere
Karspüle 2, 37073 Göttingen, Germany

Huwe, B.
Soil Physics Group, University of Bayreuth, 95440 Bayreuth, Germany,
e-mail: bernd.huwe@uni-bayreuth.de

Illig, J.
Institut für Zoologie, TU Darmstadt, Schnittspahnstraße 3, 64287 Darmstadt,
Germany

Iost, S.
Institute of Soil Science and Site Ecology, Dresden University of Technology,
Pienner Strasse 19, 01737 Tharandt, Germany

Jeske, V.
Spezielle Botanik, Mykologie und Botanischer Garten, Eberhard-Karls-
Universität Tübingen, Auf der Morgenstelle 1, 72076 Tübingen, Germany

Kessler, M.
Albrecht von Haller Institute of Plant Sciences, Department of Systematic
Botany, University of Göttingen, Untere Karspüle 2, 37073 Göttingen, Germany

Klinger, T.
Institute of Soil Science, Faculty of Forest, Geo and Hydro Sciences, Dresden
University of Technology, P.O. 1117, 01735 Tharandt, Germany

Knuth, J.
Department of Soil Geography/Soil Science, Geographic Institute, Johannes
Gutenberg University, Johann-Joachim-Becher Weg 21, 55128 Mainz,
Germany

Kottke, I.
Spezielle Botanik, Mykologie und Botanischer Garten, Eberhard-Karls-Universität Tübingen, Auf der Morgenstelle 1, 72076 Tübingen, Germany,
e-mail: Ingrid.Kottke@uni-tuebingen.de

Krashevska, V.
Institut für Zoologie, TU Darmstadt, Schnittspahnstrasse 3, 64287 Darmstadt, Germany

Küppers, B.I.L.
Institute of Botany and Botanical Gardens, University of Hohenheim, Garbenstraße 30, 70599 Stuttgart, Germany

Küppers, M.
Institute of Botany and Botanical Gardens, University of Hohenheim, Garbenstraße 30, 70599 Stuttgart, Germany,
e-mail: kuppers@uni-hohenheim.de

Lehnert, M.
Albrecht von Haller Institute of Plant Sciences, Department of Systematic Botany, University of Göttingen, Untere Karspüle 2, 37073 Göttingen, Germany

Lehmann, J.
Department of Crop and Soil Sciences, Cornell University, 909 Bradfield Hall, Ithaca, NY 14853, USA

Leuschner, C.
Plant Ecology, Albrecht von Haller Institute for Plant Sciences, University of Göttingen, Untere Karspüle 2, 37075 Göttingen, Germany,
e-mail: cleusch@uni-goettingen.de

Lischeid, G.
Ecological Modelling, Bayreuth Centre for Ecology and Ecosystem Research (BayCEER), University of Bayreuth, Dr.-Hans-Frisch-Straße 1–3, 95440 Bayreuth, Germany

Mahecha, M.D.
Max Planck Institute for Biogeochemistry, Hans-Knöll-Strasse 10, 07745 Jena, Germany, e-mail: miguel.mahecha@bgc-jena.mpg.de

Makeschin, F.
Institute of Soil Science and Site Ecology, Faculty of Forest, Geo and Hydro Sciences, Dresden University of Technology, Piennerstr. 19, 01737 Tharandt, Germany, e-mail: makesch@forst.tu-dresden.de

Mandl, N.
Albrecht von Haller Institute of Plant Sciences, Department of Systematic Botany, University of Göttingen, Untere Karspüle 2, 37073 Göttingen, Germany,
e-mail: nmandl@uni-goettingen.de

Maraun, M.
Institut für Zoologie, TU Darmstadt, Schnittspahnstrasse 3, 64287 Darmstadt,
Germany, e-mail: maraun@bio.tu-darmstadt.de

Martinez, A.
Department of Plant Physiology, Bayreuth Centre for Ecology and Ecosystem
Research (BayCEER), University of Bayreuth, Universitätsstrasse 30, 95447
Bayreuth, Germany

Matezki, S.
Institute of Plant Systematics, University of Bayreuth, Universitätsstrasse 30,
95447 Bayreuth, Germany

Matt, F.
Institut of Zoology II, University of Erlangen, Staudtstrasse 5, 91058 Erlangen,
Germany; and Estación Científica San Francisco Av. Pio Jaramillo
A. y Venezuela, esq. P.O. Box: 11-01-332, Loja, Ecuador,
e-mail: fxmatt@biologie.uni-erlangen.de

Mette, T.
Bayerische Landesanstalt für Wald und Forstwirtschaft, Abteilung und
Sachgebiet: Waldbewirtschaftung - Waldbau, Am Hochanger 11 85354 Freising,
Germany

Mosandl, R.
Institute of Silviculture, TU-München, Am Hochanger 13, 85354 Freising,
Germany, e-mail: mosandl@forst.tu-muenchen.de

Moser, G.
Plant Ecology, Albrecht von Haller Institute for Plant Sciences, University of
Göttingen, Untere Karspüle 2, 37075 Göttingen, Germany

Motzer, T.
Institute of Botany and Botanical Gardens, University of Hohenheim,
Garbenstraße 30, 70599 Stuttgart, Germany; and Department of Physical
Geography, University of Mannheim, L9, 1–2, 68131 Mannheim, Germany

Müller-Hohenstein, K.
Department of Biogeography, University of Bayreuth, Universitätsstrasse 30,
95440 Bayreuth, Germany

Nebel, M.
Staatliches Museum für Naturkunde Stuttgart, Rosenstein 1, 70191 Stuttgart,
Germany

Niemann, H.
Department of Palynology and Climate Dynamics, Albrecht von Haller Institute
for Plant Sciences, University of Göttingen, Untere Karspüle 2, 37073 Göttingen,
Germany

Norton, R.A.
College of Environmental Science and Forestry, Faculty of Environmental and Forest Biology, State University of New York, 1 Forestry Drive, Syracuse, NY 13210, USA

Nöske, N.M.
Botanic Garden and Botanical Museum Berlin–Dahlem, Free University of Berlin, Königin-Luise-Strasse 6–8, 14195 Berlin, Germany

Oberwinkler, F.
Spezielle Botanik, Mykologie und Botanischer Garten, Eberhard-Karls-Universität Tübingen, Auf der Morgenstelle 1, 72076 Tübingen, Germany

Oesker, M.
Institute of Botany and Botanical Garden, University of Hohenheim (210), Garbenstraße 30, 70599 Stuttgart, Germany, e-mail: mathiasoesker@web.de

Ohlemacher, C.
Institute of Botany and Botanical Gardens, University of Hohenheim, Garbenstraße 30, 70599 Stuttgart, Germany

Paulsch, A.
Institute for Biodiversity, Drei-Kronen-Gasse 2, 93047 Regensburg, Germany, e-mail: paulsch@biodiv.de

Paulsch, D.
Gustav-Schönleber-Strasse 3, 76187 Karlsruhe, Germany, e-mail: detlevpaulsch@hotmail.com

Pazmiño, L.
Biologia Molecular, Universidad Técnica Particular de Loja, Cayetano Alto, Loja, Ecuador

Peters, T.
Institute of Geography, FA University of Erlangen, Kochstrasse 4/4, 91054 Erlangen, Germany

Piechowski, D.
Systematische Botanik und Ökologie (Biologie V), University of Ulm, Albert-Einstein-Allee 11, 89081 Ulm, Germany

Pohle, P.
Institute of Geography, University of Erlangen–Nuremberg, Kochstrasse 4/4, 91054 Erlangen, Germany, e-mail: ppohle@geographie.uni-erlangen.de

Preußing, M.
Staatliches Museum für Naturkunde Stuttgart, Rosenstein 1, 70191 Stuttgart, Germany

Quizhpe, M.
Soil Physics Group, University of Bayreuth, 95440 Bayreuth, Germany

Reudenbach, C.
Philipps Universität Marburg, Fachbereich Geographie, Deutschhaustrasse 10, 35032 Marburg, Germany, e-mail: c.reudenbach@staff.uni-marburg.de

Richter, M.
Institute of Geography, University of Erlangen, Kochstrasse 4/4, 91054 Erlangen, Germany, e-mail: mrichter@geographie.uni-erlangen.de

Röderstein, M.
Plant Ecology, Albrecht von Haller Institute for Plant Sciences, University of Göttingen, Untere Karspüle 2, 37075 Göttingen, Germany

Rollenbeck, R.
Department of Ecoclimate, Institute of Bioclimatology and Immission Research, Technical University of Munich, Germany; and Laboratory for Climatology and Remote Sensing (LCRS), Faculty of Geography, University of Marburg, Deutschhausstrasse 10, 35032 Marburg, Germany, e-mail: rollenbeck@lcrs.de

Roos, K.
Department of Plant Physiology, Bayreuth Centre for Ecology and Ecosystem Research (BayCEER), University of Bayreuth, 95440 Bayreuth, Germany

Sandmann, D.
Institut für Zoologie, TU Darmstadt, Schnittspahnstrasse 3, 64287 Darmstadt, Germany

Scheu, S.
Institut für Zoologie, TU Darmstadt, Schnittspahnstrasse 3, 64287 Darmstadt, Germany

Schmitt, A.
Institute of Soil Science and Soil Geography, BayCEER (Bayreuth Center for Ecology and Environmental Research), University of Bayreuth, 95440 Bayreuth, Germany

Schmitt, D.
Institute of Botany and Botanical Gardens, University of Hohenheim, Garbenstraße 30, 70599 Stuttgart, Germany

Setaro, S.
Spezielle Botanik, Mykologie und Botanischer Garten, Eberhard-Karls-Universität Tübingen, Auf der Morgenstelle 1, 72076 Tübingen, Germany

Soethe, N.
Department of Plant Nutrition and Fertilization, Humboldt University of Berlin, Albrecht Thaer Weg 4, 14195 Berlin, Germany,
e-mail: Nathalie.Soethe@gmx.de

Stimm, B.
Institute of Silviculture, TU-München, Am Hochanger 13, 85354 Freising, Germany, e-mail: stimm@forst.tu-muenchen.de

Suárez, J.P.
Biologia Molecular, Universidad Técnica Particular de Loja, Cayetano Alto, Loja, Ecuador

Süßenbach, D.
Federal Environmental Agency, Wörlitzer Platz 1, 06844 Dessau, Germany

Valarezo, C.
Area Agropecuaria y de Recursos Naturales Renovables, Programa de Agroforestería, Universidad Nacional de Loja, Ciudadela Universitaria Guillermo Falconí, Loja, Ecuador

Weber, M.
Institute of Silviculture, TU-München, Am Hochanger 13, 85354 Freising, Germany, e-mail: m.weber@forst.tu-muenchen.de

Werner, F.A.
Department of Systematic Botany, Albrecht von Haller Institute of Plant Sciences, University of Göttingen, Untere Karspüle 2, 37073 Göttingen, Germany

Wilcke, W.
Department of Soil Geography/Soil Science, Geographic Institute, Johannes Gutenberg University, Johann-Joachim-Becher Weg 21, 55128 Mainz, Germany, e-mail: Wolfgang.Wilcke@uni-mainz.de

Yasin, S.
Center for Irrigation, Water resources, Land and Development, Universitas Andalas, Kampus Limau Manis, Padang 25163, Indonesia

Zech, W.
Institute of Soil Science and Soil Geography, University of Bayreuth, Universitätsstrasse 30, 95440 Bayreuth, Germany

Zeilinger, J.
Soil Physics Group, Institute of Plant Physiology, University of Bayreuth, 95440 Bayreuth, Germany; and P.O. Box 11-01-890, Loja, Ecuador

Zimmermann, B.
Institute of Geoecology, University of Potsdam, 14415 Potsdam, Germany

Zimmermann, M.
Department of Animal Ecology I, University of Bayreuth, 95440 Bayreuth, Germany

Zimmermann, R.
Institute of Botany and Botanical Gardens, University of Hohenheim, Garbenstraße 30, 70599 Stuttgart, Germany

Chapter 1
The Ecosystem (Reserva Biológica San Francisco)

E. Beck(✉), F. Makeschin, F. Haubrich, M. Richter, J. Bendix, and C. Valerezo

1.1 The Research Area in Southern Ecuador

The investigated area, termed the "Reserva Biologíca San Francisco" (RBSF) is located in the Cordillera Real, an eastern range of the South Ecuadorian Andes (Fig. 1.1), which is the weather divide between the humid Amazon ("Oriente") and the dry Inter-Andean region. In southern Ecuador the Andes are not as high as in the central and northern part of the country, but the topography of that area, called the Huancabamba depression, is more complicated. Here, a mountain junction culminates in the "Nudo de Loja" at 3800 m a.s.l., from which the Inter-Andean Sierras stretch towards SW, S and SE, all interrupted and dissected by valleys and basins. In spite of its dominance of the south Ecuadorian Andes the Cordillera Real only partly forms the watershed between the Pacific and the Atlantic Ocean. The map (Fig. 1.1) shows the looping course of the Rio Zamora, whose springs are south of Loja on the western slope of that mountain range. But north of Loja it turns eastwards and finally joins the Rio Marañon, a tributary of the Amazon. The valley of the Rio San Francisco, which is the main research area of the study presented here, belongs to the eastern escarpment of the Cordillera Real. Though still accompanied by pre-cordillera ranges, it is in principle exposed towards the Amazon basin.

The core area of the ecosystem study, the RBSF (3 ° 58' 30" S and 79 ° 4' 25" W, about 11.2 km^2), is located halfway between the province capitals Loja and Zamora in the deeply incised valley of the eponymous river (Fig. 1.2). It extends from 1800 m to 3160 m a.s.l. The logistic centre is the research station "Estacion Científica San Francisco" (ECSF), situated at the bottom of the valley at 1850 m a.s.l. and close to the communication road between Loja and Zamora. The RBSF comprises two manifestations of the ecosystem "tropical mountain rain forest", mostly undisturbed natural forest covering the NNW-facing orographically right-hand slopes of the valley and its anthropogenic replacement ecosystems on the opposite side of the valley, where the forest has been cleared by slash and burn. Pastures which are still in use or have already been abandoned extend almost up to the crest of the mountains. One charm of the study is therefore the direct comparison of the two contrasting phenotypes of the same ecosystem. For a better understanding of the RBSF, a short introduction to nature and landuse in southern Ecuador is presented in the following.

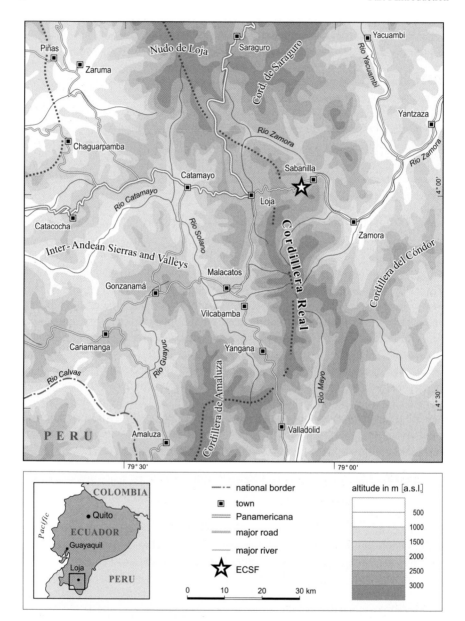

Fig. 1.1 Topographic map of southern Ecuador and location of the Research Station "Estacion Científica San Francisco" (ECSF)

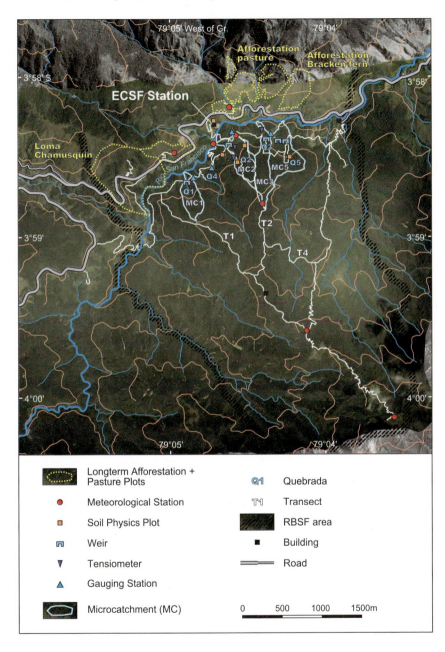

Fig. 1.2 Research area and location of the permanent plots of the study at the Reserva Biologíca San Francisco (RBSF)

1.2 Geology

Ecuador comprises three geomorphological and geological regions: the coastal plain (la Costa) in the west, the Amazon lowlands (el Oriente) including the Sub-Andean foothill zone (e.g. Cordillera del Condor) in the east and the Andes (la Sierra) in the center. The coastal plain consists of tertiary (Eocene) and quaternary sediments (Baldock 1982) whereas the Oriente is composed of a peri-cratonic foreland and a back-arc sedimentary basin, in which marine paleozoic, mesozoic and cenozoic sediments were deposited on the margins of the stable Guyana shield. The Sub-Andean zone is structurally linked to the Andes and comprises folded and slightly metamorphic mesozoic sediments like black slates, calcareous phyllites and quartzites, covered by tertiary sediments of conglomerates, shales and sandstones. Three plutonites mould parts of the western Sub-Andean zone along the major fault, which forms the tectonic margin of the Cordillera Real. One of them is the Zamora batholith, into which a part of the research area extends, consisting of leuco-granidiorites and hornblende granodiorites.

The northern and central Andean regions (la Sierra) comprise the Cordillera Occidental (Western Cordillera), the Inter-Andean basin and the Cordillera Real (Eastern Cordillera). In the West, cretaceous and eocene andesitic volcanics are overlaid by younger marine sediments and volcaniclastics. Neogene to quaternary volcanic deposits form some of the major strato-volcanoes. The Inter-Andean basin is filled with quaternary sediments and pyroclastic deposits. The eastern Cordillera consists mainly of paleozoic metamorphic rocks probably formed during a Caledonian orogenic event (Baldock, 1982).

According to Litherland et al. (1994) lithologies present in the region include a thick sequence of Paleozoic semipelites (Chiguinda unit), pelitic schists and paragneisses (Agoyan unit), amphibolites (Monte Olivo) and ortho- and paragneisses (Tres Lagunas granites, a Sabanilla unit) overlaid by mesozic metavolcanics, quartzites, slates, pelitic schists and marbles. The research area is located in the Chiguinda unit with contacts to the Zamora batholith (Fig. 1.3). The dominating metasiltstones, sandstones and quartzites of this unit are interspersed with layers of phyllite and clay schist which in concordance with the mountain range strike N–S. This complex (comprising 30 km) consists mainly of products of low-grade metamorphism which border the highly metamorphous and (again 30 km wide) Sabanilla unit in the East of the RBSF (M. Chiaradia, personal communication).

1.3 Geomorphology

The famous Ecuadorian mountains, such as the volcanoes Chimborazo, Cotopaxi or Antisana, with altitudes up to and above 6000 m a.s.l. are concentrated in north and central Ecuador. In contrast, volcanoes are absent in the Huancabamba depression, and the crest of the Ecuadorian Cordillera Real does not exceed 2800–3400 m a.s.l. Except for the breach of the Rio Zamora (about 2000 m a.s.l.) the lowest region of

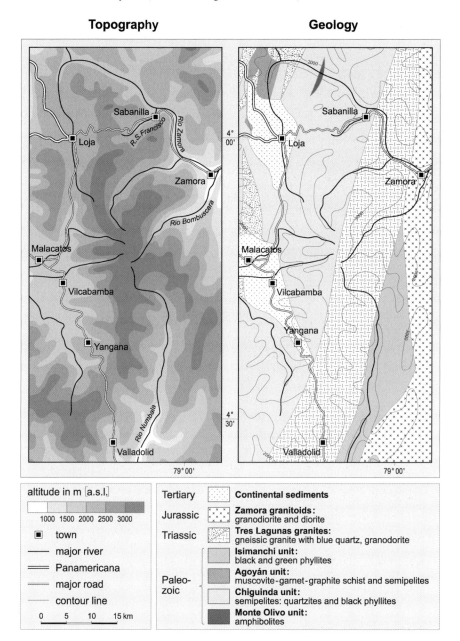

Fig. 1.3 Topography and geology of the Cordillera Real (after Litherland et al. 1994)

the Cordillera Real is the "El Tiro" pass (2750 m a.s.l.) east of Loja. Nevertheless, even here features of Pleistocene glaciation are obvious, such as head walls behind cirques containing tarns (e.g. "Lagunas de los Compadres"), smoothened bedrock with truncated spurs, and lateral and terminal moraines forming a typical nunatak landscape. The uppermost part of the Cordillera Real, although subjected to extremely wet conditions exhibits almost no indications of recent geomorphologic processes, due to the shelter by the dense páramo vegetation. The geomorphologic structure of the study area itself is quite complex. While quartzite dikes form narrow ridges declining from S to N and from SSE to NNW, respectively, the deeply incised V-shaped valleys indicate the occurrence of schist and phyllites (Sauer 1971). Due to the rugged terrain, the slopes face in all directions, with highest percentages stated for N (28%), NW (18%), SW (15%) and NE (12%). GIS-based analyses of inclination distribution prove a share of only 2% of slopes with angles less than 10 °, while the majority (57%) has angles between 25 ° and 40 °, and those steeper than 40 ° still cover 19% of the area. The steepness of the forest-covered slopes combined with the perhumid climate of the research area promotes an extraordinary frequency of landslides, partially associated with mudflow activities (Hagedorn 2002). Especially where the slopes have been scratched by human activities, e.g. by road construction, processes of fast mass movement such as rock-, earth- and landslides are dramatically enforced.

East and west of the rugged Cordillera Real the valleys are less steep and interspersed by wide basins filled with alluvial deposits. In the humid pre-cordillera east of the study area erosion and denudation processes are similar to those of the core region. In the semi-humid to semi-arid western zone torrential downpours result in gullies and debris transport, with the danger of riverbed changes by overflow and sedimentation of the adjoining floodplain. The sediments derive from sheet erosion from the slopes which has been accelerated by cultivation for hundreds of years.

1.4 Soils

The soils vary considerably between the warm and humid eastern side, the cold perhumid crest area and the warm semi-humid to a semi-arid area in the rain-shaded West of the Cordillera Real (Valarezo 1996).

Soils around the drier basins in the West often exhibit an enrichment of clay and are saturated with exchangeable cations: albic luvisols and eutric cambisols frequently occur up to 2200 m a.s.l. With increasing precipitation between 2000 m and 3200 m a.s.l. dystric cambisols, dystric planosols and gleysol gain importance (Schrumpf et al. 2001). Histosols and umbric regosols are widespread above 3000 m characterized by a high content of plant fibers, especially in the Páramo belt. Soil reaction ranges from pH 5.5 in the valleys to pH 3.0 in the Páramo.

On the humid eastern side of the Cordillera Real humic alfisols, humic acrisols and dystric leptosols are frequent between 1000 m and 2000 m a.s.l among the clay soils. Between 1500 m and 2800 m terric histosols prevail as water saturation lasts for several months (i.e. May to August at RBSF). Landslide areas within that region are characterized by umbric regosols and dystric cambisols. The latter are typical for the mountain forests between 1800 m and 2800 m a.s.l., but higher up dystric regosols are more abundant. Due to the high rainwater input most of the soils above

1800 m a.s.l show podzolic features like bleaching of the A horizon. Many of the soil profiles show reddish or black placic subsurface horizons indicating redoximorphic processes.

The stone contents of the A horizons vary considerably between more than 80% on fresh landslides and less than 10% in undisturbed mature forests with mighty organic layers. The turnover times of the organic material under closed forest canopies range between less than 8 years in the lower Zamora area and up to more than 15 years at higher altitudes. The generally low ECEC values decrease with elevation. However, the proportion of exchangeable cations increases as the kaolinite:illite ratio decreases with altitude. Plant-available N decreases with altitude concomitantly with increasing C/N values, low nutrient content of the litter, unfavourable thermal conditions and temporary water logging. Reaction of the mineral is strongly acidic (pH 4.5) and even decreases (pH 3) with altitude from the mountain forest to the Páramo. The soils, due to the diversity of the parent material and the altitudinally varying climatic and hydrologic factors display a high degree of small-scale heterogeneity (Wilcke et al. 2002) which is mirrored by the vegetation.

1.5 Climate

The climate of Ecuador is dominated by the tropical trade wind regime which is well established in the mid- and higher troposphere, with strong easterlies all over the year. The surface wind field is locally and regionally modified by the complex topography of the Andes and the thermal land-sea contrast at the Pacific coast. As outlined in Chapter 8, the south-eastern part of the country encompassing the RBSF area is also mainly influenced by easterlies, but westerlies occasionally occur especially in austral summer.

The major factors governing air temperature in Ecuador are topography, terrain altitude and ocean temperatures off the coast. The altitudinal gradient starts with the "tierra caliente" (hot land = annual average between 25 °C and 19 °C) below 1100 m, followed by the "tierra templada" (temperate land) between 1100–2200 m (19-13 °C), the "tierra fria" (cold land, 13-6 °C) up to 3800 m, the "tierra helada" (frost land, 6-0 °C) up to 4800 m and finally the "tierra nevada" at altitudes >4800 m a.s.l. The RBSF belongs to the tierra templada but extends into the tierra fria in which the thermal regime of the eastern escarpment differs significantly from the area west of the Cordillera oriental causing an asymmetric distribution of the "Tierras" (Richter 2003, see Fig. 1.4).

Spatial patterns of cloud frequency in Ecuador are illustrated in Fig. 1.5 and reveal two nearly contour-parallel bands of maximum cloud frequency (up to 85%) along the western and eastern Andean escarpments. The region of high cloudiness of the western band abruptly ends in the south Ecuadorian Huancabamba depression where the western Cordillera of the Andes attenuates and the inner Andean basin broadens. In that region the south Pacific anticyclone, in combination with divergent coast-parallel winds from the South, suppresses convection (Lettau 1976). On the

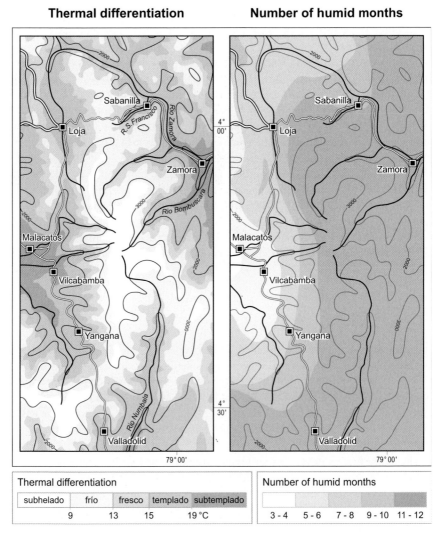

Fig. 1.4 Mean annual temperature and number of humid months for southern Ecuador based on epiphyte distribution as indicator for hygric conditions (Richter 2003), soil temperature and data from 29 meteorological stations

eastern slopes, cloudiness increases approaching the foot-hills of the Cordillera Real, with a maximum above >1800 m a.s.l. This zone is wider and more clearly established in the southern provinces of Ecuador and thus, in the RBSF area.

A detailed analysis of the spatio-temporal distribution of rainfall in Ecuador is presented in Bendix and Lauer (1992). It could be shown that rainfall formation is especially heavy at the eastern Andean slopes where the easterlies are lifted by forced convection, leading to intense condensation. Conspicuously, the zone of intense

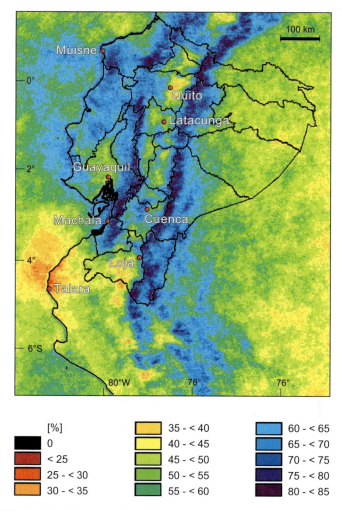

Fig. 1.5 Relative cloud frequency for Ecuador (2002–2003) and adjacent areas derived from NOAA-AVHRR data (from Bendix et al. 2004)

precipitation coincides quite well with the line of high cloudiness (Fig. 1.5) and is well known as the Andes-occurring System (AOS; refer e.g. to Bendix et al. 2006b). In contrast to the eastern slopes, the inter-Andean basins receive generally less than 1000 mm rain per year, but show a distinctive patchy structure of precipitation fields. Particularly dry "islands" are orographically isolated basins sometimes with annual precipitation below 500 mm as e.g. the Catamayo Valley in southern Ecuador.

The small-scale structure of rainfall (and cloudiness) can lead to marked climatic gradients over a short distance, which holds especially true for southern Ecuador in the vicinity of the RBSF area: The horizontal distance between the driest and the wettest point is less than 30 km (383 mm year^{-1} vs >6000 mm year^{-1}, from Catamayo to the Cordillera Real). As a consequence, the "coldest" and "hottest" spots are also found close together and the number of humid months changes from 12 per year east of the crest of the Cordillera oriental to less than four in the area of Malacatos (Fig. 1.4). A detailed analysis on the climate of the RBSF area and the adjoining region is presented in Chapters 8 and 20.

1.6 Biogeography of Southern Ecuador and the RBSF

The vegetation of Ecuador generally follows the three major landscape complexes, the drier "Costa" with semi-deciduous and deciduous forests and savannas, the evergreen Amazon rain forest in the "Oriente" and the vegetation of the Andes. Due to the altitudinal gradient and the varying climatic situations, the Andes comprise alone eight of the total 15 vegetation types recorded for Ecuador (Harling 1986; Patzelt 1996). The perhumid montane broad-leaved forest and the upper montane or Elfin forests (Ceja Andina) are well developed on the eastern escarpment of the Andes and in the northern part of the western Cordillera, but attenuate towards the drier South. The higher parts of the inter-Andean basins show the dichotomy of a grass-rich Parámo in the northern and central parts replaced by a shrubby Parámo type (Yalca) in the South. Two types of treelines have been differentiated as a consequence of the Huancabamba depression. While in the very north and south of that depression the treeline climbs to 4000 m a.s.l. and even higher, the south Ecuadorian treeline is located between 2800 m and 3300 m a.s.l. The upper treeline (around 4000 m) is formed by only two woody genera, namely *Polylepis* and *Gynoxis*, whose life-forms change with increasing altitude from single stemmed 10–15 m trees to multi-stemmed shrubs. In contrast, the lower tree line in southern Ecuador exhibits an extraordinary species diversity, indicating a kind of a timberline ecotone which is affected not by temperature or dryness but by the permanent strong wind and exceptional high moisture causing water-logging of the soils. In agreement with this interpretation, the two genera forming the higher tree line do not contribute to the species pattern of the lower tree line. The natural vegetation of the RBSF, as part of the Cordillera Real, is the evergreen lower and upper mountain rain forest. As there is no clear borderline between these two types, the latter may also be addressed as or at least merges with the "Ceja Andina".

Bordering the Podocarpus National Park in the Cordillera Real the research area contributes substantially to the biodiversity hotspot in Ecuador (Barthlott et al. 2005), for which a total of 16 000–20 000 species of vascular plants has been reported (Gentry 1977; Jørgensen and Ulloa Ulloa 1994; Jørgensen and Leon-Yanez 1999). The hotspot character is especially true for the divergence zone of the ranges south of the "Nudo de Loja" with its highly complex structures from xeric to hygric vegetation types within a short distance, where a rapid genetic interchange is guaranteed (Gentry and Dodson 1987). A general trait of the area is its extraordinary high endemism which is typical for southern Ecuador (Quizhpe et al. 2002).

A map of the potential natural vegetation of the research area, derived from hygric and thermic parameters shows the distribution of the seven principal plant formations mentioned above for the Andes of Ecuador (Fig. 1.6). There are many more in reality because not only climatic differentiation but also vegetation history and multiple disturbance regimes cause the extraordinary plant diversity.

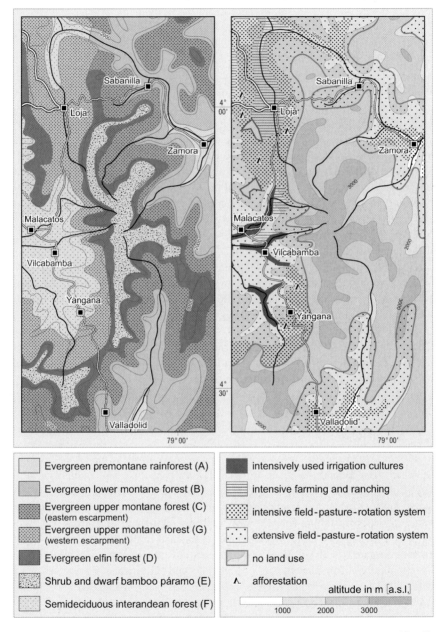

Fig. 1.6 Potential natural vegetation of the Cordillera Real (*left*) and land-use pattern with relics of the natural vegetation (*right*)

Like most parts of the Cordillera Real, the deeply incised valley of the Rio San Francisco is covered by an evergreen lower (<2150 m a.s.l.) and upper broad-leaved mountain rain forest which extends up to the tree line between 2700 m and 3000 m a.s.l. Higher up, a sub-páramo shrubland ("Yalca", "Páramo arbustivo") comes across. Classification of the mountain forest, due to its heterogeneity and the lack of a clear altitudinal zonation is difficult. According to Balslev and Øllgard (2002) the entire forest may be addressed as "bosque siempreverde montano" which is characterized by a richness of vascular and cryptogamic epiphytes. In contrast to the fuzzy altitudinal zonation, clear azonal formations can be differentiated: (a) the dense forests of the ravines and (b) the more open woodlands of the crests, which are especially rich in palms (Homeier et al. 2002). Another azonal peculiarity of the upper montane forest of the San Francisco valley is the *Purdiaea nutans* forest (2150–2600 m a.s.l., see Chapter 19). Despite the fuzzyness of the individual forest units, five types were differentiated in the RBSF on a floristic basis (Homeier 2004; Bussmann 2001a; Müller-Hohenstein et al. 2004), two in the ravines and three on the crests. In contrast to the rain forest of the lowlands, the canopy of the mountain forests of the San Francisco valley rarely exceeds 20 m in height; bigger trees are rare and bound to ravines (Müller-Hohenstein et al. 2004).

While the flora of the area has been described at least to some extent, knowledge of the fauna is still very incomplete. In addition to moths and some groups of soil microarthropods, bats and birds are rather well known. All these groups are high in species diversity. In Ecuador, more than 1600 bird species have been registered (Ridgely and Greenfield 2001; Suárez 2002). The Podocarpus National Park is one of the most important bird areas of the country with more than 560 bird species recorded (Rasmussen and Rahbek 1994). As it overlaps with the research area, bird diversity in the RBSF is also very high (see Chapter 11.1). The most famous mammals of the area are agoutis, the Andean tapir and the Andean bear, which as herbivores may considerably damage the vegetation.

1.7 Land Use

Land use types depend mainly on the local climate (Fig. 1.6): The arable land in the valley bottoms east of the Cordillera Real allows two harvests per year in a field rotation system. In the drier western basins only one harvest of field crops is common, combined with a fallow break lasting several years. Orchards with small plantations of banana, coffee and a variety of other fruit trees are widespread in the humid Zamora Valley, whereas irrigated fields of sugar cane characterize the nutrient-rich alluvial soils ("vegas") in the much drier vicinity of Malacatos and Vilcabamba. Pasture systems in all humid parts provide fodder for reasonable herds of cattle which contrasts to limited grazing facilities for sheep and goats in the drier areas west of the cordillera.

Several concepts of land use became apparent from different types of homegardens, animal-husbandry systems and pasture management carried out by the various ethnic groups of the research area, like the Saraguros, the Shuar and the newcomers ("Colonos"; Pohle 2004; see Chapter 3). While the north-facing slopes of this valley are still covered by forest, the forests on the south-facing slopes were largely converted to pasture land some 12–30 years ago (Werner et al. 2005). As the Colonos use fire for clearing the forest and also for maintaining their pastures (Paulsch et al. 2001), fire-adapted weeds invade the areas which, except geomorphologic flattenings and troughs with compacted and waterlogged soils, can hardly be used for more than 10 years (Hartig and Beck 2003). Attempts to reforest the abandoned areas have been made using *Pinus patula* or various species of *Eucalyptus*. Although these exotic species initially grow reasonably well they later suffer considerably, mainly from mineral deficiency (Brummitt and Lughada 2003). Therefore several attempts have been made to use indigenous tree species for afforestation (Weber et al. 2004; see Chapters 33 and 34.

Chapter 2
Mountain Rain Forests in Southern Ecuador as a Hotspot of Biodiversity – Limited Knowledge and Diverging Patterns

G. Brehm(✉), J. Homeier, K. Fiedler, I. Kottke, J. Illig, N.M. Nöske, F.A. Werner, and S.-W. Breckle

2.1 Introduction: Why Do We Need Biodiversity Inventories?

Highly complex ecosystems such as the tropical mountain rain forest in southern Ecuador probably harbor tens of thousands of species that interact with each other. It is impossible to understand an ecosystem without knowing the composition of its community. Such knowledge cannot be achieved without the examination of all major groups of animals, fungi, plants, and bacteria. For example, insects such as leaf beetles, ants, or hymenopteran and dipteran parasitoids have a high impact on forest ecosystems (Moutino et al. 2005; Soler et al. 2005), but have not been studied at the RBSF so far. The question of how many species there are on earth is still unresolved. Estimates range from four to 30 million species (e.g. Novotny et al. 2002). Ultimately, only counting and naming species can answer this question.

A hot debate about the methodological approaches, i.e. the usefulness and efficiency of 'traditional' taxonomy versus DNA barcoding approaches is still going on (e.g. Meyer and Paulay 2005; Will et al. 2005). Barcoding techniques provide useful sets of new characters for species descriptions and phylogenies, especially in cryptic species or taxa otherwise poor in morphological characters (e.g. Hajibabaei et al. 2006). However, species definition based on DNA sequence data can be problematic. For example, ribosomal genes can show intraspecific variation in the multinucleate Glomeromycota. The sequence types obtained from the vegetative stage can rarely be related precisely to either morphological or biological species (Sanders 2004). In our opinion, the only strategic way forward can reside on a synthesis of 'classic' and 'modern' approaches with regard to sampling campaigns, application of up-to-date information technology (Godfray 2005), and thorough taxonomic and systematic work. This might include in vitro cultivation, in the case of fungi and bacteria.

Traditional descriptive disciplines such as taxonomy, systematics and natural history provide names, phylogenies, and life history data as a service for other fields of science and their applications. However, these disciplines have suffered a great loss of capacity in past decades (e.g. Breckle 2002; Gotelli 2004). Many species identifications from the RBSF are not the primary result of research efforts in taxonomy and systematics. Rather, the recently published species checklists (Liede-Schumann and Breckle, 2008) are essentially a by-product of ecological and physiological research. This explains why many lists are still far from being complete (see below) and why only a few species have been newly described from the area thus far. Many (probably many thousands) of species in the RBSF alone are new to science and await description, but specialists are lacking for many taxa. Only some vertebrate groups are well known: All currently recorded bird and bat species are described (Matt 2001; Muchhala et al. 2005; Paulsch 2008). In other groups, however, the proportion of described species drops: from bryophytes (98%; Gradstein et al. 2008), trees (>5 cm dbh, 90%), lichens (85%; Nöske et al. 2008), to geometrid moths (63%; Brehm et al. 2005), oribatid mites (60%; Illig et al. 2008; Niedbala and Illig 2007) to fungi (5%; Haug et al. 2004; Setaro et al. 2006a;, Suarez et al. 2006; Kottke et al. 2007).

2.2 Inventory Coverage in the RBSF

After eight years of research in southern Ecuador, our knowledge about certain taxa is excellent. The area is one of the most thoroughly investigated neotropical montane forests (see the contributions to this volume). Geometrid and pyraloid moths, and oribatid mites or bush crickets (Tettigoniidae) have never been studied quantitatively in other Andean forests before (Süßenbach 2003; Brehm et al. 2005; Braun 2008; Illig et al. 2008). None of the mycorrhizal fungi has previously been known from the northern Andean forests (Haug et al. 2004, 2005; Setaro et al. 2006a; Suarez et al. 2006; Kottke et al. 2007). However, there is little reason to rest on the laurels. Our knowledge about species richness in the area remains biased and incomplete (Fig. 2.1).

The Bacteria or Archaea have not been investigated in the study area. So far, the only group of single-cell organisms studied are the Testacea (Krashevska 2008; Table 2.1). Due to their dominant role in material and energy flows in ecosystems, plants have received relatively much attention, and groups such as vascular and non-vascular cryptogams have been inventoried across the whole elevational range (see Chapter 10.1). However, botanical collections have focused on the existing trail system that is biased towards ridge sites. Bryophytes and ferns have been collected systematically across the whole altitudinal range (e.g. Kürschner and Parolly 2004a; see Chapter 19; Table 2.1).

Lichens have been sampled intensively at least in the lowermost parts (up to 2100 m) of the area (e.g. Nöske 2005). Moreover, life forms such as trees (Homeier 2004), climbers (S. Matezki, personal communication) and epiphytes (Werner et al. 2005) have been treated in extensive ecological studies, whereas terrestrial shrubs and especially herbs still remain poorly known. Concerning the large group of fungi, only mycorrhiza forming groups were investigated. The molecular diversity of the Glomeromycota forming mycorrhizas with trees and two basal groups of Basidiomycota (Sebacinales, Tulasnellales) forming mycorrhizas with orchids, ericads,

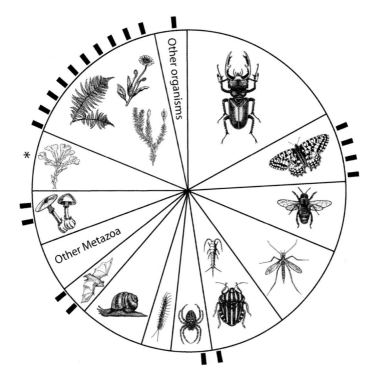

Fig. 2.1 Proportions of described species of the major taxa at a global scale and coverage of inventories at RBSF (see Table 2.1). Taxa (clockwise): Coleoptera, Lepidoptera, Hymenoptera, Diptera, other Insecta, Chelicerata, other Arthropoda, Mollusca, Vertebrata, other Metazoa, Fungi, Stramenopilata + Haptophyta, Embryophyta, other organisms. More than two-thirds of all known species are animals, and more than half of all organisms are arthropods – likely more. The figure covers organisms from all habitats, including marine ecosytems. However, only the brown algae (marked with an *asterisk*) are not expected to be present at the RBSF, whereas all other taxa are. Groups not sorted and identified at species level (e.g. Nematoda) are not included. Only a small proportion of the expected total richness has been covered so far. Data source: Lecointre and Le Guyader (2006). Illustration by G. Brehm

and liverworts, respectively, were studied but are far from being complete (Haug et al. 2004, 2005; Setaro et al. 2006a; Suarez et al. 2006; Kottke et al. 2007; Table 2.1). Saprophytic and plant parasitic fungi remain to be investigated.

Table 2.1 provides an overview of available species lists. Inventories have been carried out on birds, bats, and parts of the arthropod clades Lepidoptera, Orthoptera, and Arachnida. Although Fig. 2.1 gives global numbers of organisms and does not provide the (unavailable) proportions of taxa in an Andean forest, it roughly estimates a reasonable scenario. Thus far, the largest gaps are represented by three of the major insect orders: The Coleoptera (beetles), the Hymenoptera (ants, wasps, bees, etc.), and the Diptera (flies). With the exception of a few selected families of arthropods (see above), no other insect group has been studied (e.g. dragonflies, homopterans). Molluscs (e.g. land snails) as well as aquatic communities as a whole have been

Table 2.1 Taxa investigated in the RBSF and adjacent areas by October 2007

Taxon	Observed species number	References
Chiroptera (bats)	21 (RBSF)	Matt (2001)
	24 (1000–2900 m)	
Aves (birds)	227 (RBSF)	Paulsch (2008), Rasmussen et al. (1994): old road Zamora–Loja
	379 (1000–2800 m)	
Geometridae (geometrid moths)	1075 (RBSF)	Brehm et al. (2005), Fiedler et al. (2007)
	1266 (1040–2677 m)	
Arctiidae (arctiid moths)	287 (RBSF, 1800–2000 m)	Hilt (2005), Fiedler et al. (2008)
	446 (1040–2677 m)	
Pyraloidea (pyraloid moths)	748 (1040–2677 m)	Süßenbach (2003), Fiedler et al. (2008)
Sphingidae (hawkmoths)	36	Fiedler et al. (2008)
Papilionoidea (butterflies)	243	Häuser et al. (2008)
Tettigoniidae (bush crickets)	101 (1000–3100 m)	Braun (2002, 2008)
Oribatida (mites)	154 (RBSF)	Illig et al. (2008)
	192 (1050–3000 m)	
Testacea	78 (RBSF)	Krashevska (2008)
	110 (1050–3000 m)	
Lichens (lichens)	323 (RBSF)	Nöske et al. (2008)
Glomeromycota	83 (RBSF and afforestation areas, 18S sequences, only four known species)	Haug et al. 2004, Kottke et al. (2007)
Ascomycota	4 (RBSF, sequences, new to science)	Haug et al. 2004
Basidiomycota (Homobasidiomycetes, Heterobasidiomycetes)	90 + 6 (RBSF, sequences, new to science)	Haug et al. 2005, Kottke et al. (2007), Setaro et al. (2006a), Suarez et al. (2006)
Bryophyta (hornworts, liverworts and mosses)	515 (RBSF)	Gradstein et al. (2008), Kürschner and Parolly (2008)
Spermatophyta (seed plants)	1208 (RBSF)	Homeier and Werner (2008)
Pteridophyta (ferns)	257 (RBSF)	Lehnert et al. (2008)

ignored. Even prominent vertebrate groups such as amphibians or mammals (except for bats; Matt 2001; see Chapter 11.2) have not been studied. A sample of studied and unexplored taxa is provided in Fig. 2.2.

A complete inventory of tropical rain forests such as the RBSF can probably never be achieved, since species-rich tropical communities must be sampled intensively and over very long periods of time. An example is provided by geometrid moths: So far, 35 238 individuals representing 1223 species have been sampled during quantitative assessments at elevations of 1040–2670 m between 1999 and 2003. However, richness estimators (Colwell 2006) indicate that, despite a great sampling effort, ca. 200 more species must be expected in the area

(Brehm et al. 2005). The species checklists of less well sampled taxa therefore often provide observed minimum numbers that do not yet allow serious estimates of the actual richness. Particularly poorly sampled areas outside the RBSF such as the Rio Bombuscaro valley (the lowermost part of Podocarpus National Park, PNP) certainly harbor far more species than the currently available lists of species suggest (e.g. Chapters 11.2, 11.3).

The use of "biodiversity indicators" in "rapid biodiversity assessments" appears to be a tempting shortcut. A good indicator group is supposed to reflect the 'complete' biodiversity. However, the use of indicators is highly problematic along altitudinal gradients because the composition of taxa changes non-linearly throughout the gradient, and patterns of alpha diversity are often discordant, even between closely related groups of organisms (Brehm and Fiedler 2003; see Chapter 11.3 in this volume; Fig. 2.3). Moreover, we have to expect diverging richness patterns in different taxa along the gradient (see Chapter 11.3), and too little is still known about the diversity patterns of most groups to allow general conclusions.

2.3 The RBSF and Podocarpus National Park as Biodiversity Hotspots

It is evident that the tropical Andes are a hotspot of biodiversity (e.g. Rahbek et al. 1995; Myers et al. 2000; Brehm et al. 2005). Using data on vertebrates and vascular plants, Brummitt and Lughadha (2003) ranked the region as the top global biodiversity hotspot. Regarding many groups of organisms, the Neotropical region is more speciose than any other region of the world. However, data from the Andes are still scarce. Two large-scale diversity patterns overlap in tropical mountains in general and in the Andes in particular:

1. Species richness of most groups of organisms peaks around the equator and declines towards the poles (Gaston 2000).
2. There is a high species turnover along altitudinal gradients and usually a peak of richness not at lowest, but at medium elevations (Herzog et al. 2005; Krömer et al. 2005; Rahbek 2005).

Many hypotheses have been formulated to explain these patterns. Some of the most frequently used explanations are:

- Evolution and biogeography;
- Climate history;
- Biotic and abiotic factors;
- Stochastic effects.

Each of the concepts is plausible to some extent, and it seems most probable that a combination is actually responsible for the observed richness patterns with all their group-specific variations (Brehm et al. 2007).

Fig. 2.2 Examples of nine (mostly unidentified) species belonging to taxa that have not yet been investigated or inventoried in the RBSF (**a–i**), and nine taxa in which inventories have been carried out or started (**j–r**). **a** Beetle (Chrysomelidae), **b** wasp (Pompilidae), **c** fly (Tachinidae), **d** *Olceclostera* sp. (Apatelodidae), **e** stick insect (Phasmatodea), **f** snail (Orthalicidae), **g** *Gastrotheca* cf. *testudinea* (Hylidae), **h** Dusky Lancehead *Bothriopsis pulchra* (Viperidae), **i** Two-toed sloth *Choloepus hoffmanni* (Megalonichidae), **j** *Pantherodes colubraria* (Geometridae), **k** *Dolicheremaeus* sp. (Oribatida), **l** *Itarissa* sp. (Tettigoniidae), **m** Cinnamon Flycatcher *Pyrrhomyias cinnamomea* (Tyrannidae), **n** *Sturnira ludovici* (Chiroptera), **o** unknown Gomeromycota, **p** *Frullanoides densifolia* (Marchantiophyta: Lejeuneaceae), **q** *Purdiaea nutans* (Cyrillaceae), **r** *Fernandezia subbiflora* (Orchidaceae). Images by: G. Brehm (**a–e, j, l, m**), J. Homeier (**f–h, n, q, r**), J. Illig (**i, k**), A. Beck (**o**), and N. M. Nöske (**p**)

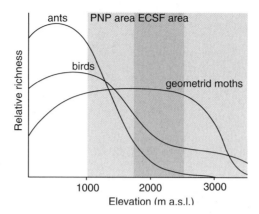

Fig. 2.3 Hypothetical curves of relative species richness of three animal taxa along an altitudinal gradient in the eastern Andes of southern Ecuador using available literature data (see text). Ant richness is expected to peak at low elevations and to decline strongly between 1000 m and 2000 m. In contrast, richness of geometrid moths and birds peaks at higher elevations and is expected to decline only at very high altitudes. Richness of all taxa at the lowest elevations is expected to be lower than at medium elevations. Illustration by G. Brehm

The diversity of the studied organisms in the tropics is usually much higher than in temperate regions, reflecting the latitudinal gradient of species richness. Some groups are exceptionally rich in the RBSF while others are not, probably because they peak at lower elevations (Fig. 2.3). Unfortunately, no hard data are available for the majority of taxa so far (see above).

Comparisons of inventories from the RBSF with other montane neotropical sites are hampered by:

1. The scarcity or absence of such inventories;
2. Differences in sampling schemes;
3. Differences in elevational range and area of the sites.

One of the most suitable approaches is therefore to compare the local (alpha) diversity. Vascular epiphytes in the RBSF are outstandingly specious. Single trees hosted up to 98 species (Werner et al. 2005), among the highest species number recorded for single trees. Six trees held a total of 225 species (Werner et al. 2005), more than have been recorded for entire Andean sites in Venezuela or Bolivia (Ibisch 1996; Engwald 1999). The RBSF harbors more than 500 species of bryophytes, probably the highest number ever recorded from a relatively small area in the tropics (Gradstein et al. 2007; Kürschner and Parolly 2007). Within the spermatophytes, the orchids are the most speciose family of the RBSF with a total of ca. 340 registered species (many more are expected; Homeier and Werner 2007). So far, this is the highest number recorded for a neotropical forest site. The genera *Stelis* (Orchidaceae) and *Piper* (Piperaceae), and the family Lauraceae show unexpectedly high species numbers (see Chapter 10.1).

Other very speciose groups include the moth family Geometridae. Brehm et al. (2005) observed 1266 morphospecies between 1000 m and 2700 m – a higher number than observed anywhere else in the world. Interestingly, the family did not show a pronounced peak of diversity but a very broad plateau (Fig. 2.3) with a high and regular species turnover (Brehm et al. 2003a, b; but see Brehm et al. 2007). In contrast, pyraloid moth richness peaked around 1000 m (Bombuscaro) and declined with increasing elevation (see Chapter 11.3). Certain taxa of plants and animals are species-rich in tropical lowlands only. Among plants, many families such as Araceae, Arecaceae, and Fabaceae show a high richness in Amazon lowland and Andean foothill forests that quickly drops as elevation increases (Jørgensen and León-Yánez 1999). Elevational richness patterns are far less well known in animals. Ants have not been investigated in the RBSF, but it is evident from qualitative observations that the area is not a hotspot for this group at all, since only a few species occur (Brehm et al. 2005; G. Brehm, J. Illig, K. Fiedler, personal observations). However, ants are obviously far more abundant and speciose at lower elevations such as in the Rio Bombuscaro valley at 1000 m. This corresponds to observations by van der Hammen and Ward (2005) and Mackay et al. (2002) in Colombia who found a pronounced decline in ant species richness along an altitudinal gradient (also, for a palaeotropical altitudinal gradient of ants, see Bruehl et al. 1999). Ant richness might peak at rather low elevations at the foothills of the Andes similarly as observed below 500 m by J.T. Longino (personal communication) in Costa Rica. However, no appropriate data along a complete altitudinal gradient are available from the tropical Andes so far.

Birds are the best known group of animals in Ecuador. The Eastern slope of the Andes is renowned for its outstanding bird diversity (Rahbek et al. 1995; Ridgely and Greenfield 2001a). Paulsch (2008; see Chapter 11.1) observed a total of 227 bird species between 1999 and 2002 in the RBSF. The number increases considerably when lower and higher elevations are included. Rasmussen et al. (1994) recorded a total of 362 bird species along the (old) Loja–Zamora road (1000–2800 m) that passes the RBSF, 292 species from the Rio Bombuscaro area (950–1300 m) and 210 species from the Cajanuma area (2500–3700 m). Only the latter number represents a near-complete list, whereas the other numbers are still underestimations (C. Rahbek, personal communication). The species richness is at the high end for Ecuador, and only some areas in Peru and Bolivia have similar bird diversity (C. Rahbek, personal communication). The elevational pattern of local bird richness in southern Ecuador is not known, but is anticipated to resemble the pattern (a foothill peak, high-elevational plateau) recorded by Herzog et al. (2005) in Bolivia.

Hypothetical curves for three selected animal taxa (geometrid moths, ants, birds) are diverging; and qualitatively similar divergences of species-richness patterns are expected to occur across the whole range of organism diversity. The curves visualize that the RBSF cannot be regarded as a hotspot of biodiversity in general. The richness of many taxa is likely to peak at lower elevations (Fig. 2.3). PNP covers a much broader elevation range (1000–3600 m) and certainly has a far higher biodiversity than the narrow elevational belt of the RBSF alone. Hence, while there is little doubt that the tropical Andes and PNP can be called hotspots of biodiversity, the RBSF is 'only' a selective hotspot.

From a conservation point of view it would be highly desirable to include lower elevations in a system of protected areas because groups showing a peak of species richness below 1000 m currently do not receive legal protection in south-eastern Ecuador. Moreover, the protection of a complete altitudinal gradient similarly as e.g. in Manú National Park in Peru is undoubtedly the best conservation strategy with regard to the threats caused by global warming. Given the dramatic decline of natural habitats in tropical Andean countries and montane forests (Svenning 1998; Hofstede et al., in press; see Chapter 4), biodiversity inventories must play an important role in selecting such areas for conservation.

2.4 Conclusion

Biodiversity inventories combined with systematic and taxonomic work ensure that trustworthy scientific names can be provided for organisms encountered during ecological or experimental work. An excellent knowledge has already been gathered in the RBSF in some groups, e.g. bats, birds, arctiid and geometrid moths, cryptogamic plants, and trees. However, large gaps still remain to be filled, e.g. in groups such as beetles, ants, wasps, bees, dipterans, most other arthropods, and mollusks. Large proportions of species are apparently new to science and particularly many arthropod species need to be described taxonomically. The RBSF is situated in the Eastern Andean hotspot of biodiversity but the species-richness of most taxa in the study area and its surroundings is still unknown. While some groups are extraordinarily diverse (e.g. geometrid moths, orchids), the richness of other taxa is low in the area and peaks far below 1800 m (the lower elevation limit of the reserve). A coordinated sampling and research approach is required to fill the most important gaps of knowledge about the biodiversity of the area in the future.

Acknowledgements Many colleagues generously provided their unpublished data. We would like to thank in particular H. Braun, C.L. Häuser, M. Lehnert, M. Maraun, F. Matt, S. Matezki, G. Parolly, D. Paulsch, and S. Liede-Schumann. M.P. Mackay, D. Paulsch, and C. Rahbek critically discussed parts of the manuscript. V. Wiese, H. Greeney, and L. Dyer (www.caterpillars.org) helped with identifications.

Chapter 3
The People Settled Around Podocarpus National Park

P. Pohle

3.1 Introduction

In southern Ecuador, a region of heterogenic ethnic, socio-cultural and socio-economic structures, profound knowledge of ethnic-specific human ecological parameters is crucial for the sustainable utilization and conservation of tropical mountain forests. In order to satisfy the objectives of environmental protection on the one hand and the utilization claims of the local population on the other hand, a detailed analysis of human ecological parameters is needed. This chapter aims to provide an introduction to the indigenous and local population of the area surrounding Podocarpus National Park in southern Ecuador. In the case of the indigenous Shuar (lowland Indians) and Saraguros (highland Indians) and the local mestizos, fundamental differences occur not only in the attitudes towards the tropical rainforest and the management of forest resources but also in wider economic and social activities, including all strategies for maintaining their livelihood. Besides differentiating the population in socio-economic terms, this chapter also deals with the regional (horizontal) distribution of ethnic groups, their traditional altitudinal stratification (vertical distribution) and recent migration trends.

3.2 The Provinces of Loja and Zamora Chinchipe

Southern Ecuador comprises the three provinces of El Oro, Loja and Zamora Chinchipe, and has a total population of more than one million people (1 007 199), 52% of whom live in the province of El Oro (525 763), 40% in the province of Loja (404 835) and 8% in the province of Zamora Chinchipe (76 601; INEC 2001). According to the census of 2001, about 86% of the population are mestizos, while indigenous Shuar and Saraguros constitute only 3% of the total population (INEC 2003). In the vertical distribution, the Shuar clearly occupy the lower altitudes, whereas the Saraguros are most prevalent in the higher altitudes, and the mestizos are strongly represented throughout the whole vertical range.

In the province of Loja, the intermediate altitudes with their temperate climate are the most densely settled and intensively cultivated areas in southern Ecuador. At altitudes where sufficient precipitation is available, intensive rain-fed farming predominates featuring a wide range of tropical and subtropical crops. In the drier valleys only irrigation-based agriculture (sugarcane, bananas) is possible. Farming is usually combined with animal husbandry; intensive ranching is also practiced. The Saraguros are the most prominent indigenous group in higher altitudes up to where agriculture is still possible.

The lower altitudes of Zamora Chinchipe province, the tropical rainforest areas, have traditionally been settled by the indigenous Shuar communities. In contrast to the ancient cultural landscape of Loja province, the province of Zamora Chinchipe represents an area of recent agricultural colonization, and exhibits the highest deforestation rates (cf. Chapter 4). The combination of a rugged mountain relief and extreme climatic conditions was for centuries a basic factor in the preservation of mountain forests in southern Ecuador – for instance, along the steep eastern mountain scarp of the Cordillera Real. But population pressure and improved technologies in road construction have opened even such ecologically unsuitable areas to agricultural colonization during the past 50 years.

3.3 The Indigenous Shuar Communities

The Shuar area of settlement traditionally lies below 1000 m a.s.l. and covers the valleys from the humid premontane forests down to the Amazonian lowland (Oriente) in the region bordering Peru. In the province of Zamora Chinchipe, the Shuar communities have settled along the Río Zamora, Río Nangaritza and Río Numpatakaime and their tributaries (Fig. 3.1). The Shuar belong to the Jívaro linguistic group (Amazonian Indians). They are the only autochthonous group of the area ever to have resisted the Incan (1463 A.D.) and Spanish (1531 A.D.) conquests (Harner 1984). They used to have a reputation as fierce warriors and defenders of their land, and as a people who hunted heads of human enemies and then shrank them (*tsantsas*; Belote and Belote 1999). They were also known for their shamanistic practices, conducted by adult males using hallucinogenic drugs (*ayahuasca*). All these specific cultural traits raised the attention of anthropological researchers centuries ago and are now the subject of a comprehensive scientific literature (cf. the bibliography of Belote and Belote 1999). During the twentieth century the practice of taking heads was all but abandoned due to pressure from the national government (Steel 1999).

The Shuar are typical forest dwellers who practice shifting cultivation, mainly within a subsistence economy (Fig. 3.2a). Their staple crop is manioc, which they plant together with taro and plantains on small rotating plots in forest gardens. In addition, they fish, hunt and gather forest products. Fishing is done with *barbasco*, a fish poison. Meat is hunted mainly with shotguns, yet many still use the blowgun

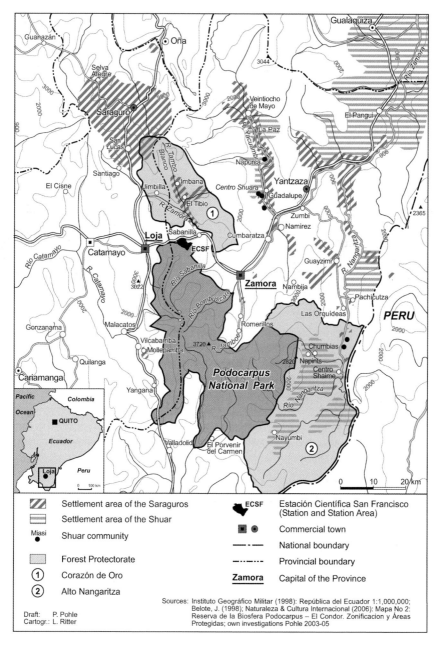

Fig. 3.1 The settlement areas of indigenous groups around Podocarpus National Park

to bring down small prey. The Shuar make extensive use of forest products, which they collect to supplement their diet or use as medicine and construction material (cf. Chapter 25). During recent decades some Shuar have also begun to raise cattle, and still others are engaged in timber extraction.

Fig. 3.2 a The people settled around Podocarpus National Park. Shuar women from Shaime taking a rest in their forest garden. Photo by A. Gerique.

Traditionally, the Shuar live in isolated houses along rivers navigable by dugout canoes. After about 8 years, houses would be moved to new locations, usually to be closer to better fish and game resources, following their decline in the previous area of settlement (Belote and Belote 1999). Nowadays most Shuar stay in one of the permanent villages (*centros*) established with the help of missionaries who entered Shuar territory during the twentieth century. Only then did the Shuar start coming into frequent contact with missionaries and colonists. In areas where competition for land use increased with the arrival of more colonists from the Sierra, some Shuar moved to more remote areas, while others adopted economic strategies of the newcomers – cattle raising and the like. With the infrastructural opening of several traditional Shuar territories (e.g. inside the Yacuambi valley), the way of life and the livelihood of many Shuar have dramatically changed during the past decades. By the 1960s, with increasing pressure on their territory coming from the colonists, the Shuar, with the help of the Salesian (Catholic) missions to the north of the Yacuambi area, began to organize to better defend their interests (Belote and Belote 1999). One result was the creation of the Shuar Federation, the first powerful indigenous organization in Ecuador (Harner 1984), and a model followed by other native groups in the country, including the Saraguros.

3.4 The Indigenous Saraguro Communities

The Saraguros, highland Indians who speak Quichua, live as agro-pasturalists for the most part in the temperate mid-altitudes (1800–2800 m a.s.l.) of the Andes (Sierra) of southern Ecuador (Fig. 3.1; Fig. 3.2b). It is assumed that they originally came from the Titicaca region in Bolivia and were resettled as workers and vassals in the Andean highlands by the Incas during their almost 70-year rule (1463–1531 A.D.) in southern Ecuador. Though they today form a single ethnic group, their ancestry is probably mixed (Belote and Belote 1997/1999). As early as the nineteenth century the Saraguros kept cattle to supplement their traditional "system of mixed cultivation", featuring maize, beans, potatoes and other tubers. By now stockbreeding has developed into the main branch of their economy (Gräf 1990). A shortage of pasture land arose at the beginning of the twentieth century, forcing the Saraguros to expand their pasturage not only into the mist forest and Páramo levels, but also particularly far into the tropical forests of the Oriente, which soon became their permanent area of settlement. In the province of Zamora Chinchipe, Saraguros are settling today along the Río Yacuambi, Río Zamora and Río Nangaritza (Fig. 3.1).

b

Fig. 3.2 b Saraguro family from El Tibio. Photo by P. Pohle.

Like other indigenous groups of Ecuador, the Saraguros have undergone cultural change: most Saraguros are now Catholic Christians who hardly speak any Quichua outside their main area of settlement (Saraguro town), but only Spanish, and increasingly seek out job opportunities away from agriculture. By the 1970s, when the Ecuadorian educational and occupational structures began to be more open to people who retained distinctive ethnic identities, the Saraguros took advantage of the changes in Ecuadorian society (Belote and Belote 1997/1999). Today they participate widely in the national educational system and have access to a wider range of occupations in different branches of the economy, whether commerce, handicrafts, business, politics, teaching, government administration, health care, justice or music entertainment. As a consequence of the economic crises in Ecuador starting in the mid-1980s, a number of Saraguros left Ecuador to seek work elsewhere, especially in Spain and the United States. Yet at least to some extent most Saraguros are still engaged in agro-pastoral activities, and raising cattle is still a major means of sustenance. Up to the present most Saraguros have maintained their distinctive ethnic identity and their pride in it.

3.5 The Saraguro Concept of Verticality

The Saraguros' traditional concept of verticality, which is quite similar to that of other high altitude populations of the world (Allan 1986), is noteworthy: To counterbalance the limited economic capacity of a single location in the mountains, they have developed an economic system in which they use various altitudinal belts, each with its own economic preconditions. Thus, the Saraguros have traditionally practiced a combined economic system comprising subsistence farming close to their villages or individual houses in the mid-altitudes of the Sierra, animal husbandry on the high altitude pastures up to the Páramo belt, and the trading of animals and animal products down in the commercial centres of the inner-Andean basins, or even as far as the Costa area. Their circular migration patterns covered a vertical range of well over 3000 m (from Páramo to coastal elevations), not to mention horizontal ranges of over 150 km (Belote 2000). Hence their traditional concept of verticality (use of natural resources in different altitudinal belts) enabled them to maintain a livelihood in a severe tropical mountain environment, where steep relief, high amounts of precipitation and poorly developed soils limit agricultural production. Today, due to the improvement in infrastructure, and especially due to the construction of the Panamericana and other more minor roads, the Saraguros' commercial centers are nowadays the towns of Saraguro and Loja. Instead of using the Páramo belt for grazing cattle, some have extended their pastures down to the Oriente – for example, the Saraguros of El Tibio, who maintain pastures in the valley of the Río Jamboe close to Zamora. Although the horizontal and vertical distances covered by the Saraguros have shrunk, their traditional combined economic system (subsistence farming, cattle rising and the selling of animal products in nearby markets) still prevails up to today.

Even though the traditional forms of life and livelihood, both of the Shuar and the Saraguros, have greatly changed under the pressure of external influences (missionary activities, agrarian colonization, etc.), both indigenous groups have been able to preserve core areas of their traditional culture, including an extensive specialized knowledge of their environment, along with numerous life-support strategies (cf. Chapter 25). Recent migration trends, however, are reversing the traditional vertical control of natural resources by the specific ethnic groups, since nowadays

Saraguros are expanding their territory far into the Oriente, while the Shuar are expanding their territory into the lower belt of the mountain rainforests of the Sierra, an area they formerly used exclusively for hunting purposes.

3.6 The Mestizos and Mestizo-Colonos

In Latin America, the term mestizo is generally used to indicate people of mixed Spanish and indigenous descent. In southern Ecuador, the term refers to the population descending from Spanish colonizers and indigenous peoples like the Saraguros, the Shuar and the pre-Incan peoples (Cañaris, Paltas, Malacatos).

Today the population of Loja and Zamora Chinchipe provinces is mainly mestizo (Fig. 3.2c). Mestizos share to a large extent a common set of values and a general cultural orientation towards the whites (*blancos*), who place considerable emphasis on their Spanish ancestry and still represent the most privileged ethnic group of Ecuador's social pyramid (US Library of Congress 2006). In the Andean highlands of Loja province, the mestizos are for the most part small-scale agro-pastoral farmers practicing subsistence agriculture. Alongside subsistence crops, small amounts of cash crops are grown, such as sugarcane, maize, peanuts and coffee (Van den Eynden 2004). In general, highland mestizos have little tendency to identify themselves as a distinct ethnic group of regional culture, except in the case of those living in the larger cities – for

Fig. 3.2 c Mestizos on their pasture in the Río Zamora valley. Photo by E. Tapia

example, the Lojanos, who have more distinctive ways of life and have been especially active in colonizing the Oriente and the Costa (Encyclopedia Britannica 2006).

In Zamora Chinchipe province mestizo-colonos are the most dominant ethnic group in terms of numbers. They are colonizers of mestizo ethnicity, said to have come from poor Spanish-speaking peasant families of the Sierra of southern Ecuador (Rudel et al. 2002). Their culture represents a blend of indigenous and Spanish colonial cultures. According to Palacios (1996), mestizo-colonos have been arriving in the Nangaritza valley since the beginning of the 1960s. Most of them originally came from highland areas (2000–3000m) in neighboring Loja province. The colonists usually make a living as farmers who cultivate subsistence and cash crops and, to varying degrees, engage in cattle farming and timber extraction. In contrast to the subsistence-oriented Shuar, the colonists engage more prominently in cash-oriented activities.

With the immigration of colonos since the 1960s, the population of the Oriente in Zamora Chinchipe province has risen sharply. The construction of new roads connecting the Andean and Amazonian areas (the Loja-Zamora road was opened in 1962), a severe drought in 1968 in Loja province (Van den Eynden 2004), and especially the national land reforms that encouraged colonization of the rainforest areas brought in ever more mestizo colonizers (Harner 1984). According to Harner, this caused serious territorial conflicts between colonizers who claimed private ownership of the land and the Shuar, who have a communal concept of land utilization and ownership. Wherever the colonists claimed land, they secured their claim by clearing it, planting pasture and establishing small herds of cattle (Rudel and Horowitz 1993). By promoting colonization, the laws also gave legal backing to a patriotic sentiment among colonists, politicians and the military that colonization would contribute to the defence of the country by establishing "a living frontier" along the disputed border with Peru (cited by Rudel et al. 2002).

3.7 Population Figures and Migration Trends

Given the differences in natural setting, the history of colonization and recent migration trends, the population figures vary significantly between the two provinces Loja and Zamora Chinchipe.

Loja province has a significantly larger total population (404 835 according to the 2001 census) and, in terms of the area of the province, a significantly higher population density (36.8 inhabitants/km^2) than the province of Zamora Chinchipe, which has a population of 76 601 and a population density of only 7.3 inhabitants/km^2 (2001 census; Table 3.1). The data clearly reflect the settlement history of both provinces, with Loja province representing an ancient cultural landscape and Zamora Chinchipe province being an area of recent agricultural colonization. The population density of Loja province has not changed dramatically during the past 39 years of population records: from 31 inhabitants/km^2 in 1962 to almost 37 inhabitants/km^2 in 2001 (INEC 1962, 2001). Sizing up the prevailing land use practices and land tenure systems, Temme (1972, p. 71) noted that the agrarian carrying

Table 3.1 Total population and population density of Loja province, Zamora Chinchipe province and Ecuador, 1990 and 2001 (INEC 1990, 2001)

	Loja province		Zamora Chinchipe province		Ecuador	
	1990	2001	1990	2001	1990	2001
Total population	384 698	404 835	66 167	76 601	9 697 979	12 156 608
Area (km^2)[a]	11 027	10 995	23 111	10 456	272 045	256 370
Population density	34.9	36.8	2.9	7.3	35.6	47.4

[a] The change in surface area between the census of 1990 and 2001 is due to the agreement signed between Peru and Ecuador concerning the borderline in the Cordillera del Cóndor

capacity of Loja province had already reached its limit in 1962 since no new land was available for agriculture development. Accordingly, it is mainly the urban population growth that has been responsible for the total population growth of Loja province from 1962 (48 751) to 2001 (183 313), while the rural population in total numbers has even decreased during these 39 years (from 236 697 to 221 522; INEC 1962, 2001). In contrast, in Zamora Chinchipe province both the urban and rural populations have increased considerably during the same time.

Further, the two provinces differ in the ethnic composition of their populations (Table 3.2). In each case, mestizos represent the major population group, at more than 80%, but Zamora Chinchipe province has a significantly larger portion of indigenous inhabitants (over 12%) than Loja province (3%). In Loja province, the resident indigenous communities are for the most part Saraguros, while in Zamora Chinchipe the dominant indigenous communities are the Shuar. There is obviously a big discrepancy between the numbers of indigenous people in the two provinces according to estimates by different authors and according to the official census. Following Van den Eynden (2004), the Shuar probably total about 20 000 people. According to Belote and Belote (1997/1999) the Saraguros number around 22 000 people. However, the official population census of Ecuador only mentions 21 725 *indígenas* of both provinces in total. It remains unclear whether the estimates were too high or whether the indigenous people changed their self-designation during the census to the mestizo majority.

Table 3.2 Ethnicity based on self-identification (census 2001; INEC 2003)

	Loja province		Zamora Chinchipe province		Ecuador	
	Total	(%)	Total	(%)	Total	(%)
Mestizo	375 558	92.8	63 729	83.2	9 411 890	77.4
Indígena	12 377	3.0	9348	12.2	830 418	6.8
Blanco	13 641	3.4	2820	3.7	1 271 051	10.5
Mulato	1987	0.5	384	0.5	332 637	2.8
Negro	1063	0.2	196	0.3	271 372	2.2
Others	209	0.1	124	0.1	39 240	0.3
Total	404 835	100	76 601	100	12 156 608	100

The vital statistics of Loja and Zamora Chinchipe provinces manifest distinct features remarkably different from that of the country as a whole (Table 3.3)[1]. While the total population of Ecuador increased rapidly during the period 1962–2001, at an annual growth rate of 2.49%, the annual population growth in Loja province (0.87%) clearly lies below the countrywide average, whereas that of Zamora Chinchipe (4.7%) ranges far above it. Between 1962 and 2001 the absolute population increased in Loja province about 1.42 times (from 285 448 to 404 835 inhabitants). The growth in the total population in Zamora Chinchipe was enormous during that period: it increased 6.7 times from 11 464 to 76 601 inhabitants (in comparison, Ecuador's population increased 2.66 times between 1962 and 2001, Table 3.3). The highest population growth rates were recorded for Ecuador and for the provinces of Loja and Zamora Chinchipe between 1962 and 1974. In this period, the populations showed an annual growth rate of 3.1% (Ecuador), 1.58% (Loja) and 9.55% (Zamora Chinchipe).

Table 3.3 Vital statistics of Loja province, Zamora Chinchipe province and Ecuador, 1962–2001 (INEC 1962, 1974, 1982, 1990, 2001)

Census (year)	Loja province		Zamora Chinchipe province		Ecuador	
	Total population	Annual population growth (%)	Total population	Annual population growth (%)	Total population	Annual population growth (%)
1962	285 448		11 464		4 564 080	
		1.58		9.55		3.10
1974	342 339		34 493		6 521 710	
		0.62		3.57		2.62
1982	360 767		46 691		8 138 974	
		0.80		4.36		2.19
1990	384 698		66 167		9 697 979	
		0.46		1.33		2.05
2001	404 835		76 601		12 156 608	

The huge increase in population in Zamora Chinchipe was due essentially to internal migration, particularly influxes from the neighboring province of Loja. In the years 1977–1982 alone, 14 306 immigrants from Loja were registered in Zamora Chinchipe (INEC 1982). Since then the gains and losses from migration have nearly balanced themselves out between the two provinces: from 1996 to 2001, Zamora Chinchipe recorded only 2853 immigrants from Loja, and Loja 2367 immigrants from Zamora Chinchipe (INEC 2001). Measured in terms of all migratory movement in all years on record during the period 1974–2001, Loja province showed a negative balance of migration, while Zamora Chinchipe province showed a positive one. The out-migration of Loja province was remarkably high in 1990: a total of 183 586 emigrants (INEC 1990). This means that in 1990 about 47.7% of

---

[1] The vital statistics in Table 3.3 need to be interpreted with caution, since the reference area changed between the census records of 1962 and 2001.

the total population left Loja province. The main destination of the emigrants from 1974 to 2001 was the province of Pichincha, containing the country's capital, Quito. The neighboring province of El Oro was another main destination of emigrants. Since the 1980s, and more so from the mid-1990s onwards, out-migration from Loja province to foreign countries has taken place. According to the *Plan Migración, Comunicación y Desarrollo* (2004), the destination of 86% of interviewed emigrants from Loja was Spain, followed by about 6% to the United States. After the introduction of visa regulations within the European Union in 2003, the foreign migratory movements have been reduced by 96%.

3.8 Factors Driving the Expansion of the Agricultural Frontier Zone

In southern Ecuador, agriculture is still the most important economic activity: In Loja province, 44% of the economically active population works in agriculture, and in Zamora Chinchipe as much as 57.9% of it is working in the primary sector (INEC 2001). Agricultural land occupies 994 854 ha in Loja province and 446 903 ha in Zamora Chinchipe province, while the number of farming units total 65 625 and 9 006 respectively (MAG, INEC, SICA 2002). It is obvious that Loja province has a high number of relatively small farms with an average size of about 15 ha, whereas Zamora Chinchipe province has a small number of farm units of a relatively large size, about 50 ha on average. However, given the structure of land tenure in Loja province, about 60% of the farms still have areas under 5 ha.

According to Van den Eynden (2004) internal migration within Ecuador only began after the country gained independence in the nineteenth century. During the colonial period the indigenous population, whether in the haciendas or in the cities, were kept in a state of dependency. They could not move around without the consent of their patrons. It was only after independence that they obtained the right to live where they wanted. Throughout the nineteenth and the first half of the twentieth century, a significant increase in the number of commercial haciendas (farms) and a continual accumulation of land by them, took place, especially in the Andean region – unlike the development in the rest of the country. In the south, the haciendas were the largest within the whole country. In 1954, for example, 0.3% of all farms occupied 50% of the land in Loja province (Van den Eynden 2004, p. 16). Apart from the haciendas, there were *minifundistas* (e.g. among the Saraguros) who owned their own small farms. From 1964 on, several land reforms attempted to redivide land by forcing landowners to sell part of their haciendas. In reality, only the most infertile and driest areas were sold, at exorbitant prices, and only very slowly (Van den Eynden 2004, p. 16). The problematic land reforms and masses of landless people caused huge migrations towards both the coastal and Amazonian areas. Today the division of land is still very irregular throughout southern Ecuador. In some areas, haciendas have been divided up, whereas in other areas landowners have maintained their large farms but reduced their size (e.g. in south-eastern Loja province).

Thus, in Loja province, the expansion of the agricultural frontier zone took place a long time ago, starting during the colonial period and reaching its limit in the first half of the twentieth century with the expansion of the hacienda system. Today there is hardly any land left to devote to agriculture. Instead, the small parcelling of land through prevailing inheritance systems (*Realerbteilung*) and the agricultural use of marginal land on steeply sloped terrain, which is disposed to erosion and increased soil depletion, have led to considerable emigration among peasant workers and their families. In contrast to the Sierra, which "suffers" from emigration, the Oriente has been registering an increased influx of persons since the 1960s. This is due to a couple of factors:

1. The agricultural land reforms, starting in the 1960s, have encouraged the exploitation of previously uncultivated areas. The colonization laws that accompanied the agrarian reforms in Ecuador have especially accelerated the colonization and expansion of the agricultural frontier in the Oriente.
2. The government has encouraged settlement by expanding the infrastructure, especially road construction, in order to secure the border with Peru.

Additionally, mining activities have enlarged the supply of jobs. At the same time, a surplus of peasant workers from neighboring Loja province began searching for new job opportunities.

Although the key driver for expansion of the agricultural frontier zone in general is still population growth, in particular cases such as those in the Oriente, other factors, including agrarian reforms, state policies, infrastructure, land tenure systems and farming strategies, may have an even higher impact. Despite the expansion of the agricultural frontier, population growth in Loja province, for one, has set in motion other processes: permanent emigration, increasing employment in non-farm activities, the fragmentation of property formerly occupied by haciendas and more.

Chapter 4
Ecuador Suffers the Highest Deforestation Rate in South America

R. Mosandl(✉), S. Günter, B. Stimm, and M. Weber

4.1 Introduction

"Deforestation" in this chapter is used according to the FAO (2001) definition: it is the conversion of forest to another land use or the long-term reduction of the tree canopy cover below the minimum 10% threshold. This implies that areas where trees have been removed as a result of harvesting or logging are not considered as "deforestation". Even if the structure or function of a forest is heavily disturbed by harvesting operations, the stand remains a forest as long as it has a tree canopy cover of more than 10% or is expected to regenerate naturally or artificially in the long run. So "deforestation" in the sense of the FAO definition does not incorporate the degradation of forests included in some other definitions, e.g. Myers (1994). Correspondingly the replacement of old-growth forests by plantations or their temporary use by shifting cultivators is not considered as destruction or deforestation. Only a complete change of land use or the destruction of forest cover which prevents its recovery to more than 10% crown cover enters the FAO statistics as "deforestation" or "forest loss".

Despite this very narrow definition of the FAO, deforestation is the most important process for a decrease in forest area (FAO 2006). Clearing the forests for agriculture or infrastructure leads to a decrease of the land category "forests" and to an increase of the land category "other land". The same effect – but to a much lesser extent – is also caused by natural disasters when the affected area is incapable to regenerate naturally (FAO 2006). An increase of forests can happen either through afforestation or by natural expansion of forests. The net change rate of forests takes account of four processes: (a) decrease by deforestation, (b) decrease by natural disasters, (c) increase by afforestation and (d) increase by natural expansion.

The deforestation rate estimated by FAO in the latest *Global Forest Resources Assessment* is the balance of these four processes (FAO 2006).

4.2 Deforestation in South America

According to *Global Forest Resources Assessment 2005* the global net change in forest area during the period 2000–2005 is estimated at −7.3 million ha/year (FAO 2006). The continent with the largest net loss of forest area in this period was South America, which suffered a net loss of forests of about −4.3 million ha/year (corresponding to a rate of −0.5% of the remaining forest area). Brazil's forests alone lost 3.1 million ha/year in this period but in relative numbers had lower deforestation rates than Ecuador, which suffers the highest rate (−1.7%) within South America. Obviously large net changes in forest area are occurring in the tropical and subtropical regions of South America (FAO 2006) with a very high biodiversity which is very likely to be reduced by the high deforestation rate. According to a prediction model of Koopowitz et al. (1994), habitat conversion caused by deforestation leads to species extinction rates that range up to 63 species/year. It is remarkable that this high extinction rate is predicted for Ecuador.

4.3 Deforestation in Ecuador

The extent of forest in Ecuador in 2005 was 10.8 million ha, which represents 39% of the land area (FAO 2006). This percentage is relatively low compared with other countries in South America, taking into account that the average forest cover rate of South America is 48% (FAO 2006). It is assumed that more than 90% of Ecuador's surface had been covered by forests originally (Wunder 2000). This implies that before human impact occurred on a large scale, the area which today we call Ecuador must have been covered by more than 25 million ha of forest. These data are confirmed by Cabarle et al. (1989), who estimated the original forest cover of Ecuador to be 26 million ha. Two major historical deforestation processes have contributed to the reduction of the forest area: first a long-lasting deforestation in the Sierra (areas with an elevation of at least 1200 m a.s.l.) in the pre-Columbian era and second a rapid forest conversion in the Costa region during the past century (Wunder 2000). The era in between these two deforestation phases, dominated by the long Spanish colonial rule, was characterized by an expansion of forests, caused by the dramatic decrease in population and also in population pressure on the forests following the Spanish conquest. After the declaration of independence in 1822 until the early twentieth century Ecuador's forest cover was largely preserved (Wunder 2000). During the cocoa boom from 1900 to the end of the 1920s and intensified during the banana boom after the Second World War (main period 1950–1965) the coastal lowland forests were cleared for agricultural crops. Cabarle et al. (1989) estimated the forest cover in 1958 was 17.5 million ha. The corresponding extent of forest must have been about 63% at that time. In 1987 the extent of forest dropped to 45% (FAO 1994). The main cause for this decrease, besides the clearing of coastal lowland forests for agricultural crops, was the opening up of the Oriente, Ecuador's Amazon region. With the oil boom of the 1970s roads were build in the Amazonian forest, which attracted agricultural colonization and timber extraction (Wunder 2000). Subsequently the reduction of forest cover continued from 43% forest cover in 1990 down to 39% in 2005 (FAO 1993, 2006).

4.4 Reasons for Deforestation

The question arises: what are the causes for the relatively high actual deforestation rate? The first idea, tracing this back to a high conversion rate of primary forests, is not confirmed by the data, because the area of primary forests remained unchanged in recent years (FAO 2006). This is certainly due to the fact that a lot of primary forests were protected. Ecuador's forest protection statistics present 21% of all forests as protected in 2002 (UNEP 2002). So it can be concluded that the main deforestation must take place in secondary forests. Granting a deforestation rate of −1.7% means a loss of 198 000 ha/year of secondary forests (FAO 2006). Not included in the statistics is an unknown area of illegally converted forests. The most reasonable explanation for these high annual losses is the change in land use. Mainly secondary forests must have been converted into agricultural land. In fact, looking at the agropastoral land-use trends in Ecuador there is a dramatic increase in pastures. From 1972 to 1985 the area of pastures increased from 2.2 million ha to 4.4 million ha and by 1989 pastures covered an area of about 6 million ha (Wunder 2000). This means an annual increase of agropastoral land of 244 000 ha during the first period of that time and 182 000 ha in the second period. Assuming that the conversion of forests into agropastoral land continued in the 1990s (on the same scale as before) then the increase in agropastoral land today is still equal to the decrease in forest cover. As cattle ranching is concentrated in the Sierra there are strong hints that the main forest losses are occurring especially in this region. Deforestation had devastating effects in parts of the coast, e.g. the elimination of over 70% of the costal mangroves (Mecham 2001) but in general in the Costa, where commercial crops are cultivated, the forest losses were lower. This is also due to the fact that, in contrast to pastures which tripled their areas from 1972 to 1989, crop lands expanded only slightly during this period (Wunder 2000).

The driving forces for the conversion into pastures are very likely rooted in socio-economic reasons. Slow growth of human capital and progressive degradation of natural capital over time is at the heart of the frustrated development experience for Ecuador and Latin America (Lopez 2003). Deforestation in Ecuador is also related to tenure insecurity (Southgate et al. 1991) and the convergence of local populations in an economic system which relies on the unsustainable exploitation of natural resources (Sierra and Stallings 1998). Long-term effects, caused by insufficient investment in education (Godoy et al. 1988), or the consequences of concentrating on short-turn returns, are further driving forces of deforestation (Wunder 2000).

Besides the high change in land use from forests to pastures there is another cause for the high deforestation rate in Ecuador: while in other countries of South

America the conversion of forests is mitigated by high reforestation efforts, no substantial areas were reforested in Ecuador during recent years. The plantation area in Ecuador is growing very slowly (FAO 2006). In the period from 2000 to 2005 it increased only by 560 ha/year (for comparison: in Chile the increase was 61 000 ha/year in the same period).

4.5 Conclusions

Identifying the causes for the high deforestation rate in Ecuador shows how to overcome this problem. There are two possible ways: first, the conversion of forests into pastures could be made unattractive, e.g. by an ecologically and economically sustainable forest management (see Chapter 26 in this volume), and second the reforestation of degraded land could be increased (Lamb 1998; see Chapter 34 in this volume). Both ways should be promoted by the government setting incentives for sustainable forest management of plantations and natural forests.

Chapter 5
Methodological Challenges of a Megadiverse Ecosystem

G. Brehm(✉), K. Fiedler, C.L. Häuser, and H. Dalitz

5.1 Introduction: What is Megadiversity?

'Megadiversity' originated as a term in the context of biodiversity conservation in the late 1980s (Mittermeier et al. 2004). It refers to countries with an extremely high level of species richness, usually found in the tropical realm, one or two orders higher in magnitude than in most temperate zone countries. In many ways countries or areas of megadiversity coincide with the slightly longer-established concept of biodiversity hotspots (Myers et al. 2000; Brummitt and Lughadha 2003; see Chapter 2 in this volume). Unlike the concept of hotspots, megadiversity also attempts to take into account degrees of endemism, phylogenetic relatedness, and other measures of diversity applied for identifying biodiversity hotspots as opposed to pure numbers of species or taxa per unit area. More than any academic differentiation, however, the term megadiversity was recently taken up and promoted at the political level, particularly under the *Convention of Biological Diversity* (CBD), as well as the *Convention on International Trade in Endangered Species of Fauna and Flora* (CITES). Following the original meeting in Cancun, Mexico, in February 2002, 15 countries formed a group of 'Like-Minded Megadiverse Countries' (LMMC) as a forum to address the specific challenges for biodiversity conservation and sustainable use faced by countries with disproportional high levels of biodiversity ('Cancun declaration 2002'). Later joined by Australia and the USA, this informal group of countries comprises many but by no means all of the recognized global biodiversity hotspots.

The Republic of Ecuador qualifies by all standards as being megadiverse, and has also been a founding member of the LMMC group. This relatively small country (approx. 284 000 km^2) harbors an outstanding variety of habitats along pronounced elevational and wet–dry gradients. The high beta diversity along these gradients, situated within the peak of species-richness at tropical low latitudes, favors an enormous biological diversity that is rivaled by only a few other regions in the world. For example, Jørgensen and Léon-Yánez (1999) listed 15.901 vascular plant species (15 306 of them native) known to occur in Ecuador, which is more than six times the number reported for Germany with its 2682 species in a 25% larger territory (357 000 km^2). Among the various habitats in

Ecuador contributing to its overall megadiverse status, the Andean forests are a significant factor which have also been identified as an area of highest priorities for biodiversity conservation (Aldrich et al. 1997; Brummitt and Lughadha 2003). Apart from the uneven spatial distribution of biodiversity, species richness is not equally distributed among taxa, and arthropods play an outstanding role (see Chapters 2, 11.3, 35 in this volume).

5.2 Methodological Challenges of Megadiversity

The scientific methodologies available for assessing and comparing megadiverse faunas and floras are essentially the same as in less diverse ecosystems. Yet, megadiversity poses three major challenges: First, the sheer number of different species makes it almost impossible to arrive at 'complete' inventories within reasonable time-scales (Novotny and Basset 2000). Second, the size and incompleteness of samples call for sophisticated analytical tools that go far beyond the simple counting of species. Third, the identification of species is critically complicated, even more so in the absence of reliable taxonomic literature and identification tools for most taxa.

The first issue is not trivial, but new methods to deal with these problems have been developed in the past decade. Mathematical procedures to estimate biodiversity burst rapidly (Colwell 2006), and empirical data as well as simulation studies now allow to select appropriate measures once a reasonable density of data has been collated (e.g. Brose and Martinez 2004). The development of 'stopping rules' allows the determination of minimum levels of sampling (Chiarucci et al. 2003). However, even the best mathematical tools are worthless unless appropriate sampling schemes are used. A freely available range of statistical estimation methods (e.g. EstimateS; Colwell 2006) can be reasonably applied only when organisms have been sampled using standardized quantitative methods with sufficient spatial and temporal replication. Even ambitious projects such as the analysis of more than 35 000 individuals of geometrid moths in the RBSF and adjacent areas (Brehm et al. 2005) still lack fully sufficient temporal and spatial replication at certain sites. DeVries et al. (1997) showed that near-complete inventories of Ecuadorian local butterfly faunas can only be achieved by large-scale sampling over many years (covering different seasons of the year), including the understorey and the canopy of the forest. It is therefore hopeless to expect meaningful richness data by sampling such habitats for a few weeks in a 'rapid assessment' style, even more so if organisms are collected qualitatively.

The second issue, i.e. to obtain meaningful comparisons between incomplete samples from diverse communities, has also been studied: Indices to calculate floral or faunal dissimilarity in such cases are now available (e.g. the NESS and CNESS family; Trueblood et al. 1994); and, by means of appropriate data transformations and sophisticated multivariate procedures (e.g. Legendre and Legendre 1998; Legendre and Gallagher 2001), it is now possible for ecologists to analyze the structure of such complex data sets.

It is the third issue which still causes the largest concern, i.e. proper and reliable sorting of samples and identification of species in cases where the flora, or fauna, is extremely rich, poorly known, or even partly undescribed. This task is often the most time-consuming in tropical community ecology. While whole organisms or samples are collected in a broad range of organisms, distance-sampling methods are applied for many groups of vertebrates. These methods require an excellent expertise that often needs to be trained over months or, more realistically, over years of fieldwork. For example, quantitative bird surveys in tropical forests require the knowledge of a vast number of similar bird voices (C. Rahbek, personal communication). Much expertise needs to be gathered in order to get 'the look' for certain patterns in other groups: Where does the well camouflaged caterpillar or the singing bush cricket sit? Where do the liverwort or *Piper* species grow? Where is the best place to net bats? In which trees do the tanagers feed or breed? A good field biologist will quickly learn, but the amount of knowledge about the many species in megadiverse communities often seems endless. While tropical fieldwork poses various problems, subsequent laboratory work on field samples is challenging in a different manner and might consume even more time. For example, during three studies of moth diversity along environmental gradients in southern Ecuador (Brehm 2002; Süßenbach 2003; Hilt 2005) the total time devoted to field work was in the order of 6–10 months each, whereas processing, sorting and identification of samples required double the time and more, even when assisted in handling the samples. Moreover, in the absence of valid literature, identification entailed repeated visits to large research collections held at natural history museums in various countries (see below). All these steps required substantial resources, not least those invested by collection curators to give advice. Therefore, any means to speed up the process of identification will be of significance for progress in tropical biodiversity research. Moreover, it is important to make the data and experiences accumulated during all larger projects internationally available and free of charge. Then, scientists who continue with similar work can build upon the knowledge that has been gained so far, even at other sites or under different institutional coverage.

5.3 Taxonomic Identification of a Diverse Insect Group: a Case Study

A bottleneck for handling megadiverse situations is the problem of obtaining reliable species identifications. Brehm et al. (Chapter 2 in this volume) and Krell (2004) discuss why proper identification is required and why the exclusive use of morphospecies is an inappropriate shortcut. Comprehensive identification literature is usually not available for most groups of arthropods and other species-rich taxa. In the case of geometrid moths, for example, most original species descriptions date from the late 1800s and early 1900s, the 'golden age' of taxonomy (Gaston et al. 1995). Unfortunately, standards for species descriptions then were usually rather low as seen from today's perspective. The species were often described in a few sentences without examination of important morphological diagnostic characters, illustrations or comparisons with other species. As a consequence, such original descriptions are usually a very poor identification tool.

The most reliable and often only way to identify geometrid species is by comparison of specimens or digital images (Fig. 5.1) with type material deposited in natural history museums (e.g. Brehm 2002; Brehm et al. 2005; Chapter 2 in this volume). However, such an approach is time-consuming and costly since much material needs to be examined in many museums. For this work, hundreds of presorted morphospecies needed to be compared with thousands of specimens in different museum collections: finally, a total of 1266 species were sorted, 63% of which were assigned to known species (Brehm et al. 2005). Since the most relevant museums were visited (most twice), this percentage actually indicates that ca. one-third of the species are likely to be new to science and need to be described taxonomically. Fortunately, type specimens of neotropical geometrid moths are rather concentrated in a few museums [the most important being the Natural History Museum (London), the United States Museum of Natural History (Washington D.C.), the Zoologische Staatssammlung (Munich), the American Museum of Natural History (New York), the Senckenberg Museum (Frankfurt), the Humboldt Museum (Berlin)]. Species identification becomes of course more difficult for those taxa in which type material is scattered in many museums around the globe.

Preliminary identification would greatly be enhanced if digital images of all type material deposited in museums became available in the Internet, since this allows one to compare samples quickly and cheaply, and to prove existing identifications. Although such projects are on their way for some insect taxa, such as butterflies (Lamas et al. 2000; Häuser et al. 2004), the vast majority of Neotropical insect type material is still not accessible via online databases. Though undoubtedly extremely useful, putting digital images of type material on the web is not a solution for all problems: Such images cannot replace proper identification literature written by experienced taxonomists. Moreover, good identification literature usually does not illustrate the often odd and worn types but proper specimens belonging to the same species. High standards in identification literature as set in the European fauna (e.g. Hausmann 2001) are still a far vision for most tropical Lepidoptera and insects in general, particularly when considering: (a) the recent retirement (without replacement) of taxonomists working on neotropical Geometridae, and (b) the general neglect of taxonomy as a fundamental biological discipline.

5.4 Imaging of Plants and Moths

High-resolution digital images of specimens and samples of many groups of organisms can nowadays easily be obtained, stored, and administered (Basset et al. 2000; for best practice, see Häuser et al. 2005). Relational databases allow easy data access, data exchange, and the publication of images and data on the Internet. Identification tools can be incorporated in such databases and are available in the

Fig. 5.1 Sixteen (out of 102 species collected in the RBSF and adjacent areas) belonging to the very species-rich geometrid genus *Eois* (*amarillada, angulata, antiopata, azafranata, basaliata, binaria, biradiata, borrata, burla, camptographata, chrysocraspedata, ciocolatina, cobardata, cogitata, contraversa, encina*; Brehm et al. 2005). No identification literature is available for most tropical insects, including *Eois*. Digital images (three megapixels, Nikon Coolpix 990) of the upper- and undersides of more than 1200 species of geometrid moths were taken and used for identification in museums and for documentation. The *Eois* images are provided on the Internet (http://www.personal.uni-jena.de/~b6brgu2). *Bars* 10mm. Images taken by G. Brehm

field for local researchers. During the course of our research in Ecuador, fresh plant specimens were scanned soon after collection. The difference between such fresh specimens and flattened, brown herbarium specimens often was considerable. Pigment colors in moths and butterflies usually fade less rapidly than in plants if the specimens are protected from light.

Appropriately spread Lepidoptera specimens display most parts of their wing patterns (Fig. 5.1), and moths can usually be photographed months or even years after collection without a significant loss of quality. Images taken in the field can provide additional and highly valuable data, e.g. about behavior. However, such images are much more difficult to obtain than images of pinned specimens. Unlike the small and secretive moths, sedentary organisms such as plants can relatively easily be photographed in the field. Combined field and herbarium images provide a rather complete visual representation of the respective individual. Close-ups of flowers or fruits provide important characters in high resolution. It is well known that images are generally easier to memorize than pure text and 'dry' sets of characters. They effectively help scientists to recognize their target organisms in the field.

5.5 Data Processing and Accessibility

In species-rich regions, fieldwork results in a huge number of collected specimens and associated meta data. Each collected specimen must be labeled with data such as country, region, habitat, elevation, geographical position, time of collection, and collector. The specimen should also be labeled with its individual database number and the assigned scientific name. All these meta data are also stored in a database together with further information (e.g. morphological characters, where and when identified, etc.) and digital images. It is evident that databases are much more efficient for the storage and analysis of data than spreadsheet software. Moreover, typing efforts and the risk of misspellings are reduced. A rather complete standard for the design of taxonomic databases is available from the Taxonomic Database Working Group (TDWG; for a description of the standards, see: www.bgbm.org/TDWG/CODATA/Schema/).

Biodiversity informatics is a rapidly growing field for the data analysis of large collections stored in natural history museums. However, many of the initiatives are related to specimen data that are already stored in collections. Data availability is often restricted to experts and/or local networks, and not all tools are intended for fieldwork at a relatively low level of taxonomic knowledge. In research networks such as in the RBSF, data accessibility between all participating groups is of crucial importance. Site-specific conditions of a research area and available facilities define how data can be shared among the researchers. If the Internet can be accessed, a web-based database is most appropriate (e.g. as provided by the Arthropods of La Selva project, available at: http://viceroy.eeb.uconn.edu/ALAS/ALAS.html). However, in smaller research stations in tropical areas that lack permanent Internet access, a local version could be the best way to share information.

5.6 'Visual Plants' – a Biodiversity Database for Non-Specialists

'Visual Plants' is a database that specializes in storing information on plants. The program can be installed on local computers and is used in the research network at the RBSF. It is also available on the Internet (www.visualplants.de). The main intention of the database is the presentation of images together with their meta data. Attached characters describe the specimen to allow a pre-selection of species based on vegetative characters, such as life form, leaf arrangement, or flower color.

A set-based query tool allows the interactive minimization of the number of specimens or field images stored in the database through the selection of character statements. Although the system is not a complete identification system, it is a useful tool, particularly for non-specialists and where other sources for identification (floras, etc.) are absent. It publishes specimens collected in the field, even if they are not identified at species level, since this might ease the discovery of new species. Hopefully, the database will be expanded to taxa other than plants in the future to provide a multi-tool for preliminary species identification of a broad range of organisms in the field.

5.7 Conclusions

Megadiverse ecosystems such as tropical montane forests in Ecuador pose major methodological challenges with their species-richness, incomplete sampling, and difficulties in the identification of species. Digital images taken from the type specimens as well as from fresh specimens can greatly ease identification. The availability of additional information, e.g. morphological characters, is essential, preferably organized in a database. However, such tools still need to be developed for most groups of organisms.

Chapter 6
Investigating Gradients in Ecosystem Analysis

K. Fiedler(✉) and E. Beck

6.1 Gradient Analyses as an Alternative to Experimentation

Natural ecosystems are usually highly complex. They are built up by a multitude of organisms which interact with each other as well as with their abiotic (i.e. physical and chemical) environment. The more kinds of organisms thrive in an ecosystem, the more complex is its internal structure. For example, food web complexity is a function of the direct and indirect interactions between organisms, and this increases in a non-linear manner with the number of players. Food web complexity, in turn, affects emergent properties of ecosystems such as water and nutrient cycles, or the dynamics and resilience of the system in an unpredictable and changing environment. Accordingly, understanding the function of natural ecosystems becomes ever more challenging with increasing species richness.

The classic, and usually most powerful, approach to understand ecosystem processes is by experimentation. Systems are manipulated in a controlled manner and with a sufficient level of replication to assess whether their responses follow predictions derived from some theory. Alternatively, systems can be artificially created with a known structure (micro- and mesocosm approaches) and their properties measured as a function of system structure or of manipulation of predictor variables. However, these approaches are only feasible for tiny subsets of an ecosystem as complex as a tropical mountain forest (Armenteras et al. 2003; Brooks et al. 2002; Chapter 4 in this volume). The resources required to manipulate entire forest plots with a meaningful number of replicates likewise preclude this approach for the most part. Experimentally assembled forest systems take a very long time to mature and hardly ever reach the complexity and diversity of natural tropical forest.

Given these constraints, the analysis of a wide variety of ecosystem parameters along environmental gradients provides a powerful approach to gain insight into many aspects of the structure, function, and dynamics of a mountain forest. Environmental gradients can be viewed as 'natural experiments' in that regard. Gradient analyses are extremely useful to detect patterns in the first part. This is particularly important in systems which are thus far largely *terra incognita*, i.e. where a pattern analysis is the initial and decisive step to generate hypotheses about causal relationships. Characteristic patterns of ecosystem parameters become visible and thereby reveal changes along gradients. For this to be successful, the gradients need to be sufficiently extensive, viz. they should cover a substantial range of the ecological parameters under study. In addition, gradient studies can also contribute much to test hypotheses that have been derived from ecological theory.

One general drawback of gradient studies is that their outcome is by necessity correlative in nature. It is therefore sometimes difficult to separate cause and effect. In particular it is a challenge to disentangle direct and indirect effects, or to assess the relative strength of effects in highly interactive systems with many feed-back pathways. The key to overcome these problems is a clear theoretical framework to build upon, with sufficient replication of measurements in space or time, a high quality of data collation, and a subsequent evaluation using multivariate statistical procedures (e.g. multiple regression models, canonical correspondence analysis, path analysis) to establish the strength and direction of links between the measured variables (Legendre and Legendre 1998). Ultimately, this correlative evidence can then be tested and validated by experiments on small accessible subsystems or by modeling studies.

In our case study of a tropical mountain forest, we chose two contrasting environmental gradients as a backbone: a natural one (elevation, with a concomitant change of many abiotic traits and organismic responses) and an anthropogenic one [the intensity of impact by humans, which affects the biota directly (forest clearing) and/or indirectly via abiotic traits].

6.2 The Altitudinal Gradient: Effects of Elevation

Altitudinal gradients have attracted much interest in ecology (Körner 2000; Lomolino 2001; Rahbek 2005). Abiotic conditions such as temperature, precipitation, humidity, and soil conditions change profoundly with increasing altitude (for tropical mountain forests, see e.g. Tanner 1977; Marrs et al. 1988; Grieve et al. 1990; Flenley 1995). As a result the net primary production of ecosystems and species richness of most organisms also decrease, but with many differences between the kinds of organisms and frequently not in a monotonic manner (Gentry 1988; Lomolino 2001; Rahbek 2005; see Chapters 10.1–10.5 and Chapters 11.1–11.4). Altitudinal gradients thus provide ideal settings to study effects of various ecological factors across small spatial scales. Given their limited spatial extent, differences in species composition between sites can confidently be attributed to responses to local environmental conditions, whereas historical factors, which play an important role in latitudinal gradients, are far less important.

The basic geophysical effect of elevation is a nearly linear decrease in ambient temperature by about 0.6 K per 100 m. Other abiotic factors, such as precipitation, wind, geomorphology (inclination), shallow or deep soil, and the hydrological situation change in a more complex, non-linear and site-specific manner. Collectively, these changes are mirrored by typical characters of the vegetation that recur in all tropical mountain areas. Traits that vary with elevation include the frequency of specialist plant life-forms such as epiphytes or vines, the root-to-shoot ratio of plant biomass, and physiological parameters of plants (CO_2-fixation, transpiration rates; see Chapters 12–17). Likewise the distribution patterns of most animal species reveal clear altitudinal preferences and limits, which can usually be related to their thermal physiology (see Chapter 11.3). Food webs may be altered by elevational shifts, for example when host plants or host animals become unavailable for specialist herbivores or parasitoids, respectively, above a critical altitudinal limit. However, recent work by Novotny et al. (2005) indicated that, for a speciose guild of insect herbivores, such constraints by the changing composition of the vegetation are less intense than assumed earlier. Organisms can thus be viewed as integrative indicators of environmental change that respond to a multitude of physical and chemical factors along altitudinal gradients.

Of special interest are cases where individual abiotic factors and responses of different groups of organisms do not go in parallel. If a measured gradient deviates from the theoretically predicted one, a reason for that must be found. For example, the decrease in air temperature is generally linear, whereas the increase in precipitation is non-linear. These kinds of disproportionalities usually indicate the effect of other, often unexpected, individual factors. Many such deviations are caused by microsite effects which mirror the geomorphologic situation rather than the influence of altitude per se. Microclimate-mediated effects on the vegetation or on soils (e.g. through site-specific litter production and biomass turnover), irregularities caused by nutrient deposition from remote areas through airstreams (Fabian et al. 2005) or the effects of strong wind, which can considerably depress the upper tree line, are features which become apparent along the altitudinal gradient.

With regard to biological diversity, a recent review by Rahbek (2005) revealed two common elevational patterns of species richness. Of many organisms, diversity decreases in a roughly monotonic manner with altitude. This is to be expected if resource limitation and thermal constraints govern species diversity in an ecosystem. In other groups the diversity–altitude relationship is hump-shaped with a maximum somewhere in the range 1000–2000 m a.s.l. There is considerable debate as to the origin and ecological reality of these mid-elevation peaks of biological diversity. According to the mid-domain concept, hump-shaped diversity relationships are expected from geometric (i.e. non-ecological) constraints on habitat area available along mountain slopes (Colwell and Hurtt 1994; Colwell and Lees 2000). Mid-domain effects also extend to functional aspects such as biotic interactions and phenological events (Morales et al. 2005; Bendix et al. 2006; Cueva et al. 2006). However, mid-domain effects failed to explain observed patterns in a number of instances (for a review, see Colwell et al. 2004) and explained only a minor fraction of data variance in others (e.g. Brehm et al. 2007).

Two other important ecological concepts to be examined along elevational gradients are temperature–size rules and the Rapoport effect. Animals usually grow larger in areas with a colder climate (for endothermic vertebrates, this is known as Bergmann's rule; see Atkinson 1994; Mousseau 1997; Ashton et al. 2000; Meiri and Dayan 2003;

Brehm and Fiedler 2004). Rapoport's rule predicts that the distributional ranges of cold-adapted species become larger at higher latitudes and elevations (Gaston et al. 1998). It is speculated that more species may then coexist in warm tropical habitats due to their narrower ranges (Stevens 1989). However, predictions derived from both these concepts need to be tested along tropical altitudinal gradients to assess their general validity. The (few) available studies do not suggest that Rapoport or mid-domain effects are sufficient to fully explain elevation patterns of tropical biodiversity. Clearly integration of further aspects such as resource limitation (Hobie et al. 1994) or non-linear spatial dynamics of biotic interactions will be necessary to unravel the causes for the variety of elevation patterns seen across the plant and animal kingdom.

6.3 Gradients in Disturbance or Land Use

Gradients of different intensities of anthropogenic disturbance (severity and frequency of impact) of the original ecosystem reflect modes and time-spans of land use. Such gradients deal with man-made ecosystems or ecosystems recovering (if possible) from former land use (succession after abandonment; see Chapters 6.1–6.7). Strictly speaking, analyses across different land-use intensities are not true gradient studies, since these 'gradients' are not universally connected by steady transitions, and habitat types are typically classified as such by the observer. Correspondingly, the indicators used to order habitats along an impact axis are sometimes not unambiguous.

What is the aim of examining disturbance gradients and what sense does it make to investigate land use along an inferred gradient of ecosystems subjected to impacts of different severity and duration? Studies of the consequences of converting the natural tropical mountain rain forest into agricultural land will: (a) contribute to the question of the stability and resilience of the original ecosystem, (b) allow an estimation of the (permanent) input which is necessary to achieve livelihood for people who convert forest into pastures or arable land, (c) support an assessment of the probability of an unaided succession into the former or a similar substitute ecosystem after the impact has finished, i.e. once the area has been abandoned, (d) be necessary for identifying levels or modes of impact where a stable land-use ecosystem can be maintained. This could be considered a 'sustainable land use'.

Likewise, such studies could be useful in developing strategies for the management and reforestation of abandoned agricultural and pasture areas by determining the remaining abiotic (soil) and biotic (soil organisms, seed bank) resources available for regeneration (e.g. Davis et al. 1999). Another important aspect of studies across land-use gradients is to identify the loss of biodiversity in anthropogenically altered habitats (e.g. Fiedler et al. 2007b). Such knowledge may be used for the purpose of biodiversity indication (e.g. Lawton et al. 1998; Schulze et al. 2004); and this is particularly urgent in biodiversity hotspots, where the ongoing conversion of natural habitats into areas used for agriculture and forestry is the most substantial threat to many unique species. In contrast, the intermediate disturbance hypothesis and other non-equilibrium models predict that a (transient) peak of biodiversity, which may even exceed the richness of the natural forest, might occur at moderate disturbance levels with regard to both the intensity and frequency of 'stress' events (Levin and Paine 1974; Connell 1978; Huston 1994). However, the positive effects of disturbance on biodiversity are heavily scale-dependent (Hill and Hamer 2004) and in tropical areas may also depend on the proximity of natural source habitats to allow for re-colonization by many forest-dependent species after disturbance.

Which premises must be fulfilled to allow land-use gradients to be established?

1. The geophysical conditions such as altitude, climate, and soils should be as similar as possible.
2. The continuous or discontinuous measures taken by humans (e.g. fertilization, pasturage, burning, planting of crops or other non-indigenous plants) must be somehow quantifiable.
3. The recent history of land use in the areas should be known and similar.

Only if these premises are fulfilled, can robust results be expected from comparative studies across different land-use systems and intensities.

6.4 Gradients of Regeneration

A conceptually different kind of gradient can be studied in succession habitats (see Chapters 27, 32, 35 in this volume). In the mountains of southern Ecuador such habitats occur in a 'natural' variant (recovery of vegetation after tree fall or landslides; see Chapters 23, 24 in this volume) and in an anthropogenic form (succession after abandonment of land-use; see Chapters 28, 31 in this volume). Both these types of succession may be considered as reverse gradients of disturbance. Unraveling parallels and differences between natural succession, human-induced succession, and land-use gradients have the potential to reveal much information about resilience and recovery potentials of mountain forest habitats. One would expect close parallels, especially if similar organisms and interactive mechanisms are the main drivers of ecosystem dynamics. In contrast, if succession after anthropogenic impact takes different routes than after natural disturbance, this hints to massive changes in the role of particular species, abiotic factors, or an interplay of these.

6.5 Conclusion

Much can be learned from studying natural and anthropogenic environmental gradients, especially in a fragile and excessively complex ecosystem such as the tropical mountain forest where experimentation faces severe limitations. This information is not only relevant to science (e.g for answering some of the most challenging questions in ecology and evolution), but also for society, e.g. in terms of sustainable development and conservation of unique threatened biodiversity. If many facets of science can be integrated in a trans-disciplinary manner, the benefit will be even greater.

Chapter 7
The Investigated Gradients

E. Beck(✉), R. Mosandl, M. Richter, and I. Kottke

7.1 Introduction

As discussed by Fiedler and Beck (Chapter 6 in this volume) gradient analysis is a powerful tool for the investigation of an ecosystem. The knowledge of the RBSF that has arisen from gradient analysis is presented in the following chapters. This chapter, however, attempts to introduce the reader to the strategy of the ecosystem study, discussing the features which can be revealed by identifying appropriate gradients. One basic requisite of the joint venture was to concentrate all projects on the same core research area, the RBSF in order to produce coherent data (Fig. 7.1). Satellite areas or monitoring stations were established to address questions of supraregional relevance. On that rationale four gradients were followed, two of which are natural gradients based on climatic factors while the other two result from human activities.

7.2 The Natural Gradients

7.2.1 The Horizontal Gradient

The horizontal gradient was investigated at an altitude of 1950 m over a distance of about 30 km. Here, the effects of humidity between the dry rain-shadow situation in the west (Vilcabamba) and the windward effect in the east (Zamora, El Libano) become obvious. Three climate diagrams (Fig. 7.2) show the most important differences. Average temperatures are about 2.6 K higher in the west than in the east and the annual fluctuations of the mean temperatures are much smaller than the diurnal

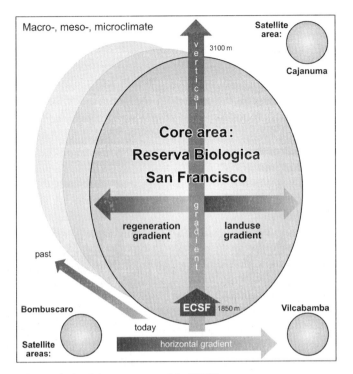

Fig. 7.1 Gradient analysis of the ecosystem of the RBSF

oscillations, a situation that is typical of the inner tropics. The most conspicuous differences between the three sites arise from the recordings of precipitation and evaporation. Annual precipitation is three times higher at El Libano than at Vilcabamba. In contrast, evaporation at Vilcabamba is more than twice as much as at El Libano. In both respects, RBSF holds an intermediate position. Seasonality, especially of air humidity and rainfall, also varies along the horizontal gradient. The wettest period in RBSF and El Libano is from April to July with 52% of the annual rainfall. Vilcabamba in the rain-shade receives 62% of its annual precipitation during a shorter humid season early in the year. The intermediate climatological situation of RBSF is also reflected by the vegetation. While the majority of the taxa found in the RBSF are of Amazonian origin, a few taxa reach their most eastern limit here. This holds e.g. for the deciduous trees like *Tabebuia chrysantha* which has its distribution centres further west in the much drier inter-Andean basin and the coastal area.

Spatial heterogeneity on a midrange scale is high. Between crests, slopes and gorges substantial changes in the vegetation were found concomitantly with differences in abiotic factors. Almost all parameters change from the more moist gorges with larger trees to the more open crests with smaller trees (Chapter 18 in this volume).

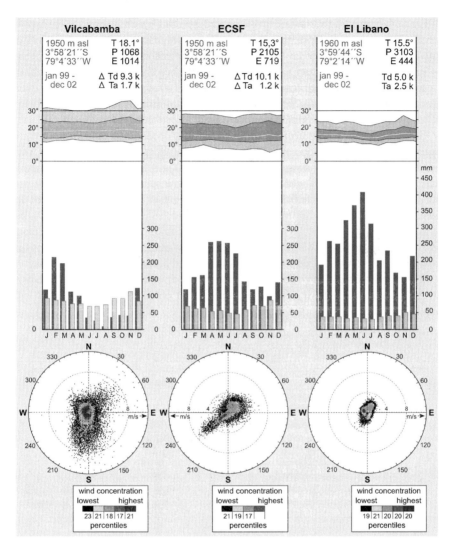

Fig. 7.2 Climate diagrams of Vilcabamba in the semiarid western part of the Cordillera Real, the RBSF (measured at ECSF) and of Zamora-El Libano at the west-eastern escarpment. All weather stations are at the same altitude. *T* Annual temperature mean, *P* precipitation, *E* evaporation, $\Delta\, Ta$ mean annual temperature range, $\Delta\, Td$ mean diurnal temperature range

7.2.2 The Vertical Gradient

The vertical gradient, as investigated in detail on the RBSF terrain, is clearly characterized by the data of three weather stations at 1960 m, 2670 m and 2930 m a.s.l. (Chapter 8 in this volume), but also by many other abiotic and biotic features, summarized in Fig. 7.3. Interestingly, the precipitation gradient does not correspond to the typical adiabatic temperature lapse rate. Rainfall is around 450 mm/100 m below 2600 m and ca. 360 mm/100 m above that level, but the uppermost part of the RBSF terrain is subjected to an extremely high rainfall which results from strong and humid easterly airflows. Páramo plants are well adapted to wind, rain and cold, and typical plant communities may be composed of up to 90 vascular plant species per 100 m^2 (Chapter 10.4 in this volume).

Soils, throughout on acid parent material, show a considerable increase of the organic top layer with increasing elevation. Obviously, conditions for litter decomposition become more and more unfavourable (Chapter 14 in this volume), mainly due to water logging, low temperatures, concomitant weak microbial activity, and decreasing nutrient content of the litter (Wilcke et al. 2002). These layers are densely rooted like in other tropical mountain forests (Grieve et al. 1990; Tanner et al. 1998; Hafkenscheid 2000) and the root/shoot ratio of plant biomass increases five-fold from 1050 m to 3060 m a.s.l. (Röderstein et al. 2005; Chapter 15 in this volume). This result indicates difficulties in nutrient acquisition by the trees, which leads to increased root production together with efficient mycorrhization (Chapter 10.5 in this volume). In principle the data suggest phosphorus limitation at lower and strong nitrogen limitation at higher altitudes.

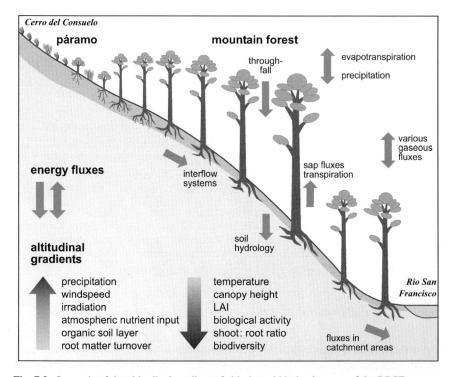

Fig. 7.3 Synopsis of the altitudinal gradient of abiotic and biotic elements of the RBSF

Litter-eating oribatid species, like other primary decomposers, are fairly rare. The decrease in number of individuals with altitude explains the slower decomposition and accumulation of organic material with increasing elevation (Chapter 11.4 in this volume).

Where not cleared, an evergreen forest extends from the valley bottom up to the tree line ecotone at about 2850 m a.s.l. This forest was originally addressed as primary mountain rain forest (Bussmann 2001b, 2002); however, later, large patches of the forest in particular below 1950 m turned out to be secondary forest which is hardly older than 50 years (Chapter 32 in this volume). Nevertheless, the main part of the forest may still be considered as primary forest.

The mean canopy height of the forest decreases from about 32 m at 1050 m a.s.l. to 9 m at 3060 m a.s.l. (Chapter 15 in this volume), and the same tendency holds for both the stem diameter (1.7–0.7 m dbh) and the LAI (Chapter 16 in this volume). The density of stems, however, increases from about 1000/ha to 8300/ha (Chapter 15 in this volume). On a larger scale, seven major zonal types of the forest have been differentiated around and in the RBSF (Chapter 10.1 in this volume), six of which are evergreen and one semi-deciduous (see Fig. 1.6 in this volume).

The decrease in tree species diversity and canopy height with elevation is a tendency that has been generally observed on tropical mountain forests (Lieberman et al. 1996; Aiba and Kitayama 1999). While such decrease holds for the terrestric plant cover, the diversity of the epiphytes increases with altitude up to about 2500 m. The mountain crest, as already mentioned, is subjected to quasi-permanent strong easterlies, giving rise to a small scale pattern of a Yalca-type bushland or "upper Andean forest" and a special kind of Páramo vegetation (Cleef et al. 2003).

Birds and bats play an important role as pollinators and fruit-dispersers (Chapter 11.2 in this volume). The functional relationship between bird communities and vegetation structure revealed a maximum of biodiversity in the narrow altitudinal range between 1900 m a.s.l. and 2000 m a.s.l., presumably due to the more open vegetation in this altitude (Chapter 11.1 in this volume). Diversity of ectothermic herbivorous insects was also expected and finally proven for moths to decrease along the elevation gradient (Chapter 11.3 in this volume). However, on a lower systematic level, diversity patterns of the extraordinary species rich moths were variable and taxonomic composition of local ensembles changed markedly with elevation.

7.3 Gradients Resulting from Human Activities

7.3.1 *Gradient of Forest Disturbance for Land Use Purposes*

In the RBSF wide areas of natural forest have been cleared for human use. These comprise rather small-scale clearings for roads and settlements but also vast areas for agricultural land. Human impacts can be arranged along an abstract gradient reflecting the distance in naturalness from the original ecosystem (Fig. 7.4). Five classes of (decreasing) naturalness depending on the intensity of land use were differentiated, starting with Class I, the untouched forest. A forest management experiment to promote the growth of valuable tree species was initiated on a reasonably sized patch (Chapter 26 in this volume), representing Naturalness Class

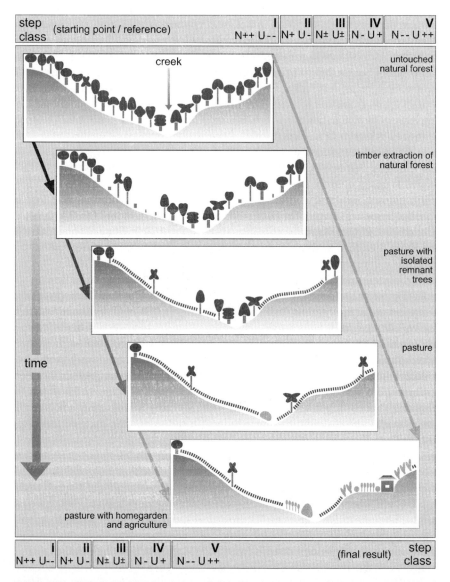

Fig. 7.4 Progressive gradient (from ++ to --) of human disturbance in the study area. Naturalness Classes (*N*) and their intensity of use (*U*, steps *I* to *V*) are defined by their supposed distance to the reference area "natural forest" and "extensive land use (home-gardens)"

II. Quite different from the natural ecosystem are the areas where forest has been converted into pastures leaving behind isolated islets of trees. This type of silvo-pastoral use of the area was ranked as Naturalness Class III. Pastures which are completely devoid of trees were classified as Naturalness Class IV, and areas of extensive agricultural use such as home-gardens or crop fields represent Naturalness Class V.

Anthropogenic habitat changes may, however, not only decrease biodiversity but even increase it, at least with respect to individual groups of organisms. While the α-diversity of epiphytic bryophytes and geometrid moths declined by ca. 30% and that of vascular epiphytes by 80% from Naturalness Class I to Naturalness Class IV, the species richness of lichens, geometrid and arctiid moths was even ca. 10–50% higher in disturbed forest or another late successional stage (Naturalness Class II) than in primary forest. A comparable positive tendency was recorded for birds (see Chapters 10.3 and 11.1 in this volume).

7.3.2 Gradients of Regeneration

If land use is abandoned or reduced in intensity, a reverse development can be expected. In principle two routes of naturalization have been observed:

1. Upon decreasing intensity of use, the naturalness of the system recovers and finally arrives at the stage of a secondary forest (Fig. 7.5). This takes place mainly on former small-scale clearings surrounded by the natural forest. Interestingly, species composition and structure of these forest show clear differences which can be related to the mode of disturbance of the natural forest (Chapter 32 in this volume).
2. Abandoned pastures, where a quasi-stable weed vegetation of bracken fern and heavily propagating bushes develops and where, because of the distance to the natural forest and the lack of seed import, no regeneration of forest has been observed (Chapter 28 in this volume). In these areas attenuation of faunal diversity is most pronounced (Chapter 35 in this volume).

Conversion of pastures into a kind of natural forest can be achieved by planting indigenous trees. Naturalness Class II seems to be achievable within a limited time span. Different tree species were tested for their usefulness in an afforestation experiment (Chapter 34 in this volume). Besides some exotic and pioneer species as trailblazers, late successional native tree species were also used.

Studying the progressive as well as the regressive gradients of human impact facilitates an insight into the degradation processes and options for rehabilitation of the tropical mountain forest. From the acquired knowledge, assessments of the mode and severity of the anthropogenic disturbance can be made and suggestions for efficient interventions can be provided.

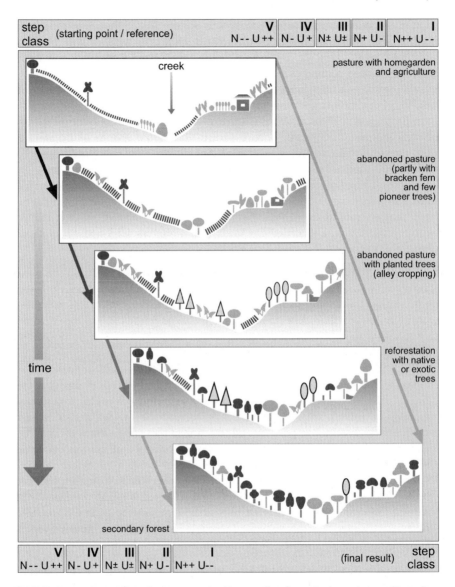

Fig. 7.5 Regressive gradient (from ++ to −−) of human disturbance in the study area. Naturalness Classes (*N*, classes *V* to *I*) are defined according to their supposed distance between the reference areas "natural forest" and extensive land use (*U*)

Chapter 8
Climate

J. Bendix(✉), R. Rollenbeck, M. Richter, P. Fabian, and P. Emck

8.1 Introduction

The altitudinal gradient of climate is a major driver for abiotic and biotic processes in the study area. Hence, a set of meteorological stations was established along an altitudinal gradient in the RBSF area (Table 8.1). The observations are representative for the climate outside the forest, which gives a meteorological background signal for micrometeorological measurements within the forest. The aim of the current Chapter is to present an inventory of the meteorological gradients in the central study area. Climatic irregularities and their impact are described in Chapters 20–23.

8.2 Data and Methods

The basic equipment consists of a series of three standard automatic meteorological stations (MS) of which the first one (ECSF MS) started operating in March 1998 (Richter 2003). The figures presented in this Chapter are based on the entire time series available for each meteorological station. The installations have been significantly extended since 2001. A set of quadratic mesh grid collectors (see Schemenauer and Cereceda 1994) for the measurement of fog and cloud water deposition (including horizontal rain) and additional totaling rain gauges was successfully installed. Data are collected at a weekly sampling interval. More sophisticated instrumentation especially for the in situ and indirect measurement of precipitation was installed at the ECSF MS. A detailed description of the ECSF meteorological station is given by Rollenbeck et al. (2007).

Table 8.1 Observation sites along the altitudinal gradient (1998–2005). *MS* Meteorological station, *RT* rainfall totaliser, *FC* fog collector, *LAWR* radar

Name	Altitude [m a.s.l.]	Type of measurement	Start of operation
Rio	1800	FC, RT	April 2002
ECSF MS	1960	MS, FC, RT	March 1998
Plataforma	2270	FC, RT	April 2002
TS1 MS	2660	MS, FC, RT	May 1998
Cerro MS	2930	MS	April 1998
Cerro del Consuelo	3180	FC, RT, LAWR	April 2002

8.3 Results and Discussion

8.3.1 Gradient of Solar Irradiance

The average hourly irradiance at the main meteorological stations reveals a clear altitudinal gradient of available solar energy (Fig. 8.1). This is especially established at noon and during the afternoon (1200–1600 hours) when upcoming clouds shelter the ground from direct irradiance.

Irradiance is especially reduced in the main rainy season in the austral winter (JJA) when satellite image interpretation points to an overall high cloud frequency of ca. 84% for the whole study area (Bendix et al. 2006a). Energy input is significantly enhanced from the late boreal autumn/winter (O-F) to boreal spring (AM), with the exception of February to March which show a very high cloud frequency with an average of up to 91% at the Cerro del Consuelo. The reduction of irradiance from the ECSF station up to the top of the Cerro del Consuelo is especially high during the austral winter (ca. −52% in August) while the lowest gradients are observed for the drier months of the year (Fig. 8.2; e.g. ca. 18% in January).

8.3.2 Gradient of Air Temperature

The average gradient of air temperature (decrease of temperature with terrain height) in the study area varies according to the time of the day and season (Fig. 8.3). Temperature differences (ΔT) between ECSF MS and Cerro MS are generally reduced during early morning hours (ΔT ca. 4–5 °C) but increase significantly during daylight, especially in the relatively drier months. November illustrates the drier season with a ΔT of ca. 9–10 °C, in comparison with the wet month of July which is characterised by a ΔT of ca. 6–7 °C. The main rainy season in the austral winter is generally the coldest time of the year.

The average gradient of air temperature between ECSF and Cerro del Consuelo of 0.61 °C·100 m^{-1} resembles the moist adiabatic lapse rate, which is similar to (but slightly higher than) other observations at the eastern Andean slopes of central

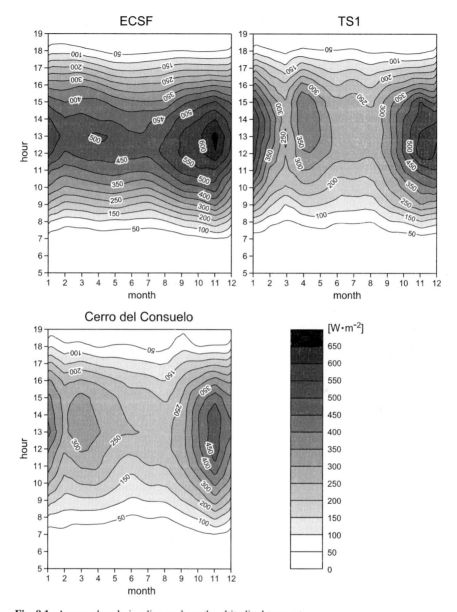

Fig. 8.1 Average hourly irradiance along the altitudinal transect

Ecuador (0.52–0.54 °C · 100 m^{-1}; Lauer and Rafiqpoor 1986; Bendix and Rafiqpoor 2001). The significant homogeneity over the year (standard deviation σ = 0.02 °C) underlines the perhumid character of the diurnal climate in the study area. However, the behaviour of temperature alters with season and terrain height. While the average

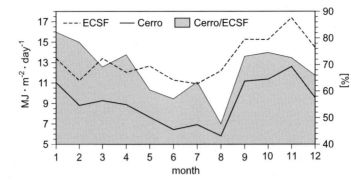

Fig. 8.2 Average daily solar irradiance at the ECSF and Cerro meteorological stations

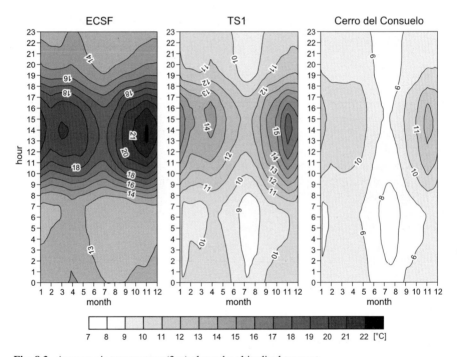

Fig. 8.3 Average air temperature (2 m) along the altitudinal transect

annual temperature gradient between TS1 and Cerro MS is 0.69 °C·100 m^{-1}, the high fluctuation over the year ($\sigma = 0.11$ °C) illustrates a change of the thermal environment with similar values in the austral winter (e.g. 0.53 °C·100 m^{-1} in July) and gradients which resembles more the dry adiabatic lapse rate in the drier months of the year (e.g. 0.88 °C·100 m^{-1} in November). This change does not occur in the lower parts of the atitudinal transect (ECSF to TS1) where the moist adiabatic gradient is fairly constant over the year (0.67 °C·100 m^{-1} in July; 0.5 °C·100 m^{-1} in

November). Ecological importance may also be attributed to the extremes of air temperature as presented in Table 8.2. As expected, the coldest temperature of the observational period is registered at the Cerro MS and the highest at the ECSF MS. Overall, the span of air temperature decreases with terrain height.

8.3.3 Gradient of Vapour Pressure Deficit

The gradient of vapour pressure deficit (VPD) illustrates the typical situation of tropical valleys crossing the eastern Cordillera of the Andes with a relatively dry valley bottom and moister slopes (Fig. 8.4). Humidity generally increases with terrain altitude, especially in the drier months of the year during daylight. The situation at the top of the Cerro del Consuelo is nearly constant over the year with humidities close to saturation point, which means a high tendency to cloud formation, with some exceptions during Veranillo del Niño events in the early austral summer (October to January).

However, the inversion of VPD in the early morning hours of the drier months (compare ECSF and TS1 in months 10–12 and months 1–3) contrasts the general decrease with altitude. VPD is close to zero at the ECSF but increases significantly to the TS1 MS. This is most likely due to the formation of low stratus clouds/radiation fog in the boundary layer which is triggered by relatively unhampered nocturnal outgoing radiation. As Fig. 8.4 shows, these clouds are burnt away by solar heating during the morning when the normal gradient of VPD is reconditioned. It should be emphasised that periods of very high VPD can occur day and night in the drier months (e.g. above 2 hPa Cerro MS), especially during a north-westerly air flow. If these periods occur during daylight, atmospheric water stress of the vegetation can be the result.

8.3.4 Gradient of Wind Field

Average wind velocity in time and space is the result of a complex interaction of larger-scale circulation patterns and interactions with topography as well as local breeze systems which are caused by the diurnal course of heating and cooling. However, obvious altitudinal behaviour of wind velocity can be deduced from

Table 8.2 Absolute minimum and maximum air temperature during the observational period along the altitudinal transect

Met. station	Min	Max	Max–min
ECSF	5.0 (December, 0800 hours)	29.1 (April, 1400 hours)	24.1
TS1	4.7 (November, 2100 hours)	25.5 (November, 1500 hours)	20.8
Cerro del Consuelo	2.6 (February, 1000 hours)	20.7 (December, 1500 hours)	18.1

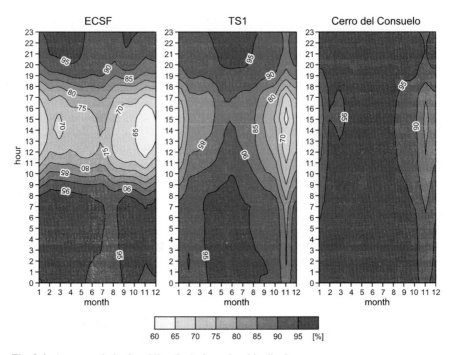

Fig. 8.4 Average relative humidity (2 m) along the altitudinal transect

Fig. 8.5. The valley stations show relatively low average wind speeds <2 m s^{-1} which are just slightly enhanced at the ECSF MS during noon/afternoon when the thermal upvalley/upslope breeze is well established. A drastic change occurs at the highest point (Cerro MS) where the average wind speed exceeds 7 m s^{-1}. It is striking that wind velocity shows a maximum during the rainy season in the austral winter (June to August).

The absolutely hourly maximum of wind velocity over the observational period reveals the same altitudinal characteristics. The valley stations (ECSF, TS1) only reach the intensity of a fresh breeze (level 5 on the Beaufort scale) while stormy winds (level 8 on the Beaufort scale) are occasionally reached at the Cerro MS (19.8 m s^{-1} at Cerro MS).

The altitudinal gradient of the prevailing wind direction clearly reveals the dynamic reason for the strong difference in wind speed (Fig. 8.6). The ECSF station is characterised by the diurnal cycle of the vally breeze system. The noon/afternoon hours are dominated by an up-slope/up-valley flow (NE/E) while the reversed nocturnal circulation system consists of cold air drainage flow from the slopes and down the valley (SE in the early morning hours). The situation is less clear at the TS1 MS where easterly directions begin to predominate. The situation completely changes at the Cerro MS where south-easterly directions prevail nearly the whole time. Only in the drier month of November (0000–0500 hours) are westerly winds established in the average wind field. The streamflow can also help to explain the

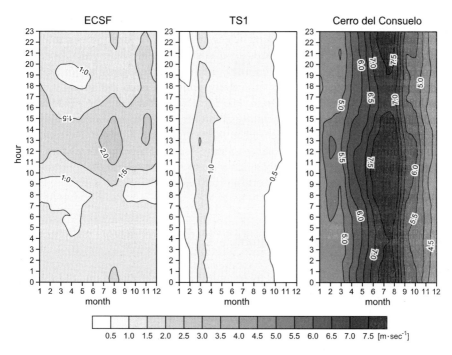

Fig. 8.5 Average wind velocity (2 m) along the altitudinal transect

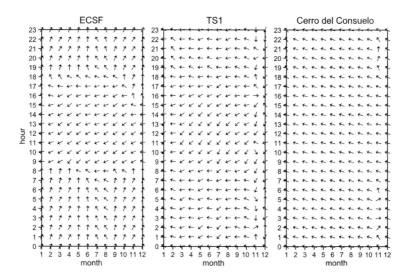

Fig. 8.6 Vectorial average of wind direction (2 m) along the altitudinal transect

altitudinal gradient of wind velocity. The Cerro MS is affected in that it is nearly unprotected from the synoptic wind which is characterised by strong mid- and upper-level easterlies from the Amazon almost throughout the whole year. Upper level westerlies from the Pacific reach the study area only occasionally late in the year.

In contrast, the ECSF MS is effectively protected from the strong synoptic circulation by the surrounding terrain. The slightly enhanced average wind velocity during the insolation period coincides with general observations of thermal breeze systems and is due to the stronger radiative forcing during daylight. The low average wind speed at the TS1 MS points to a transition zone between the weaker upper branch of the valley/up-slope circulation and the synoptic flow which however, is significantly weakend due to the topographical protection of the site.

8.3.5   Gradient of Precipitation

Rainfall is a major ecological and hydrological factor, keeping in mind that a considerable quantity of ecologically relevant water can be provided through direct deposition of cloud and fog water on the vegetation (e.g. Ataroff 1998). Fig. 8.7 illustrates that the main rainy season at all stations of the transect lasts from April to August even if rainfall is high all through the year.

Hence, the study area generally belongs to the regime type of the eastern Andean slopes as described by Bendix and Lauer (1992). It is striking that the lowest stations tend to an earlier maximum (April to May) while the peak time at the higher stations (TS1, Cerro) is shifted towards the austral winter (May to July). The figures shows a clear vertical gradient of rainfall with relatively low values at the valley bottom and a strong increase to the top of the Cerro del Consuelo. The average gradient yields an increase of $220\,\text{mm} \cdot 100\,\text{m}^{-1}$. However, precipitation is nearly constant between ECSF and plataforma as well as between TS1 and Cerro del Consuelo. The real steep gradient occurs between 2270 m and 2660 m a.s.l. (Plataforma to TS1) with an increase of $663\,\text{mm} \cdot 100\,\text{m}^{-1}$. Along the total vertical distance of 1320 m from Cerro to ECSF, rainfall decreases by about 54%.

By assessing the gradient of precipitation it should be emphasised that conventional rain gauge measurements have well known disadvantages, especially in windy environments which can lead to a severe underestimation of rainfall (see e.g. Groisman and Legates 1994; Serra et al. 2002). A main source of error is horizontally advected raindrops which are not adequately captured by the gauge. It becomes clear from Fig. 8.5 that an underestimation is most likely at least for the stations at Cerro del Consuelo due to the high average wind speed. Cloud/fog water deposition is also a function of wind velocity which determines the throughflow rate of cloud/fog droplets (through vegetation or a fog collector). The fog collectors (FC) used in the current study samples both, horizontal rain and cloud/fog water deposition which act as an ecologically relevant water increase. Fig. 8.8 illustrates the great importance of this source of water in the mountain crest area, while it becomes nearly neglectable at the valley bottom (ECSF FC).

Again, the peak of intake of additional atmospheric water to the ecosystem can be observed between April and June, with the steepest gradient between TS1 FC and Cerro FC (275 mm · 100 m^{-1}). Table 8.3 reveals that the additional amount of atmospheric water most probably not collected by the rain gauges yields 41.2% for the Cerro del Consuelo. This would mean a yearly total of 6701 mm of ecologically available water input from the atmosphere.

Of interest are the rain rates along the altitudinal transect. Hourly average rain rates are generally low for all stations. This shows that most of the precipitation originates from orographic cap clouds formed due to the smooth stable ascent of air parcels over mountains, maybe with embedded shallow convection (Lin 2003; Bendix et al. 2006b). Generally, the rain rate decreases from Cerro del Consuelo to the ECSF. This gives evidence that most rainfall is formed in the crest area and rain droplets evaporate on their way to the valley bottom. The highest rain rates

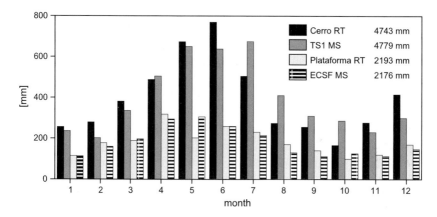

Fig. 8.7 Average monthly rainfall as collected from each meteorological station (MS; see Table 8.1) and rainfall totaliser (*RT*) along the altitudinal transect

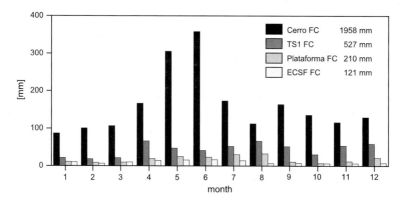

Fig. 8.8 Average monthly water intake to each fog collector (*FC*; see Table 8.1) along the altitudinal transect

Table 8.3 Water surplus due to horizontal rain and cloud/fog water deposition along the altitudinal transect

Station	Rain (mm)	Horizontal deposition (mm)	Sum (mm)	Water increase (%)
ECSF	2176	121	2297	5.6
Plataforma	2193	210	2403	9.6
TS1	4779	527	5306	11.0
Cerro del Consuelo	4743	1958	6701	41.2

occur in the main rainy season but two times a day with enhanced intensities can be observed:

1. An afternoon maximum (1300–1600 hours) related to convective events due to local thermal forcing;
2. An early morning maximum which is most probably a result of mesoscale dynamics as described by Bendix et al. (2006b). However, strong convective showers occasionally occur at all stations, with maximum rain rates of up to 36.2 mm h^{-1}.

8.4 Conclusions

The current section illustrates that specific altitudinal gradients occur for all meteorological elements. Generally, the ecological conditions due to weather become more and more unfavorable with increasing terrain height. This is exemplarily emphasised by the strong gradient of insolation between the valley bottom and the crest at the Cerro del Consuelo of −52% in the main rainy season. The high cloudiness in the crest area is reflected by a relative humidity of the air which is close to saturation most of the year. A remarkable gradient is recognised for wind speed which is clearly related to the prevailing wind direction. A significant jump to high values occurs between the lower meteorological stations (ECSF, TS1) and the crest site (Cerro MS). Together with partially steep gradients of precipitation and temperature in the upper part of the profile, the slope of the natural forest can be subdivided into three vertical zones:

1. The valley bottom (Rio to plataforma) which is sheltered from the synoptical circulation by the forelying Cordillera del Consuelo. Hence, the upper Rio San Francisco Vally in the vicinity of the ECSF station is situated in a lee situation from the prevailing easterly trades. The valley stations are clearly influenced by local circulation patterns with the typical reversal of streamflow between day and night.
2. The synoptic situation becomes more and more prominent in altitudes where the shelter effect of the Cordillera del Consuelo weakens. Hence, the level between plataforma and TS1 can be considered to be a transition zone to the third level where the effects of unsheltered synoptical circulation dominates.

3. This is the crest zone at the Cerro del Consuelo. Precipitation is drastically increased, especially the water input by horizontal rain and cloud/fog water deposition. High cloudiness and rainfall but also the high wind velocities of the nearly undisturbed strong easterlies lead to an especially unfavourable thermal environment.

The decrease of rainfall towards the valley bottom clearly shows that rainfall formation mainly takes place on a larger scale due to the barrage effects of the easterlies at the windward site of the Cordillera del Consuelo. An additional local trigger for precipitation (especially in the afternoon hours) is the thermal breeze system of the valley. It should be stressed that gradients vary over the year. The main rainy season in the austral winter reveals the greatest gradients in precipitation and wind velocity, while the drier months are characterised by greater altitudinal gradients in solar irradiance, air temperature and relative humidity.

Chapter 9
Soils Along the Altitudinal Transect and in Catchments

W. Wilcke(✉), S. Yasin, A. Schmitt, C. Valarezo, and W. Zech

9.1 Introduction

In the literature, it is well documented that there are systematic changes in soil properties and thus in soil fertility along altitudinal gradients in tropical mountains (e.g., Marrs et al. 1988; Grieve et al. 1990; Schrumpf et al. 2001). Grieve et al. (1990) reported increasing soil organic matter concentrations along an altitudinal transect from 100 m to 2600 m a.s.l. in Costa Rica. Furthermore, it has been reported for several forests that C/N ratios increase with increasing altitude (Edwards and Grubb 1982; Schrumpf et al. 2001). Mineralization rates are negatively correlated with the C/N ratio and the lignin and polyphenol concentrations (Tian et al. 1995). Consequently, N mineralization and nitrification rates decrease with increasing altitude (Marrs et al. 1988). For the study area in Ecuador, it was shown that the pH and N concentrations decrease and C/N ratio and hydromorphic properties increase with increasing altitude between 1850 m and 3050 m a.s.l. in a previous study (Schrumpf et al. 2001).

9.2 Methods

This chapter presents the properties of soils studied along a transect ranging from 1050 m a.s.l. in the lowland tropical rain forest via the lower montane forest between 1850 m and 2190 m to the Páramo regions above 2500 m. Within this long transect the section between 1850 m and 2500 m was studied at a higher intensity ("short transect"). Furthermore, between 1850 m and 2200 m the spatial distribution of soil properties was assessed in even greater detail at 47 sampling sites for an area of about 30 ha belonging to three catchments covering 8–13 ha (a map of the transects and catchments is shown in Fig. 1.2, in Chapter 1). Further forest soil data are reported in Chapters 13–15, 17. Soil data for pastures are presented in Chapter 31.

We determined bulk density, texture, pH, potential and effective cation-exchange capacities (CEC_{pot}, CEC_{eff}), little crystalline iron oxides (Fe), and crystalline pedogenic iron oxides (Fe_d) with standard methods (Schlichting et al. 1995). Total polyphenol concentrations in the organic horizons were determined with the Folin-Ciocalteau method (Box 1983). The amount of lignin in the organic horizons was estimated as vanillyl + sinapyl + coumaryl (=VSC) lignin according to Kögel (1986). Total concentrations of C and N in organic layer and mineral soil samples were determined with an elemental analyzer after grounding in a ball mill. To determine total metal concentrations, organic layer and mineral soil samples were digested with concentrated HNO_3. Metal concentrations in all extracts were measured with flame atomic absorption spectrometry. Total P concentrations in extracts were determined with ICP-OES.

9.3 Results and Discussion: Relationship Between Altitude and Soil Properties

9.3.1 The Long Transect

The most obvious and consistent finding was that organic matter increasingly accumulated on top of the mineral soil, forming thick organic layers with increasing altitude (Fig. 9.1). At the same time, C/N ratios decreased systematically, resulting in an increasingly poorer N availability with altitude (Fig. 9.2). Similar results are reported by Iost et al. (Chapter 14 in this volume). The decreasing N availability with altitude is reflected by an increasing belowground:aboveground biomass ratio (see Chapter 15) and decreasing foliar N concentrations (see Chapter 17). These results confirm previous work at other tropical mountain sites (e.g., in Costa Rica; Marrs et al. 1988; Grieve et al. 1990). However, from Figs. 9.1 and 9.2 it is evident that there is considerable scatter in the data concerning the relationship between altitude and organic matter accumulation and turnover. This is attributable to the high soil heterogeneity at the small scale (Wilcke et al. 2002).

9.3.2 The Short Transect

The short transect comprised eight study sites (numbered I–VIII from the lower to upper positions), of which the two lowermost (Sites I and II, at 1960 m and 2070 m a.s.l., respectively) were in a slope position and six were on the ridge (Sites III–VIII). At each study site three replicate soils were sampled.

Most soils were Humic Cambisols (Sites I, III, IV) or Dystric Cambisols (Sites II, V, VI). At Site VII, there were Dystric Planosols and at Site VIII Terric Histosols (FAO, UNESCO 1997). As we did not observe systematic differences in the properties of the soils in slope and in ridge positions, we bulked all results and evaluated the relationship between altitude and soil properties irrespective of the topographic position.

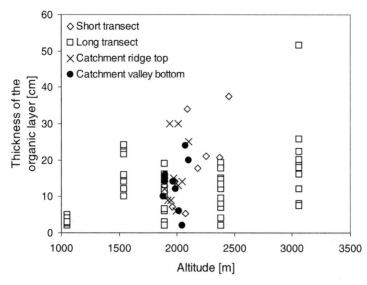

Fig. 9.1 Relationship between altitude and the thickness of the organic layer along the short and long transects and in Catchment 2 (see Fig. 1.2 in Chapter 1) in ridge top and lower slope positions. Part of the data was provided by Susanne Iost and Franz Makeschin

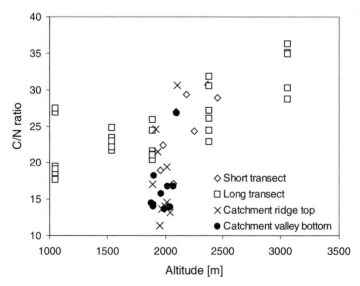

Fig. 9.2 Relationship between altitude and the C/N ratios in the organic layer along the short and long transects and in Catchment 2 (see Fig. 1.2 in Chapter 1) in ridge top and valley bottom positions. Part of the data was provided by Susanne Iost and Franz Makeschin

Table 9.1 Mean element concentrations in the O horizons ($n = 3$, mean of Oi, Oe, and Oa horizons weighted according to the thickness of these horizons) at eight sites along the "short altitudinal transect" in south Ecuador

	Altitude (m a.s.l.)	C	Ca	K	Mg	N	P
		(g kg^{-1})					
Slope							
Site I	1960	390	3.6	3.5	1.4	21	8.7
Site II	2070	345	6.9	5.7	1.9	20	10
Ridge							
Site III	1980	443	1.5	1.8	1.2	20	7.7
Site IV	2090	485	0.51	1.1	0.64	18	5.7
Site V	2180	465	1.7	1.1	0.84	16	5.2
Site VI	2250	466	0.56	1.3	0.95	19	6.1
Site VII	2370	413	0.56	1.2	0.61	13	4.1
Site VIII	2450	356	0.18	1.1	0.34	12	3.4

The mean concentrations of all macronutrients (N, P, K, Ca, Mg) in the three replicate soils per site of the whole organic layer decreased with increasing altitude (Table 9.1). The correlation between altitude and mean element concentrations in the organic layer was significant for all elements except for Ca and K. For the latter two elements, the correlation was only significant after one outlier (Site II) was removed. Thus, soil fertility decreased with increasing altitude.

Mean C concentrations of the whole organic layer (consisting of the three morphologically different Oi, Oe, and Oa horizons – the Oi horizon consists of fresh litter, the Oe of fragmented litter, and the Oa of humified material without visible plant structures) were not related with altitude, but the C concentrations of the Oi horizon showed a positive correlation ($r = 0.86$). Increasing C concentrations in the Oi horizon with increasing altitude are probably attributable to decreasing nutrient concentrations of the litter. The latter is likely because of decreasing nutrient concentrations in leaves with increasing altitude (see Chapter 17).

In contrast to the concentrations, the mean storages of all studied elements were independent of altitude, reflecting the reciprocal trends of increasing organic layer thickness and decreasing nutrient concentrations with increasing altitude.

Thus, decreasing nutrient concentrations with increasing altitude were compensated by increasing organic layer thickness. Consequently, the mass of nutrients per unit area was independent of altitude but the availability decreased with increasing altitude because of decreasing concentrations. This assumption is supported by the fact that root biomass and the proportion of root biomass of total above- and belowground biomass increased with increasing altitude in our study area probably as response to decreased nutrient availability (Röderstein et al. 2001; Soethe et al. 2006; see also Chapter 15).

The mean VSC lignin concentrations in the organic layer of the three replicate soils per site correlated significantly negatively with increasing altitude (Fig. 9.3). The decrease in VSC lignin concentrations in the organic layer with increasing altitude suggests that lignin accumulated relative to other organic compounds at the

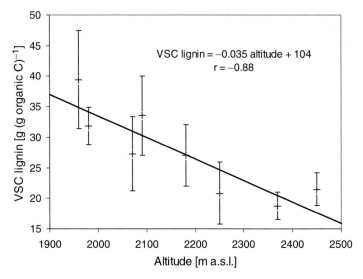

Fig. 9.3 Relationship between altitude and mean VSC lignin concentrations in the organic layer. *Error bars* indicate standard deviation ($n = 3$)

lower study sites because of a generally higher mineralization rate. However, lignin did not seem to contribute to reduced organic matter turnover with increasing altitude, which would have been shown by increasing lignin concentrations with increasing altitude. As furthermore the polyphenol concentrations were independent of altitude, we conclude that there are no indications that the decreased organic matter turnover rates at higher altitude are attributable to the concentrations of lignin or polyphenols. Thus, our results do not support the hypothesis of Bruijnzeel et al. (1993) that reduced nutrient supply at higher altitudes is related to increased polyphenol production of the plants and resulting higher recalcitrance of the organic matter. Instead, frequent waterlogging and decreasing temperatures are likely to be the main reason for the increasing thickness of the organic layer with increasing altitude. Furthermore, increasingly unfavorable nutrient supply with increasing altitude might contribute to a reduced organic matter turnover.

In the A horizon, the texture became coarser with increasing altitude. There were close significant correlations between altitude and sand ($r = 0.75$) and clay concentrations ($r = -0.80$). This went along with a decrease in mean CEC_{eff} (−0.84), mean CEC_{pot} (−0.85), and mean Fe_o (−0.80) and Fe_d (−0.82) concentrations. However, among the texture, CEC_{eff}, CEC_{pot}, Fe_o, and Fe_d of the B horizons none was correlated with altitude. In the A and B horizons, pH values [A horizons pH (KCl), $r = -0.66$; B horizons pH (H_2O), $r = -0.67$ were significantly negatively correlated with altitude. As there is no systematic change in texture of the B horizons along the altitudinal transect, we attribute the decrease in clay concentrations of the A horizon to pedogenetic processes. Acidity increased with increasing altitude [from pH (KCl) 3.4 in the A horizon at the lowermost site to pH (KCl) 3.0 at the uppermost site] as well as the duration of waterlogging indicated by qualitatively increasing occurrence of mottling and the decrease in oxalate-soluble Fe (Fe_o, from $0.77\,g\,kg^{-1}$ in the A horizon at the lowermost site to $0.10\,g\,kg^{-1}$ in the A horizon at the uppermost site) and dithionite-citrate-soluble Fe concentrations (Fe_d, from $0.40\,g\,kg^{-1}$ to $0.07\,g\,kg^{-1}$) of the A horizons with increasing altitude. Therefore, ferrolysis might be responsible for clay destruction. Ferrolysis is caused by the production of H^+ as a consequence of Fe(II) oxidation after the Fe(II) ions released by reduction of Fe(III) oxides have replaced base metal ions at the cation exchanger surfaces. Furthermore, reductive dissolution of clay-sized Fe(III) hydroxides and vertical and lateral leaching of the released Fe might also contribute to the loss of clay. The negative trends in CEC_{eff} and CEC_{pot} with altitude are attributable to the decrease in clay concentrations.

In summary, our results demonstrate that the obvious decrease in organic matter turnover with increasing altitude is attributable to increasing soil wetness, reduced temperatures, and deteriorated nutrient supply although nutrient storages were independent of altitude. Decreased organic matter turnover rates, however, are not related to increasing concentrations of recalcitrant lignin and polyphenol. A similar deterioration of living conditions for microorganisms and growth conditions for plants with increasing altitude is reported in Chapters 14, 15, and 17, respectively. The latter Chapters discuss biotic responses to changing soil conditions with altitude.

9.3.3 The Catchments

The quantitatively dominating soils in all catchments were Humic Cambisols occupying 28, 60, and 28% of the area in catchments 1, 2, and 3. All soils are shallow, loamy-skeletal with high mica content (Yasin 2001). Hydrological and biogeochemical properties, processes, and budgets in these three catchments are described in Chapters 12 and 13.

In the 47 sampled soils, the thickness of the whole organic layer (Oi, Oe, Oa horizons) ranged between 2 cm and 43 cm, with a mean of 16 cm. The thickness increased with increasing altitude giving Histosols (mainly Folic Histosols) above ca. 2100 m, whereas lower down Cambisols dominated. At most profiles, Oi, Oe, and Oa horizons could be distinguished. The organic layers were densely rooted. The bulk density of the organic horizons increased with increasing depth over $0.08–0.23\,g\,cm^3$ and abruptly changed to about $1\,g\,cm^3$ in the mineral soil. The thick and densely rooted organic layers are common under tropical mountain forest (Grieve et al. 1990; Tanner et al. 1998; Hafkenscheid 2000).

As shown in Table 9.2, the total concentrations of most elements in the organic horizons were similar to those under a Venezuelan tall mountain forest with high cloud incidence on moderately fertile soils derived from sedimentary rocks north of our study region, except that N and P concentrations were higher in our region

(Steinhardt 1979). The concentrations of elements in the organic horizons in our region were higher than in four lower mountain rain forests on fertile soils derived from volcanic rocks in New Guinea (Edwards 1982).

The mean K, N, and P concentrations in the organic layer increased with depth (i.e., Oi < Oe < Oa), whereas those of C, Ca, and Mg decreased (Table 9.2). In all the organic horizons, the mean C, Ca, Mg, N, and P concentrations were higher than in the underlying A horizons, whereas those of K were lower. The mean C/N [Oi: 30, with standard error (s.e.) 1.9; Oe: 20, s.e. 1.0; Oa: 16, s.e. 0.6] and C/P (Oi: 710, s.e. 65; Oe: 505, s.e. 35; Oa: 459, s.e. 41) decreased in the order, Oi > Oe > Oa. The abrupt decrease in C concentrations between the Oa and A horizons and the high C concentrations in the Oa horizons indicate that mixing of the organic layer with the mineral soil was limited probably because of weak biological activity. If we assume that soil organic matter roughly contains 50% C, then we can estimate the contribution of mineral soil material (silicates and oxides) to the total mass of the horizon, on average, as increasing in the order: Oi (8%) < Oe (14%) < Oa (26%). Thus, although there was little biological activity, there was still some mixing of organic matter and mineral soil.

Element turnover times in the organic layer are the combined result of mineralization, leaching, and mixing of mineral soil with the organic layer. The depth distribution of the elements in the organic layer indicates the relative importance of these three processes. The reason for the increase in K concentrations in the order: Oi < Oe < Oa < A horizon was probably an increasing contribution from the mineral soil. In contrast, the decrease in the concentrations of all macronutrients except for K indicates that these elements did not accumulate during decomposition of the organic matter probably because they were leached into the mineral soil or taken up by the vegetation. The decrease in C concentrations and the increase in N and P concentrations with increasing depth giving decreasing C/N and C/P ratios is the result of the release of CO_2 during mineralization and the immobilization of N and P.

The chemical properties of the organic and the A horizons varied greatly (Table 9.2). The pH of the organic horizons spanned a wide range of buffer systems and was more variable than that in the A horizons (pH 3.9–5.8; Wilcke et al. 2001a). There were close and significant correlations between pH and concentrations of Ca ($r = 0.83$) and Mg ($r = 0.84$) in the organic horizons (data of Oi, Oe, Oa horizons pooled; Fig. 9.4). Thus, the concentrations of Ca and Mg in the organic horizons (= Ca_O, Mg_O) may be predicted from the pH by regression ($Ca_O = 4.8$ pH $- 17$; $Mg_O = 1.3$ pH $- 4.4$).

The wide range of pH of the organic horizons was unexpected because all soils were derived from acid parent material. Possible reasons for this heterogeneity are differences in litter quality among the many different plant species and a heterogeneous distribution of redox potentials. Temporal waterlogging might result in the production of alkalinity and thus an increase in pH compared with permanently well aerated soils.

The results demonstrate that the conditions for plant growth were heterogeneous, and local shortages of nutrients might not affect the whole forest. Further, because there was a large storage of all plant nutrients, the shortages, if present, may arise

Table 9.2 Mean and range (in brackets) of selected physical and chemical properties of the organic horizons, with comparable data from the literature. *ND* Not detected

Horizon	Thickness (cm)	pH in H$_2$O	C	Ca	K	Mg	N	P
					Total concentrations (g kg^{-1})			
Oi	2.5 (0.5–7.0)	5.0 (3.9–7.4)	463 (251–529)	10 (1.0–26)	2.6 (0.5–7.9)	3.4 (0.30–7.3)	17 (7.5–28)	0.87 (ND–4.6)
Oe	4.1 (0.5–25)	4.5 (3.5–7.1)	428 (262–516)	8.0 (0.55–23)	2.8 (0.65–9.9)	2.4 (0.34–6.3)	23 (12–32)	0.95 (ND–1.5)
Oa	9.1 (0.5–32)	3.8 (3.1–6.7)	370 (105–503)	4.8 (0.19–21)	4.0 (0.64–13)	1.5 (0.13–6.7)	23 (8.3–32)	0.97 (0.39–2.2)
Steinhardt (1979)[a]			–	10	2.1	2.5	11	0.72
				6.2	1.5	1.5	14	0.79
Edwards (1982)[b]			–	1.5–1.6	0.16–0.20	0.23–0.28	1.2–2.2	0.06–1.44

[a] Mean concentrations in Oi (upper values) and Oa horizons (lower values) of a Venezuelan lower mountain cloud forest.
[b] Oi horizons in a New Guinean lower mountain rain forest (concentrations: range of the means of three litter fractions from four forest types; storage: range of four forest types).

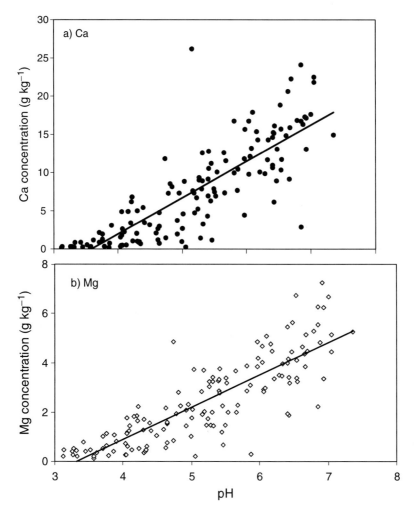

Fig. 9.4 Relationships between the concentrations of Ca (**a**) and Mg (**b**) and pH in the O horizons. The *lines* represent the regression lines of the concentrations of Ca on pH (Ca = 4.8 pH − 17, $r = 0.83$) and of the concentrations of Mg on pH (Mg = 1.3 pH − 4.4, $r = 0.84$)

because these nutrients are not readily available. A similar conclusion was drawn for Malaysian mountain forests by Bruijnzeel et al. (1993).

Part of the variation in soil properties is related to systematic effects such as the topographic position. Therefore, we studied ten soils along each of two transects at ca. 1 m distance from the stream draining the catchment between 1880 m and 2100 m a.s.l. ("valley bottom") and on the ridge between 1890 m and 2110 m a.s.l. ("ridge top"). These soils were included in the overall set of 47 soils and located only in one of the catchments (MC2; see Fig. 1.2 in Chapter 1).

The ridge top soils had, on average, a significantly lower pH in H_2O (organic layer: pH 4.0; A horizons: pH 4.2) than the valley bottom soils (pH 4.6/4.9, paired-differences test, $P<0.05$). The mean CEC_{eff} in the A horizons was significantly higher in the ridge top soils than in the valley bottom soils because of the higher organic matter concentrations (Fig. 9.5). The mean base saturation in the A horizons was significantly lower in the ridge top soils than in the valley bottom soils. In the valley bottom soils, mean total concentrations of Ca (O horizons: 8.7 g kg^{-1}/A horizons: 0.67 g kg^{-1}), Mg (2.5/0.94), and P (1.0/0.61) were higher than in the ridge top soils (Ca: 4.3/0.37; Mg: 1.5/0.70; P: 0.96/0.50); most differences between valley bottom and ridge top soils were significant. Enhanced acidification, lower base saturation, and lower total element concentrations of the ridge top than of the valley bottom soils were the result of element leaching from ridge top to valley bottom positions. The reason is probably the higher water input than in the sheltered valley bottom positions, and slightly lower soil temperatures.

Our results demonstrate that topographic position explains part of the spatial variability in soil properties under tropical mountain rain forests. This is attributable to differences in climatic conditions but also to the position in the landscape where water and matter transport occurs from upper to lower areas. Another systematic reason for soil heterogeneity was the landslides because of their effect on soil fertility (Wilcke et al. 2003; see also Chapter 24).

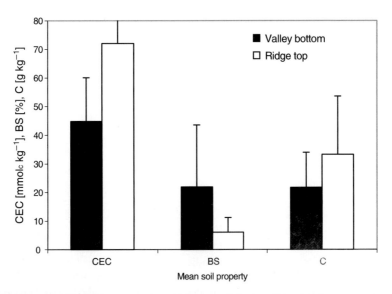

Fig. 9.5 Mean cation-exchange capacity (*CEC*), base saturation (*BS*), and C concentrations (*C*) of ten soils from the valley bottom and ridge top positions, respectively. *Error bars* indicate standard deviations

9.4 Conclusions

In summary, it can be clearly stated that soil properties change systematically with altitude along all studied shorter and longer transects. Soils become richer in organic matter, more acid, and less fertile as indicated by decreasing nutrient concentrations with increasing altitude.

In contrast, total storages of nutrients do not decrease with increasing altitude because of the increasing accumulation of thick organic layers on top of the mineral soil being a large nutrient reservoir. However, even at similar altitude soils are highly heterogeneous because of the heterogeneous substrates for soil formation, varying hydrologic conditions, element redistribution on a landscape level as a consequence of the steep and variable morphology, and potentially also because of drawbacks of the enormous diversity of plant species. Part of the soil heterogeneity at similar altitudes can be explained by systematic variations in different topographic positions. Ridge top soils have thicker organic layers and are more acid and depleted in nutrients leached to lower positions than valley bottom soils.

Acknowledgments We thank Richard Cuenca and Hector Valladarez and numerous Ecuadorian and German student helpers for their contributions to this work.

Chapter 10.1
Potential Vegetation and Floristic Composition of Andean Forests in South Ecuador, with a Focus on the RBSF

J. Homeier(✉), F.A. Werner, S.R. Gradstein, S.-W. Breckle, and M. Richter

10.1.1 Introduction: Andean Forests

A decrease in overall plant species diversity and forest stature with increasing elevation is a general trend in moist tropical mountain areas (e.g., Aiba and Kitayama 1999; Cavelier 1996; Gentry 1988, 1995; Grubb 1977; Lieberman et al. 1996; Richter 2001) and was confirmed for the RBSF (Homeier 2004).

However, the mechanisms responsible for the altitudinal change in both plant diversity and forest structure in tropical mountain forests remain poorly understood (see Chapters 10.3, 10.4 in this volume).

The altitudinal gradient, as well as the topographical gradient at a smaller scale, combines sets of environmental factors such as soil hydrology, nutrient supply, temperature and wind exposure, which are difficult to disentangle. The environmental abiotic settings of the study area are described by Bendix et al. (Chapter 8), Richter and Moreira-Muñoz (2005) and Wilcke et al. (Chapter 9).

Ecuadorian woody plant species richness declines with increasing elevation from the Amazonian lowlands, which harbour the world's highest alpha diversity in trees (Balslev et al. 1998), to upper montane forests (Jørgensen and León-Yánez 1999). Comparing Neotropical forests, Gentry (1988) found no pronounced differences in woody plant diversity for sites below ca. 1500 m, but a general decrease above. In contrast, epiphyte species richness peaks in montane forests: around 1000–2500 m a.s.l. in vascular epiphytes (Ibisch et al. 1996; Küper et al. 2004; Krömer et al. 2005) and between 2500–3000 m a.s.l. in bryophytes (Wolf 1993). Family patterns of species richness are fairly contrasting, but many are hump-shaped with maxima at middle elevations (Kessler 2002a). The same is true for elevational patterns of endemism (Kessler 2002a). In general, highest percentages of endemism for the Ecuadorian vascular plant flora are found in the Andean forests above 1500 m a.s.l. (Jørgensen and León-Yánez 1999; Kessler 2002a). The Podocarpus National Park in particular is known for its high endemism with more than 200 species restricted to this area (Valencia et al. 2000).

10.1.2 Results and Discussion

10.1.2.1 Potential Vegetation of the Podocarpus National Park Area

A classification of the plant formations was compiled using information from former publications, maps and our own records (Aguirre et al. 2003; Diels 1937; Grubb et al. 1963; Grubb and Whitmore 1966; Kessler 1991; Madsen and Øllgaard 1993; Ulloa and Jørgensen 1995; Valencia et al. 1999; Balslev and Øllgaard 2002; Lozano 2002; Richter 2003). Seven major types of potential natural vegetation are distinguished for the wider study area (see Fig. 1.6 in Chapter 1 in this volume). We omit the description of a variety of additional main vegetation types in south-western Ecuador (west of easternmost Loja province and the upper Rio Catamayo drainage) such as coastal dry forests or cloud forests, being not of major relevance for our study site in terms of floristic interchange. There is of course a great variety of vegetation subtypes, since not only the floristic composition or climatic gradients drive the differentiation of communities, but also edaphic factors, diverging traits of vegetation history and multiple disturbance regimes. Not all of these subtypes can be treated in the context of this short overview.

Currently, much of the province of Zamora–Chinchipe and most of the neighbouring province of Loja is stripped of its natural vegetation. For many centuries, but especially in the course of the past decades, large parts of the study area were converted. The main current threats for the forests of the region are extraction of timber, escaping fires and especially conversion into cattle pasture. With the construction of further roads, deforestation expands and the last remote territories are opened up to colonization. Quite likely, only a few patches within protected areas such as the Podocarpus National Park will survive the coming decades at the current rate of conversion. The vegetation types are specified in Table 10.1.1.

10.1.2.2 Floristic Composition of the RBSF

To date, 1208 species in 425 genera and 131 families of native spermatophytes have been recorded from the RBSF (Homeier and Werner 2007). This list remains fragmentary; sampling is incomplete, especially as non-woody plants and higher altitudes have only been collected opportunistically. We estimate the 1000 ha reserve harbours some 1500–1700 species of seed plants, or about 10% of the entire spermatophyte flora of Ecuador.

Both in terms of physiognomy and floristic composition, two steep local gradients promote the coexistence of a wide array of vegetation types and microhabitats. With increasing altitude, height and closure of the canopy quickly decline, coupled with high species turnover. At an even smaller spatial scale, corresponding changes take place along the ravine–slope–ridge gradient (Homeier 2004; see Chapter 18 in this volume).

Table 10.1.1 The seven vegetation types depicted in Fig. 1.1.6 and some of their characteristic vascular plant species (altitudinal ranges of the types strongly depend on local conditions)

A Evergreen premontane (= submontane) rainforest around Zamora, ca. 800–1300 m a.s.l.; tierra subtemplada, 12 humid months.
  This forest can attain up to 40 m height. The few characteristic species include: *Ceroxylon amazonicum*, *Euterpe precatoria*, *Iriartea deltoidea* and *Oenocarpus bataua* (Arecaceae), *Ochroma pyramidale* (Bombacaceae), *Terminalia amazonia* and *T. oblonga* (Combretaceae), *Grias peruviana* (Lecythidaceae), *Clarisia racemosa*, *Ficus* spp and *Sorocea trophoides* (Moraceae), *Otoba glycicarpa* (Myristicaceae), *Piper* spp (Piperaceae), *Palicourea guianensis*, *Psychotria* spp (Rubiaceae), *Pouteria* sp. (Sapotaceae).

B Evergreen lower montane forest ca. 1300–2100 m a.s.l. on the eastern escarpment of the Cordillera Real; tierra subtemplada and templada, 12 humid months.
  Canopy height up to 30 m. Characteristic species: *Alzatea verticillata* (Alzateaceae), *Chamaedorea pinnatifrons*, *Dictyocaryum lamarckianum* and *Wettinia maynensis* (Arecaceae), *Piptocoma discolor* and *Mikania* spp (Asteraceae), *Tabebuia chrysantha* (Bignoniaceae), *Vismia tomentosa* (Clusiaceae), *Cyathea caracasana* (Cyatheaceae), *Inga acreana* and other *Inga* spp (Fabaceae), *Nectandra lineatifolia*, *N. membranacea* and *Ocotea aciphylla* (Lauraceae), *Miconia imitans*, *M. punctata* (Melastomataceae), *Cedrela montana* (Meliaceae), *Ficus* spp, *Morus insignis* and *Sorocea trophoides* (Moraceae), *Piper* spp (Piperaceae), *Heliocarpus americanus* (Tiliaceae).

C Evergreen upper montane forest on the eastern escarpment above Sabanilla, ca. 2100–2700 m a.s.l.; tierra fresca, 12 humid months.
  Canopy heights attain up to 25 m. Characteristic species: *Ilex rimbachii* (Aquifoliaceae), *Hedyosmum* spp (Chloranthaceae), *Clethra revoluta* (Clethraceae), *Clusia ducu* and *Tovomita weddeliana* (Clusiaceae), *Weinmannia pinnata* and other *Weinmannia* spp (Cunoniaceae), *Cyathea bipinnatifida* (Cyatheaceae), *Purdiaea nutans* (Clethraceae), *Bejaria aestuans* (Ericaceae), *Alchornea grandiflora* (Euphorbiaceae), *Macrocarpaea revoluta* (Gentianaceae), *Eschweilera sessilis* (Lecythidaceae), *Licaria subsessilis*, *Ocotea benthamiana* and *Persea ferruginea* (Lauraceae), *Graffenrieda emarginata*, *G. harlingii* and *Tibouchina lepidota* (Melastomataceae), *Myrica pubescens* (Myricaceae), *Myrsine coriacea* (Myrsinaceae), *Calyptranthes pulchella* and *Myrcia* spp (Mytraceae), *Podocarpus oleifolius* and *Prumnopitys montana* (Podocarpaceae), *Dioicodendron dioicum* and *Palicourea* spp (Rubiaceae), *Matayba inelegans* (Sapindaceae), *Drimys granadensis* (Winteraceae).

D Evergreen elfin-forest (ceja de montaña) on the eastern and western escarpment of the Cordillera Real up to the timberline, ca. 2700–3100 m a.s.l.; tierra fría, 12 humid months.
  This forest type forms the timberline. Canopy height up to 6–8 m, rarely higher. Characteristic species: *Ilex* spp (Aquifolicaeae), *Puya eryngioides* (Bromeliaceae), *Hedyosmum cumbalense* and *H. scabrum* (Chloranthaceae), *Clethra ovalifolia* (Clethraceae), *Clusia elliptica* (Clusiaceae), *Weinmannia cochensis*, *W. loxensis* and *W. rollottii* (Cunoniaceae), *Gaultheria reticulata* (Ericaceae), *Escallonia myrtilloides* (Grossulariaceae), *Orthrosanthus chimborazensis* (Iridaceae), *Persea ferruginea* and *Ocotea infravoveolata* (Lauraceae), *Gaiadendron punctatum* (Loranthaceae), *Graffenrieda harlingii* (Melastomataceae), *Myrteola phylicoides* (Myrtaceae), *Hesperomeles ferruginea* (Rosaceae), *Styrax foveolaria* (Styracaceae), *Symplocos sulcinervia* (Symplocaceae), *Gordonia fruticosa* (Theaceae), as well as impenetrable bamboo of *Chusquea falcata* and *C. scandens* (Poaceae).

E Shrub and dwarf bamboo páramos in the crest region of the Cordillera Real above timberline, ca. 3100–3700 m a.s.l.; tierra subhelada, 12 humid months.
  Also referred to as jalca in a physiognomical sense. This vegetation attains heights of up to 2 m. Characteristic species: *Gynoxis* spp (Asteraceae), *Puya eryngioides* and *Puya nitida* (Bromeliaceae), *Hypericum decandrum* (Clusiaceae), *Rhynchospora vulcani* (Cyperaceae), *Bejaria resinosa*, *Disterigma pentandrum*, *Gaultheria erecta*, *G. reticulata* and *Vaccinium floribundum* (Ericaceae), *Escallonia myrtilloides* (Grossulariaceae), *Brachyotum andreanum* (Melastomataceae), *Neurolepis asymmetrica*, *N. elata*, *N. laegaardii* and the bamboo species

(continued)

Table 10.1.1 (continued)

Chusquea neurophylla (Poaceae), *Monnina arbuscula* (Polygalaceae), *Valeriana microphylla* and *V. plantaginea* (Valerianaceae).

F Semideciduous interandean forest in the valley region west of the Cordillera Real around Malacatos, Vilcabamba and Yangana, ca. 1400–2400 m a.s.l.; tierra templada and fresca, 6–8 humid months.
Canopy height up to 15 m, emergents occasionally up to 20 m. Characteristic species: *Cacosmia rugosa*, *Verbesina lloensis* and *Zexmenia helianthoides* (Asteraceae), *Tabebuia chrysantha* (Bignoniaceae), *Ceiba insignis* (Bombacaceae), *Cordia macrocephala* (Boragninaceae), *Buddleja lojensis* (Buddlejaceae), *Capparis scabrida* (Capparidaceae), *Croton wagneri* (Euphorbiaceae), *Acacia macracantha*, *Anadenanthera colubrina* (Fabaceae), *Lepechinia mutica* and *Scutellaria volubilis* (Lamiaceae), *Adenaria floribunda* (Lythraceae), *Cantua quercifolia* (Polemoniaceae), *Coccoloba ruiziana* (Polygonaceae), *Oreocallis grandiflora* and *Roupala obovata* (Proteaceae), *Dodonaea viscosa* (Sapindaceae), *Streptosolen jamesonii* (Solanaceae).

G Evergreen upper montane forest on the western escarpment above Loja and Yangana, ca. 2400–2800 m a.s.l.; tierra templada and fresca, 8–11 humid months.
Canopy height up to 20 m. Characteristic species: *Hedyosmum racemosum* (Chloranthaceae), *Clusia alata*, *C. elliptica* (Clusiaceae), *Weinmannia glabra*, *W. macrophylla* (Cunoniaceae), *Persea ferruginea* (Lauraceae), *Graffenrieda harlingii* (Melastomataceae), *Cedrela montana*, *Ruagea hirsuta* (Meliaceae), *Myrica pubescens* (Myricaceae), *Ardisia* sp., *Myrsine andina* (Myrsinaceae), *Myrcianthes rhopaloides* (Myrtaceae), *Prumnopitys montana* (Podocarpaceae), *Oreocallis grandiflora* (Proteaceae), *Cinchona macrocalyx* and *C. officinalis*, *Palicourea heterochroma* (Rubiaceae).

The diversity of vegetation types is further increased by tracts of secondary vegetation (mostly at 1800–1950 m a.s.l.) in different stages of recovery from clear-cutting and the construction of a pipe system for the close-by hydroelectric plant, respectively.

Finally, the study area constitutes a well developed mosaic of natural successional stages due to tree fall gaps (see Chapter 23) and landslides, which are conspicuously common throughout the whole area (see Chapter 24). However, successional habitats of natural or anthropogenic origin are not treated in this chapter. The timberline is poorly defined and highly variable in altitude (see Chapter 10.4); the forest decreases in stature and opens gradually giving way to shrub páramo at ca. 2600–2700 m a.s.l. along exposed ridges and considerably higher in major ravines (ca. 3000 m a.s.l.).

The most important seed plant families are Orchidaceae (337 spp), Rubiaceae (65 ssp), Bromeliaceae (64 ssp) and Melastomataceae (55 spp; Homeier and Werner 2007). So far, more than 280 tree species (diameter at breast height exceeding 5 cm) are known from the RBSF; the most speciose families are Lauraceae, Melastomataceae and Rubiaceae. *Graffenrieda emarginata* (Melastomataceae; see Homeier 2005a), *Purdiaea nutans* (Clethraceae; see Homeier 2005b), *Alchornea grandiflora* (Euphorbiaceae) and *Podocarpus oleifolius* (Podocarpaceae) are some of the most common and characteristic tree species of the RBSF forest.

About 140 climber species are known to occur in the RBSF, the most important families being Asteraceae, Apocynaceae *s.l.* and Araceae.

More than 400 angiosperm epiphytes (the largest families being Orchidaceae, Bromeliaceae, Piperaceae) have been registered. Werner et al. (2005) found up to 98 species of vascular epiphytes on single trees – one of the highest numbers recorded worldwide (see Schuettpelz and Trapnell 2006).

Fern and fern allies with 257 species (Lehnert et al. 2007) and bryophytes with more than 515 species (Gradstein et al. 2007; Parolly et al. 2004) are equally speciose. The number of bryophyte species is the highest recorded from such a small tropical area and constitutes about one-third of the entire bryophyte flora of Ecuador (Léon et al. 2006).

Four main forest types were distinguished for the lower reaches of RBSF by their tree species composition (Fig. 10.1.1; Homeier 2004; Homeier et al. 2002). These were confirmed by studies on the syntaxonomy of epiphytic bryophytes (Parolly and Kürschner 2004b) and on climber species composition (Matezki, in preparation). The less accessible vegetation types such as ravine forest above 2200 m a.s.l. remain to be studied in detail.

The following description lists some structural features and characteristic plant species for each of the four forest types and for the shrub páramo that occurs at the highest elevations of the RBSF.

10.1.2.2.1 Forest of the San Francisco Valley and Major Ravines Below ca. 2200 m a.s.l. (Forest Type I, Subtype of B. Evergreen Lower Montane Forest)

The tallest and most speciose forest grows on gentle lower slopes of the San Francisco river and within major ravines where the canopy attains 20–25 m with emergents occasionally approaching heights of 35 m (Fig. 10.1.2). Megaphyllous shrubs (e.g., Piperaceae) and herbs (Cyclanthaceae, Heliconiaceae, Zingiberaceae) abound in the understorey. Terrestrial and bole climbing Araceae are both common and speciose. Understorey epiphytes are predominantly skiophilous ferns, especially Aspleniaceae (*Asplenium auritum, A. theciferum*), Dryopteridaceae (*Elaphoglossum* spp, *Polybotria* spp), Hymenophyllaceae (*Trichomanes angustatum, T. diaphanum, T. reptans*) and Vittariaceae (*Polytaenia brasilianum, P. lineatum*) as well as *Guzmania confusa* and *G. morreniana* (Bromeliaceae). Due to low light levels generalistic and heliophilous epiphytes (e.g., Ericaceae, Tillandsia spp, Orchidaceae) are virtually confined to higher strata.

Several important families of moisture loving herbs show a relatively low diversity and abundance at RBSF and are largely restricted to the understorey of this forest type (e.g., Begoniaceae, Campanulaceae, Gesneriaceae).

Characteristic trees include *Piptocoma discolor* (Asteraceae), *Tabebuia chrysantha* (Bignoniaceae), *Hyeronima asperifolia* and *Sapium glandulosum* (Euphorbiaceae), *Nectandra linneatifolia* and *N. membranacea* (Lauraceae), *Merania* sp., *Miconia punctata* and other *Miconia* spp (Melastomataceae), *Inga*

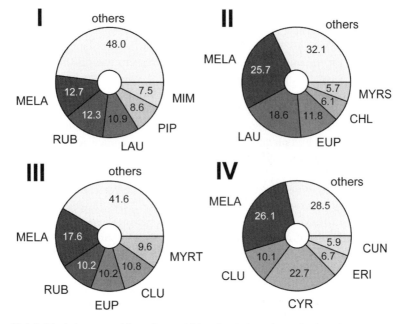

Fig. 10.1.1 Floristic composition of trees (dbh ≥5 cm) throughout the four forest types (*I–IV*) from the RBSF described in detail. *Numbers* indicate the respective percentages of total tree species richness (after Homeier 2004). *CHL* Chloranthaceae, *CUN* Cunoniaceae, *CLU* Clusiaceae, *CYR* Cyrillaceae, *ERI* Ericaceae, *EUP* Euphorbiaceae, *LAU* Lauraceae, *MELA* Melastomataceae, *MIM* Mimosaceae, *MYRS* Myrsinaceae, *MYRT* Myrtaceae, *PIP* Piperaceae, *RUB* Rubiaceae

spp (Mimosaceae), *Morus insignis*, *Naucleopsis sp. nov.*, and *Ficus* spp (Moraceae), *Prumnopitys montana* (Podocarpaceae), *Micropholis guyanensis* (Sapotaceae) and *Heliocarpus americanus* (Tiliaceae).

Predominating shrubs include *Saurauia* spp (Actinidiaceae), *Chamaedorea pinnatifrons* (Arecaceae), *Piper* spp (Piperaeae), *Faramea* spp, *Palicourea* spp and *Psychotria* spp (notably *P. tinctoria*; Rubiaceae).

Bryophytes are very abundant and speciose in this forest type and cover tree trunks and branches with thick mats. Characteristic epiphytes in the forest understorey and lower canopy include numerous species of the liverwort genus *Plagiochila* (e.g., *P. bryopteroides*, *P. deflexirama*, *P. diversifolia*, *P. heterophylla*, *P. montagnei*, *P. oresitropha*, *P. punctata*, *P. stricta*, *P. superba*), furthermore the feather-like *Bryopteris filicina* and *Porella crispata*, the pendent *Radula voluta* and *Taxilejeunea pterigonia* and robust mosses such as *Porotrichum expansum*, *Porotrichodendron superbum*, *Prionodon densus*, *Pterobryon densum* and various pendent Meteoriaceae. Tree bases harbour large mats of *Trichocolea tomentosa* and *Rhizogonium spiniforme* and tree canopies are rich in species of *Frullania* (e.g., *F. brasiliensis*, *F. intumescens*, *F. kunzei*, *F. pittieri*, *F. arecae*). Thalloid liverworts of *Symphyogyna* (*S. aspera*, *S. brasiliensis*, *S. brogniartii*) and *Riccardia andina* are characteristic on forest soil.

Fig. 10.1.2 Profile diagram of forest type I (Forest of the San Francisco valley and major ravines below ca. 2200 m a.s.l., subtype of B. Evergreen lower montane forest) at the RBSF (1960 m a. s.l., depth of the profile is 5 m). Trees with diameters exceeding 5 cm: *1 Talauma* sp., *2 Ficus citrifolia*, *3 Morus insignis*, *4 Sapium glandulosum*, *5* Lauraceae, *6 Inga* sp. 1, *7 Prumnopitys montana*, *8* Myrsinaceae, *9 Miconia theaezans*, *10 Ilex* cf *amboroica*, *11 Ocotea* sp., *12* Lauraceae, *13 Micropholis guyanensis*, *14 Miconia quadriporata*, *15 Inga* sp. 2, *16 Ficus tonduzii*, *17 Psychotria tinctoria*, *18 Psychotria* cf *alba*, *19 Psychotria montivaga*, *20 Palicourea luteonivea*

The general appearance of the forest becomes considerably wetter at higher altitudes due to lusher epiphytic vegetation (i.e., increasing abundance of epiphytic bryophytes and vascular plants).

10.1.2.2.2 Forest Along Ridges and Upper Slopes from ca. 1900 m to 2100 m a.s.l. (Forest Type II, Subtype of B. Evergreen Lower Montane Forest)

This forest type attains 15–20 m on slopes but as little as 8–10 m along the most exposed ridges (Fig. 10.1.3). A characteristic feature of this forest type is the thick raw humus layer. Araceae are poorly represented; characteristic species are the bole climbing *Anthurium cutucuense* (shady upper slope forest) and the mostly terrestrial *A. grubbii* (open ridges). The herb and shrub layer is dominated by *Elaphoglossum* spp (Dryopteridaceae), *Cyathea bipinnatifida* (Cyatheaceae) and *Guzmania* spp (Bromeliaceae). Epiphytes are strikingly more abundant and speciose in the understorey than in forest type I. Epiphyte species characteristic for upper slope forest are *Hymenophyllum lobatoalatum* (Hymenophyllaceae), *Acronia microcardia, A.* cf *bivalvis, Pleurothallis lilijae* (Orchidaceae); characteristic in ridge forest are *Masdevallia gnomii* and congeners or *Specklinia* sp. nov. (Orchidaceae). Shared character species include *Mezobromelia capituligera, Guzmania killipiana* and *G. sanguinea* (Bromeliaceae) and *Brachycladium* spp (Orchidaceae).

In respect to tree species composition, forest type II is also quite different from the above-mentioned type I. Some frequent and characteristic species are *Alzatea verticillata* (Alzateaceae), *Dictyocaryum lamarckianum* and *Wettinia aequatorialis* (Arecaceae), *Weinmannia pinnata, W. sorbifolia* and *W. spruceana* (Cunoniaceae), *Abarema killipii* (Fabaceae), *Hyeronima moritziana* (Euphorbiaceae), *Ocotea aciphylla* (Lauraceae), *Graffenrieda emarginata* and *Miconia calophylla* (Melastomataceae), *Podocarpus oleifolius* (Podocarpaceae) and *Matayba inelegans* (Sapindaceae). While forest type I is strongly dominated by mesophyllous and straight trunk trees, ridge trees tend to be crooked and microphyllous (Paulsch 2002; see Chapter 10.3 in this volume). Among climbers the genus *Mikania* (Asteraceae) is well represented.

The bryophyte flora of this forest type is quite different from that of the previous one. *Plagiochila* and robust mosses characteristic of forest type I are less conspicuous, instead the epiphytic layers are dominated by members of *Bazzania* (e.g., *B. cuneistipula, B. hookeri, B. longistipula, B. phyllobola, B. stolonifera), Lepidozia cupressina* and the mosses *Leucobryum antillarum* and *Campylopus huallagensis*. Characteristic species of *Plagiochila* include the delicate *P. aerea* and *P. punctata* and the robust *P. tabinensis*. Ground-dwelling species are more abundant than in the previous forest type and include the mosses *Campylopus huallagensis, Leucobryum giganteum, Pyrrobryum spiniforme* and others.

10.1.2.2.3 Forest of Ridges and Upper Slopes at ca. 2100–2250 m a.s.l. (Forest Type III, Subtype of C. Evergreen Upper Montane Forest)

With increasing altitude, the canopy successively opens and trees do not surpass 12 m (ridge) and 15 m (slope) in height, respectively (Fig. 10.1.4). Epiphytes are abundant and highly diverse, last but not least close to the ground where they constitute

Fig. 10.1.3 Profile diagram of forest type II (Forest along ridges and upper slopes at ca. 1900–2100 m a.s.l., subtype of B. Evergreen lower montane forest) at the RBSF (2050 m a.s.l., depth of the profile is 5 m). Trees with diameters exceeding 5 cm: *1 Alchornea grandiflora, 2 Hyeronima moritziana, 3 Weinmannia pinnata, 4 Clusia* sp., *5 Matayba inelegans, 6 Graffenrieda emarginata, 7 Ocotea benthamiana, 8 Ocotea* sp., *9 Persea* sp., *10 Lauraceae, 11 Hedyosmum anisodorum, 12 Miconia* cf *calophylla, 13 Podocarpus oleifolius, 14 Rudgea* sp., *15 Sloanea* sp., *16 Ilex* cf *amboroica*

a blend of moisture-demanding (e.g., *Hymenophyllum ruizianum* (Hymenophyllacae), *Semiramisia speciosa* (Ericaceae)), generalistic and (especially on ridges) heliophilous species (*Tillandsia laminata, T.* cf *aequatorialis* (Bromeliaceae), *Melpomene moniliformis* (Grammitidaceae) and *Utricularia jamesonii* (Lentibulariaceae) along with numerous orchid taxa).

Common tree species are *Hedyosmum translucidum* (Chloranthaceae), *Clusia* cf *ducuoides, Clusia* spp and *Tovomita weddeliana* (Clusiaceae), *Weinmannia haenkeana* and *W. ovata* (Cunoniaceae), *Purdiaea nutans* (Clethraceae), *Alchornea grandiflora* (Euphorbiaceae), *Endlicheria oreocola, Licaria subsessilis, Ocotea benthamiana* and *Persea subcordata* (Lauraceae), *Eschweilera sessilis* (Lecythidaceae), *Graffenrieda emarginata* (Melatomataceae), *Calyptranthes pulchella* and *Myrcia* sp. (both Myrtaceae) and *Podocarpus oleifolius* (Podocarpaceae).

Fig. 10.1.4 Profile diagram of forest type III (Forest of ridges and upper slopes at ca. 2100–2250 m a.s.l., subtype of C. Evergreen upper montane forest) at the RBSF (2210 m a.s.l., depth of the profile is 5 m). Trees with diameters exceeding 5 cm: *1 Graffenrieda emarginata, 2 Clusia* sp. 1, *3 Podocarpus oleifolius, 4 Calyptranthes pulchella, 5 Alchornea grandiflora, 6 Myrcia* sp., *7 Hyeronima moritziana, 8 Clusia* sp. 2, *9 Dioicodendron dioicum, 10 Eschweilera sessilis, 11 Palicourea angustifolia, 12 Palicourea loxensis, 13 Faramea glandulosa, 14 Hedyosmum translucidum, 15 Purdiaea nutans, 16 Siphoneugena* sp., *17 Matayba inelegans*

The bryophyte flora of this forest type is rather similar to that of the previous one but is somewhat richer in heliophytic species such as the moss *Macromitrium trichophyllum* and the liverwort *Herbertus pensilis*, due to the more open and lower forest canopy (Parolly and Kürschner 2004b). The latter authors also recorded the moss *Leptotheca boliviana* and the liverwort *Bazzania aurescens* (earlier identified as *B. tricrenata*) as characteristic of this forest type.

10.1.2.2.4 Ridge Forest at ca. 2250–2700 m a.s.l., up to the Timberline (Forest type IV, Subtype of C. Evergreen Upper Montane Forest)

With increasing altitude the canopy further decreases in cover and height (5–10 m). Larger trees are typically scattered, closed canopy is restricted to saddles (Fig. 10.1.5). *Purdiaea nutans* (Clethraceae) monodominates this vegetation type up to ca. 2700 m. The forest ground is densely covered by terrestrial bromeliads, especially *Guzmania paniculata*.

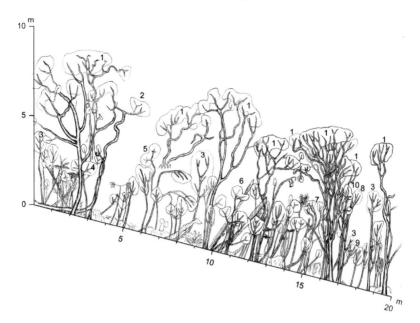

Fig. 10.1.5 Profile diagram of forest type IV (Ridge forest at ca. 2250–2700 m a.s.l., subtype of C. Evergreen upper montane forest) at the RBSF (2450 m a.s.l., depth of the profile is 5 m). Trees with diameters exceding 5 cm: *1 Purdiaea nutans, 2 Persea feruginea, 3 Graffenrieda harlingii, 4 Symplocos bogotensis, 5 Ternstroemia cleistogama, 6 Weinmannia fagaroides, 7 Freziera campanulata, 8 Clusia sp., 9 Weinmannia ovata, 10 Styrax trichostemon*

Obviously related to the high level of exposure to light, the understorey features a great abundance of canopy epiphytes such as *Racinaea dielsii*, *Vriesea fragrans*, *V. lutheri* (Bromeliaceae), *Maxillaria adendrobium*, *M. aggregata*, *Specklinia rubens*, *Trichosalpinx lenticularis* and *T. scabridula* (Orchidaceae). More hygrophilous plants such as *Racinaea seemannii* (Bromeliaceae), *Hymenophyllum ruizii* (Hymenophyllaceae) or *Lepanthes mucronata* (Orchidaceae) are restricted to particularly sheltered spots, like dense tangles.

Characteristic trees and shrubs are *Ilex rimbachii*, *I. scopulorum*, *I. teratopis* and *I. weberlingii* (Aquifoliaceae), *Weinmannia elliptica*, *W. fagaroides* and *W. loxensis* (Cunoniaceae), *Licaria subsessilis*, *Ocotea infravoveolata* and *Persea ferruginea* (Lauraceae), *Gaiadendron punctatum* (Loranthaceae), *Brachyotum campanulare*, *Graffenrieda harlingii* (Melastomataceae), *Cybianthus marginatus* (Myrsinaceae), *Calyptranthes pulchella* and *Myrcia* spp (Myrtaceae), *Podocarpus oleifolius* (Podocarpaceae), *Freziera karsteniana* (Theaceae) and *Drimys granadensis* (Winteraceae).

Undoubtedly, the most remarkable bryophyte feature of the *Purdiaea* forest is the abundance of the rare liverworts *Pleurozia paradoxa* and *P. heterophylla* (Pleuroziaceae; see Chapter 19 in this volume). *Pleurozia heterophylla* (only three known localities worldwide) grows on trunks and branches, whereas *P. paradoxa* grows on tree bases; the latter extends upwards into páramo. Because of the low and open canopy, the epiphytic community of the *Purdiaea* forest is rich in Macromitriaceae (*Groutiella, Macromitrium, Schlotheimia*; ten species), which are typical canopy mosses of montane forest (Gradstein et al. 2001). Further characteristic epiphytes include members of the genus *Herbertus* (*H. divergens, H. juniperoideus, H. sendtneri*), *Scapania portoricensis, Lepicolea pruinosa, Bazzania bidens, B. longistipula, Plachiochila bifaria, P. pachyloma, P. tabinensis* and the mosses *Holomitrium sinuosum* and *Bryohumbertia filifolia* (Parolly and Kürschner 2004b). Canopy twigs habour numerous tiny liverworts of Lejeuneaceae. The terrestrial layer can be very conspicuous and is made up of species of *Bazzania* (e.g., *Bazzania gracilis, B. hookeri, Bazzania* sp. nov.), *Syzygiella* (e.g., *S. anomala, S. tonduzana*), *Lepidozia* (*L. caespitosa, L. macrocolea, L. squarrosa*) and in wet places *Sphagnum*.

10.1.2.2.5 Subpáramo ("Ceja de Montaña") Above the Timberline at ca. 2700 m (3000 m) to 3150 m a.s.l. (Subtype of D. Evergreen Elfin Forest)

The vegetation above the timberline is largely dominated by shrubs (Fig. 10.1.6). Some characteristic species are *Loricaria* sp. (Asteraceae), *Clethra ovalifolia* (Clethraceae), *Weinmannia cochensis* (Cunoniaceae), *Desfontainia spinosa* (Loganiaceae), *Brachyotum* spp, *Miconia* spp (Melastomataceae), *Myrica pubescens* (Myricaceae), *Styrax foveolaria* (Styracaceae) and *Symplocos sulcinervia* (Symplocaceae).

Terrestrial herbs become more speciose at these altitudes; some common species are *Dorobaea* sp. (Asteraceae), *Puya eryngioides, P. roseana* and *Guzmania lychnis* (Bromeliaceae), *Paepalanthus meridensis* (Eriocaulaceae), *Calamagrostis macrophylla, Chusquea falcata* and *Neurolepis elata* (all Poaceae) and *Arcytophyllum* spp (Rubiaceae).

The bryophyte flora of the subpáramo is very distinct although still poorly studied. Characteristic trunk-inhabiting species include the endemic moss *Macromitrium perreflexum* (abundant!) and the liverworts *Herbertus acanthelius, H. pensilis, Jamesoniella rubricaulis, Lepicolea pruinosa, Leptoscyphus gibbosus* and *Pleurozia paradoxa*. Several of these may also occur on soil. Twigs are inhabitated by a unique community of tiny liverwort species including four that were described as new to science from the RBSF: *Physotheca autoica* (new monotypic genus), *Plagiochila cucullifolia* var. *anomala, Diplasiolejeunea erostrata* and *D. grandirostrata*. The terrestrial layer is well developed and made up of species of *Riccardia* (notably *R. fucoidea*), *Isotachis* (*I. lopezii, I. muticeps*), *Sphagnum* (species still unidentified), *Nowellia curvifolia*, the robust mosses *Polytrichadelphus aristatus* and *Rhacocarpus purpurascens* and some of the trunk epiphytes mentioned above.

Fig. 10.1.6 Profile diagram of the subpáramo [Subpáramo (ceja de montaña) above the timberline at ca. 2700 m (3000 m) to 3150 m a.s.l., subtype of D. Evergreen elfin-forest] at the RBSF (2750 m a.s.l.). Trees with diameters exceeding 5 cm: *1 Podocarpus oleifolius, 2 Clusia* cf *ducuoides, 3 Weinmannia loxensis, 4 Calyptranthes* cf *pulchella, 5 Schefflera* sp., *6 Geonoma* sp., *7 Myrica pubescens, 8 Meriania rigida, 9 Macleania* sp., *10 Weinmannia fagaroides, 11 Miconia* cf *jahnii, 12 Weinmannia elliptica, 13 Graffenrieda harlingii*

10.1.2.2.6 The Forests of the RBSF in the Neotropical Context

Compared with many other neotropical montane forests the RBSF is peculiar in several aspects. The canopy along upper slopes and ridges is remarkably low and open across the entire elevational gradient. This points to a limitation of plant growth as a result of unfavourable environmental conditions.

The average canopy height of 6–8 m found on ridges at 2300–2500 m a.s.l. (Homeier 2004) is very low compared with other montane forests in Ecuador even at considerably higher elevations (e.g., ridge forest at Yanayacu Biological Station in North Ecuador attains 15 m at 2900 m a.s.l.; J. Homeier and F.A. Werner, personal observation). The timberline is very low (2700–3000 m a.s.l.) and the genus *Polylepis* (Rosaceae), usually constituting the timberline of other Andean forests, is missing in South Ecuador (see Chapter 10.4 in this volume).

Comparable neotropical mountain areas with similar stunted, microphyllous forests include the Cordillera del Condor (located just east of Podocarpus NP), the Guyana shield (including the Venezuelan tepuis), and northern Peru (Alto Mayo). Apart from structural features these areas share certain plant taxa mainly restricted to sandstone. The most notable example is the tree *Purdiaea nutans* (Clethraceae), which dominates the upper slopes and ridges of the RBSF forests above 2100 m a. s.l. (see Chapter 19 in this volume). Other species are *Quiina yatuensis* (Quiinaceae), a tree species until recently only known from Venezuelan Guyana, or the fern *Pterozonium brevifrons* (Pteridaceae).

At the family level, however, floristic composition is quite similar to other Neotropical montane forests, with Lauraceae, Melastomataceae and Rubiaceae being the most speciose tree families as, e.g., reported by Webster and Rhode (2001) for Maquipucuna reserve, North-west Ecuador (1100–2800 m a.s.l.) or by Dorr et al. (2000) from Guaramacal, Venezuela (1500–2800 m a.s.l.).

The plant species richness at RBSF is high. Several taxa show surprisingly high species numbers compared with other neotropical montane forests, although comparable floristical data from neotropical montane forests are still scarce.

Exceptionally high local species richness has been confirmed for some taxa such as the orchids in general (C.H. Dodson, personal communication) and the genus *Stelis* in particular with ca. 40 species (C. Luer, personal communmication), Lauraceae with more than 40 species (H. van der Werff, personal communication) and *Piper* with over 30 species.

Small-scale environmental heterogeneity boosts the development of floristically and structurally distinct vegetation types and thus contributes to the outstanding plant species richness of the area as well as the region as a whole. In addition to the processes typical for primary forests, such as gap dynamics (see Chapter 23 in this volume) and the common landslides (see Chapter 24), steep climatic gradients (see Chapter 8) allow a wide array of trees to coexist and form a mosaic of physiognomic distinct vegetation units. These, in turn, amplify the abiotic heterogeneity to create an even wider magnitude of microhabitats and regeneration niches for other life forms. The combination of these processes together with the high topographic diversity and the related edaphic patchiness (see Chapter 9) maintain a high environmental heterogeneity throughout space and time.

10.1.3 Conclusions

In conclusion, the extraordinarily broad matrix of environmental conditions in the RBSF and its wider surroundings fosters high beta diversity allowing the coexistence of habitat specialists at the smallest spatial scale. The geographical position of the reserve in the ecotone between the hyperhumid eastern Andean slope and the dry interandean forests of South Ecuador on the one hand and the central and the northern Andes on the other hand provides access to an immense species pool communicated by the unique regional depression of the Andes.

Chapter 10.2
Past Vegetation and Fire Dynamics

H. Niemann(✉) and H. Behling

10.2.1 Introduction

In order to understand the landscape history of the Podocarpus National Park region in the southeastern Ecuadorian Andes, we started our first field campaign in March 2005 to explore and core sediment archives. We collected several sediment cores from lakes, peat bogs and soils within the Podocarpus National Park region, including the RBSF for our palaeoecological studies focused on the development in the late Pleistocene before about 10 000 years BP and in the Holocene about 10 000–0 years BP.

We also installed 41 pollen traps to collect the modern pollen rain for one year on an altitudinal gradient between 1810 m and 3200 m a.s.l. by 50-m steps of elevation in the RBSF. Despite the importance to understand the landscape history only a few pollen records are available from the Ecuadorian Andes (e.g., Bush et al. 1990; Colinvaux 1988, 1997; Graf 1992; Hansen et al. 2003).

Within the framework of the DFG Research Group the following questions and aims were addressed:
- How was the vegetation development and dynamic as well as the climate dynamic in the Podocarpus National Park region since the late Pleistocene?
- Is it possible to separate drier and wetter periods?
- When and how did intensive fires occur in the Podocarpus National Park region? Were fires natural or of anthropogenic origin?
- When did the Podocarpus National Park region come under human influence and how strong was the human impact?

Our studies provide important background information to understand the stability and dynamics of modern Mountain Rainforest and Páramo ecosystems in southeastern Ecuador. Here we provide our first results on late Quaternary vegetation, climate and fire dynamics from the southeastern Andes in Ecuador. First results are available from the RBSF (1990 m, 2520 m, 3155 m a.s.l.) and the El Tiro Pass (2810 m a.s.l,), both core sites are located next to the main road between Loja and Zamora, near the northern frontier of the Podocarpus National Park.

The distance from the El Tiro Pass to Loja is about 10 km and the distance from RBSF to Loja is about 20 km. The small bog at the El Tiro Pass (3° 50' 25.9" S, 79 °08' 43.2" W) is about 800 m distant from the summit which is at about 3200 m elevation, located in a small depression of 30 m width and 60 m length. The maximum depth of the bog is 1.3 m.

10.2.2 Materials and Methods

The peat deposit at the El Tiro Pass in the Subpáramo zone was cored using a manual core (Russian corer) with 40 mm core diameter. The total length of the core was 127 cm. Sections of 50 cm length were extruded on-site with split PVC tubes, wrapped in plastic film and stored under dark and cold (+4 °C) conditions before processing.

At the RBSF, soil deposits were collected at three different locations by excavation. Cores were stored in 50 cm, split PVC tubes and wrapped with plastic film. The samples are taken in different altitudes in the Lower Mountain Rainforest zone on a small plateau at 1990 m elevation (T2/250), in the Upper Mountain Rainforest zone on a small plateau, at 2520 m elevation (Refugio) and in the Páramo zone in a small depression next to the summit at 3155 m elevation (Cerro de Consuelo).

For accelerator mass spectrometer (AMS) radiocarbon dating, five subsamples (organic material and charcoal fragments) were taken from the El Tiro core and two subsamples each (organic materials, charcoal fragments) were taken from the three RBSF soil cores. So far, two AMS radiocarbon dates are available for the El Tiro bog core and two for the RBSF soil cores (T2/250, Refugio). Radiocarbon ages were calibrated with CalPal (Weninger et al. 2004).

For pollen and charcoal analysis, 64 subsamples (0.25 cm^3) were taken at 2 cm intervals along the El Tiro Pass sediment core. From the three soil cores from the RBSF, a total of 73 subsamples (0.25 cm^3) were taken at 2 cm intervals along the cores. All samples were processed with standard analytical methods (Faegri and Iverson 1989). A minimum of 300 pollen grains were counted for each sample. The pollen sum includes trees, shrubs and herbs and excludes fern spores and aquatic pollen taxa. Pollen identification relied on the reference collection from the second author with more than 2000 neotropical species and literature (Behling 1993; Hooghiemstra 1984). Pollen and spore data are presented in pollen diagrams as percentages of the pollen sum. Carbonized particles (10–150 µm) were counted on pollen slides and presented by concentration (particles/cm^3). The software TILIA, TILIAGRAPH and CONISS was used for illustrations of the pollen and spore data (Grimm 1987).

10.2.3 Results and Discussion

10.2.3.1 The Bog Core from the El Tiro Pass

The first results of pollen and charcoal analysis of the El Tiro Pass bog core at 2810 m elevation are shown in Fig. 10.2.1. The core base (126 cm) is dated at 16 517±128 years BP (19 530–20 140 cal years BP) indicating deposits of late Pleistocene age. The upper core part (at 31 cm) is dated of 1828±55 years BP (1700–1830 cal years BP) indicating deposits of late Holocene age. The pollen percentage diagram shows a selected number of important pollen and spore taxa, out of about 90 taxa, which have been identified so far.

Abundant herb pollen, especially Poaceae with a decreasing trend as well as frequent *Plantago* (Plantaginaceae), mark the late Pleistocene period (127 – ca. 87 cm core depth). The identified *Plantago* pollen are *P. rigida* and *P. australis*, but this still needs to be investigated. Pollen of Cyperaceae, Asteraceae and *Valeriana* (Valerinaceae) are also frequent during this period. Tree and shrub pollen such as Melastomataceae, *Weinmannia* (Cunoniaceae), *Hedyosmum* (Chloranthaceae), *Podocarpus* (Podocarpaceae), *Ilex* (Aquifoliaceae) and *Myrsine* (Myrsinaceae) are found only in low amounts. Pteriophyta spores and spores of *Huperzia* (Lycopodiaceae) are rare. The charcoal concentration in the analyzed sediment samples for this period is low.

The pollen assemblage document that the El Tiro Pass which today is naturally covered with Subpáramo, was covered with Grass Páramo during the late Pleistocene. The pollen data from Holocene mire in the Chingaza National Park at 3730 m elevation, in the eastern Cordillera of middle Colombia in the Grass Páramo zone, showed a high occurrence of *Plantago rigida* pollen. Cushions of *P. rigida* dates from early Holocene (9000 years BP) until today and experienced several periods of expansion and re-expansion. The expansion of the *P. rigida* cushion took place under cool and humid conditions. Under warmer or drier conditions the expansion ends (Bosman 1994). Fires were rare during the late Pleistocene in the El Tiro region. This has also been found in the pollen record from Lake Surucucho (3200 m elevation) at the Amazonian flank of middle Ecuador (Colinvaux 1997).

A marked increase of tree and shrub pollen of Melastomataceae, *Weinmannia*, *Hedyosmum* and a decrease of herb pollen, especially Poaceae, Cyperaceae, Asteraceae, *Valeriana* and *Plantago* are observed during the transition of the late Pleistocene to the Holocene period (approx. 87 cm to ca. 75 cm core depth). Pteriophyta (monolete, trilete) spores increases and *Huperzia* decreases at the end of this period. Charcoal concentration increases little. The data from the pollen record of Lake Surucucho (3200 m), at the Amazonian flank of southern Ecuador, reveal an increase in *Weinmannia* and *Hedyosmum* at 10 300 years BP (Colinvaux 1997). The data from pollen record of Laguna Chorreras at an elevation of 3700 m (Cajas National Park, western Cordillera) report a decrease in *Huperzia* sp. spores during the transition of the late Pleistocene–Holocene

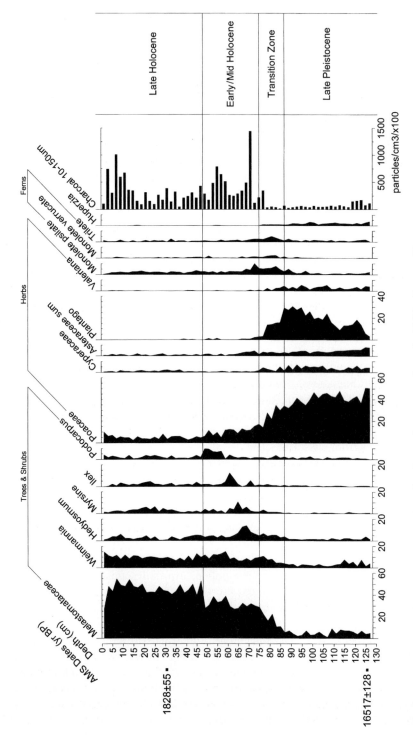

Fig. 10.2.1 Pollen percentage diagram of selected pollen and spore taxa from the El Tiro Pass. yr Years

period. *Huperzia* sp. and *Lycopodium* sp. characterize the upper cold wet Páramo (Hansen et al. 2003).

During the early to mid-Holocene period (approx. 75 cm to ca. 48 cm) Melastomataceae pollen increases markedly. Pollen of *Weinmannia, Hedyosmum, Myrsine, Ilex* and *Podocarpus* show asynchronous maxima between 70–50 cm core depth. This succession may indicate the formation of the Upper Mountain Rainforest vegetation at the study site. It is possible that the Upper Mountain Rainforest vegetation coheres with a warmer climate, as indicated by the stronger decomposed organic material during this period. This suggestion has to be examined in detail when other additional records are available from the region. Monolete and trilete Pteriophyta spores decreases at the beginning of this period. Charcoal concentration increases markedly, indicating that fires became quite frequent during this period. It is difficult to decide whether they are of natural or anthropogenic origin. First human activity in the region of Loja is dated by around 4000 years BP (Guffroy 2004), but human activities may have occurred in the dry Inter-Andean valley much earlier. For example, in the Sabana de Bogota (Colombia), the presence of Amerindians could be established from 12 500 years BP onward and possibly even before that time (Van der Hammen 1978). It is somewhat speculative, but it might be possible that fires on the El Tiro mountains slopes may originate from anthropogenic fires by hunting activities in the savanna vegetation of dry Loja basin since the early Holocene.

During the mid- to late Holocene period (ca. 48–0 cm) Melastomataceae pollen become dominant, while other tree and shrub pollen taxa such as of *Hedyosmum, Myrsine, Ilex* and *Podocarpus* are slightly less frequent. The pollen data of this period suggest the formation of a relatively stable Subpáramo vegetation of the study site since that time. Fires remain frequent during the mid-Holocene period and become more frequent during the late Holocene, before decreasing in the very late Holocene in the El Tiro region, probably at lower elevations, suggesting an increase of human activities near or at the El Tiro slopes. The pollen record from Lake Surucucho (3200 m), at the Amazonian flank of middle Ecuador, shows a strong increase of fires during the late Holocene (Colinvaux 1997). This may suggest that the increase of fire frequency in the El Tiro Pass record derives from human activity.

10.2.3.2 The Soil Core from the RBSF – T2/250

The first results of pollen and charcoal analyses from the RBSF T2/250 soil core at 1990 m elevation are shown in Fig. 10.2.2. It is the lowermost of the three RBSF sediment cores, taken at a slope position on a small plateau in the Lower Mountain Rainforest zone. Macroscopic charcoal in the lowermost part of the core (33 cm) is dated at 915±38 years BP (793–897 cal years BP) indicating deposits of late Holocene age. The pollen percentage diagram shows a selected number of

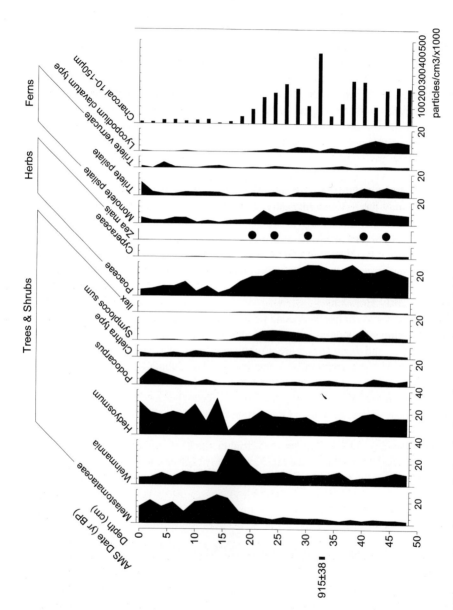

Fig. 10.2.2 Pollen percentage diagram of selected pollen and spore taxa from the RBSF – T2/250

important pollen and spore taxa, out of about 93 taxa, which have been identified so far.

At a core depth of 48–16 cm, tree and shrub pollen of *Hedyosmum*, *Weinmannia* and *Symplocos* (Symplocaceae) are relatively frequent, *Podocarpus*, *Ilex*, Melastomataceae and *Clethra* type (including pollen of *Purdiaea nutans* (Clethraceae) pollen are relatively rare. Pollen of *Hedyosmum*, *Ilex* and *Symplocos* decreases, while Melastomataceae and *Weinmannia* pollen increase. Herb pollen, especially from Poaceae, is quite frequent (about 20%). Pollen of *Zea mais* (Poaceae) is found in five samples during this period. Pteriophyta (monolete and trilete) and spores of the *Lycopodium clavatum* type are relatively high. Charcoal concentration is high with a marked decrease at the end of this period. The high number of Poaceae pollen, as well as *Zea mais*, coupled with the strong frequency of fires, as well as the occurrence of macroscopic charcoal in the sediments, indicates human activity near the core site.

At a core depth of 16–0 cm, tree pollen of Melastomataceae and *Hedyosmum* are relatively frequent. *Weinmannia* pollen increases markedly at the beginning of the upper core section. Tree pollen of *Podocarpus* is rare, but increases at the end of the upper core section. Pollen of *Ilex*, *Symplocos* and the *Clethra* type pollen is rare. Herb pollen is less frequent and *Zea mais* pollen is absent. The charcoal concentration is very low. The lower distribution of Poaceae species and the missing *Zea mais*, coupled with the very low frequency of fires, indicate the absence of human activity near the core location. The vegetation development after the absence of human disturbance indicates a local regeneration of the forest, primarily by *Weinmannia*, Melastomataceae, followed by *Hedyosmum* and *Podocarpus*.

10.2.3.3 The Soil Core from the RBSF – Refugio

The pollen and charcoal analyses of the RBSF Refugio soil core at 2520 m elevation are shown in Fig. 10.2.3. It is the middle of the three RBSF sediment cores, taken at a ridge position with a small plateau in the Upper Mountain Rainforest zone. Macroscopic charcoal in the lowermost part of the core (34.5 cm) is dated at 854±45 years BP (730–870 cal years BP), indicating deposits of late Holocene age. The pollen percentage diagram shows a selected number of important pollen and spore taxa, out of about 72 taxa identified so far.

At a core depth of 44–35 cm, tree and shrub pollen of Melastomataceae, *Weinmannia* and *Clethra* type (including pollen of *Purdiaea nutans*) are frequent. Pollen of *Podocarpus*, *Ilex*, *Hedyosmum*, *Clusia* (Clusiaceae) and Myrtaceae are relatively rare. Herb pollen such as Poaceae is represented in relatively low amounts. Pteriophyta spores, especially the *Cyathea conjugate* type, are frequent in the lower core section. Charcoal concentration is moderate with a very high maximum at 38–36 cm. At 35–0 cm, tree pollen of especially Melastomataceae and *Weinmannia* increase, while other tree pollen decrease. *Zea mais* is found in two samples at a core depth of 35–33 cm. Spores of *Cyathea conjugata* type are now absent. Spores of

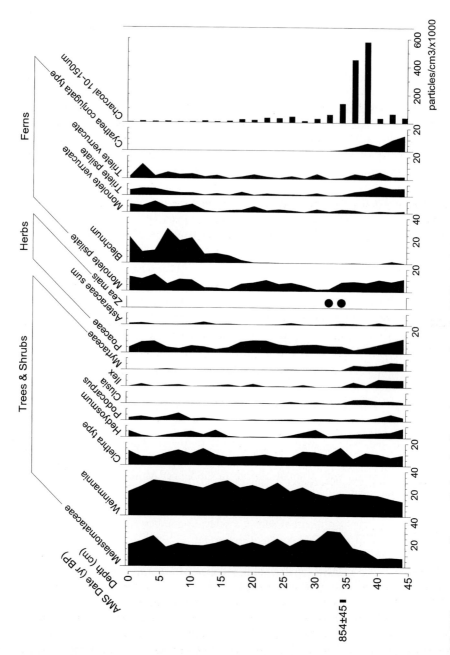

Fig. 10.2.3 Pollen percentage diagram of selected pollen and spore taxa from the RBSF – Refugio

Blechnum (Blechnaceae) become very frequent in the upper core section. Charcoal concentration is low and becomes even lower in the upper core section.

The floristic composition (higher occurrences of *Podocarpus, Ilex, Hedyosmum, Clusia,* Myrtaceae) is changed markedly by the strong increase in Melastomaceae species. These changes correlate with the marked decrease in charcoal particles and date at 845±45 years BP. Apparently the decrease in fire frequency is responsible for this vegetation change. The occurrence of frequent fires at 845±45 years BP or before and the occurrence of a single corn pollen also suggest some human activities near the study site. The local extinction of the tree fern *Cyathea,* probably *C. conjugate,* might be also related to the frequent burning in the past. Interesting is the local establishment of *Blechnum* during the recent times at the study site. The relatively frequent pollen of the *Clethra* type originate at the Refugio site probably almost from the tree *Purdiaea nutans,* while *Clethra* itself is rare (Gradstein, personal communication). *Purdiaea nutans* is a rare tree in the northern Andes, but is exceptionally frequent at the Refugio site of the RBSF research area (Gradstein et al. 2007; Homeier 2005; see Chapter 19 in this volume). Frequent fires in the past may have contributed to the establishment the *Purdiaea* forest in the Refugio area.

10.2.3.4 The Soil Core from the RBSF – Cerro de Consuelo

The first data of the RBSF Cerro de Consuelo soil core at 3155 m a.s.l. are shown in Fig. 10.2.4. It is the uppermost of the three RBSF sediment cores, taken at a small depression near the summit in the Páramo zone. The pollen percentage diagram shows a selected number of important pollen and spore taxa, out of about 88 taxa identified so far.

The pollen record in general shows no marked changes except the higher occurrence of *Podocarpus, Hedyosmum, Ilex, Myrsine* and Ericaceae pollen at a core depth of 49–37 cm. At 39–22 cm the charcoal concentration is markedly higher than in the lower and upper core parts. The decrease in *Podocarpus, Hedyosmum, Ilex, Myrsine* and Ericaceae after the increase in the frequency of fires may suggest that the tree line shifted to lower altitudes; however, this still has to be studied in detail.

10.2.4 Conclusions

The first palaeo-environmental results interfered from the 127 cm long and 16 517±128 year BP radiocarbon-dated sediment core from the El Tiro Pass at an elevation of 2810 m indicate that Grass Páramo was the main vegetation type in the Podocarpus National Park region during the late Pleistocene period. Grass Páramo was rich in Poaceae and *Plantago*. It is possible that, during the early to mid-Holocene period, an Upper Mountain Rainforest vegetation type developed at the study site, suggested by the succession stages of *Weinmannia, Hedyosmum, Myrsine, Ilex* and *Podocarpus*. The stronger decomposition of the accumulated organic material

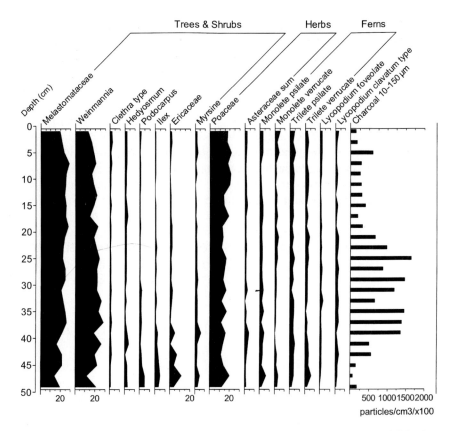

Fig. 10.2.4 Pollen percentage diagram of selected pollen and spore taxa from the RBSF – Cerro de Consuelo

suggests a warmer climate than today. The modern Subpáramo vegetation became established during the mid- to late Holocene period at the El Tiro Pass. Fire were rare during the late Pleistocene period, but became quite frequent during the Holocene, probably in the lower El Tiro mountain region. The comparison with other Ecuadorian records suggests that the fires are of anthropogenic origin.

The first data from the three studied soil cores of the RBSF research area indicate that fires were common in the research area during the late Holocene. Frequent fires and a relatively high occurrence of grass pollen (Poaceae), as documented in the lower part of RBSF-T2/250 soil core, show early human activities in the research area. Past fires have markedly influenced the floristic composition of the Mountain Rainforest and vegetation changes are found after the reduction or absence of fires.

The two radiocarbon dates from different elevations (915 ± 38 years BP at 1990 m and 854 ± 45 years BP at 2520 m) show that the decrease in human activity happened about 800–500 years BP. The absence of human activity in the study region during the past centuries might be related to a reduction in the human population. The Inca occupied the area between Loja and Zamora in middle of the fifteenth century during their invasion northwards. It was the strategy of the Inca to settle defeated populations in other regions of their empire. This happened with the Palta, an indigenous group settling in southeastern Ecuador, as well as in the study region, from 4000 years BP (Guffroy 2004). In the sixteenth century the Inca were defeated by the Spanish conquests. In the mid-seventeenth century the Spanish conquests lost their control over the region between Loja and Zamora in fights with the indigenous population. After some centuries with low human activity a resettling started in the mid-twentieth century, mostly by Mestizes Colonos, settlers from other parts of Ecuador. The basis for the fast resettling was the construction of the road between Loja and Zamora during this time. Many parts of the regenerated mountain rainforest (indicating political instability in the study region) were disturbed again by slash and burn farming and for pasture during recent decades.

Acknowledgements Special thanks to Achim Bräuning and Felix Matt for helping us in the logistical support of the field work and for Bräuning's fruitful discussion on past environmental changes.

Chapter 10.3
Forest Vegetation Structure Along an Altitudinal Gradient in Southern Ecuador

A. Paulsch(✉), D. Piechowski, and K. Müller-Hohenstein

10.3.1 Introduction

Approaches to classify vegetation are numerous but can be divided into two major groups: floristic approaches and physiognomic/structure-based approaches (Beard 1973), both leading to equal results when compared (Brocque and Buckney 1997; Webb et al. 1970; Werger and Sprangers 1982). Floristic systems (e.g. Braun-Blanquet 1928) base on plant species and species composition to classify vegetation within a hierarchical system. They are well established in temperate zones with their surveyable number of plant species (e.g. Dierschke 1994, Sautter 2003). The higher plant diversity in the tropics is well known and especially true for our investigation area (Henderson et al. 1991; Jørgensen and León-Yánez 1999). All floristic surveys in the tropics have to focus on some selected taxa (e.g. Kessler 2002; Krömer et al. 2006), functional groups like trees (e.g. Condit et al. 2002; Phillips et al. 2003) or vascular epiphytes (e.g. Werner et al. 2005) or create a flora for a locally distinct area (e.g. Mori et al. 1997, 2002; Ribeiro et al. 1999). Physiognomic approaches are based on structural parameters of parts of plants (e.g. leaves) or whole plants (e.g. architecture) and can be used without a detailed knowledge of the local flora. The most important structure-based classification systems by Ellenberg and Mueller-Dombois (1967) and Holdridge et al. (1971) are widely used to name vegetation units (formations) of bioms and sub-bioms worldwide (e.g. UNESCO 1973). However, these classifications are not adapted to smaller areas, especially where human influence has largely disturbed the natural vegetation and replaced it by vegetation units, where structural characteristics appear which were not known before (lopped trees, grazed bushes, plantations with one species all of the same age). Several structural approaches are applicable on smaller scales but none of them is useable in the investigation area due to improper characters (Orshan 1986), or the usage of floristic parameters (Halloy 1990; Parsons 1975) or due to the restrictiveness in classic forestry parameters (Condit 1998; Proctor et al. 1988). For a detailed comparison see Paulsch (2002) and Paulsch and Czimczik (2001).

Here we present results of a new structural approach on the basis of systems by Orshan (1986), Parsons (1975), Richards et al. (1940) and Werger and Sprangers

(1982). We follow Barkman (1979) in defining vegetation structure as the "horizontal and vertical arrangement of vegetation". The set of 102 structural parameters per stratum used in our system is applicable on a plot level concerning the vegetation, the plant community or a closed mosaic of different plant communities. Additionally, we investigate how the forests examined on plot level are matched into the highly fragmented, anthropogenically altered landscape. We follow the definition of landscape by Forman (1997): "a landscape is a mosaic where the mix of local ecosystems or land-uses is repeated in similar form over a kilometers-wide area". The objective on this level is to choose and characterize a representative section of the man-made landscape by structural attributes. These attributes are of outstanding importance once the chosen section is regarded as an ecosystem, providing space (an assemblage of biotops) for all living creatures (biocenoses).

Beside the fast and easy application of a structure-based classification, functional relationships on the ecosystem level can be derived. Ewel and Bigelow (1996) stated that: "it is the mix of life-forms, not the mix of species, that exerts major controls over ecosystem functioning", where life-forms = structural elements.

Special structures are responsible for the uptake of water or nutrients, others are decisive for photosynthesis and again others influence or determine the microclimate. Additionally important are biocenotic connections between animals and plant or vegetation structures. Dziedzioch (2001) showed many relations between more than 20 hummingbird species and different forest types in the investigation area (at plot level).

10.3.2 A New Structure-Based Classification System

The land use patterns of 12 farms (*fincas*) were recorded on the landscape level for parts of the San Francisco valley between Zamora and the ECSF (Fig. 10.3.1). The size and shape of patches were noted, based on the concepts of Forman (1997) and Wiens et al. (1993). The number of neighboring land-use units and the sharpness of boundaries were observed for each of the 245 units as well as e.g. the degree of fragmentation, the size and shape of different vegetation communities, the density of borderlines and the types of linear vegetation units (Table 10.3.1; Müller-Hohenstein et al. 2004) and the density, diameter at breast height (dbh), height and cover of the remaining trees (Paulsch et al. 2001). The grouping of patches into classes of similar characteristics was done with cluster analyses.

The study area reached along an altitudinal gradient from the Bombuscaro section of the Popocarpus National Park via the San Francisco valley and the area of the ECSF to the Cajanuma Section of the Podocarpus Park. Some 34 plots were installed in the Bombuscaro section (960–1090 m a.s.l.) and 19 on the property of the adjacent Finca Copalinga (1030–1580 m a.s.l.). The ECSF area hosted 102 plots (1820–2650 m a.s.l.) and 11 plots were investigated at the Cajanuma section (2750–3100 m a.s.l.).

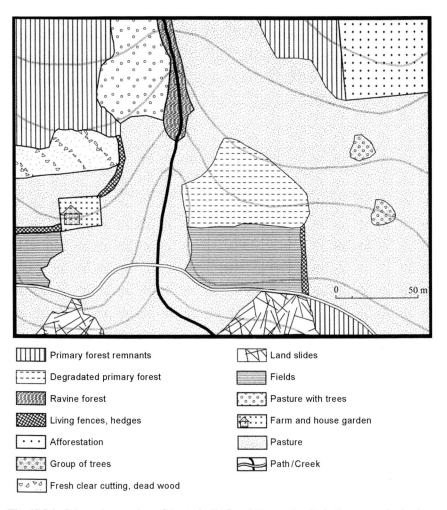

Fig. 10.3.1 Schematic overview of the typically found structural units in the area under land use in the San Francisco valley

Characteristics of forest stands were documented on these 166 plots of 20×20 m each. These plots covered an altitudinal gradient from 960 m to 3100 m. For each stratum of woody vegetation within these plots two groups of character classes were investigated: the first group contained eight parameters on a ratio scale, such as dbh, height of stratum or distance between stems. Data were collected as an average value for the stratum. The second group contained 94 parameters per stratum on an ordinal scale. This scale followed Webb et al. (1970) and had the four categories: absent, rare, abundant and dominant.

These parameters included e.g. leaf size, leaf shape, degree of epiphytic coverage, crown shape and crown development (Table 10.3.1). The complete catalogue

Table 10.3.1 Structural parameters on different levels

Level	Structural parameters
Landscape level: mosaic of vegetation units, natural or man-made	Size and shape of patches, degree of fragmentation, length and density of boundaries, linear structures, land use induced structures like gardens, pastures with or without trees, shade trees, etc.
Plot level: vegetation mosaic in closed stands	Stratification, total cover, distance between stems, distance between crowns, diameter at breast height, abundance of epiphytes, lianas, leaf size, leaf length, leaf width, leaf shape, leaf consistency, thickness and roughness of bark, angle of branches, ramification and branching pattern, crown shape, crown development, height of tree, stem diameter, coverage of mosses, ferns, orchids and bromeliads on stem and branches, etc.

of parameters and the classes within each parameter are given by Paulsch (2002) and Piechowski (2003). Statistical interpretation was carried out by the use of cluster analyses and the Ward algorithm, with the squared Euclidean distance as a distance measure.

To find relationships between the total of 204 parameters for the highest and the lowest strata and altitude the Spearman rank correlation coefficient (r_s) was calculated. To avoid a table-wide false discovery rate only probabilities of $P<0.001$ were considered.

10.3.3 Results and Discussion

Landscape level classification In the majority, the 245 land-use units of the 12 fincas investigated were not regularly shaped but followed geo-morphological features of the terrain (e.g. small valleys, landslides) or human-induced boundaries (e.g. slash and burn destruction of primary forest). Figure 10.3.1 gives a schematic overview of the typical structural units in the area under land use in the San Francisco valley. Pasture is the dominant form of land use and pasture patches can be further differentiated by the presence of trees or groups of trees, by the amount of dead wood still present or the degree of fern (*Pteridium arachnoideum*) invading the grass-dominated patches.

Regular-shaped gardens are found directly around the farm buildings, providing the farmers with fruit, vegetables and medical plants. In some places afforestations with *Pinus* sp. and *Eucalyptus* sp. can be found. Landslides occur typically in the context of road construction. Forest remnants can be found especially along little creeks (*quebradas*) and recently cleared patches border the primary forests at the upper level of the properties. The landscape pattern described by the structural analysis shows attributes of a just recently colonized region: boundaries follow geomorphological structures instead of linear property boundaries, pastures still include dead trees not yet rotten and farms are irregularly distributed in a matrix of primary forest. It can be concluded that the landscape undergoes a dynamic process of land-use changes that will lead to more loss of primary forest (see Chapters 3 and 4 in this volume). The structural classification is an appropriate means of documenting this process.

10.3.3.1 Plot Level Classification and Altitudinal Correlations

The 166 plots were grouped by cluster analyses into 21 forest types of different degree of human interference. Table 10.3.2 assigns the nine primary forest types to the altitudinal range of their occurrence.

The grouping of the plots into structural forest types was based on a multitude of parameters and it is impossible to name only a handful of individual key parameters that would justify the grouping (Paulsch 2002). In order to handle the structural forest types as descriptive units the very obvious parameters of leaf size and topographic situation were used in name-giving, despite the fact that the types are distinguished by the varying occurrence of more than 100 parameters. Thus, the nine types of Table 10.3.2 can be described as follows.

The primary forest of the ridge of Cajanuma was an elfin forest with small stunted trees, with diagonal stems, a high coverage of epiphytic mosses and vascular epiphytes and a high percentage of *Weimannia* species (see Chapter. 10.4). The canopy stratum was only 10–15 m high, 5–25 trees per plot covered 40–50% and the crown distance was 1–3 m. Trees had a dbh of 40 cm and mostly irregular or umbrella-shaped crowns. The simple or compound leaves were micro-mesophyll, had a semi-sclerophyll consistency and were held horizontally. Bamboo was omnipresent, overgrowing young trees and filling gaps.

The canopy stratum of *the primary ravine forest of higher altitudes* was 20–25 m high and 5–25 trees per plot covered 50–60%. The dbh was 40 cm, the irregular or umbrella-shaped crowns had a distance of 1–3 m. The simple and malacophyll leaves were held horizontally or at an angle of 45 ° and had mesophyll or micro-mesophyll sizes. Most individual trees hosted some vascular epiphytes, connecting elements were rare or absent, climbers conspicuous. The percentage of standing dead trees was higher than average and stilt roots were abundant.

The primary ravine forest of lower altitudes contained the highest and thickest trees of the whole investigation area and had the highest percentage of emergent trees, partly the economically valuable *Prumnopitys montana* (Humb. & Bonpl. Ex Willd.) de Laub (Podocarpaceae). In the canopy stratum, 1–5 individual trees covered 40–50%. The irregular or umbrella-shaped crowns had a distance of <1 m.

The canopy reached an average height of 25–30 m with trees of 60 cm dbh but single emergent trees grew 35–40 m high and had a dbh of 130 cm. The simple and malacophyll leaves were of micro-mesophyll or mesophyll size and were held horizontally or up to an angle of 45 °.

The microphyll ridge forest had the widest altitudinal range of all forest types in the investigation area. It was characterized by a low and sparse canopy layer and a dense undergrowth stratum, where grasses, species of Cyclanthaceae and ground-living bromeliads filled the space between the woody plants. Sunlight on the ridges was so intense and the canopy so sparse that all kinds of epiphytic plants could develop even in the lower stratum. Compared with the canopy strata of all other forest types ground cover, dbh, height of stratum, height of lowest branch and height of main ramification were under-represented. The high percentage of semi-sclerophyll leaves corresponded with the dominance of *Purdiaea nutans* Planch. (Cyrillaceae) in the canopy stratum (see Chapter 19). A high percentage of species of the Clusiaceae was responsible for the over-representation of semi-succulent leaves (see Chapters 10.1, 10.4).

Compared with the other forest types, the height of canopy, height of lowest branch and height of main ramification in *the macrophyll ridge forest* were under-represented and the distance between crowns and stems less than average. Ground cover was higher and more crowns than in other forest types were restricted by their neighbors. Bamboo as a connecting element was over-represented in the canopy stratum. The over-representation of macrophyll semi-sclerophyll and bicolored leaves corresponded with the high percentage of species of Melastomataceae (genera *Miconia, Graffenrieda*) in the canopy stratum (see Chapters 10.1, 10.4).

Palm trees were over-represented in the canopy stratum of *the megaphyll ridge forest* corresponding with the over-representation of funnel-shaped crowns, a crown volume restricted to the top, a high number of trees without branches and a high percentage of megaphyll compound leaves.

Compared with the canopy strata of other forest types, *the mesophyll ridge forest* was over-represented by diagonal stems, fan-shaped crowns, dead branches, epiphytic ferns, bamboo and connecting lianas. Climbers were conspicuous.

In *the forest with dense canopy* the crowns of neighboring trees grew into each other, the number of trees in the canopy stratum was higher than in other types and climbers and lianas were highly abundant.

The low ridge forest was characterized by a sparse canopy of less than 20 m height, a dbh of 20–30 cm, semi-sclerophyll leaves and a high abundance of vascular epiphytes.

For a closer description of which parameters are most abundant or characteristic for each type, see Paulsch (2002) and Piechowski (2003).

The distribution of structural forest types does not follow a clear altitudinal gradient throughout the investigation area. Instead, two main gradients determine vegetation structure: the degree of destruction mainly caused by humans and the relief. Especially in the lower part of the investigation area between 960 m and 1500 m and in the lower part of the ESCF area (1800–2000 m) the human influence by selective logging or cutting and burning causes different vegetation types at the same altitudinal level. In the mid-elevation part of the ECSF area (2000–2300 m) the distribution of forest types is mainly caused by the heavily inclined relief and the small scale mosaic of deep valleys and steep ridges. Only at an altitude between 2400 m and 3100 m, which is above the steep valleys (in the higher part of the ECSF area, in the Cajanuma section of the Podocarpus National Park) the forest vegetation forms a kind of altitudinal belts.

Webb et al. (1970) stated that the number of features which could be described in a structural approach is theoretically unlimited and therefore a choice has to be made. The catalogue of features used in the study presented is based on the results of an intense testing of the applicability of several approaches known from the literature (for more details, see Axmacher 1998). From the tested systems, those features were combined which were really present in the investigation area and detectable with field methods in tropical forest vegetation. The catalogue was extended by features concerning connecting elements like lianas and bamboo and features describing the distribution of epiphytes because of the importance of these life-forms, particularly for montane forests (Bogh 1992; Finckh and Paulsch 1995; Grubb et al. 1963; Nadkarni 1984). The resulting structural classification approach offers an easy and quick way with only a minimal training requirement to map the vegetation of a distinct area like e.g. a national park. In contrast, a floristic classification would be very time-consuming and require the knowledge of a team of experts for all the different plant taxa.

Highly significant changes ($P \leq 0.001$) in the occurrence of parameters with increasing altitude are e.g. crown development, leaf consistency, leaf size, occurrence of drip-tips and leaf angle: with increasing altitude the percentage of semi-sclerophyll ($r_S = 0.401$) and semi-succulent leaves ($r_S = 0.332$) increases as well as the occurrence of nanophyll ($r_S = 0.294$) and nano-microphyll ($r_S = 0.445$) leaves, while the occurrence of larger leaves (mesophyll $r_S = -0.345$) decreases. Not only the size and consistency of leaves changes, but also the angle in which they are presented. With increasing altitude leaves are less often presented horizontally ($r_S = -0.576$), more at steeper angles ($r_S = 0.681$). All these findings might be explained as adaptations to extreme amounts of radiance at higher elevations that occur in the rare occasions of cloudless days and lead to the danger of drought stress, especially on the less profound soils on ridges. The decreasing number of drip-tips with increasing altitude ($r_S = -0.696$) might be related to different precipitation patterns: at the lower elevations heavy rains occur regularly, while at higher elevations a main part of the moisture comes from clouds and slight drizzle. Other changes relate to the shape of the trees as a whole: with increasing altitude a higher percentage of crowns in the canopy layer were fully developed ($r_S = 0.658$) and not restricted by neighbors ($r_S = -0.527$). At lower elevations and especially in ravines occurs hard competition for light and trees will fill every gap in a way that allows only some individuals to develop full crowns, whereas at higher elevations the lack of other resources (e.g. nutrients) is determining tree growth. So trees are dispersed more scarcely and crowns in the canopy layer do not outgrow each other in their competition for light.

These examples show that structural features and their changes can be related to the change in environmental conditions.

10.3.3.2 Comparison with Existing Zonation Models

The expression 'montane rain forest' is used in different contexts (e.g. Beard 1955; Ellenberg 1975; Gradstein and Frahm 1987; Grubb et al. 1963; Lauer and Erlenbach 1987; Richards 1952). Table 10.3.2 places nine of the primary structural forest types according to their altitudinal range and the relief position. It can be observed that the altitudinal zonation of structural forest types does not contradict the established zonation systems and confirms a change in forest formations at about 2200 m a.s.l.; moreover, it allows further distinction of forest types which would be assigned to one single formation in the traditional classification systems.

10.3.3.3 Functional Relationships

The classification of structural forest types was meant as a basis for the investigation of functional relationships between vegetation and other components of the forest ecosystem. Dziedzioch (2001) studied species composition and phenology of hummingbird-visited plants in a 1-ha transect in the lower part of the ECSF investigation area. Her data revealed that the distribution of hummingbird-visited plant species and plant individuals reflected the structural forest types. As a consequence, the classification of structural forest types can now be used as an instrument to obtain a very rapid overview of habitat suitability for a certain functional group, i.e. hummingbirds or their food plants. D. Paulsch used the structural forest types in the ECSF area as a basis for his investigations of bird communities and also found relationships (see Chapter 11.1 in this volume). The structural classification system thus provides a solid basis for further investigation of functional relationships.

Table 10.3.2 Structural forest types

Altitude (m a.s.l.)	Structural classification system	
4000	–	
	Shrub and grass páramo	
3000	Primary forest Cajanuma	
2000	Primary ravine forest at higher altitude	Microphyll ridge forest
	Primary ravine forest at lower altitude	Macrophyll ridge forest
		Megaphyll ridge forest
		(Mesophyll ridge forest)
1000	Forest with dense canopy	Low ridge forest
0		

10.3.4 Conclusion

The newly developed and tested structural classification approach offers an easy and quick way with only a minimal training requirement to map the vegetation of a distinct area like e.g. a national park. The altitudinal zonation of the resulting structural forest types does not contradict the established zonation systems and confirms a change in forest formations at about 2200 m a.s.l. in the investigation area. Moreover, it allows further distinction of forest types which would all be assigned to one single formation in the traditional classification systems. Investigation of functional relationships shows that the classification of structural forest types can be used as an instrument to obtain a very rapid overview of habitat suitability for a certain functional group, i.e. hummingbirds or their food plants. Thus, the system provides a solid basis for further investigation of functional relationships.

Chapter 10.4
Vegetation Structures and Ecological Features of the Upper Timberline Ecotone

M. Richter(✉), K.-H. Diertl, T. Peters, and R.W. Bussmann

10.4.1 Introduction: Neotropical Timberline Ecotones and the Special Case of Cordillera Real

The position of the upper timberline varies considerably along the neotropical cordilleras. Reviewing a north–south profile, differences between western, central, and eastern chains and/or escarpments become obvious. This is best demonstrated by an example from northern Chile and north-western Argentina at the southern limit of the Neotropics around 23 °S: While tree stands are missing completely on the western escarpment of the high Atacama, apart from small groups in creek and salar habitats (Richter and Schmidt 2002), the eastern part around Jujuy shows a *Polylepis* timberline at around 4000 m a.s.l. (Kessler 1995). At 550 km further north, open *Polylepis* woodland climbs up to 4800 m a.s.l. with smaller treelets reaching even 5100 m on Sajama and Parinacota near the border of Bolivia and Chile (Jordan 1983; Hoch and Körner 2005).

Polylepis, sometimes joined by *Gynoxis* trees in the Central Andes of Peru and in central Ecuador, is the most prominent member of South American tree- and timberlines. In some cases low soil temperatures with mean air temperatures of between 5.5 °C and 7.5 °C are seen as a main ecological trigger explaining their uppermost occurrence in an altitudinal belt. This comes close to the 7 °C mean for soil temperatures at 50 cm soil depth as postulated by Walter (1973; see also Körner 1999). In contrast to this finding, Bendix and Rafiqpoor (2001) give proof that thermal growth conditions can drop temporarily to 1.9 °C with a yearly average temperature of about 4.25 °C within a *P. incana* forest at Papallacta in northern Ecuador at 4060 m a.s.l.

In the northernmost part of the neotropical mountains Holarctic elements form the timberline. *Pinus* and *Cupressus* are the uppermost tree genera at about 4000 m a.s.l. between Central Mexico and Guatemala (Beaman 1962; Ern 1974; Lauer 1978; Veblen 1978). Between the northern coniferous and the southern *Polylepis* timberline the transitory cordillera in Costa Rica and western Panama is characterized by a relatively low treeline at about 3400 m a.s.l. Here a few woody genera such as *Quercus*, *Vaccinium*, and *Viburnum* from the northern hemisphere meet

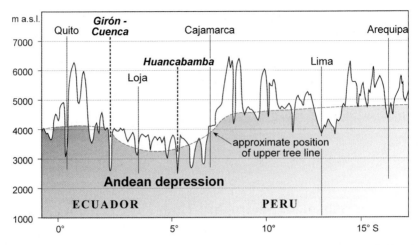

Fig. 10.4.1 Position of the upper tree line and lowest glacial stands within the Neotropical section of the Andes (western escarpment of the main chain)

Drimys, *Clusia*, *Miconia*, *Myrsine*, *Schefflera*, *Prumnopitys*, and *Weinmannia* from the south (Kappelle et al. 1995; Weber 1958). At this point it becomes obvious that tree diversity of the "low neotropical timberline", containing a mixture of geo-elements, is much richer than of the "high neotropical timberlines" further north and south.

A second, less known low neotropical timberline interrupts the high one of *Polylepis* stands in the northern part of the Andes (Venezuela, Columbia, northern-central Ecuador) and in the Central Andes (central-southern Peru, Bolivia, northern Chile, north-western Argentina). This is between 3 °S and 7 °S, within the so-called Amotape–Huancabamba Floristic Zone (Weigend 2002) of the Andean Depression, once again presented by an upper timberline rich in tree species (Fig. 10.4.1). Very few peaks of the Andean chain in this region transgress 4000 m a.s.l. and only few stands of *P. weberbaueri* are known from the south of this section (6 °S at around 3300 m; Baumann 1988). This timberline depression results from the absence of the usually superposing *Polylepis* belt, which is combined with extremely high precipitation in the eastern chain of the cordilleras in northern Peru and southern Ecuador, since the genus tends to avoid perhumid mountain regions (Kessler 1995).

10.4.2 Study Area and Methods

Three study areas of different character are located in the northern half of the Cordillera Real, which forms part of the northern Huancabamba Depression. Towards the tectonic rupture of Girón–Cuenca as the northern borderline of the depression zone, only one adjacent mountain complex follows north of the Rio

Zamora breach, which is the Nudo de Loja (syn. Cordillera de Saraguro). The southward continuation of the 60 km north–south running Cordillera Real is the Cordillera de Sabanilla or Cordillera de Amaluza. In these adjoining mountain chains the uppermost timberlines exhibit elevations similar to those in the Cordillera Real (between 2600 m and 3400 m a.s.l.). The minimum and maximum extensions raise the question as to whether the driving force for treeless sites in the timberline ecotone is always the same.

Two of the three study areas are located at lower altitudes corresponding to "low timberline sites" within the vast zone of the "low Neotropical timberline". One is Cerro del Consuelo in the upper part of the RBSF terrain, between 2730 m and 3040 m a.s.l. (Fig. 10.4.2: upper sketches) and the second is Paso El Tiro on the road from Zamora to Loja, between 2790 m and 2820 m a.s.l. The third area corresponds to a "high timberline site" within the zone of the "low Neotropical timberline", situated between 3160 m and 3360 m a.s.l.

For floristic research eight transects of 50×2 m, corresponding to a method recommended as a sampling method for projects in the tropics by the Missouri Botanical Garden (2007), were established in each area. Each of the eight transects at different altitudes was divided in ten sections of 5×2 m, which were sampled individually during fieldwork. All plant species were collected and life form distributions were recorded, including values of percentage coverage for similarity analysis. Collected specimens were identified in the Herbario de la Universidad Nacional de Loja.

Climate data presented are drawn from a network of 14 automatic weather stations installed since 1997. An extraordinarily high rainfall input within the timberline belt of the Cordillera Real can be verified, as confirmed by the data from four automatic weather stations at 2670 m (TS1 in RBSF), 2830 m (El Tiro), 2930 m (Cerro del Consuelo), and 3400 m (Páramo de Cajanuma). Five-year means of annual precipitation range between 1500 mm year^{-1} (El Tiro) and 5000 mm year^{-1} (TS1; Emck 2006). According to radar extrapolation (Rollenbeck, personal communication; see Chapter 8 in this volume) even this amount might be exceeded in some remote mountain areas of the surroundings.

10.4.3 Results and Discussion

10.4.3.1 Floristic Composition and Plant Diversity at Three Timberline Sites

The two lower areas show a rather regular timberline pattern with a gradual shift from slope forests towards dwarf forests and herbaceous formations on top of the ridges. In contrast, the Cajanuma area at higher altitude presents an irregular timberline structure. In the latter case, a mixture of dense elfin forests and uniform *Chusquea* bamboo stands, *Neurolepis* dwarf bamboo patches and mixed shrub

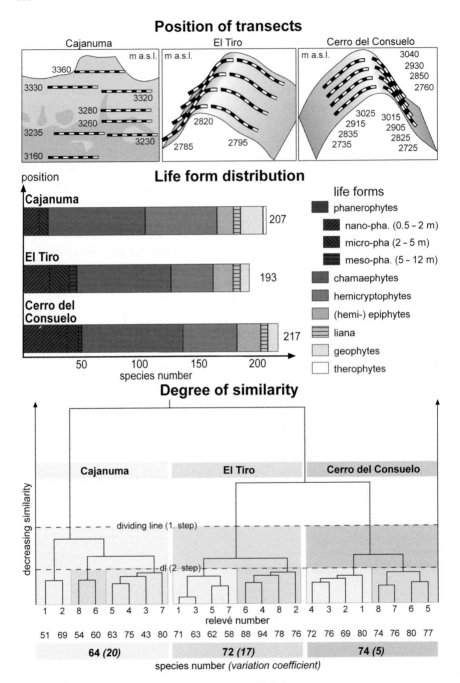

Fig. 10.4.2 Comparative overview of the community similarity and plant formation structure within the three timberline study areas. *Above* Position of the eight transects within each area. *Center* Life form numbers for the three areas and species numbers. *Below* (Dis-)similarity of the species composition between each transect and the three areas [presence/absence data, Jaccard index; *bottom*: species number (and *variation coefficient*) per transect]

páramos was found interspersed with tree stands, forming a complicated ecotone complex. Consequently, some of the physiognomic aspects of the transects vary considerably. Some of them are species-poorer bamboo formations without any trees, while others reveal species-richer fog forests and include all types of life forms available. Disturbances cause differently sized gaps in this area [landslides, soil creeping, possibly also rare wildfires under irregular climatic conditions, and even burrowing mammals such as agoutis (*Solenodon* spp)]. All transects show a multi-layered coverage of vascular plants ranging between 95% and 205%. Lowest values correspond to some of the dwarf bamboo páramos and highest to mixed shrub páramos with small trees containing epiphytes, while elfin forests hold an intermediary position. In addition, cryptogams cover 20–90% of the surface and supplementary trunk and branch envelopes occur too.

The gradual transition from dense forests towards open woodland and scattered trees, which is a typical feature of many *Pinus* and *Polylepis* timberlines in drier Neotropical mountains, does not exist in any of the research areas. In contrast, the perhumid climate of the low timberline ecotones investigated here explains not only a much denser coverage but also the presence of additional life forms such as lianas and hemi-epiphytes. Interestingly, most (165 species = 39% of the total) are dwarf shrubs up to 50 cm in height, while grasses and herbs hold the second position (99 species = 23% of the total, seealso Fig. 10.4.2, centre). Among the 76 species of phanerophytes (18% of the total), the crude climatic conditions lead to a decreasing number of taller subclasses, as 44 nano-forms are followed by 23 micro- and only nine meso-forms. The latter does not occur in the cooler and wetter high timberline area of Cajanuma (Table 10.4.1).

The total of 39 timberline tree species found is surprisingly high, compared to the "high neotropical timberline" and especially with respect to sub- or extratropical equivalent ecotones. This is even true for the lower amounts per area shown in Table 10.4.1, where the limited number of shrubs and small trees for Cajanuma once again hints to harsher growth conditions. Six woody genera are among the ten first species-rich taxa, with *Miconia* (first rank = 26 spp), *Weinmannia* (second rank = 11 spp), *Symplocos* (joint fourth = 9 spp), *Ilex* (joint fourth = 9 spp), *Brachyotum* (joint fourth = 9 spp), and *Disterigma* (ninth rank = 7 spp). The differences in numbers of tree species between El Tiro and Cerro del Consuelo trace back to the broader altitudinal range of the latter research area, which covers a vertical distance of 310 m (Fig. 10.4.2, top).

Table 10.4.1 Species number of woody life forms on the eight transects (800 m^2) in the three areas investigated

Tree life forms	Number of species		
	Cajanuma (3200 m a.s.l.)	El Tiro (2800 m a.s.l.)	Cerro del Consuelo (2850 m a.s.l.)
Nano-phanerophytes (0.2–2.0 m)	16	22	37
Micro-phanerophytes (2–6 m)	12	15	19
Meso-phanerophytes (6–18 m)	0	7	2

In difference to the number of shrub and tree species, no significant variations between the total amounts of vascular species are notable between the three areas, with 205 ± 12 species per $800\,m^2$ plot found. The total of 422 species encountered in the three areas likewise reflects the tropical aspect of vegetation. Still more remarkable is the fact that only 30 species (i.e. 7.1%) are shared by all three areas. Species numbers per transect range between 43 (Cajanuma, dwarf bamboo páramo) and 94 (El Tiro, mixed formation extending from herbaceous to forested stands) with a mean of 70 species per transect. The highest variation was found at Cajanuma (20%). This is easily explained by the great diversity of plant formations. The reason for the considerable difference between the deviation rates of El Tiro and Cerro del Consuelo (17% compared with 5%) in contrast, traces back to different wind impacts (see Section 10.4.3.2). Windward and leeeffects are much stronger at El Tiro (see Section 10.4.3.2) than at Cerro del Consuelo and hence create more pronounced differences in species composition. This explains a clear separation of the four western from the eastern transects in the dendrogram. All sites of higher species richness (even valid for the short sections) are located on the exposed eastern sides with its mixed formation structures, while the forested western sides have lower species numbers (uneven relevé numbers in the dendrogram, Fig. 10.4.2, bottom). This contrast between the two aspects is lacking at Cerro del Consuelo where wind is of minor importance.

The dendrogram in Fig. 10.4.2 shows that the degree of species dissimilarity within an area is highest at Cajanuma, where the dividing line separates three assemblies instead of two at El Tiro and Cerro del Consuelo. The between-area similarity rate underlines the diverging character of Cajanuma. This might be explained not only by environmental factors in this elevated study area but also by its location west of the main crest-line. The shorter and straighter pass-way between Cerro del Consuelo above and El Tiro at the western margin of the same San Francisco Valley might facilitate seed interchange.

The floristic composition of the three areas indicates similar as well as diverging traits (Fig. 10.4.3). With regard to the family structure, the five first ranked families are the same in all study sites: Asteraceae, Bromeliaceae, Ericaceae, Melstomataceae, and Orchidaceae account for the most frequent species, while in the next four positions only Poaceae still form a prominent group in each of the areas. The ranking of the first five families varies in the sense that the worldwide-distributed Asteraceae and Ericaceae dominate at the higher reaches while the tropical group of Melastomataceae is strongest in the two lower timberline ecotones. *Miconia* (Melastomataceae), the species richest genus overall, is less present at the highest area of distribution. *Tillandsia* (Bromeliaceae), the third-ranked genus of the total species pool, is always well distributed while *Weinmannia* as the second-ranked genus shows more variety among its area members. Apart from the woody genera *Miconia*, *Weinmannia*, and *Brachyotum*, the two climber groups of *Mikania* and *Bomarea* are conspicuous associates in each of the three areas.

Communities can be separated by some frequent woody species in the timberline ecotone. *Hedyosmum cumbalense* and *H. scabrum* as well as *Weinmannia pubescens* and *W. rollottii* are important timberline members in Cajanuma and are not registered in the other two areas. Instead, the latter harbor *Clusia elliptica* and *C. ducuoides*,

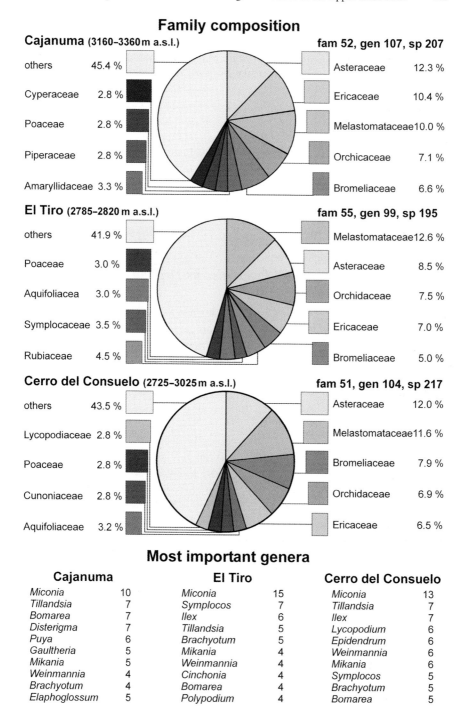

Fig. 10.4.3 Percentage of the contribution of the first nine most important plant families, and species numbers of the first most important genera per study area

Hedyosmum translucidum, Miconia jahnii, Ocotea sp., *Schefflera* sp. 1, and *Weinmannia fagaroides*. In all three areas *Clethra revoluta, Cybianthus marginatus, Graffenrieda harlingii*, and *Miconia radula* play an important part.

Differences in the distribution of shrubs were obvious. Frequent species only occurring in Cajanuma were *Brachyotum andreanum* and *B. campii, Gynoxis cuicochensis, Macleania rupestris*, and *Miconia ligustrina*, among others. Restricted to the two lower timberlines were *Brachyotum campanulare* and *B. rotundifolium, Cyathea frigida, Gaiadendron punctatum, Geissanthus andinus, Gynoxis fuliginosa, Ilex scopulosum, Myrsine andina, Symplocos canescens*, and *Weinmannia elliptica*. Abundant in all areas were *Freziera minima, Gaultheri amoena, G. erecta, G. reticulata, Gynoxis miniphylla, Miconia tinifolia*, and *Symplocos clethrifolia*. *Chusquea falcata* and the dwarf species *C. neurophylla* and *Neurolepis elata* were prominent species of all three sites.

10.4.3.2 Climate and Soil Properties as Driving Forces

The two weather stations of the Cajanuma area are located at 3240 m a.s.l. within the timberline ecotone. One provides data from a mixed elfin forest located on a slope with coarse debris, the other from a 120 m nearby dwarf bamboo páramo on a flat ridge with loamy soil above solid rock. Long-term climate data are available for temperature and relative humidity (200 cm, 30 cm), wind speed and direction (400 cm, 30 cm at the forest site, 250 cm at the páramo stand), global irradiation and emission (30 cm within the forest, 200 cm at the páramo), and soil moisture (−10 cm, −50 cm).

Fig. 10.4.4 gives an idea of two ecologically decisive weather types, i.e the rainfall situation (by far predominant) and the fog situation, both derived from the quasi-permanent easterlies, and the short radiation periods during October and November originating from western airflows (so-called "veranillo"). Both situations are highlighted by taking November 2005 as an ideal example. The first half of this month presents rainy weather and the second sunny weather. Rainy weather is typical for a season of at least 10 months between December and October; and even during the short "dry" period fogs and rain prevail. The precipitation type is marked by a rather constant but not strong input whereas thunderstorms with downpours are rare in the crest area, which is mostly enveloped in shallow cloud caps.

Fig. 10.4.4 Comparison of air temperature (amplitude at 200 cm and 30 cm above ground), irradiance and emission (daily sum w h m^{-2} at 200 cm above ground), relative humidity (amplitude at 200 cm and 30 cm above ground), soil moisture in vol% at −10 cm and −50 cm (values at 4:00 p.m.), maximum and mean wind speed per day (based on hourly means; at 200, 400, 30 cm above ground) over an open dwarf bamboo stand on an exposed side ridge and in a slope forest (120 m nearby) within the timberline ecotone at Cajanuma at 3240 m and 3230 m a.s.l., respectively. Precipitation measurements were taken on the side ridge at 100 cm above ground

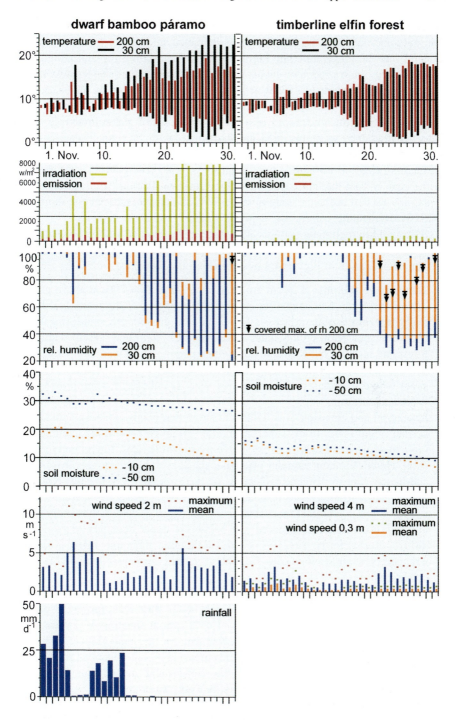

A comparative interpretation of both weather situations leads to an explanation of the main microclimatic triggers. They are characterized by long-time equal conditions of narrow temperature and humidity amplitudes during fog and rainfall and extraordinarily high irradiation inputs during sunshine. In the latter context record values were recorded: up to $1.832\,\text{w}\,\text{m}^{-2}$ at the nearby Páramo climate station (3400 m a.s.l.; March 2000) and up to $1.756\,\text{w}\,\text{m}^{-2}$ at Cajanuma–Mirador (2960 m a.s.l.; December 1999). Such occasions of a very high excess of global irradiance above potential irradiance were measured and explained by Emck (unpublished data) as "white-screen" effects of reflection by transparent and opaque clouds. They give a reason for the temperature maxima on plant surfaces exceeding 50 °C – just contrary to the equally cool thermal conditions during most of the year. Thus, plants at the timberline are best adapted to long-time wet surroundings but must endure short-time stress by overheating, extreme or "super-irradiance", and desiccation too.

Although microclimatic features at the timberline ecotone within the Cordillera Real do not differ considerably from general principles, some noteworthy characterizations are given in Fig. 10.4.4 by comparing the two sites.

1. Temperature. Minima at 30 cm are nearly the same as at 200 cm above ground during and some few days after rainfall. After drying of the upper soil layers slight freezing at soil surfaces is found on bare ground. No significant differences are seen between the stand temperature at 30 cm and 200 cm within the forest and also not between forest and open páramo at 200 cm. Temporary heat stress is restricted to the uppermost part of the canopies and to open herbaceous strata where it explains xeric structures such as pubescent, scleromorphic, or waxy leaf surfaces.
2. Irradiation and emission. It is hard to imagine that plants at the Cajanuma timberline might suffer from "super-irradiance" although 84% of the daytime during a year are rainy or foggy (calculated by sunshine rates; Emck, unpublished data). However, many vascular plant species show reddish colors on fresh leafs and hence, hint to carotene protecting against excessive UV-B rates. Not only global "super-irradiance" is responsible for any of the protective features. Since UV radiation is scattered more than total shortwave radiation, maximum absolute UV intensities are observed just below the upper boundary of clouds by diffuse fluxes (Barry 1992). Since the Cajanuma mountain ridge is mostly wrapped by shallow cloud caps the timberline area consequently underlies such conditions.
3. Relative humidity. According to the thermal and solar circumstances for the exceptional dry period during the last part of the time-scale presented in Fig. 10.4.4, relative humidity is also marked by a special situation of dryness with minimum values below 30%. This is even true for the stand climate at 200 cm within the elfin forest, where once again similar amplitudes like that in the open environment become obvious. Only the near-ground air and thus the undergrowth does not suffer in the same way by water vapor deficits.
4. Soil moisture: Soil water contents differ considerably between the two sites. Soils of the open páramo on the ridge develop on solid rocky layers, which on

their part cause long-lasting water stagnation. Consequently, especially for the lower soil strata directly above the rock surface, moisture content is considerably higher than in the coarse porous debris layer beyond the forests. In contrast, thick humus layers in elfin forests guarantee that the soils there never dry up strongly. Thus elfin forests create their own water regime by a mighty uppermost organic stratum that conserves moisture also during the (rarely more than two weeks long) dry periods of Veranillo del Niño. Furthermore, the immense water storage capacities of thick epiphytic bryophyte packages on tree branches are verified as a general feature of elfin forests (Müller and Frahm 1998; Kürschner and Parolly 2004b).

5. Wind: Diurnal means of wind velocity as well as maximum gust speeds of the páramo stand (2 m above ground) are around twice as high as above the forests (0.5 m above canopy surface or 4 m above ground, respectively). Within the forest the air-flow is once again much weaker, sometimes reaching only one-fifth of the above-canopy wind speed. Interestingly, the wind direction in open stand differs considerably from the direction above the forest canopy. While the open stand was prevalently marked by air flows from the south-west (39%) and from the north-west (34%) the forest site only 60 m off showed a clear dominance of the north-west (63%, only 9% from the south-west). A weather station on the crest-line at 3400 m a.s.l. measured the easterly airflow with an input of 45% from the north-east, while 31% from the north-west traces back to the special October–November phenomenon of sunny weather during the last two weeks of observation. An interpretation of the different directions hints at turbulence effects at the lee side where easterlies are often turned into western winds that are weaker than on the crest side. However, the wind velocities vary considerably on the lee side depending on the orographic position. During the observation time shown in Fig. 10.4.4, the mean wind speeds were: on the main crest 14.5 km h^{-1} (max. gust 49.7 km h^{-1}), at 160 m lower down on the side ridge with the páramo site still 11.3 km h^{-1} (max. gust 40.3 km h^{-1}), and over the forest site only 5.8 km h^{-1} (max. gust 20.9 km h^{-1}). Apart of an effect of friction caused by an irregular canopy surface, the location of the forest on a less exposed slope between two side ridges provides a more effective screen.

Wind as a decisive force for the position of timberline is proved by measurements on Paso El Tiro at around 2810 m a.s.l. (Fig. 10.4.5). Of course, mean wind and gust speeds are highest on the crest itself. Only 8 m beyond the ridge, mean velocity is reduced to 58% on the leeside. Interestingly, wind speed is still lower on the windward side at 8 m and 13 m east of the crest. This phenomenon traces back to a rather plain topography in front of the ridge but beyond a steeper slope, which lifts up the upwind from the east. The consequence is a patchy structure of isolated but dense tree stands in flat hollows or on wind-protected flats on the windward escarpment below the crestline. In contrast, dense forest cohorts climb just up to the rim of the same crest, forming a sharp timberline on the leeside.

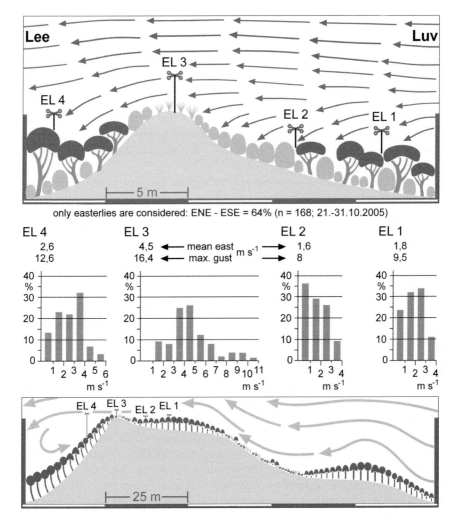

Fig. 10.4.5 Wind speed classes on Paso El Tiro during an easterly period (21–31 October 2005). The evaluation is restricted to the prevailing wind direction (sector ENE–ESE = 64% of the observation period). Air-stream lines give an idealized interpretation of friction effects caused by the local topography and the vegetation type; an overview of the surrounding profile is given in the *bottom sketch*. The two sketches (*top, bottom*) give an impression of the sharp timberline at the leeside of the El Tiro crest and a nearby wind-protected isolated tree stand on the windward side

10.4.4 Conclusion: Too Much Wind and Too Much Water

One of the most surprising results is the high species number of trees at the timberline ecotones of the Cordillera Real. This is all the more astonishing under the aspect of climatic conditions (researchers active in the area sometimes raise the question why any of all those organisms might like to live under such circumstances...). This taxonomic richness is not only an attribute of the input from the humid tropical lowlands but also results from the dizzying kaleidoscope of semi-arid to perhumid environments in the neighborhood (Young et al. 2002; Richter and Moreira-Muñoz 2005).

Soil temperature is not a decisive factor for the delimitation of tree growth in this region of tropical timberline. It shows rather high average values without decisive variations around 11 °C at 2930 m a.s.l. in Cerro del Consuelo and around 8.8 °C at 3400 m a.s.l. on the main ridge above Cajanuma. Further long-time comparative measurements of soil temperatures at −50 cm at 2650 m a.s.l. in RBSF hint to an average mean reduction of 2 K in a dense forest as against to open stands. Taking this value of diminution in forest habitats as well as a vertical thermal gradient of 0.58 K per 100 m into account, mean soil temperatures of 7 °C under forests should be reached at about 3650 m to 3750 m a.s.l. This corresponds to an elevation between 400 m and more than 800 m above the highest position of existent timberlines in the Cordillera Real. This study indicates that the postulated cooling impact as a main trigger for tropical timberlines is questionable and obsolete.

Instead, wind has to be considered as a much more important factor to explain the absence of trees on exposed ridges, as shown for the El Tiro area. However, this is not the case at the Consuelo site where wind is of lesser importance, while rainfall and fog show an exuberant impact underlined by dense *Sphagnum* cushions or by the presence of a not yet identified *Drosera* spec. Consequently, "low timberlines" (or "local timberlines", in analogy to the expressions "climatic" and "local snow-line") are caused by strong wind and/or soil moisture impact. In the case of the Cordillera Real this observation also holds true for some elevated U-shaped valleys such as the upper Sabanilla Valley where an ensemble of cold-air drainage and water-logging results in *Sphagnum*-rich bogs. In the same region many ridge sites also suffer from water-saturated soils that inhibit tree growth, which creates local timberlines expanding along both sides of crests. In some areas, this symptom starts above 2600 m a.s.l. On lower elevations the same ridges disappear under a coherent forest cover as they become steeper and thus water drainage is more efficient apart from lower precipitation and higher evaporation rates towards inferior parts of V-shaped valleys.

To a minor extent, temperature might play an additional part. It explains for example that the two "low local timberlines" have floristically more in common than either of them have with the comparatively more isolated flora of the "high local timberline" at Cajanuma. Most apparently the latter traces back to each of the mentioned factors, i.e. wind and soil-logging on the ridge sites combined with a quasi-permanent cooling effect, which however cannot explain the lack of trees alone. Consequently, the presence or absence of tree stands and forests within all timberline ecotones in the Cordillera Real depends on a complicated synergetic bundle of driving factors. These are given by strong wind and/or high soil-water content, in some areas to a limited degree supported by cool climate.

Acknowledgement We gratefully acknowledge the help of Bolívar Merino at Herbario Loja for assistance with the identification of the collected plant material.

Chapter 10.5
Mycorrhizal State and New and Special Features of Mycorrhizae of Trees, Ericads, Orchids, Ferns, and Liverworts

I. Kottke(✉), A. Beck, I. Haug, S. Setaro, V. Jeske, J.P. Suárez, L. Pazmiño, M. Preußing, M. Nebel, and F. Oberwinkler

10.5.1 Introduction

Most land plants constitute obligate symbiosis with soil fungi forming mycorrhizae (= fungal roots) in the case of vascular plants and mycorrhiza-like associations in the case of liverworts and hornworts. The fungi significantly improve nutrient uptake in the mostly nutrient-limited conditions in nature and supply minerals to the plants (Read and Perez-Moreno 2003). Glomeromycota, Hymenomycetes of Basidiomycota, Pezizales, some Leotiales, and other Ascomycota are the most important mycorrhizal fungi forming structurally distinctive associations, as determined by evolution and environmental constraints (Kottke 2002; Brundrett 2004). The tropical mountain rain forest in South Ecuador is not only exceptionally diverse in tree species, but also a hotspot for ericads, orchids, ferns, and liverworts (see Chapters 2 and 10.1 in this volume). Diverse mycorrhizal associations were, therefore, expected. Very restricted information was previously published from similar forests (Alexander and Högberg 1986). We, therefore, started with morphological and ultrastructural investigations of the mycorrhizae of the 220 tree species, the ericads, some abundant and mostly endemic, epiphytic pleurothallid orchids, and some epiphytic ferns, and we also investigated some mycorrhiza-like associations of liverworts. Here we consider the mycorrhizal state and the occurrence of the distinct mycorrhizal associations along the altitudinal and horizontal gradients, and we compile a number of new structural features.

10.5.2 Mycorrhizal State of the Diverse Plants

All the investigated 115 tree species were found to form mycorrhizae, the vast majority displaying arbuscular mycorrhizae with highly diverse Glomeromycota (Kottke et al. 2004; Kottke and Haug 2004; Kottke et al. 2007). Colonization of tree roots was prominent. Quantification of fungal colonization of ten individuals of *Inga acreana* (Mimosaceae) proved a high frequency of arbuscules and vast spreading of mycelium inside and outside the roots, while the numbers of appressoria and vesicles were variable, and the number of spores linked to the root surface was low (Fig. 10.5.1). The picture was typical for the majority of the trees under survey and independent of whether roots were sampled from the pure humus layer on the mountain ridge or from the mineral soil in the ravines.

Three nyctaginaceen species growing scattered on the steep slopes of small ravines and the San Francisco River formed ectomycorrhizae with Basidiomycota and Ascomycota, but the number of fungi and the level of mycorrhization varied between the tree species (Haug et al. 2005). *Graffenrieda emarginata* (Melastomataceae), the most frequent tree on the mountain ridge, but missing in the ravines, formed arbuscular mycorrhizae and simultaneously ectomycorrhizae with an ascomycete on the same individual roots (Haug et al. 2004). The bushy ericads

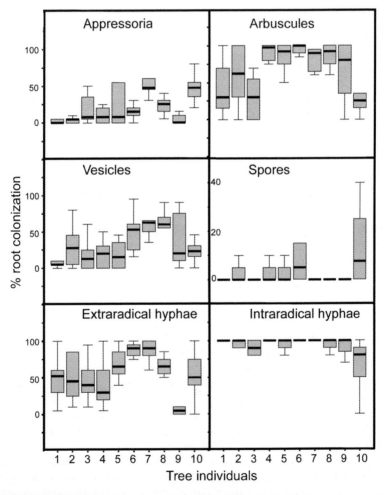

Fig. 10.5.1 Fungal colonization in arbuscular mycorrhizae of *Inga acreana*, ten investigated trees, 10 cm of roots each. Box plot diagrams show median, 25% and 75% percentiles, minima and maxima; SPSS ver. 10.7

in open habitats formed mycorrhizae of the known ericoid type, but the hemiepiphytic, neotropical ericads in the pristine forest displayed a special type of mycorrhizae formed with Sebacinales and ascomycetes (Setaro et al. 2006a, b). Almost all the last-order rootlets of ericads collected from the upper humus layer or the tree bark were well colonized, while the roots of epiphytic orchids were only colonized by mycorrhizal fungi where they contacted the tree bark. The orchid mycorrhizae displayed typical hyphal coils in the cortical tissue formed by Tulasnellales and Sebacinales (Basidiomycota; Suárez et al. 2006; Kottke et al. 2007). Roots of epiphytic Grammitidaceae (Pteridophyta) were associated with Ascomycota, while the epiphytic Polypodiaceae were found non-mycorrhizal (Pazmiño 2006). Terrestrial liverworts belonging to basal systematic groups (Marchantiales, Pallaviciniaceae) were associated with Glomeromycota displaying similar structures as in arbuscular mycorrhizae (Nebel et al. 2004). Thalli of Aneuraceae (liverworts) sampled from fallen stems and rocks were associated with Tulasnellales displaying the same structures as the orchid mycorrhizae (Kottke et al. 2007).

The strong mycotrophic dependence of trees and ericads was already known for temperate forests, tropical lowland forests, and tropical dry forests (Janos 1980; Metcalfe et al. 1998; Wubet et al. 2003). The plants in the tropical mountain rain forest and the bushy páramo occur at conditions of low nutrient availability (see Chapters 13, 17), and dependence on mycorrhizae is reflected in the high degree of fungal colonization.

10.5.3 New and Special Features of Mycorrhizae

10.5.3.1 Arbuscular Mycorrhizae of Alzatea verticillata

The arbuscular mycorrhizae of *Alzatea verticillata* (Alzateaceae) and other trees were covered by conspicuous large amounts of the mycelium of Glomeromycota. Comparative morphology of the supraradical mycelium distinguished a number of consistent, previously not described characters (Fig. 10.5.2a–d). Coarse, irregularly branched hyphae (Fig. 10.5.2a), branched appressoria-like plates with thick walls and irregular outline (Fig. 10.5.2b), and fine branched appressoria-like plates with hyaline outline (Fig. 10.5.2c) were observed. Supraradical vesicles of regular or irregular shape (Fig. 10.5.2c and d) also occurred on the root surface. Penetration points, where hyphae breach the cell wall to colonize the cortical cells, were only observed at more simple appressoria (Figs. 10.5.2a, b, c). Most of the distinct characters of the supraradical mycelium were also observed on mycorrhizae of *Podocarpus oleifolius*, *Critoniopsis floribunda*, *Micropholis guyanensis*, *Tabebuia chrysantha*, *Nectandra acutifolia* and several melastomataceen species of the pristine forest (Beck et al. 2007).

Two mycorrhizal types distinguished on *A. verticillata* displayed intercellular appressoria-like structures between the inner cortical cell layers, reminiscent of the Hartig net of ectomycorrhizae, while the outer cortical cells contained hyphal coils (Fig. 10.5.3; Beck et al. 2005).

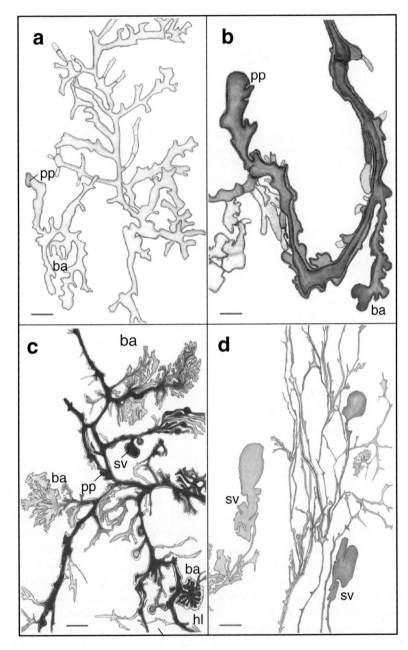

Fig. 10.5.2 a-d Supraradical mycelium observed on selected arbuscular mycorrhizae of *Alzatea verticillata*. For explanation see text. *ba* Branched appressoria-like structure, *hl* hyaline wall layer, *pp* penetration point, *sv* supraradical vesicle. *Bar* 20 μm

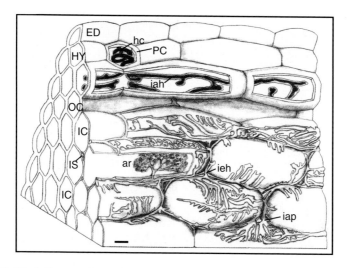

Fig. 10.5.3 Block diagram of a special mycorrhiza of *Alzatea verticillata*, displaying intracellular hyphal coils in the outer cortical layer and the passage cells of the hypodermis, intracellular arbuscules, and fine, branched appressoria-like structures between the inner cortical cell layers. *ED* Epidermal layer, *HY* hypodermal layer, *IC* inner cortical layer, *IS* intercellular space, *OC* outer cortical layer, *PC* passage cell, *ar* arbuscule, *hc* hyphal coil, *iah* intracellular hypha, *iap* intercellular appressorium, *ieh* intercellular hypha. *Bar* 10 μm. First published by Beck et al. (2005)

10.5.3.2 *Ectomycorrhizae of Guapira and Neea (Nyctaginaceae), and Graffenrieda emarginata (Melastomataceae)*

Three members of the Nyctaginaceae, two *Neea* species and one *Guapira* species, were found to form ectomycorrhizae (Fig. 10.5.4). While *Neea* sp. 1 showed ectomycorrhizae on short roots with five different fungal partners (Fig. 10.5.4b–d), *Neea* sp. 2 and *Guapira* sp. had long roots which were only partially covered by a hyphal mantle (Fig. 10.5.4a; Haug et al. 2005). Root segments without hyphal mantle showed root hairs (Fig. 10.5.4a). Only one morphotype was detected, respectively. The morphotypes of *Neea* sp. 2 and *Guapira* sp. exhibited plectenchymatous hyphal mantles with loose arrangement of the hyphae in the outer and compact arrangement in the inner layers. The *Guapira* mycorrhizae showed a prominent Hartig net on and between the epidermal cells (Fig. 10.5.5a). Loose hyphae and intracellular infections in the root hairs and in the epidermal cells were observed in the root segments without hyphal mantle (not shown). A prominent Hartig net was developed between root-hair-like outgrowths of the epidermal cells in the *Neea* sp. 2 mycorrhizae (Fig. 10.5.5b, asterisks). The molecular identification of the fungal partners of the *Neea* sp. 2 and the *Guapira* morphotype revealed one *Thelephora/Tomentella* species respectively (Haug et al. 2005). The described situation of long root systems only partly transformed into ectomycorrhizae by just one fungus, not suppressing root hair formation, and partly penetrating into cells has not

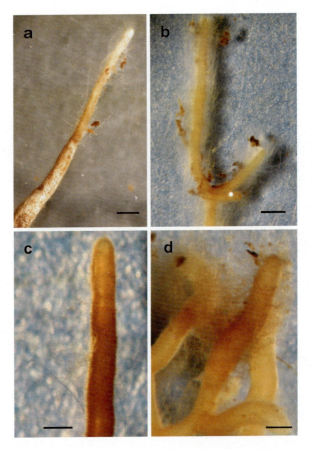

Fig. 10.5.4 Micrographs of ectomycorrhizae of Basidiomycota with nyctaginaceen trees: **a** *Tomentella/Thelephora* sp. – *Guapira* sp., **b** *Lactarius* sp. – *Neea* sp. 1, **c** *Tomentella/Thelephora* sp. 1 – *Neea* sp. 1, **d** *Tomentella/Thelephora* sp. 2 – *Neea* sp. 1. For explanations see text. Bars 1 mm in **a** and **b**, 0.5 mm in **c** and **d**. First published by Haug et al. (2005)

been described from any other plant family so far. We hypothesize that the situation displays an early evolutionary step in ectomycorrhiza formation in the Nyctaginaceae. Other ectomycorrhizal nyctaginaceen species are known from Peru and Venezuela (Alexander and Högberg 1986; Moyersoen 1993), but the Nyctaginaceae are the only family within the Caryophyllales that form ectomycorrhizae, others form arbuscular mycorrhizae or are non-mycorrhizal.

The rootlets of *Graffenrieda emarginata* were associated with arbuscular mycorrhizal fungi in the inner cortical layers and simultaneously, on the same individual roots, a superficial Hartig net was formed by a septate fungus (Fig. 10.5.5c). The hyphae of the brownish fungus invaded between the epidermal cells, but were blocked at the hypodermal layer (Fig. 10.5.5c). The ultrastructure of this ectomycorrhiza forming fungus indicated an ascomycete, and molecular phylogenetic studies

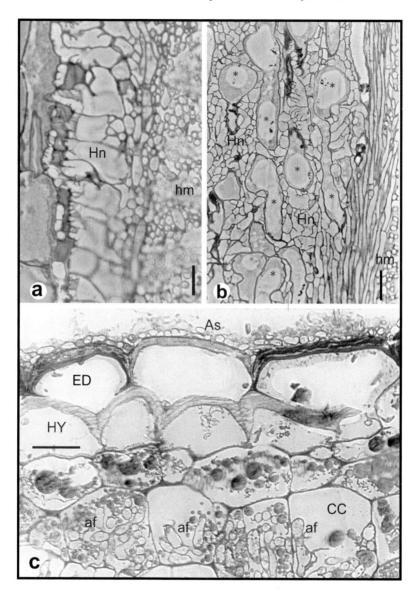

Fig. 10.5.5 Longitudinal sections through ectomycorrhizae of: **a** *Guapira* sp. and **b** *Neea* sp.2, and **c** simultaneous Hartig net formation with an Ascomycete and arbuscular mycorrhizae with Glomeromycota by *Graffenrieda emarginata*. *CC* Cortical cell layer, *ED* epidermal layer, *HY* hypodermal layer, *af* arbuscular mycorrhizal fungi, *hm* hyphal mantle, *Hn* Hartig net; asterisks indicate root hairs. *Bars* 15 μm in **a** and **b**, 30 μm in **c**. **a, b** First published by Haug et al. (2005), **c** first published by Haug et al. (2004); reproduced with permission of NRC

identified the fungal symbiont as a new taxon within the *Hymenoscyphus ericae* aggregate (Haug et al. 2004). Related ascomycetes associated with ericads were shown in previous studies (Read and Perez-Moreno 2003) to have access to organic nitrogen. If this feature counts also for the fungus associated with *G. emarginata* the fungus could give additional nutrient support to this tree species rooting in the pure humus layer of the mountain ridge and this would probably explain why *G. emarginata* is the most frequent tree there.

10.5.3.3 Mycorrhizae of Andean Clade Ericads

Anatomy, ultrastructure, and molecular investigation of the mycorrhizae of the monophyletic "Andean clade" ericads (Kron et al. 2002; Powell and Kron, 2003) showed a new subcategory of mycorrhizae, the cavendishioid mycorrhiza (Fig. 10.5.6; Setaro et al. 2006a, b). The hyphal sheath consists of several layers of fine hyphae growing in parallel to the root axis (Figs. 10.5.6, 10.5.7a). The hyphae form a superficial Hartig net-like structure penetrating between the radial walls of the single layer of cortical cells (Figs. 10.5.6, 10.5.7c). The hyphae penetrate into the cortical cells and form intracellular coils therein (Figs. 10.5.6, 10.5.7a, arrowheads). Inside the host cell, the hyphae almost double in diameter (Fig. 10.5.7c).

The predominant fungi were hymenomycetes displaying dolipori with imperforate, slightly curved to straight parenthesomes (Fig. 10.5.7b, c, arrowheads). Molecular investigations revealed that the fungi were diverse species of the order Sebacinales Group B (Weiß et al. 2004; Setaro et al. 2006a). Ascomycetes were observed in few roots and preliminary molecular identification connected the sequences to Leotiomycetes.

The cavendishioid mycorrhiza is structurally similar to the arbutoid mycorrhiza, but enlargement of intracellular hyphae was not observed in the latter association

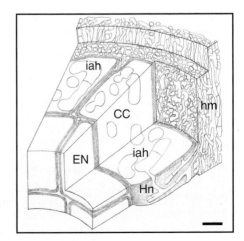

Fig. 10.5.6 Block diagram of the cavendishioid mycorrhiza. *CC* Single cortical cell layer, *EN* endodermal layer, *iah* intracellular hypha, *hm* hyphal mantle, *Hn* Hartig net. *Bar* 4.5 µm. First published by Setaro et al. (2006a)

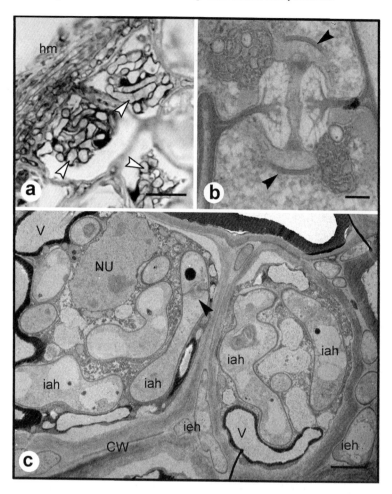

Fig. 10.5.7 a Light micrograph of the cortical layer with intracellular hyphal coils (*arrowheads*) and hyphal mantle of the cavendishioid mycorrhiza. **b** ultrastructure of the fungal porus with imperforate parenthesomes (*arrowheads*). **c** ultrastructure of intracellular and intercellular colonization of the epidermal layer by hymenomycetes; note enlargement of intracellular hyphae. *CW* Plant cell wall, *NU* plant nucleus, *V* plant vacuole, *hm* hyphal mantle, *iah* intracellular hypha, *ieh* intercellular hypha. *Bars* 10 μm in **a**, 0.15 μm in **b**, 2.5 μm in **c**. First published by Setaro et al. (2006a)

(Münzenberger et al. 1992). Additionally, ericads with arbutoid mycorrhizae belong to the Arbutoideae and the Pyrolaceae (Monotropoideae) and associate with ectomycorrhizal fungi (Molinia and Trappe 1982), whereas cavendishioid mycorrhizal ericads belong to the "Andean clade", a crown group of Vaccinioideae (Kron et al. 2002) forming mycorrhizae with ericoid mycorrhizal fungi. An independent, parallel evolution of the cavendishioid mycorrhiza and the arbutoid mycorrhiza is derived from these data justifying the new term cavendishioid mycorrhiza (Setaro et al. 2006b).

10.5.3.4 Mycorrhizae of Epiphytic Ferns

The studied epiphytic species of the family Grammitidaceae (*Micropolypodium* sp, *Grammitis paramicola, Lellingeria subsessilis, Melpomene assurgens, Cochlidium serrulatum, Melpomene pseudonutans, Ceradenia farinosa*) formed mycorrhizae while all investigated epiphytic species of the family Polypodiaceae (*Campyloneurum amphostenon, Niphidium carinatum, Pleopeltis percussa*) lacked mycorrhizae. In the Grammitidaceae the hyphae entered via the root hairs without the formation of appressoria (Fig 10.5.8). The hyphae formed loose coils in the epidermal and cortical cell layers with occasional short branching, spreading from one cell to another. The hyphae were regularly septate and brownish. Ultrastructural studies confirmed the colonization by Ascomycota (data not shown). A similar type of mycorrhiza has so far only been described for epiphytic ferns from Costa Rica (Schmid et al. 1995; Kottke 2002).

10.5.3.5 Appressoria on Rhizoids of *Jensenia spinosa* (Pallaviciniaceae, Liverworts)

While vascular plant roots are invaded by Glomeromycota via appressoria formed on the root surface (see above), in *Jensenia spinosa* (Pallaviciniaceae; liverworts) we observed the formation of appressoria on rhizoids from where

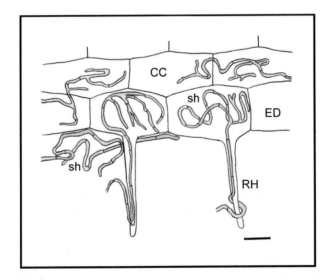

Fig. 10.5.8 Scheme of the mycorrhiza formed between Ascomycota and epiphytic ferns of the family Grammitidaceae. *CC* Cortical cell layer, *ED* epidermal layer, *RH* rhizoid, *sh* septate hypha. Bar 10 µm

hyphae enter and grow through the rhizoids to colonize the creeping axis. Appressoria are fairly distinct (Fig. 10.5.9) and this is related to hyphal diameter, hyphal staining with methyl blue, hyphal wall structures, and the formation of coils, arbuscules, and vesicles in the liverwort creeping axis (Kottke, unpublished observations).

Formation of appressoria on rhizoids of liverworts appears to be restricted to Glomeromycota as Ascomycota and hymenomyceteous Basidiomycota invade rhizoids directly, without the formation of appressoria (not shown). The formation of appressoria and invasion via the rhizoids is supposedly a basal form of symbiotic colonization of plants by Glomeromycota which was replaced by appressoria formation and penetration through the root surface in vascular plants in later evolution.

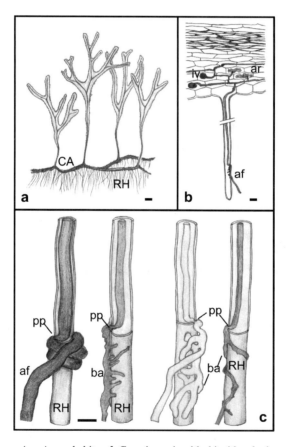

Fig. 10.5.9 a *Jensenia spinosa*, habitus. b Creeping axis with rhizoid and arbuscular mycorrhizal fungi. c Diverse fungal appressoria on rhizoids and penetration of hyphae into rhizoids. *CA* Creeping axis, *RH* rhizoid, *af* arbuscular mycorrhizal fungus, *ar* arbuscule, *ba* branched appressorium, *iv* intracellular vesicle, *pp* penetration point. *Bars* 1 mm in **a**, 5 μm in **b** and **c**.

10.5.4 Conclusion

These investigations revealed for the first time the omnipresence of mycorrhizae in a tropical mountain rain forest of the Northern Andeans and gave hints to evolutionary constraints in the diverse plant and fungal lineages, which appeared even more important than relation to the environment. The evolutionary constraints account especially for the new, cavendishioid type of mycorrhiza in the Andean clade ericads, the ectomycorrhizae found with the nyctaginaceen species, and the family-specific mycorrhization or lack of mycorrhizae in epiphytic ferns and the associations of liverworts with specific fungal groups (Kottke and Nebel 2005).

Arbuscular mycorrhizae were previously related to mineral soil while ectomycorrhizae were thought to be better adapted to soils rich in organic matter (Read and Perez-Moreno 2003). Contrary to this situation in temperate biomes, trees on the steep mountain ridge formed the majority of fine, absorptive, arbuscular mycorrhizal rootlets in a thick layer (30–50 cm) of pure organic matter, but arbuscular mycorrhizae were also found in the mineral soil on the steep slopes of ravines (Kottke et al. 2004). No change of dominance from arbuscular mycorrhizae to ectomycorrhizae occurred along the altitudinal gradient, in contradiction to the situation described for temperate and boreal biomes by Read and Perez-Moreno (2003). Instead, in the tropical, pristine forest, ectomycorrhizae were only observed in nyctaginaceen tree species, a neotropical plant family with no close relation to other ectomycorrhiza-forming plant families. A previously unknown combined association of arbuscular mycorrhizal fungi and Ascomycota was found with *Graffenrieda emarginata* mycorrhizae, the *Hymenoscyphus* sp. probably giving extra support to this most frequent tree species in the investigation area because of the supposed ability to mobilize organic nitrogen. Thus, mycorrhizae of trees in the pristine tropical mountain rain forest are fairly more diverse and the ecological relations are quite distinct from the situation in the temperate biomes, a finding confirmed by other authors (Alexander and Lee 2005).

Acknowledgements The authors thank Lorena Endara, Jürgen Homeier, Markus Lehnert, James Luteyn, and Alfons Schäfer-Verwimp for help with identification of the plant species.

Chapter 11.1
Bird Species Distribution Along an Altitudinal Gradient in Southern Ecuador and its Functional Relationships with Vegetation Structure

D. Paulsch(✉) and K. Müller-Hohenstein

11.1.1 Introduction

The Tropical Andes harbour about 2780 bird species representing 28% of the bird species of the world (BirdLife International and Conservation International 2005). In Ecuador more than 1600 bird species are registered (Ridgely and Greenfield 2001b; Suárez 2002) giving the country one of the highest bird species per area ratio world-wide. Birds play an important role in the tropical mountain ecosystem as they are essential plant-pollinators and fruit-dispersers and form a part of the food chain. In order to understand the ecosystem functioning of tropical mountain forests, the investigation of bird communities as part of the local fauna is indispensable. Along an altitudinal gradient from 1800 m a.s.l. (valley of the Río San Francisco) up to 2650 m a.s.l. (upper border of mountain rain forest) functional relationships between bird communities and vegetation structure were investigated. The study was based on the classification system of tropical mountain forests developed by Paulsch (2002).

11.1.2 Methods

Bird species were recorded by point-counts (Bibby et al. 1995; Gregory et al. 2004) at 30 investigation sites distributed in the different structural forest types, as described by Paulsch (2002) and Paulsch et al. (Chapter 10.3 in this volume). The sites were visited regularly over a one-year period (between 2000 and 2002) at times of main bird activity, which are early morning at sunrise and late afternoon before sunset. All bird species which could be observed or heard during 30 min within a radius of 50 m were recorded. Unknown bird songs were tape-recorded and compared later with voice-recordings by Moore and Lysinger (1997) and Krabbe (2000).

At 15 of these sites birds were also captured with mist-nets in order to record secretive living species which are otherwise hard to observe (Karr 1981). At all sites mist-nets were set up for three successive days. The black mist-nets were

6 m long and about 3 m high and had a mesh width of 16 mm. Due to the steep terrain and the dense undergrowth, only three mist-nets could be used at each site. Mist-nets were controlled every hour from sunrise to sunset. Captured birds were determined to species; and weight, length of wings, bill, tarsus, tail and overall length were noted to allow comparisons when the bird was recaptured (Gosler 2004). Individuals were marked by coloured rings which allowed recognition not only in the case of recapture but even if the birds were observed with binoculars. Hummingbirds (Trochilidae) were marked by shortening the tail-feathers.

The data on bird species distribution were completed by all bird observations 'by chance' in the investigation area.

11.1.3 Results and Discussion

11.1.3.1 Altitudinal Distribution of Bird Species and Feeding Guilds

A total number of 213 bird species occurring in an altitude between 1800 m and 3150 m a.s.l. in the research area was recorded. Mist-netting caught 327 individuals in 68 species during 2700 net-hours. Of these individuals 32 were recaptured and 27 colour-ringed individuals were observed again with binoculars. A further 14 species were observed by other scientists (Dziedzioch 2001; Braun and Matt personal communication; Schmid personal communication) working in the RBSF, raising the total number of bird species to 227. This number fits well with the results of other investigations. Jiménez and Lopez (1999) named 310 bird species between 1500 m and 2900 m a.s.l. for the area 'San Francisco' situated in the direct neighbourhood west of the RBSF. Along the 'Loja–Zamora Road' (1000–2800 m a.s.l.) Rahbek et al. (1995) found 362 bird species, with the highest number of 317 species between 1500 m and 2500 m a.s.l.

A weighted mean of the altitudinal distribution of each bird species was calculated by summarising all the altitudes in which the individuals occurred and grouped in classes for 100 m of altitude (Fig. 11.1.1). Only data for the 150 bird species recorded by point-counts or mist-netting are considered in this figure. Overflying individuals and bird species living exclusively near water were excluded.

As Fig. 11.1.1 shows, 64 bird species, which is more than 42% of the total species number, occur at altitudes between 1800 m and 2000 m a.s.l., with the highest proportion (47 species) between 1900 m and 2000 m a.s.l. Forest types under human influence, semi-open and open successional stages are found at this elevation in the investigation area. These vegetation types are extraordinary rich in birds as the habitats of forest species, forest edge species and the species of the more open areas overlap here. With increasing altitude the recorded number of bird species decreases. Different kinds of primary ridge forests and primary ravine forests are

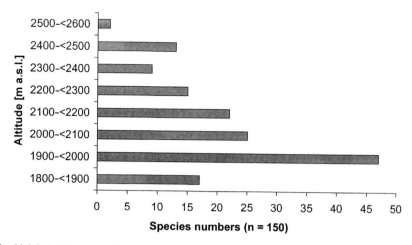

Fig. 11.1.1 Distribution of bird species numbers along the altitudinal gradient

found between 2000 m and 2200 m a.s.l. (Paulsch 2002; Chapter 10.3 in this volume) harbouring 47 bird species, followed by 39 species at higher elevations of a ridge forest type dominated by *Purdiaea nutans* which reaches its timberline at 2650 m a.s.l. (Homeier 2004; Chapter 19 in this volume).

Chavez (2001) found quite similar numbers, with 137 and 120 bird species in two investigation areas situated in the Andean cloud-forest of north-west Ecuador at an altitude between 1800 m and 2200 m a.s.l. In the RBSF 125 bird species were recorded by point-counts and mist-netting in this altitudinal range. Herzog et al. (2005) observed 63 bird species in an mountain rainforest of Bolivia between 1800 m and 1999 m a.s.l. and 76 species between 2000 m and 2249 m a.s.l compared with 100 and 105 species in these altitudes in the RBSF. Between 2250 m and 2499 m a.s.l. the species numbers of the Bolivian rainforest (58 species) and the RBSF (68 species) are more similar (Herzog et al. 2005).

The most important feeding guilds in the investigation area are formed by insectivorous, frugivorous and nectarivorous species (Fig. 11.1.2). These groups play an essential role for the ecosystem functioning as hummingbirds are important plant-pollinators and frugivorous bird species disperse fruits and seeds over great distances.

Bird species of two families belong to the guild of nectarivores in the research area. These are the hummingbirds (Trochilidae) and six species of honeycreepers (the genera *Diglossa*, *Diglossopis* of the Thraupidae) which rob nectar by perforating the flower-bottoms but also feed on insects and fruits. In the investigation area 29 species of hummingbirds were recorded which pollinate the flowers of a variety of ornithophilous plants inside the mature forest, at the forest edge and in the more open areas. Referring to Dziedzioch (2001), most plant species which are visited by hummingbirds belong to epiphytes and phanerophytes of the families Bromeliaceae, Orchidaceae and Ericaceae.

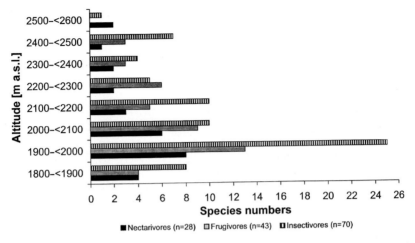

Fig. 11.1.2 Distribution of nectarivorous, frugivorous and insectivorous bird species numbers along the altitudinal gradient

Of the recorded bird species in the investigation area, 57 are frugivores (Fig. 11.1.2). The larger bird species like doves, parrots, guans and trogons forage mainly on bigger fruits, whereas the smaller fruits are dispersed mainly by tanagers, thrushes, vireos and a few species of the mainly insectivorous tyrants. Most bird species in the research area (100 species) belong to the guild of the insectivores (Fig. 11.1.2). The 70 species recorded by point-counts and mist-netting along the altitudinal gradient derive from 12 families with the tyrants (Tyrannidae) as the one with most insectivorous species.

The altitudinal distribution of the 23 hummingbird species and five species of honeycreepers, 43 species of frugivores and the insectivorous species all recorded by point-counts and mist-netting in the investigation area is shown in Fig. 11.1.2. Calculation of the centres of distribution and grouping in altitudinal classes follow the description given for Fig. 11.1.1.

The distribution pattern of the species of each of the three guilds along the altitudinal gradient is very similar to the distribution of all bird species shown in Fig. 11.1.1. A total of 12 nectarivorous, 17 frugivorous and 33 insectivorous species have their centre of distribution at 1800–2000 m a.s.l. Nine species of nectarivores, 14 species of frugivores and 20 species of insectivores were recorded with their altitudinal centre of distribution between 2000 m and 2200 m a.s.l. In the higher elevations between 2200 m and 2600 m a.s.l. the *Purdiaea nutans* forest is characterised by a low and open canopy favouring a rich flora of cryptogamic plants (see Chapter 19 in this volume). In the undergrowth stratum epiphytic orchids and bromeliads are more abundant than in any other undergrowth of the forest types in the investigation area (Paulsch 2002). Only seven nectarivorous, 12 frugivorous and 17 insectivorous bird species have their centre of distribution in this altitude. The distribution centres of two more species of hummingbirds and one species of honeycreepers lie in the páramo vegetation around 2700 m a.s.l. (Braun and Matt, personal communication).

Referring to the feeding guilds, Chavez (2001) found 15/12 nectarivores, 36/30 frugivores and 61/57 insectivores in his two investigation sites. In the RBSF 26 nectarivores, 32 frugivores and 63 insectivores were recorded by point-counts and mist-netting between 1880 m and 2200 m a.s.l. These numbers are very similar to those recorded by Chavez (2001), with the exception of the nectarivores showing a much higher number of hummingbirds in the RBSF. Dziedzioch (2001) states the very high diversity of plant species visited by hummingbirds and the heterogeneous vegetation mosaic composed of different kinds of structural forest types and the adjacent secondary vegetation types as reasons for the occurrence of so many hummingbird species. Referring to the 23 species of hummingbirds Dziedzioch (2001) described, only eight species were seen regularly inside the mature forest, whereas all other species were recorded in forest-edge situations. Our investigation found only one hummingbird species (*Phaethornis syrmatophorus*) as a typical forest species, four species were recorded only at the forest edge and 18 species occurred in the forest as well as at the forest edge. The hummingbird community of the montane rainforests between 1900 m and 2100 m a.s.l. shows only a small number of specialists and a high proportion of generalists caused by a high degree of temporal and spatial dynamics of the nectar resources (Dziedzioch 2001).

11.1.3.2 Bird Species in Different Structural Forest Types

The investigation by Paulsch (2002) led to 12 different structural forest types in the RBSF. Four of these structural types were chosen for the investigation of bird communities by point-counts and mist-netting; and three different types of ridge forests occurring in relatively small areas were united to one ridge forest type and two more types were added (Table 11.1.1). Due to human influence in the lower parts of the investigation area and the heavily inclined relief forming a small scale mosaic of deep valleys and steep ridges in mid-elevations, the distribution of structural forest types does not follow a clear altitudinal gradient in all parts of the investigation area (Chapter 10.3 in this volume). In the following section bird species numbers are given as recorded in each structural forest type (Table 11.1.1).

The bird species richness of these structural forest types is shown in Fig. 11.1.3 as the mean values of all bird species recorded by point-counts and mist-netting at the different investigation sites. Only one investigation site for point-counts could be set up and visited regularly for 'Primary ravine forest at higher altitude' due to its inaccessibility inside the forest type.

Mean values of species richness are highest in the two structural forest types 'Ravine forest under human influence' and 'Forest edge of lower altitude', with 38 species each. These two structural types are found in an altitude between 1810 m and 1985 m a.s.l. and are characterised by their location at or near to the forest edge, making these vegetation types extraordinarily rich in birds. The structural forest types 'Primary ravine forest at lower altitude', 'Primary ridge forest at lower altitude' and 'Primary ridge forest at higher altitude' are less diverse in species. The

Table 11.1.1 Numbers of bird species in different structural forest types (after Paulsch 2002)

Structural forest type	Altitude (m a.s.l.)	Number of bird species
Ravine Forest under human influence	1810–1940	64
Forest edge at lower altitude	1870–1985	67
Primary ravine forest at lower altitude	1880–2120	57
Remnants of ravine forest at lower altitude	1920–2100	42
Primary ridge forest at lower altitude	2000–2210	62
Primary ravine forest at higher altitude	2180–2290	18
Primary ridge forest at higher altitude	2170–2650	68

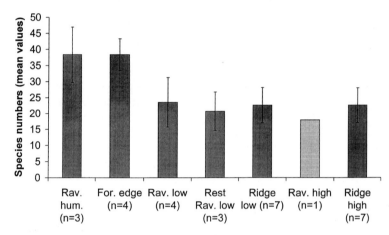

Fig. 11.1.3 Bird species richness (mean values) in different structural forest types. *Rav. hum.* 'Ravine forest under human influence', *For. edge* 'Forest edge of lower altitude', *Rav. low* 'Primary ravine forest at lower altitude', *Rest Rav. low* 'Remnants of ravine forest at lower altitude', *Ridge low* 'Primary ridge forest at lower altitude', *Rav. high* 'Primary ravine forest at higher altitude', *Ridge high* 'Primary ridge forest at higher altitude'

total species numbers are rather equally distributed in these forest types (Table 11.1.1). A lower species richness is found in the forest type 'Primary ravine forest at higher altitude' which might, however, be higher due to the rarity of visits to this forest type. The 'Remnants of ravine forest at lower altitude' also show a lower species richness due to their isolation from intact forest on the opposite side of the Río San Francisco valley.

Those differences in species composition between bird communities of primary tropical forests and forest fragments are described by many authors (e.g. Kattan et al. 1994; Waltert et al. 2004). Especially large, sparsely distributed or specialised species are very sensitive to forest fragmentation due to the loss of food resources, possible nesting sites and other ecological factors important for a suitable habitat (Turner 1996). A comparison of the species richness of the primary ravine forest in the RBSF and the three remnants of ravine forests on the opposite site of the Río San Francisco valley showed that the total numbers of bird species declined from 57 in the primary forest to 42 in the remnants (Table 11.1.1). The composition of the bird communities changed dramatically as only three of the 12 typical forest species recorded in the primary ravine forest could be found in these fragments. However, more than one-third of the bird species in the forest remnants belonged to those species typical for the forest edge or more open habitats, which made up only 7% in the primary ravine forest.

11.1.3.3 Functional Relationships Between Bird Species Distribution and Vegetation Structure

The habitat requirements of those bird species which are typical for one of the structural forest types were more intensively investigated, respecting their altitudinal ranges (Table 11.1.2).

Functional relationships between bird species distribution and vegetation structure could be worked out for five of these species (*Pyrrhomyias cinnamomea*, *Basileuterus tristriatus*, *Thamnophilus unicolor*, *Pseudocolaptes boissonneautii*, *Synallaxis unirufa*). Certain structural parameters which are typical for one or another structural forest type (Paulsch 2002) fulfil the special habitat requirements of these bird species regarding breeding or feeding behaviour.

The Streaked Tuftedcheek (*Pseudocolaptes boissonneautii*) for example is a typical species of the evergreen mountain forest and the elfin forest, occurring at an altitude between 1700 m and 3200 m a.s.l. (Remsen 2003). Its diet consists mainly of insects and their larvae which are found from mid-storey up to the forest canopy (Remsen 2003). This bird species is an epiphyte-specialist searching for arthropods inside bromeliads, mosses and other epiphytic vegetation on stems and branches (Remsen 2003). Old holes of woodpeckers in dead trees or branches are used for nesting (Ridgely and Greenfield 2001b). In the RBSF the Streaked Tuftedcheek was recorded at an altitude between 2090 m and 2550 m a.s.l.: 62% of the recorded individuals occurred in the 'Primary ridge forest of lower altitude' and 38% in the 'Primary ridge forest of higher altitude'. Paulsch (2002) showed that these two types of ridge forest offer more dead stems and branches than the types of ravine forest, thus favouring the building of holes by woodpeckers which can later be used for breeding by the Streaked Tuftedcheeks. The higher strata of the ridge forests show a higher number of epiphytic ferns and vascular epiphytes (especially bromeliads) than the ravine forests (Paulsch 2002). This offers the Streaked Tuftedcheek more opportunities for foraging in the ridge forests types.

11.1.4 Conclusions

The bird communities of the RBSF are characterised by a high species richness, peaking in those structural forest types situated at the forest edge between 1900 m and 2000 m a.s.l. At this altitude an extraordinarily high number of hummingbird

Table 11.1.2 Bird species typical for one structural forest type

Species	Family	Altitudinal range (m a.s.l.)	Structural forest type
Chlorospingus parvirostris	Thraupidae	1860–2000	Ravine forest at lower altitude
Pyrrhomyias cinnamomea	Tyrannidae	1870–2070	Forest edge at lower altitude
Basileuterus tristriatus	Parulidae	1870–2075	Ravine forest at lower altitude
Ochthoeca cinnamomeiventris	Tyrannidae	1880–2070	Ravine forest at lower altitude
Momotus aequatorialis	Momotidae	1880–2000	Ravine forest at lower altitude
Thraupis cyanocephala	Thraupidae	1880–2075	Forest edge at lower altitude
Thamnophilus unicolor	Thamnophilidae	2040–2210	Primary ridge forest at lower altitude
Hemitriccus granadensis	Tyrannidae	2040–2370	Primary ridge forest at lower (higher) altitude
Pseudocolaptes boissonneautii	Furnariidae	2090–2550	Primary ridge forest at lower (higher) altitude
Synallaxis unirufa	Furnariidae	2170–2550	Primary ridge forest at higher altitude
Chlorospingus ophthalmicus	Thraupidae	2240–2550	Primary ridge forest at higher altitude
Chlorornis riefferii	Thraupidae	2260–2550	Primary ridge forest at higher altitude

species is found, which are very important for the ecosystem as plant pollinators. Due to the occurrence of the *Purdiaea nutans* forest between 2200 m and 2650 m a.s.l. at the RBSF, the bird community found here differs from those of other locations at the same altitude. At Cajanuma for example a cloud-forest dominates at this altitude, harbouring a quite different bird community (Andrade 1996).

Comparison of the bird communities of the primary ravine forest in the RBSF and the three forest remnants on the opposite site of the Río San Francisco valley shows a declining number of species due to forest fragmentation, especially typical forest species. If a high species richness is to be maintained it is thus indispensable to protect the primary mountain rainforests of Ecuador, which might be a difficult task as they show an even higher deforestation rate than the lowland forests (Doumenge et al. 1995).

Functional relationships between bird species distribution and vegetation structure can be shown for different bird species. The habitat requirements of further species must be intensively investigated to learn more about the value of certain structural parameters for these species and for the whole bird communities of the different structural forest types.

Chapter 11.2
Seed Dispersal by Birds, Bats and Wind

F. Matt(✉), K. Almeida, A. Arguero, and C. Reudenbach

11.2.1 Introduction

The study of altitudinal transects in tropical high mountains furthers the understanding of adaptation and coupling strategies of fauna and flora within the same biogeographical zone under comparable geological and climatologic conditions (Körner 2000). The neotropical chiroptera are especially well suited for such investigations, as they are embedded in this natural experiment of speciation and niche differentiation (Voss and Emmons 1996). The bat species are of unique diversity within the vertebrates because of their diverse food spectrum, their highly diverse life-forms, their high abundance, and their complex mobility patterns. Chiroptera supposedly take an ecological key role in the tropical forests. In the tropical lowland rain forest 70–98% of tree species produce seeds and fruits that are spread by animals, and most single-tree locations are due to seed dispersal by animals (Howe and Smallwood 1982; Janzen 1982). Various plant species depend on bats as pollinators and seed dispersers, and about 100 of the 140 species of neotropical Phyllostomidae feed partly or entirely on fruits (Fleming et al. 1988). In French Guiana, for instance, about 93% of all pioneer species are spread by birds or bats (Charles-Dominique 1986). However, current knowledge on autecology and synecology of the neotropical bats relies mainly on lowland-forest studies with a focus on the description of pan-neotropical species (Bonaccorso 1979; Fleming 1988; Brosset and Charles-Dominique 1990; Francis 1990 Willig et al. 1993; Brosset et al. 1996; Kalko et al. 1996; Thies 1998). Only a few studies exist that analyze the shifting of species formation in relation to habitat fragmentation and degradation (Fenton et al. 1992; Cosson et al. 1999; Sampaio 2001). Comprehensive research studies of the chipterofauna in tropical mountain rain forests are scarce, and when performed, they usually provide only basic descriptions and species lists (Graham 1983, 1990; Munoz-Arango 1990; Patterson et al. 1998; Sanchez-Cordero 2001). The current study was performed to derive deeper synecological knowledge of the unknown chiroptera coenose along the altitudinal gradient in a tropical mountain rain forest in the area of Loja. The study focuses on frugivorous bats and their role as seed dispersers in disturbed areas since frugivorous bats are comparatively frequent and easy to trap, their excrements are simple to analyze, and they offer a precise interaction with plants.

The sampling of data from highly mobile fauna is a common problem in species inventories. In this study, two established methods were employed: Fisher's (Williams) alpha-diversity (Fisher et al. 1943) and rarefaction methods (Hurlbert 1971).

Data sampling was performed from 1997 to 1999 and 2003 to 2004, covering altitudes from 900 m to 3400 m a.s.l. in the central area of the RBSF.

11.2.2 Results and Discussion

11.2.2.1 Spatial Range and Food Spectrum of Frugivorous Bats

Based on 195 successful nightly trapping campaigns a total of 1014 bat individuals belonging to 20 species were captured. They could be assigned to the three major food guilds: frugivorous, nectarivorous, and insectivorous chiroptera (Table 11.2.1).

The correlations between observed and expected species abundance at the three altitudes are highly significant according to the hypothesis of a logarithmic abundance distribution. The Fisher alpha-diversity of the bat communities follows the altitudinal gradient. According to the 95% confidence intervals, the alpha-diversity reaches 5.2 ± 2.0 at Rio Bombuscaro, 3.7 ± 0.7 at RBSF, and 2.9 ± 1.1 at Cajanuma/El Tiro. The Hurlbert rarefied species numbers were significantly higher in the Rio Bombuscaro plot than in the RBSF plot. This supports the results of the Fisher alpha-diversity and suggests that the high species number ($n=20$) of the RBSF area is a direct effect of the large number of trapped individuals ($n=834$). A comparison based on 89 traps as the largest common number of trapped individuals yielded 15 species in the area of Rio Bombuscaro, 13 species in the RBSF area, and ten species in the Cajanuma and El Tiro regions.

Table 11.2.1 Trapped chiroptera, ordered by guilds and altitude

Trapping plot	Altitude (m a.s.l.)	Total individuals	Frugivorous bats			Nectarivorous bats			Insectivorous bats		
			%	Individuals	Species	%	Individuals	Species	%	Individuals	Species
Bombuscaro	900–1000	89	73	65	9	9	8	1	18	16	5
RBSF	1800–2000	834	83	691	9	4	34	3	13	109	8
Cajanuma El Tiro	2800–3000	91	51	46	4	8	7	2	42	38	4

11.2.2.2 Seed Dispersal Into Disturbed Areas Occurs by Wind and Frugivorous Bats and Birds

Frugivorous bats are the dominating guild with 83% of all net-trapped chiroptera individuals at 1800–2000m a.s.l. The nine frugivorous species identified (Table 11.2.2) feed on at least 51 species of forage plants. The significant prevalence of seeds from either shrubs (Piperaceae, Solanaceae, Araceae) or trees (*Ficus* sp., Cecropiaceae) in the bat excrements divides the bats into understorey and canopy frugivores (Table 11.2.2; Bonaccorso 1979).

The analysis of seed records shows an inhomogeneous distribution (Table 11.2.2).

One-third (33%) can be assigned to seed morphotypes 1–14, representing the seeds of yet unknown Piperaceae species, 20% to four *Ficus* species (type 31), and another 20% to the Solanaceae. The remaining seed morphotypes make up the individual fractions below 4%. Fig. 11.2.1 highlights the marked differences in annual distribution dynamics for each of the six frugivores most frequently trapped during one year.

Carollia brevicauda is a residential all-season bat species with a moderate but continuous appearance. In contrast, the chiroptera *Sturnira bidens* is a seasonal species with a single peak in July. The bats *S. erythromos* and *S. ludovici* display no regular annual patterns but a highly variable occurrence during all seasons. Both of the canopy guild bats *Platyrrhinus infuscus* and *Artibeus phaeotis* present a synchronous trend with a maximum in July and a minimum from April to May. This seems to be directly related to the availability of mature fruits like *Cecropia* and *Ficus*.

Analysis of the range of activity of the chiroptera was performed by placing identification rings on all trapped frugivorous bats. The six most frequent frugivorous species (Table 11.2.3) were chosen for further analysis.

Interpretation of the range of bat activities as derived by calculating the distances between trapping (marking location) and re-trapping sites was normalized by the frequencies of all distances between all trapping locations (Fig. 11.2.2, black bars). All the species show a significant divergence of the trapping re-trapping distances from the expected distances (χ^2 test).

Carollia brevicauda shows the best correlation of both distance distributions. The mean distance of two re-trapping locations is 1530m. It can be expected that the spatial activity of *C. brevicauda* extends the trapping locations.

The spatial activity of *S. erythromos* is significantly smaller and within the range of expectations: 72% of all re-trapped individuals were re-trapped within a distance of less than 200m from the place of their first capture, 54% exactly at the same trapping location. The maximum distance between two trapping locations was 395m.

S. bidens shows a similar spatial behavior as *S. erythromos*: 45% of all re-trapped individuals were re-trapped within a distance of less than 200m from the place of their first capture, 15% at the same trapping location. The maximum distance between two trapping locations was 992m.

Table 11.2.2 List of 51 identified seed morphotypes based on 493 samples of bat excrement of the frugivorous chiroptera species *Sturnira erythromos*, *S. ludovici*, *S. bidens*, *Carollia brevicauda*, *Platyrrhinus infuscus*, *P. lineatus*, *Artibeus phaeotis* and *A. hartii* at the the RBSF plot (1800–2000 m a.s.l.)

Family	Genus	Species	Fraction of total seeds
Piperaceae	*Piper*	sp. 1	ca. 33%
		cf. *nebuligandens*	
		sp. 2	
		sp. 3	
		sp. 4	
		sp. 5	
		sp. 6	
		sp. 7	
		sp. 8	
		sp. 9	
		sp. 10	
		cf. *scutilumsum*	
		sp. 11	
		sp. 12	
Solanaceae	*Solanum*	cf. *trichneuron*, cf. *versabile*	ca. 20%
		sp. 1	
		sp. 2	
		sp. 3	
	Physalis	*peruviana*	
	Solanum/ Cyphomandra	sp. 4	
	Solanum	sp. 5	
	Larnax	*psilophyta*	
	Solanum	sp. 6	
		sp. 7	
		sp. 8	
	Solanum/ Lycopersicum	sp. 9	
Actinidaceae	*Saurauia*	*bullosa*	
		laxifolia	ca. 4%
		peruviana	
Araceae	*Anthurium*	sp. 1	ca. 4%
		sp. 2	
Ericaceae	Unknown	Unknown	ca. 1%
Clusiaceae	*Vismia*	*tormentosa*	ca. 2%
Moraceae	*Ficus*	sp. 1	ca. 20%
		sp. 2	
		sp. 3	
		sp. 4	
Lobeliaceae	*Centropogon*	cf. *densiflorus/granulosus*	<1%
Flacourtiaceae	Unknown	sp.	<1%
Staphyleaceae	*Turpinea*	sp.	ca. 1%
Rosaceae	*Rubus*	sp.	<1%
Cecropiaceae	*Cecropia*	*montana*	ca. 4%
		sp.	
Unknown	Unknown	Unknown	ca. 11%

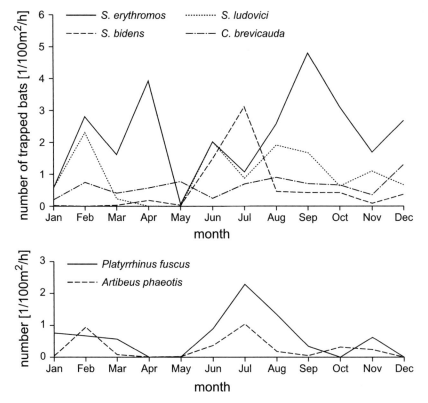

Fig. 11.2.1 Probability of trapped bats shown by species and month. The species specific trapping probability is calculated for 100 m² net area per hour and month

Table 11.2.3 Frequencies of the six most frequent re-trapped frugivores in the trapping area of RBSF. On average 14% of the identified chiroptera were trapped at least once again

Species	Number of identified individuals	Number of re-trapped individuals
Sturnira erythromos	185	22 (12%)
Sturnira ludovici	90	9 (10%)
Sturnira bidens	21	8 (38%)
Carollia brevicauda	43	16 (37%)
Plathyrrinus infuscus	89	7 (8%)
Artibeus phaeotis	26	2 (8%)
Total	454	64 (14%)

The pattern of the trap/re-trap distance distribution is more difficult to interpret for *S. ludovici*. This species shows a heterogeneous spatial behavior. About 30% of the re-trappings were at the former trapping locations but the other 70% were re-trapped at distances ranging from 400 m to 1000 m. The maximum distance between two trapping locations was 958 m.

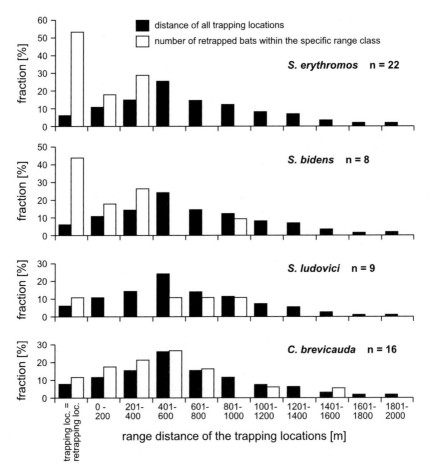

Fig. 11.2.2 Frequency distribution of distances between trapping (marking) and re-trapping locations of four frugivorous bat species at the RBSF (1800–2000 m a.s.l.) The *black bars* represent the distribution of ranges between all trapping locations itself, whereas the *white bars* indicate the distribution of re-trapping within a specific range class

11.2.2.3 Seed Dispersal

To obtain a better understanding of the seed dispersal processes and the role of the different seed dispersers (bats, birds, wind), a field experiment was performed. The nine registered frugivorous bat species and additionally the most important frugivorous birds (approx. 110 registered species) were investigated with respect to their seed transport capabilities. Additionally, wind-borne seed dispersal was considered. Seed and excrement traps were installed on three areas covered by different succession vegetation types. The experiment areas were two landslides (approx. 1200 m^2

and 1400 m^2), one mostly covered by herbs and surrounded by forest, the other densely covered by shrubs on a regenerating pasture without adjacent forest (2500 m^2). In total 209 excrement traps (1×1 m, separated into 100 traps on landslides, 109 traps on shrub plots) and 20 seed traps (wind-borne seed dispersa;)(60×60 cm, ten on landslides, ten on shrub plots) were installed on each type of investigation area evenly distributed in a rectangular raster of 5×5 m and controlled at 0600 hours and 1800 hours on ten consecutive days in each month during one year.

The analysis of all samples yielded seeds of 128 plant species. Of these, 67 plant species are dispersed by birds, 33 by bats, and 31 by wind. The density of seed rain into the landslide plots was 2.72 seeds/m^2 by wind, 0.93 seeds/m^2 by birds, and 0.093 seeds/m^2 by bats. In the shrub plot the wind spread 1.8 seeds/m^2, birds defecated 1.6 seeds/m^2, and bats 0.004 seeds/m^2.

According to the food preferences of the frugivorous bats, they are the main dispersers of Araceae, Cecropiaceae, Moraceae, Piperaceae, and Solanaceae (Table 11.2.4). In total, 33 species could be identified that were dispersed only in the shrub free areas of the disturbance plots. Only one species was found in the shrubby parts of both plots. The origin of bat-dispersed seeds was from more distant areas.

The species dispersed by birds are very distinct from those spread by bats. Basically they belong to the families Ericaceae, Melastomataceae, and Rubiaceae. The diversity of seeds and the density of the seed rain by birds are higher than by bats. However, it can be observed that succession plots surrounded by forests received a 'seed rain' richer in species than the more distant shrub areas. The majority of the bird-dispersed plant species originate from the plot itself or its immediate vicinity.

The seeds dispersed by wind belong to the families Asteraceae, Poaceae, and Tiliaceae. In total, 31 species were identified on both disturbance plots. Of these, 17 species were found only on landslides, five species only on shrub areas, and nine species on both sub-types of disturbance areas.

An important finding is that in most cases the seeds deposited by bird excrements originate from the trees and bushes adjacent to the record location. This indicates both a fast digestion process and that the majority of defecating activities related to the observed birds are performed while they are sitting on trees or bushes and not in flight. Hence pre-existing vegetation patterns are a basic precondition that birds will be able to contribute to a re-colonization of the disturbed areas. In contrast, frugivorous bats mostly defecate in flight; only one bat-related excrement record could be reported on shrub areas. The list of species whose seeds are exclusively dispersed by frugivorous bats to the bare soil or herb-covered parts of the landslide areas (Table 11.2.4) allows the conclusion that the frugivorous bats are remote seed dispersers. Their contribution is the first remote disposal of pioneer species on shrub- and tree-free areas. Hence it can be assumed that they play a key role in starting early succession stages and introducing the spatial setup patterns of bushes and trees onto disturbed areas.

Table 11.2.4 Preferred plant species as dispersed by frugivorous bats, birds and wind seperated by plot type. *B* Birds, *C* Chiroptera, *H* hangslide, *S* shrub-plot. *B* Birds, *C* Chiroptera, *H* hangslide, *S* shrub-plot

Familiy	Species	Disperser	Plot type
Actinidaceae	*Saurauia bullosa*	C	H
	Saurauia laxifolia	C	H
Araceae	*Anthurium amoenum*	C	H
	Anthurium effusipatha	C	H
	Anthurium longirostrum	C	H
	Anthurium sp. 1 (planas)	B/C	H
	Anthurium sp. 2	C	H
	Anthurium sp. 3	C	H
	Anthurium sp. 4 (zig-zag)	C	H
	Anthurium truncicolum	C	H
Cecropiaceae	*Cecropia gabrielis*	B/C	H/S
	Cecropia polyphlebia	B/C	H
Clusiaceae	*Vismia tomentosa*	C	H
Ericaceae	*Cavendishia* sp. (javier)	C	H
	Cavendishia zamorensis	B/C	H
Moraceae	*Ficus* sp. 1 (q1)	C	H
	Ficus sp. 2 (entrada a la estación)	C	H
	Ficus sp. 3 (q1 manchas rojas)	B/C	H
	Ficus sp. 4 (Felix muestra)	C	H
	Ficus sp. 5 (tomate pegajoso)	C	H
	Ficus sp. 6 (nuevo)	C	H
	Ficus subandina	B/C	H
Piperaceae	*Piper* cf. *ecuadorense*	C	H
	Piper sp. 1 (paraguas)	C	H
	Piper sp. 3 (hoja lustrosa)	C	H
	Piper sp. 4 (hoja lanza)	C	H
	Piper sp. 5 (gigante)	C	H
Solanaceae	*Solanum aphyrodendron*	C	H
	Solanum aserolanatum	C	H
	Solanum sp. 1 (chara)	C	H
	Solanum sp. 2 (fx261)	C	H
	Solanum sp. 3 (frutos chicos)	C	H
	Solanum sp. 4 (fx216)	C	H

As a result, bird guano and the corresponding seeds are concentrated at the borders of the disturbance areas near the shrub islands, while bat-dispersed seeds can be found preferably in the center or on clear areas within the disturbed plots.

11.2.3 Conclusions

On the basis of these findings, it can be concluded that the frugivorous bat contribution to the diversity structure and richness of succession and disturbance areas is significant. This affects both the species (Table 11.2.4) and the dispersal patterns.

It can be clearly noticed that the diversity of seed insertion, due to the contribution of the frugivorous bats, is significantly higher on landslides than on the shrubby control area. Regarding the case of the landslides, this may also be affected by the close vicinity of the forest that provides an optimal spectrum of fruit plants, compared with the abandoned shrubby range land that offers only moderate food resources.

Although the diversity of bats is not exceptionally high, these animals play an outstanding ecological role in pollination and fruit/seed dispersal.

Chapter 11.3
Variation of Diversity Patterns Across Moth Families Along a Tropical Altitudinal Gradient

K. Fiedler(✉), G. Brehm, N. Hilt, D. Süßenbach, and C.L. Häuser

11.3.1 Introduction: Altitudinal Gradients and Insect Diversity

Comparative and experimental investigations along the steep altitudinal gradient that is characteristic for Andean ecosystems have formed one of the central paradigms of research at the RBSF in southern Ecuador summarized in the present volume (Lomolino 2001; Rahbek 2005; see Chapter 6 in this volume). For ectothermic insects the increasingly harsher climatic conditions at higher elevations alone impose severe constraints (McCoy 1990; see Chapter 8). Herbivorous insects find fewer plant species and less plant biomass available as nutrient resources at high elevations (see Chapter 10.1). Herbivorous insects are one of the most diverse guilds in terrestrial ecosystems, and within this guild phytophagous beetles and the Lepidoptera (moths and butterflies) are especially species-rich. Available studies on Lepidoptera (the taxonomically best known large insect group) often revealed a mid-elevation bulge of species richness (Holloway and Nielsen 1999; Rahbek 2005). However, many variations occur to this general pattern. For example, on Mt. Kilimanjaro, diversity in the moth family Geometridae is surprisingly low and decreases but slightly throughout the montane forest belt at 2000–3000 m (Axmacher et al. 2004). In the Ecuadorian Andes, in contrast, diversity in the same moth family is exceptionally high and remains almost constant at 1000–2700 m (Brehm et al. 2003b). Yet, these Andean geometrid ensembles (for terminology, see Fauth et al. 1996) change fundamentally with regard to species composition, representation of higher taxa, and body sizes (Brehm and Fiedler 2003, 2004a; Brehm et al. 2003a, b). In this chapter we extend and summarize our investigations of moth diversity in this Andean altitudinal gradient by integrating data on two further, very species-rich moth taxa, viz. Pyraloidea and Arctiidae.

The three taxa were selected because:

1. They represent large proportions of moth diversity globally (numbers of described species: Geometridae: 21 000; Pyraloidea: 16 000; Arctiidae: 11 000; Scoble 1992) as well as in the Neotropical region (numbers of described species: Geometridae: 6450; Pyraloidea: 4562; Arctiidae: 5931; Heppner 1991; Munroe et al. 1995; Scoble 1999).

2. They differ markedly from one another in many bionomic traits, body size, and host plant affiliations.

Pyraloidea are usually small (forewing length mostly <30 mm), slender and delicate (Munroe and Solis 1999). Pyraloid larvae are concealed feeders which bore inside stems or fruits, or live in webs or leaf rolls (Munroe and Solis 1999). Concealed feeders are typically more specialized with regard to host plants (Gaston et al. 1992). Few pyraloids are chemically defended by sequestered plant metabolites or toxic gland secretions. Most Geometridae species are also rather delicate moths, but are generally larger in size than pyraloids. The naked caterpillars are mostly ectophagous on foliage, with exceptions in the Geometrinae and the speciose Eupitheciini many of which are hemi-endophagous in inflorescences. As in the Pyraloidea, aposematic coloration and sequestration of plant toxins are rare in the Geometridae. In contrast, Arctiidae moths are larger (fore wing length often >20 mm, especially in the subfamily Arctiinae) and more robust (for Ecuadorian samples, see Hilt 2005). Many arctiids are aposematically colored and chemically well defended (Häuser and Boppré 1997; Kitching and Rawlins 1999; Weller et al. 1999). Arctiid larvae possess a dense cover of setae, feed externally on a wide range of plants, and a substantial fraction of them appears to be less specialized with regard to host plants (Kitching and Rawlins 1999; Holloway et al. 2001).

We here address three questions:

1. How does species diversity change in all moths, in the three moth clades, and in their larger subordinated taxa, along an altitudinal gradient?
2. Does the gross taxonomic composition of moth ensembles (i.e. representation of subfamilies and tribes) change with altitude, or do certain taxa contribute a rather invariant share at all elevations?
3. How does species composition of the moth ensembles change with altitude?

11.3.2 Methods: Sites, Sampling and Data Processing

We sampled moths at 22 sites situated at 1040–2677 m a.s.l. (two replicate sites at each level of altitude; for locality data and a map, see Brehm and Fiedler 2003; Brehm et al. 2005) at the northern border of the Podocarpus National Park and in the adjacent Reserva Biológica San Francisco. Chapters 1, 8, 10.1, and 10.3 provide details on the topography, climate, and vegetation of the study area along the altitudinal gradient.

Moths were attracted to weak artificial light sources (2×15 W) at ground level and manually sampled (Axmacher and Fiedler 2004; Brehm and Axmacher 2006). Assuming an attraction radius of 50–100 m of the light sources, the total sampling area amounted to just 0.17–0.69 km^2. Light-trap samples do not perfectly represent actual populations but rather reflect the activity of species attracted to light (Muirhead-Thomson 1991; Butler et al. 1999). However, for nocturnal Lepidoptera this sampling method is unrivalled in terms of specimen and species numbers that can be gathered (for a critical evaluation of the methodology, see Schulze and Fiedler 2003; Wirooks 2005).

Sampling was restricted to the evening hours between 1830 hours and 2130 hours, local time. Hence 'late night species' were not covered. We collected moths during three field periods (April to May 1999, October 1999 to January 2000, October to December 2000; for catch dates, see Süßenbach 2003) and pooled replicate catches from each site into one sample. Two to four nightly catches were combined for the Geometridae (which occurred in larger numbers), whereas up to nine catches per site were collated for the less abundant Arctiidae and Pyraloidea (Table 11.3.1).

Specimens were taxonomically identified using published literature and reference collections (see Acknowledgments; Table 11.3.1). In view of the tremendous species richness of moths (see Chapter 2 in this volume), this process provided a great challenge on its own (see Chapter 5). Species lists are available by Süßenbach (2003), Brehm et al. (2005), and Hilt (2005). Voucher specimens are deposited in the State Museum of Natural History, Stuttgart, Germany.

As a robust measure of local diversity we used Fisher's α of the logseries (e.g. ; Hayek and Buzas 1997; Beck et al. 2002; Axmacher et al. 2004). Species numbers are less informative for rich communities of mobile organisms, unless near-complete inventories are achieved (Gotelli and Colwell 2001). Logseries-type rank-abundance distributions are common in nature (Engen and Lande 1996; Hubbell 2001). To extend comparisons to lower taxonomic scales with smaller sample sizes, we calculated taxon ratios based on the data shown in Table 11.3.2 (Brehm and Fiedler 2003).

We ordinated moth samples by non-metric two-dimensional scaling based on Bray–Curtis similarities (Legendre and Legendre 1998). Raw data were fourth-root transformed to alleviate dominance effects. Ordinations based on other similarity measures (Sørensen index, CNESS index for various values of the sample size parameter m) as well as application of other ordination techniques (correspondence analysis, detrended correspondence analysis) all yielded very similar results (e.g. Brehm and Fiedler 2004b). Two-dimensional ordinations were deemed sufficient due to their low stress (stress is a measure of poorness-of-fit between the configuration in reduced ordination space and the original similarity matrix; Clarke 1993).

Relationships between diversity measures or with environmental variables were assessed using standard correlation and regression techniques. The match between matrices of faunal similarity was assessed using permutation tests for matrix rank correlations. If multiple tests of significance were performed, we used Benjamini and Hochberg's (1995) approach to control for a table-wide false discovery rate.

11.3.3 Results and Discussion

11.3.3.1 Local Diversity of Moth Ensembles

We recorded 28 743 moths representing 2129 species (or 'parataxonomic units' in the case of moths that we could not formally identify; Krell 2004) of the three focal taxa. Geometridae were by far the most abundant and speciose family. Values of Fisher's α for all 22 sampling sites cumulated across the gradient were: 250.1 (Geometridae),

Table 11.3.1 Numbers of moth species (S) and individuals (N) at the 22 light trapping sites along the altitudinal gradient in the Ecuadorian Andes. 'identified' refers to the fraction where taxonomic determinations were achieved at species level

Site	Altitude (m a.s.l.)	Arctiidae S	N	Geometridae S	N	Pyraloidea S	N	All combined S	N
1a	1040	52	108	134	410	220	696	406	1214
1b	1040	149	712	247	976	315	1352	710	3039
2a	1380	59	155	200	623	166	892	425	1670
2b	1380	54	159	158	393	135	528	347	1080
3a	1800	59	175	203	496	92	368	354	1039
3b	1800	59	153	171	384	94	331	324	868
4a	1850	60	192	178	473	78	250	316	915
4b	1875	90	354	225	649	155	442	470	1445.0
5a	2005	75	434	202	618	114	323	391	1375
5b	2005	49	186	190	429	80	232	319	847
6a	2112	49	140	217	782	87	269	353	1191
6b	2113	61	187	177	440	85	257	323	884
7a	2180	50	135	292	1200	89	315	431	1650
7b	2212	52	290	273	1116	72	415	398	1822
8a	2290	49	138	200	683	75	506	324	1327
8b	2308	48	202	144	384	77	455	269	1041
9a	2375	50	173	201	725	77	586	328	1484
9b	2387	46	128	259	981	68	617	374	1726
10a	2524	55	123	191	596	70	296	316	1015
10b	2558	65	135	192	447	62	268	319	850
11a	2671	34	82	167	473	65	306	266	861
11b	2677	40	100	209	660	86	640	335	1400
Total		371	4461	1013	13 939	748	10 344	2132	28 744
Identified (%)		64.7	63.8	64.7	72.2	10.3	19.5	45.6	51.9

185.1 (Pyraloidea), 96.1 (Arctiidae), and 531.6 (all three moth taxa combined). This high diversity was not uniformly distributed along the altitudinal gradient. The expected decline in the diversity of ectothermic herbivores occurred in the Pyraloidea and among all moths combined (Fig. 11.3.1; statistics: Table 11.3.3).

We observed no such decrease in the Geometridae. Brehm et al. (2003b) showed that, within the Geometridae, diversity of species belonging to the subfamily Larentiinae even increased at higher altitudes, whereas Ennominae diversity did not change along the gradient. In the Arctiidae the altitudinal decrease was just marginally significant. Analyzing the two subfamilies of Arctiidae separately revealed that only diversity of Lithosiinae decreased with increasing altitude, whereas in the far more species-rich Arctiinae diversity did not change consistently with elevation. There was also no significant pattern in the two largest Arctiinae clades (viz. Phaegopterini, Ctenuchini–Euchromiini; the latter termed 'Ctenuchini' in the following for the sake of brevity), but sample sizes were low, limiting the power of tests. Within Pyraloidea, the altitudinal decline of diversity was stronger in the family Crambidae rather than in Pyralidae. The subfamily Pyraustinae was by far the largest

Table 11.3.2 Higher classification of Geometridae, Arctiidae, and Pyraloidea with numerical representation of the groups in the cumulative sample along the altitudinal gradient (1040–2677 m a.s.l.). Systematics: Scoble (1999) and Pitkin (2002) for Geometridae; Munroe et al. (1995) for Pyraloidea; Jacobson and Weller (2002) for Arctiidae

Taxon	Species	Individuals	Taxon	Species	Individuals
Geometridae	1013	13939	Pyraloidea	748	10344
Desmobathrinae	1	2	Crambidae	638	9829
Ennominae	502	6636	Crambinae	35	596
Geometrinae	57	715	Schoenobiinae	5	128
Larentiinae	381	5721	Cybalomiinae	3	16
Oenochrominae	3	23	Glaphyriinae	8	162
Sterrhinae	69	842	Scopariinae	18	644
			Musotiminae	29	322
Arctiidae	371	4461	Midilinae	1	3
Lithosiinae	68	1941	Acentropinae	28	258
Arctiinae	303	2520	Odontiinae	6	293
Arctiini	7	13	Evergestiinae	2	18
Phaegopterini	169	1139	Pyraustinae	503	7389
Ctenuchini	116	1313			
Pericopini	11	55	Pyralidae	105	475
			Pyralinae	4	4
			Chrysauginae	12	56
			Galleriinae	10	180
			Epipaschiinae	18	50
			Phycitinae	61	185

and strongly dominated this pattern. Other pyraloid taxa were too small to allow for meaningful calculation of diversity statistics. The altitudinal decrease in diversity was 5–10 times steeper (t-test, $P<0.05$) within Pyraloidea, Crambidae, and all moths combined, than in Arctiidae or Lithosiinae.

Local diversity of Geometridae and Arctiinae was largely unrelated to diversity patterns in any other of the moth groups examined (Table 11.3.4). Diversity patterns were more similar between Pyraloidea, all Arctiidae, Lithosiinae, and all moths combined.

11.3.3.2 Changes in Taxonomic Composition with Altitude

Representation of moth subfamilies and tribes along the altitudinal gradient varied considerably within the Arctiidae, Pyraloidea (Table 11.3.5), and Geometridae (Brehm and Fiedler 2003). Among Arctiidae, species proportions of Phaegopterini increased and Ctenuchini decreased with altitude. Lithosiinae had an increasingly higher share of individuals and species at high elevations, whereas Ctenuchini and Pericopini contributed less to Arctiidae ensembles at higher altitudes. Ctenuchini comprised a sizeable proportion of the arctiid ensembles at 1040–1875 m (>35% of species and individuals), but accounted for usually ca. 20% and less at higher elevations.

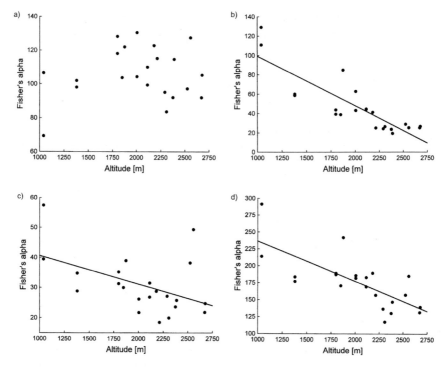

Fig. 11.3.1 Relationship between altitude and local diversity (measured as Fisher's α) of various moth groups. **a** Geometridae, **b** Pyraloidea, **c** Arctiidae, **d** all these moths combined. The linear least squares regression line is included where significant. For statistics, see Table 11.3.3

Table 11.3.3 Results of linear regressions of alpha diversity (measured as Fisher's alpha) against altitude. Given are regression coefficients b, their standard errors SE(b), the coefficient of determination R^2, and the error probability P (significant results printed in italics; * significant after correction for a table-wide false discovery rate at $P=0.05$). $N=22$ sites.

Taxon	b	SE(b)	R^2	P
Geometridae	0.0050	0.0071	0.0243	0.488
Pyraloidea	−0.0512	0.0071	0.7212	*<0.0001**
Crambidae	−0.0495	0.0058	0.7835	*<0.0001**
Arctiidae	−0.0095	0.0039	0.2285	*0.024*
Arctiinae	0.0035	0.0038	0.0415	0.363
Phaegopterini	0.0051	0.0028	0.1427	0.083
Ctenuchini	−0.0026	0.0023	0.0567	0.286
Lithosiinae	−0.0033	0.0010	0.3364	*0.005**
All moths combined	−0.0592	0.0126	0.5250	*0.0001**

Within the Pyraloidea, at higher altitudes the family Pyralidae increased and the Crambidae decreased in their proportions of species, but this was not significant for fractions of individuals. At the subfamily level, Scopariinae, Odontiinae, Galleriinae, and Phycitinae tended to be more prevalent at higher altitudes, whereas Pyraustinae, Acentropinae, Musotiminae, and Glaphyriinae were more important at lower elevations. Schoenobiinae exclusively occurred in low numbers at medium elevations (1800–2300 m). Only three of the more abundant pyraloid subfamilies did not show any clear altitudinal pattern (Crambinae, Chrysauginae, Epipaschiinae).

Table 11.3.4 Correlation of local diversity between moth groups ($N=22$ sites), based on Fisher's alpha. Within each cell, the upper entry is Pearson's rP, the lower entry gives the corresponding P value. n.s. not significant ($P>0.05$). All significant results (printed in italics) remained so after correction for a table-wide false discovery rate of $P=0.05$.

	Pyraloidea	Arctiidae	Arctiinae	Lithosiinae	All moths combined
Geometridae	−0.177	0.123	−0.169	−0.130	0.287
	n.s.	n.s.	n.s.	n.s.	n.s.
Pyraloidea	–	0.641	0.002	0.767	0.874
		0.001	n.s.	*<0.001*	*<0.001*
Arctiidae	–	–	0.567	0.800	0.777
			0.006	*<0.001*	*<0.001*
Arctiinae	–	–	–	0.325	0.016
				n.s.	n.s.
Lithosiinae	–	–	–	–	0.751
					<0.001

Table 11.3.5 Relationship between subfamily or tribe proportions with altitude (Spearman's rank correlation rS, $N=22$ sites). The arctiid tribe Arctiini and four pyraloid subfamilies were disregarded because of low representation (Cybalomiinae, Midilinae, Evergestiinae, Pyralinae; cf. Table 11.3.2). Significant results ($P<0.05$) printed in italics.; * significant after correction for a table-wide false discovery rate of $P=0.05$.

	Individuals		Species	
Taxon	rS	P	rS	P
Arctiidae				
Lithosiinae	0.630	*0.002**	0.472	*0.026*
Arctiinae	−0.630	*0.002**	−0.472	*0.026*
Phaegopterini	0.345	0.115	0.696	*<0.001**
Ctenuchini	−0.838	*<0.001**	−0.867	*<0.001**
Pericopini	−0.455	0.033	−0.405	0.0618
Pyraloidea				
Crambidae				
Pyraustinae	−0.327	0.138	−0.558	*0.007**
	−0.335	0.128	−0.437	*0.042*
Acentropinae	−0.917	*<0.001**	−0.765	*<0.001**
Crambinae	−0.390	0.073	0.320	0.147
Schoenobiinae	−0.252	0.258	−0.178	0.429
Glaphyriinae	−0.532	*0.011**	−0.191	0.395
Scopariinae	0.784	*<0.001**	0.823	*<0.001**
Musotiminae	−0.858	*<0.001**	−0.400	0.065
Odontiinae	0.720	*<0.001**	0.710	*0.001**
Pyralidae				
Chrysauginae	0.390	0.073	0.529	*0.011**
	0.268	0.228	0.380	0.081
Galleriinae	0.528	*0.012**	0.508	*0.016**
Epipaschiinae	−0.331	0.132	−0.159	0.479
Phycitinae	0.426	*0.048*	0.652	*0.001**

11.3.3.3 Species Turnover Along the Altitudinal Gradient

Ordinations of all moth taxa revealed very similar altitudinal gradients of species turnover (Fig. 11.3.2). No clustering of ensembles was detectable, as would be expected if moth ensembles of particular altitudinal zones were to form distinct associations. The gap between sites at altitudinal levels 1 and 2 versus levels 3 and higher in the ordination diagrams reflects the larger altitudinal distance between sampling sites (Table 11.3.1). For the seven moth taxa with sufficient sample sizes (Arctiidae, Arctiinae, Phaegopterini, Ctenuchini, Lithosiinae, Pyraloidea, Geometridae), as well as for all moths combined, there was a highly significant (and linear) correlation between the first ordination axis and altitude (rP = 0.877–0.975, $P<0.001$). Correlations of species turnover with mean air temperature during nightly catches at the sampling sites were also all very close (rP = 0.885–0.980, $P<0.001$). Finally, similarity matrices for all these moth groups were highly correlated to each other (rS = 0.838–0.986, $P<0.001$).

11.3.3.4 Moth Diversity, Elevation, and Thermal Ecology

The diversity of the three moth (super-)families was exceptionally high at the scale of individual sampling sites and impressively so if viewed for the entire study landscape. With far more than 2100 moth species representing the three studied families

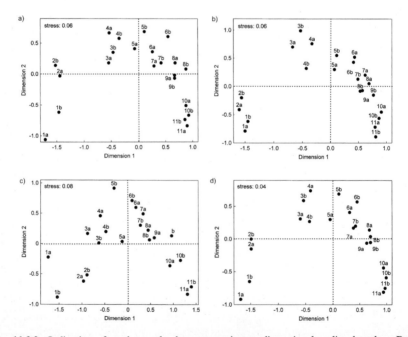

Fig. 11.3.2 Ordination of moth samples by non-metric two-dimensional scaling based on Bray-Curtis similarity matrices (raw abundances fourth-root transformed). **a** Geometridae, **b** Pyraloidea, **c** Arctiidae, **d** all these moths combined. In all cases moth samples are ordinated according to altitude

sampled from a small area, the montane forest belt of the Eastern Cordillera in the southern Ecuadorian Andes qualifies as a global hotspot of moth diversity (Myers et al. 2000; Brummitt and Lughadha 2003; Brehm et al. 2005; see Chapter 2 in this volume). In the two 'macro-moth' families Arctiidae and Geometridae, ca. 64% of the species and ca. 70% of the individuals could be formally identified thus far (see Chapter 5). The remainder includes mostly undescribed species whose taxonomic evaluation is now under way (Brehm 2004, 2005; Pitkin 2005). The identification rate was much lower for the taxonomically less known Pyraloidea. Still, our parataxonomic sorting in this clade followed by evaluation aided through taxonomists should allow for meaningful analyses of gross patterns (Krell 2004).

As was expected for ectothermic herbivorous insects, the total moth diversity significantly declined in the montane forest zone of southern Ecuador at 1040–2677 m elevation, even though at the highest sites the moth diversity was still remarkably high (Fisher's α: 120–185). Such a decrease in diversity is concordant with the cooler, more humid environmental conditions, and lower resource diversity at higher altitudes.

However, this overall decline in moth diversity did not uniformly recur at smaller systematic scales. Importantly, exceptions from that rule were not confined to small lineages with 'aberrant' ecological requirements, but occurred in large species-rich taxa. Brehm et al. (2003b) showed that diversity of Geometridae was constantly high along the investigated altitudinal gradient, with one large subfamily not responding to altitude (Ennominae), while in the other large subfamily (Larentiinae) diversity even increased towards higher elevations. With the subfamily Arctiinae we found one more speciose moth group whose diversity did not decline with altitude in the studied elevational range. Rather, in this subfamily diversity was lowest at intermediate altitudes (2000–2200 m; α: 15–25), whereas ensembles at higher and lower elevations were almost equally diverse (2290–2677 m, α: 30–50; 1040–1875 m, α: 25–45).

The reasons why Geometridae and Arctiinae do not follow the common trend of declining diversity at higher elevations might possibly differ. Geometridae moths are known to fly at unusually low thoracic temperatures, often just marginally above ambient temperature (Heinrich and Mommsen 1985; Utrio 1995, Rydell and Lancaster 2000). This physiological specialization should make them less sensitive to the linear decline in ambient temperatures from 22 °C mean annual temperature at the lowest to 12 °C at the highest sampling sites (see Chapter 8 in this volume). Thus, due to their flight physiology even delicate, slender, small geometrids may be able to remain active at low temperatures, and this might also allow the smaller and less robust Larentiinae to dominate at higher altitudes (Brehm and Fiedler 2004a). Arctiinae, in contrast, usually require higher thoracic temperatures (above 30 °C) to commence flight activity (Rydell and Lancaster 2000). However, many Arctiinae are robust moths with a dense thoracic hair cover. This is particularly true for the Phaegopterini and Arctiini, whereas Ctenuchini or Pericopini are often more slender, less hairy, and are thus less robust flyers. We therefore hypothesize that their better isolation against thermal losses, rather than an adaptation of flight metabolism to lower temperatures, allows Phaegopterini moths to thrive in large numbers along the entire altitudinal gradient.

Moth taxa where we observed a clear decline in diversity with elevation are rather delicately built, and less well isolated by hair-like thoracic scales. This applies to the majority of Pyraloidea, but also to the arctiid subfamily Lithosiinae. The proportional decrease of Ctenuchini and Pericopini relative to all arctiid moths also fits here. The Pyraloidea showed the overall largest reduction of diversity from the warm lower montane forest to the cool cloud forest zone. While at the lowest sites at Bombuscaro (1040 m) pyraloid ensembles were exceptionally diverse ($\alpha > 100$), ensembles at elevations above 2200 m were much less rich ($\alpha < 35$).

11.3.3.5 Moth Diversity and Host Plants

Most moths have herbivorous larvae, which tend to be specialized with regard to host plant use (Scoble 1992; Holloway et al. 2001). One would therefore predict that moth diversity mirrors plant (i.e. resource) diversity. Plant diversity in Ecuador is extremely high (Madsen and Øllgaard 1994, Jørgensen and León-Yánez 1999), yet floristic data for our study area are still incomplete (see Chapter 10.1 in this volume). Knowledge of the host plant relationships of most recorded moth species is very fragmentary (Brehm 2002; Süßenbach 2003; see also Dyer et al. 2007; Janzen and Hallwachs 2007). This limits the search for relationships between plant and insect diversity. In the study area the forest at elevations between 2100–2700 m largely consists of mono-dominant stands of the tree *Purdiaea nutans* (see Chapter 19), but this was hardly mirrored by moth diversity patterns. Only among Pyraloidea (especially Crambidae, Pyraustinae) was the diversity almost constant and low (α: 15–25) throughout the *Purdiaea* forest zone. Whether this finding reflects reduced resource availability in this unusual cloud forest remains to be tested.

Pertinent host plant data are yet unavailable for most moth species that we have recorded. Certain changes in community patterns can nevertheless be linked to resource affiliations at a higher taxonomic scale. Most clades among the Lepidoptera have characteristic feeding specializations which are often conserved even across continents (Powell et al. 1999; Holloway et al. 2001). Pyraloid larvae are frequently endophagous or shelter-building and tend to be more specialized than ectophagous caterpillars (Gaston et al. 1992). Moreover, smaller-sized moth species generally tend to be more specialized with regard to host plants (Loder et al. 1998). Thus, higher host plant specialism might contribute to the steeper decline of pyraloid diversity as opposed to the more polyphagous arctiids, whose diversity changed much less in response to altitude.

Within the Arctiidae, Ctenuchini and Pericopini (many aposematic species with complex chemical defence) decreased considerably at higher elevations, whereas Phaegopterini (robust, often cryptically colored, putatively more polyphagous species) proportionally increased. The larvae of Lithosiinae often feed on algae, lichens, and liverworts, although certain Neotropical genera have angiosperm host plants (Dyer et al. 2007; Janzen and Hallwachs 2007; F. Bodner, M. Zimmermann, and K. Fiedler, unpublished observations). Despite the abundance of lichens in the cloud forest zone, Lithosiinae diversity significantly decreased with elevation. However, at high altitudes Lithosiinae usually comprised more than 50% of all arctiid individuals in our samples, suggesting that the few species adapted to the upper cloud forest zone have ample resources there.

Four pyraloid subfamilies (Scopariinae, Odontiinae, Galleriinae, Phycitinae) proportionally increased at higher elevations. Scopariinae larvae often feed on mosses or ferns; in the tropics the subfamily is most prevalent in montane habitats (Munroe and Solis 1999). Thus their commonness in the upper cloud forest zone was expected. The host plant relationships and habitat requirements of Neotropical Odontiinae, Galleriinae, and Phycitinae are too incompletely known (Munroe and Solis 1999) to relate their unexpected proportional increase at high elevations to life history traits.

The pyraloid subfamilies Acentropinae, Glaphyriinae, and Musotiminae proportionally decreased with altitude. Acentropinae larvae often feed on aquatic plants (Munroe and Solis 1999). Our light-trapping sites at elevations of 1040–2000 m were close to streams, so the distributional pattern of Acentropinae in our samples could indeed reflect an association with freshwater habitats. Musotiminae are fern feeders (Munroe and Solis 1999). It was thus surprising that moths of this subfamily became scarce at elevations above 2100 m, while ferns were still abundant and diverse (Kessler 2001). The feeding habits of Glaphyriinae larvae are so diverse that no inferences can be made at subfamily level.

11.3.3.6 Taxonomic Composition, Scale, and the Problem of Biodiversity Indication

Brehm et al. (2003a, b) and Brehm and Fiedler (2003) observed that taxonomic composition significantly changed along the altitudinal gradient within the moth family Geometridae, depending on taxonomic scale (family, subfamily, tribe). Our analyses revealed that similar variation recurs also in two other large moth clades. Patterns partially contrasted from the tribe (e.g. Ctenuchini: altitudinal decrease) nested in a subfamily (Arctiinae: no clear altitudinal change) which is again nested in a family (Arctiidae: weak altitudinal decline). The gross taxonomic composition of moth ensembles also profoundly changed with altitude. It was thus surprising that species turnover was highly concordant across all moth groups. Patterns of beta diversity were also invariant to taxonomic scale. All our ordinations revealed the same, recurrent pattern: a steady altitudinal change of species composition, with no evidence for the existence of faunal zones. In all cases ordinations nearly perfectly mirrored the altitude of sampling sites. This suggests that, apart from all idiosyncrasies in host-plant use, physiology or habitat requirements, one major variable must govern species turnover in all investigated moth groups in much the same way. Given the large differences in feeding habits and behaviour between the study groups, the most likely candidate for a master variable regulating species turnover among ectothermic invertebrates in the montane forest zone is temperature. Indeed, correlations between scores of samples in ordination space and temperature were all highly significant.

Over the past ten years, many efforts have been devoted to the establishment of 'biodiversity indicators' (e.g. Schulze et al. 2004). The main idea behind this search is that wherever complete inventories can almost never be achieved (as with rich tropical arthropod faunas), or where results are required within short periods of time (e.g. urgent conservation decisions), it would be ideal to restrict monitoring and assessment effort to a few taxa that can easily be investigated. Such indicators should be easy to sample, they should represent the same proportion of the fauna at all study sites (Pearson 1994; Beccaloni and Gaston 1995), identification should be easy, and conclusions drawn from such indicators should be transferable to other groups of organisms.

Among the moths investigated here, Arctiidae (especially Phaegopterini) appear to be good candidates at first sight. They can easily be sampled by attracting them to light, and species identification is possible to an unusually high degree (as compared to other tropical invertebrates). Yet, at a closer look the suitability of arctiids as biodiversity indicators becomes questionable. Their diversity patterns are poorly correlated with those seen in other moths. Also, within the Arctiidae the discordant diversity patterns at lower taxonomic scales and variable taxon ratios further undermine their versatility. These same arguments apply to every other moth group we have studied here or previously (Geometridae: Brehm and Fiedler 2003). More generally, correlations among diversities of various organisms were sometimes found (Schulze et al. 2004), but were also absent or erratic in many other studies (Lawton et al. 1998; see also Simberloff 1998). Our study along an altitudinal gradient is particularly intriguing for the problem of biodiversity indication, since a common environmental 'master variable' (viz. temperature) was supposed to concomitantly reduce diversity in ectothermic herbivorous moths in much the same way. But, while this influence does exist with regard to species turnover, it does not uniformly exist with regard to local species diversity.

11.3.4 Conclusions and Perspectives

Our study of moths along an altitudinal gradient in the Ecuadorian Andes revealed remarkable idiosyncrasies in patterns of alpha diversity across various moth groups and taxonomic scales. These render extrapolations from one exemplar group to others, or upscaling to the entire guild of herbivorous insects, impossible. In contrast, species turnover was extremely concordant across groups with very different life-history characters, suggesting that temperature – and not the availability of plant resources – plays the leading role in governing community change. It would have been desirable to extend sampling to lower as well as higher elevations to capture altitudinal patterns more completely (Rahbek 2005). At present, therefore, we cannot ascertain whether moth richness would become even higher at lower elevations, or whether some groups at least have true mid-elevation peaks (Brehm et al. 2007).

We still know little about the functional role the exceptionally high diversity of moths plays in Andean mountain forests. The elevational decline of Arctiidae and Pyraloidea diversity is mirrored by a concomitant decrease in abundances in these same taxa (Pearson correlations of mean log-transformed catches per night, after controlling for effects of canopy closure as a proxy for vegetation density; Arctiidae: $rP = -0.591$, $P = 0.006$; Pyraloidea: $rP = -0.742$, $P = 0.0002$). Thus these two moth groups become, as a whole, increasingly less important as herbivores or as prey organisms for insectivores or parasitoids at higher elevations. Yet individual species or subordinated taxa may still be significant in the food webs at high altitudes, such as the lichen moths or Scopariinae pyraloids that occur in substantial numbers above 2400 m a.s.l. Moreover, in the especially diverse family Geometridae, the decline in moth abundance was far less pronounced and not significant ($rP = -0.332$, $P = 0.15$). Overall, moths form an abundant and diverse component of food-webs throughout the mountain forest belt, even though their precise functional role awaits to be quantified.

Hence, what is most urgently desired is a better understanding of the functional interrelationships between moths, their host plants and natural enemies to allow for more precise interpretation of diversity patterns. Large-scale rearing programs similar to those initiated by Dyer et al. (2007) or Janzen and Hallwachs (2007) are a necessary next step. Since Andean moth communities are among the richest that have been documented globally thus far, but are under severe threat through continuous deforestation (Doumenge et al. 1995; see Chapter 4), the time-window for improving our understanding of these organisms may close soon.

Acknowledgments We thank all taxonomists and curators for access to reference collections under their care and assistance with identifications: Axel Hausmann and Andreas Segerer (Zoologische Staatssammlung, Munich); Daniel Bartsch (State Museum of Natural History, Suttgart); David J. Carter, Martin Honey, Malcolm J. Scoble and Linda M. Pitkin (The Natural History Museum, London); Bernard Landry (Muséum d'Histoire Naturelle, Geneva); Matthias Nuss (State Museum of Zoology, Dresden); Wolfgang Speidel (Zoological Museum Alexander Koenig, Bonn); Wolfram Mey (Zoological Museum of the Humboldt University, Berlin). Giovanni Onore (Pontifícia Universidad Católica, Quito) and NCI (Loja) gave logistic and administrative support to our work. Teresa Baethmann, Doreen Fetting, Adrienne Hogg, Eva Mühlenberg, Frank Ruge, Rita Schneider, and Annick Servant assisted with setting and data-basing moths. We thank Christian H. Schulze for constructive comments on an earlier manuscript version. The Ministerio de Medio Ambiente del Ecuador granted research permits. Dedicated to the late Clas M. Naumann in appreciation of his continuous support extended to the senior author.

Chapter 11.4
Soil Fauna

M. Maraun(✉), J. Illig, D. Sandmann, V. Krashevska, R.A. Norton, and S. Scheu

11.4.1 Introduction

Tropical mountain rain forests form biodiversity hotspots for a number of animal and plant taxa (Küper et al. 2004; see Chapter 2 in this volume). Only a few taxa, such as parasitic Hymenoptera, appear not to conform to the general rule of increasing diversity from temperate to tropical regions (Gauld et al. 1992). However, one important component of the animal community, the soil fauna, has rarely been studied in tropical rain forests, especially in mountain regions. The reasons for this presumably are that: (a) soil animals have to be extracted from a 3-dimensional medium by heat or other methodology, (b) there are few keys for determination and (c) a large number of species is still not described.

11.4.2 Methods

We investigated the density and diversity of the soil microarthropod community in a tropical mountain rain forest in southern Ecuador. We focused on microarthropods (mainly oribatid mites and collembolans), since at our study site the density of soil macrofauna, such as earthworms, diplopods and isopods, was low, which may be due to low pH and low resource (litter) quality. Due to the low density of large decomposer animals we expected soil microarthropods to significantly affect decomposition processes. Changes in density and community structure of the soil mesofauna were investigated along an altitudinal gradient spanning over 2000 m. To evaluate the trophic structure of the soil food web we studied natural variations in stable isotope ratios ($^{15}N/^{14}N$; $^{13}C/^{12}C$) of dominant soil fauna taxa. Knowledge on the trophic structure of the decomposer food web is necessary for understanding the limiting factors of soil animal species and the role of decomposer species for decomposition processes. By manipulating carbon and nutrient availability we evaluated limiting factors for soil microorganisms and investigated the role of microorganisms as food resources for animal decomposers. Finally, by investigating the decompositon of leaf litter enclosed in litterbags of different mesh size we evaluated the role of decomposer animals for litter decomposition at different altitudes.

The studies were carried out at the RBSF (see Chapter 1 in this volume). The region is covered with little-disturbed mountain rain forest. The climate is semi-humid, the average annual precipitation at 1900 m altitude is about 2000 mm and the average annual temperature is 16 °C. The soil types are mainly Aquic and Oxaquic Dystropepts (Schrumpf et al. 2001) and the pH ranges between 4.0 and 4.5 (Wilcke et al. 2002). The study sites were located at 1000 m (Bombuscaro), 1850 m (Estacion Scientifica San Francisco, ECSF) and 3100 m a.s.l. (Cajanuma).

11.4.3 Results and Discussion

11.4.3.1 Soil Microarthropod Density, Diversity and Reproductive Mode

The density of oribatid mites was about 34 400 individuals (ind)/m^2 at 1000 m (Fig. 11.4.1). This density is similar to that of other tropical forests (rainforest in Australia: 43 000 ind/m^2; Plowman 1981) and to limestone forests of the temperate zone (35 000 ind/m^2 at a base rich Danish beech forest; Luxton 1981), but much lower than that of acidic deciduous forests of the temperate zone where oribatid mite densities may reach 180 000 ind/m^2 (Maraun and Scheu 2000) or more (Persson et al. 1980). This is rather surprising since the lower density of oribatid mites (and also that of other microarthropods) in base rich as compared to acidic temperate forests presumably is due to the activity of earthworms and other large decomposers which remove the litter layer (Migge 2001; Eisenhauer et al. 2007), but as stated above these large decomposers are rare at our study site.

Surprisingly, the density of oribatid mites further decreased with increasing altitude. At 2000 m the density of oribatid mites was only 20 000 ind/m^2 and at 3100 m a.s.l. only 5400 ind/m^2. This decline was particularly unexpected since the thickness of the organic layers increases with altitude and in temperate forests the density of microarthropods increases with the thickness of the organic layers (Maraun and Scheu 2000).

At 1850 m we found about 150 oribatid mite species, including several new species (Niedbala and Illig 2007). This is higher than in most temperate forests where species numbers usually range between 30 and 80 (Luxton 1975; Persson et al. 1980; M. Maraun, unpublished data). Possibly, high plant diversity and the associated high litter diversity is responsible for the high species number of soil microarthropods, but factors driving oribatid mite species diversity are generally little understood (Anderson 1975; Maraun et al. 2003a). In temperate forest ecosystems it increases from open to forest habitats but is little affected by tree species diversity (Migge et al. 1998; Hansen 2000).

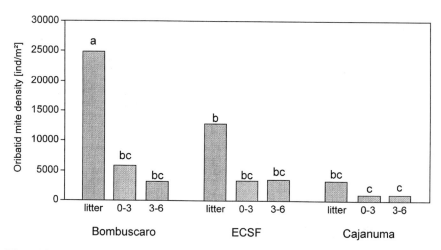

Fig. 11.4.1 Oribatid mite densities at Bombuscaro (1000 m), ECSF (1850 m) and Cajanuma (3100 m) in the litter and upper soil layers (0–3 cm, 3–6 cm depths). *Bars* sharing the same *letter* are not significantly different (Tukey's HSD test; $P < 0.05$). *ind* Individuals

Compared with temperate forests, the number of parthenogenetic taxa of oribatid mites at the studied tropical mountain rain forest is small. In temperate forests about 60% of the species and up to 90% of the individuals of oribatid mites reproduce via parthenogenesis (Maraun et al. 2003b), whereas at the studied tropical forest (3100 m) only 23% of the species and about 35% of the individuals reproduce via parthenogenesis (V. Eissfeller and M. Maraun, unpublished data). These findings are consistent with the hypothesis that due to more pronounced biotic interactions, such as competition, the proportion of sexually reproducing taxa increases towards the tropics (Sanders 1968; Glesener and Tilman 1978). However, densities of oribatid mites at the studied tropical forests are low and therefore intra- and interspecific competition likely are of little importance.

11.4.3.2 The Soil Food Web

The structure of food webs, i.e. the trophic relationships between animals, plants and microorganisms, has mainly been studied in temperate ecosystems. It has been suggested that compared to temperate forests the number of trophic levels in tropical forests is increased (Reagan et al. 1996). However, recent studies indicate that the number of trophic levels in tropical food webs in fact is similar to that in temperate systems (Kupfer et al. 2006).

We investigated the trophic structure of the soil food web of the studied mountain rain forest at 1850 m a.s.l. by analysing natural variations in stable isotope ratios ($^{15}N/^{14}N$; $^{13}C/^{12}C$). The results indicate that the soil food web of the studied tropical rain forest spans about four trophic levels (Fig. 11.4.2) which is similar to that in temperate forests (Scheu and Falca 2000). In the studied food web primary decomposers, i.e. litter feeding species, were rare, potentially reflecting low litter quality (Illig et al. 2005). A large number of 'decomposer' animals were in fact predatory or necrophagous, suggesting that various putative decomposer soil animal species (especially in Oribatida) feed on other soil invertebrates, in particular nematodes or on animal carcasses (Schneider et al. 2004).

Another interesting pattern was that parthenogenetic oribatid mite species had lower ^{15}N values than sexually reproducing species. This suggests that parthenogenetic reproduction prevails in primary decomposers but not in predators. Potentially, predominance of parthenogenetic reproduction in primary decomposers reflects that these species do not co-evolve with their resources, i.e. dead organic material, and therefore sexuality can be abandoned more easily (Hamilton 2001). In contrast, species at higher trophic levels may be confronted with co-evolutionary arms races with their prey and need to reproduce sexually to avoid going extinct (Red Queen hypothesis; Hamilton 1980).

In addition to the ^{15}N/^{14}N ratios in a number of soil animal species we also measured the ^{13}C/^{12}C ratio. The data support the conclusion that primary decomposers indeed are rare in the studied forest. As suggested earlier (Scheu and Falca 2000) decomposer animals constitute two trophic guilds, differing significantly in ^{15}N signatures, i.e. primary and secondary decomposers. Schmidt et al. (2004) proposed that the combined analysis of the ratios of ^{15}N/^{14}N and ^{13}C/^{12}C facilitates identification of primary and secondary decomposers since the latter not only are more enriched in ^{15}N but also less depleted in ^{13}C. In fact, in our study δ^{13}C signatures (cf. Lajtha and Michener 1994) of most decomposer taxa were less depleted by about 1.5–2.5 δ units compared with leaf litter of *Graffenrieda emarginata* (Fig. 11.4.3). Surprisingly, we did not find any mesofauna taxon classified as primary decomposer by δ^{15}N and δ^{13}C signatures. The only primary decomposer species we identified was Diplopoda sp. 1 and this species was rare. Signatures of δ^{15}N and δ^{13}C of Diplopoda sp. 1 were very similar to that of leaf litter of *G. emarginata*. This is surprising since consumers generally are enriched in ^{15}N by on average 3.4 δ units (Post 2002). Similar δ^{15}N signatures of primary decomposers and their food resources have been found previously. It has been suggested that ^{15}N fractionation in decomposer animals generally is low compared to herbivores and predators (Vanderklift and Ponsard 2003).

Signatures of δ^{13}C of predators also spanned a range of more than two units. In some predators, such as Staphylinidae, Pselaphidae and *Amerioppia* spp, the δ^{13}C signatures were close to those of primary decomposers. In other predators, such as Uropodina sp. and Gamasina sp. 1 and sp. 2, they were close to those of secondary decomposers. This suggests that at our study site, and potentially in soil food webs in general, there are two predator guilds, one predominantly feeding on primary decomposers, the other predominantly on secondary decomposers. Generally, the secondary decomposer channel appears to be more diverse, supporting the view that predators in soil predominantly feed on secondary decomposers rather than on primary decomposers (Scheu and Falca 2000).

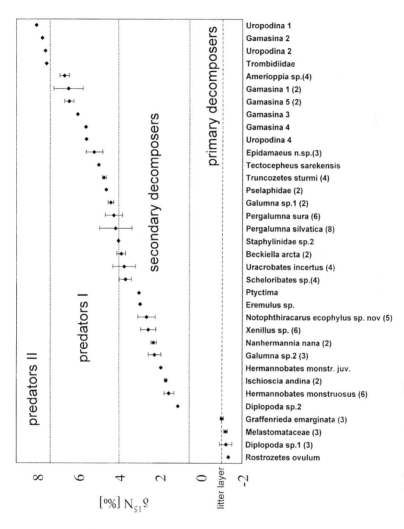

Fig. 11.4.2 Variation of $\delta^{15}N$ signatures of soil animals extracted from *Graffenrieda emarginata* leaf litter material from the ECSF site (1850 m a.s.l.). Single measurements and means of 2–8 replicate measures (*numbers in brackets*) with SD. If no standard deviation is shown only a single sample was analysed. Species were ordered by increasing $\delta^{15}N$ signatures

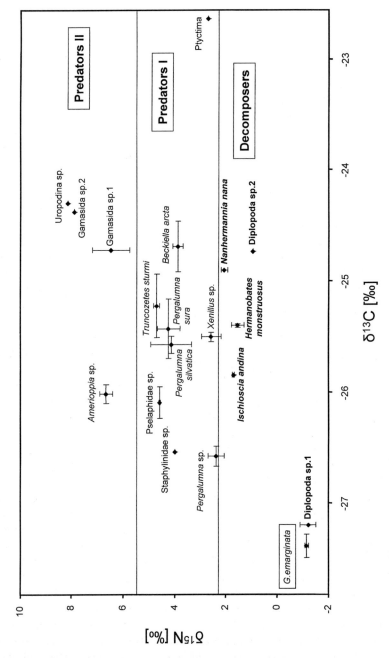

Fig. 11.4.3 Variation in $\delta^{13}C$ and $\delta^{15}N$ signatures of soil animals extracted from *Graffenrieda emarginata* leaf litter material from the ECSF site (1850 m a.s.l.). If no standard deviation is shown only a single sample was analysed Decomposer taxa in bold

11.4.3.3 Decomposition and Microarthropod Colonization of Litter

Litter decomposition in tropical mountain rain forests is slower than in lowland tropical rain forests (Heneghan et al. 1999), which may be due to lower temperatures in mountain rain forests or due to low litter quality. We investigated leaf litter decomposition of two abundant tree species (*G. emarginata, P. nutans*) at the studied tropical mountain rain forest (Illig et al. 2007a, b). In addition, we studied the role of the soil mesofauna for the decomposition of these two litter types. To exclude mesofauna the litter was enclosed in litterbags of 48 µm mesh; litter accessible to mesofauna was enclosed in 1 mm mesh. Litter mass loss, content of C and N, basal respiration, microbial biomass and the colonization of the litterbags by microarthropods were analyzed after exposure in the field for 2, 6 and 12 months at 1850 m and 2270 m. Both litter types decomposed slowly (average of 68.5% of initial dry weight remaining after 12 months) and, generally, *P. nutans* litter decomposed slower than *G. emarginata* and mixed litter (Fig. 11.4.4). The decomposition rates were similar to those of oak–pine litter in a forest stand in Japan (Kaneko and Salamanca 1999), sweet chestnut and beech litter in two deciduous woodland forests in England (Anderson 1973) and *Quercus coccifera* litter in Northern Greece (Argyropoulou et al. 1993). However, compared with tropical lowland forests decomposition rates were low; in these systems only about 30% of the initial dry weight remained after one year (Paoletti et al. 1991; Heneghan et al. 1999; Franklin et al. 2004).

One of the reasons for the slow decomposition may be low litter quality, i.e. low nutrient concentrations, in particular that of nitrogen (Enriques et al. 1993). Indeed, nitrogen concentrations of the leaf litter (F material) of *P. nutans* and *G. emarginata* were lower (C/N ratio of 73.6 and 41.9, respectively; J. Illig, unpublished data) than those of the F material of European beech (C/N ratio of 21; Maraun and Scheu 1996). Each of the litter materials (*G. emarginata, P. nutans*, mixed litter) decomposed faster at 1850 m than at 2270 m (averages, respectively, 60% and 76% of the initial litter mass remaining after 12 months). Lower decomposition rates at higher altitudes likely were due to lower temperatures.

At 1850 m altitude the mixture of *G. emarginata* and *P. nutans* litter decomposed significantly faster than both single litter types (Fig. 11.4.4) indicating that combining the two litter types accelerates decomposition processes ('non-additive effect'). The reason for this non-additive litter decomposition may be the higher humidity in mixed litter (Wardle et al. 2003) or the mixing of litter of different resource quality which may stimulate the decomposition of low quality litter, i.e. litter low in nitrogen (Smith and Bradford 2003; Quested et al. 2005).

The most abundant microarthropods in the litterbags were oribatid mites, followed by Collembola, Gamasina, Uropodina, Prostigmata and Astigmata. Each of these taxa was more abundant at 1850 m than at 2270 m. Oribatid mites in the litter bags constituted 37 morphospecies and were dominated by *Scheloribates* sp., *Pergalumna sura* and *Truncozetes sturmi*. Species composition was similar in both litter types, supporting previous findings that the structure of soil decomposer microarthropod communities is little affected by litter type (Walter 1985; Migge et al. 1998; Hansen 2000).

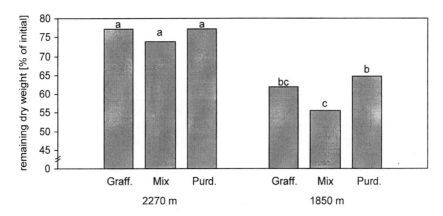

Fig. 11.4.4 Decomposition (measured as remaining dry weight) of leaf litter of *Graffenrieda emarginata* (*Graff.*), *Purdiaea nutans* (*Purd.*), and a mixture of both (*Mix*) exposed in the field for 12 months at two different altitudes [m a.s.l.]. *Bars* sharing the same *letter* are not significantly different (Tukey's HSD test; $P < 0.05$)

Soil microarthropods generally did not significantly affect litter decomposition ($F = 0.04$; $P = 0.84$). This is consistent with results of our food web analysis suggesting that litter feeding soil meso- and macrofauna are scarce. However, higher density and diversity of secondary decomposers as compared to primary decomposers suggest that soil animals may affect decomposition processes at later stages of decay. Therefore, exposure of litter for 12 months, as done in the present study, may not have been long enough to evaluate the role of decomposers for litter decomposition in the studied tropical mountain rain forest. Long-term studies are needed to prove if this in fact is the case. Furthermore, soil microfauna should also be investigated in these studies since microfauna groups, such as Nematoda and Testacea, are abundant at the studied mountain rain forest and their role in decomposition processes may well exceed that of meso- and macrofauna.

11.4.3.4 Bottom-Up Forces in the Soil Food Web

Animal, plant and microbial communities are structured by bottom-up (i.e. resources) and top-down forces (i.e. predators), but the relative importance of both is debated (Hairston et al. 1960; McQueen et al. 1989; Hunter and Price 1992; Moran and Scheidler 2002). For soil systems Hairston et al. (1960) suggested that especially saprophagous animals and microbial decomposers, such as bacteria and fungi, are regulated by the availability of resources, i.e. the input of dead organic matter. Scheu and Schaefer (1998) confirmed the importance of resource limitation for earthworms in a temperate deciduous forest. However, decomposer animals are not only regulated by bottom-up forces; indirect effects, such as bioturbation by earthworms, may override bottom-up effects, especially in soil microarthropods (Maraun et al. 2001; Salamon et al. 2006). However, the latter are presumably of little importance in the studied mountain rain forest since large decomposer animals are virtually lacking.

By increasing the supply of resources to decomposer biota we expected to be able to uncover the role of resource limitation in the studied decomposer system. First, in a short-term laboratory experiment we studied the role of nutrients (N, P) and carbon (C, glucose) added as C, CN, CP and CNP for the growth of soil microorganisms. The study evaluated whether microbial growth in the litter layer was limited by nutrient availability. Second, by adding nutrients (N, P) and carbon (C, glucose) in all combinations (control, C, N, P, CN, CP, NP, CNP) repeatedly to the soil of the studied rain forest for one year we investigated the nutrient and carbon limitation of microorganisms and soil animals under natural conditions in the field.

In the short-term laboratory experiment, microbial growth (measured as the slope of the respiration curve after resource addition) was measured (Scheu and Parkinson 1994). We hypothesized: that (a) carbon addition will not stimulate microbial growth since it is primarily limited by nitrogen and phosphorus (Enriques et al. 1993) and (b) nitrogen addition (together with carbon) stimulates microbial growth in the studied rain forest since nitrogen is assumed to be more important than phosphorus for plants and microorganisms, whereas in lowland rain forests the opposite may be true (Tanner et al. 1998). In agreement with our first hypothesis, carbon addition stimulated little microbial growth, particularly at higher altitudes (Fig. 11.4.5), indicating that nutrient limitation increases with altitude. However, in contrast to our second hypothesis, the addition of carbon together with nitrogen also only little affected microbial growth indicating that nitrogen is not limiting microbial growth. Unexpectedly, the addition of phosphorus (together with C or CN) generally strongly increased microbial growth, and this effect increased with altitude indicating that phosphorus limitation increases with altitude. Overall, the results indicate that:

1. Microbial growth is mainly limited by the availability of phosphorus.
2. The availability of nutrients (esp. phosphorus) declines with altitude.

In the long-term field experiment the amount of resources added was equivalent to about five times the annual input of C, N and P with aboveground plant litter. We hypothesised that the additional resource supply predominantly affects lower trophic levels, i.e. microorganisms and decomposer animals, and propagates to higher trophic levels but with decreasing intensity due to the dampening of bottom-up forces at higher trophic levels in soil systems (Salamon et al. 2006). Indeed, fungal biomass (measured as ergosterol content) and density of Astigmata strongly increased after addition of carbon (Fig. 11.4.6). The density of Entomobryidae, Onychiuridae, Ptyctima, Desmonomata and juvenile Oribatida and also predatory taxa (Uropodina, Gamasina) increased after N and C application, although the response was not uniform, and some taxa remained unaffected (e.g. Arachnida, Prostigmata, Formicidae, some subgroups of oribatida). Predatory taxa generally responded little to resource additions. Overall, the results support the hypotheses that: (a) additional resources mainly affect lower trophic levels (b) carbon and (less pronounced) also nitrogen and phosphorus are important limiting factors for basal trophic taxa, in particular fungi and decomposer soil mesofauna and (c) resource addition effects propagate only slightly to higher trophic levels, i.e. to predatory taxa.

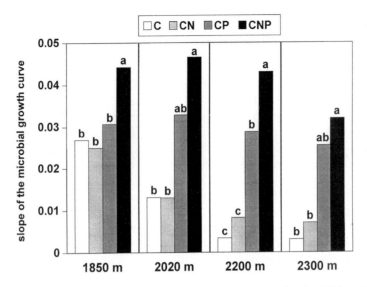

Fig. 11.4.5 Microbial growth in F litter material from the ECSF site after the addition of carbon (*C*), carbon + nitrogen (*CN*), carbon + phosphorus (*CP*) and carbon + nitrogen + phosphorus (*CNP*; see text for details). *Bars* sharing the same *letter* are not significantly different (Tukey's HSD test; $P < 0.05$)

11.4.4 Conclusions and Prospect for Future Studies

In the studied mountain rain forest the densities of soil macrofauna and soil microarthropods (oribatid mites, collembolans) were low. As in temperate forests, low densities of decomposer soil macrofauna may be caused by low soil pH. However, these low densities of soil microarthropods are surprising since, in temperate and boreal forests resembling the studied mountain rain forest with respect to low pH and thickness of organic layers, the density of soil microarthropods is high. In contrast to density, the diversity of soil microarthropods was high compared with temperate and boreal forests. High diversity might be due to the presence of a high number of microhabitats (e.g. litter types).

The natural variation in the stable isotope ratios ($^{15}N/^{14}N$, $^{13}C/^{12}C$) of soil animals indicates that the number of trophic levels and the structure of the soil food web resembles that of temperate forest ecosystems; however, there appear to be fewer species functioning as primary decomposers. As reported from temperate forests, secondary decomposer animals were more enriched in ^{13}C than primary decomposers, and this may allow separation of consumer chains based on primary and secondary decomposers.

Litter decomposition was generally slow and decreased with altitude. High moisture and low litter quality may limit the decomposition process. The acceleration of litter decomposition in mixed litter material, as observed for the litter of *P. nutans* and *G. emarginata*, suggests that non-additive effects significantly contribute to decomposi-

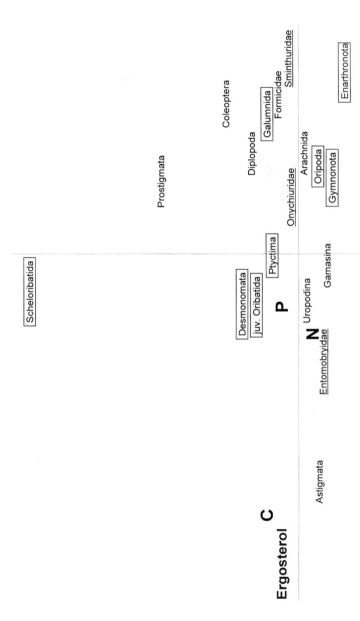

Fig. 11.4.6 Canonical correspondence analysis plot of fungi (indicated by ergosterol) and soil animal taxa after one year addition of carbon (*C*), nitrogen (*N*) and phosphorus (*P*) separately and in combination (CN, CP, NP, CNP). Collembola are *underlined*; oribatid mites are enclosed in *boxes* (eigenvalue of axis 1 = 0.28; eigenvalue of axis 2 = 0.12)

tion processes. This might be particularly important in tropical forest ecosystems with high tree species diversity as in the studied tropical mountain forest.

The short-term addition of carbon (glucose) and nutrients (N, P) to litter material (F layer) indicated that phosphorus limits microbial growth and that the nutrient deficiency increases with altitude. The long-term (one year) addition of carbon (glucose) and nutrients (N, P) indicated that in particular carbon limits soil fungi and Astigmata whereas the densities of decomposer taxa (Entomobryidae, Onychiuridae, Ptyctima, Desmonomata, juvenile Oribatida) and some predatory taxa (Uropodina, Gamasina) increased after combined application of C and N. Most predatory taxa did not respond to the addition of resources, indicating that bottom-up forces are dampened at higher trophic levels. The results suggest that carbon and (less pronounced) nitrogen are limiting elements for basal trophic taxa in the long term.

Acknowledgement We thank Melanie Pollierer for commenting on earlier versions of the manuscript.

Chapter 12
Water Relations

W. Wilcke(✉), S. Yasin, K. Fleischbein, R. Goller, J. Boy, J. Knuth,
C. Valarezo, and W. Zech

12.1 Introduction

Rainfall in tropical mountain forests ranges between several hundreds and several thousands of millimeters per year (Cavelier et al. 1997; Bruijnzeel 2001). Many of these forests receive considerable cloud water inputs (Bruijnzeel 2001). During its passage through the forest, the quantity of the water entering the forest with rainfall and clouds changes because of increasing evaporation and transpiration during percolation through the ecosystem (Likens and Bormann 1995; Bruijnzeel 2001). An important water loss is the evaporation of water intercepted in the canopy. Measurements of interception loss of tropical mountain forests also vary over a broad range (Cavelier et al. 1997). The portion of throughfall and stemflow reaching the stream after its passage through the soil depends on water flow paths. The occurrence of rapid interflow is a frequent phenomenon on steep forested hill slopes (Mulholland et al. 1990; Bonell et al. 1998) and is mainly attributable to the high significance of macropores (Buttle and McDonald 2000). Therefore, the assessment of water and element fluxes in the soil requires the due consideration of fast interflow.

The catchment approach, i.e. the measurement of all input, output, and internal water fluxes within a well defined and usually not too large water catchment can be considered as classic in ecosystem research (Likens and Bormann 1995; Matzner 2004). The advantage of the catchment approach is that it allows for the determination of evaporation and transpiration fluxes which are otherwise not easily accessible by budgeting all input and output fluxes. In an ideal catchment, all output fluxes can be measured at a weir located at the outlet of the catchment. This requires the catchment to be watertight and also implies that all surface flow is captured at the outlet. Furthermore, the catchment should only receive water input from the atmosphere via precipitation but not via surface or groundwater flow from neighboring catchments. Usually, the assessment of these prerequisites is complicated because it is difficult to determine the water-conducting properties of the subsoil and underlying rock and because topographic catchment borders are not necessarily the same as hydrological ones. Watershed approaches have been used at several locations in the temperate zone (e.g., Hauhs 1985; Likens and Bormann 1995; Matzner 2004) but rarely in tropical rain forests. One of the few studies in forests similar to ours was conducted in Puerto Rico (McDowell and Asbury 1994).

12.2 Methods

General information about the precipitation and its distribution along the whole altitudinal transect can be found in Chapters 8, 20, 22 in this volume. Here, we focused on three small catchments between 1900 m and 2200 m a.s.l. The small catchments (termed microcatchments, MC) were on 30–50 ° steep slopes on the north-facing flank of the San Francisco valley. Microcatchment (MC)1 had a size of 8 ha, MC2 of 9 ha and MC3 of 13 ha (see Fig. 1.2 in Chapter 1). Microcatchments 2 and 3 are entirely forested, whereas the upper part of MC1 has been used for agriculture until about 10 years ago. This part is currently undergoing natural succession and is covered by grass and shrubs. Soil details of the three study catchments are given by Wilcke et al. (Chapter 9).

To measure incident precipitation, throughfall, stemflow, and stream flow, we used standard devices described in detail by Fleischbein et al. (2005, 2006). Briefly, precipitation was measured with Hellmann-type collectors, surface flow with a manually calibrated V-shaped weir. Measurements were made in at least weekly intervals since April 1998. The equipment was arranged along ca. 20-m transects and on clearings adjacent to the catchments shown in Fig. 1.2 in Chapter 1.

To assess the canopy interactions, we used weekly data of incident rainfall, throughfall, and stemflow between April 1998 and April 2001. In 2001, the Leaf Area Index (LAI), inferred from light transmission, and epiphyte coverage were determined (Fleischbein et al. 2005). To characterize the hydrological properties of the canopy we determined the storage capacity of the leaves (S) and of the trunks and branches (S_t), and the fractions of direct throughfall (p) and stemflow (p_t) with a regression approach (Rutter et al. 1975; Crockford and Richardson 1990; Hörmann et al. 1996). Furthermore, we investigated the controls of interception loss by relating canopy density and epiphyte coverage of trees with interception losses.

To model surface water flow, we used TOPMODEL (Beven et al. 1995; Kinner and Stallard 2004). A model description and a description of model parameterization, calibration, and validation are given by Fleischbein (2004) and Fleischbein et al. (2006).

Evapotranspiration was calculated as the difference between incident precipitation and surface flow. We determined the interception loss and transpiration separately by budgeting canopy fluxes and whole catchment fluxes, respectively, neglecting direct evaporation from the soil and stream surfaces.

12.3 Results and Discussion

12.3.1 Canopy Interactions

Mean annual incident rainfall at the three gaging stations between 1998 and 2001 – the period used for determination of canopy properties – was 2320–2560 mm

Table 12.1 Water budget of three approx. 10-ha large microcatchments (MC1–MC3) under lower mountain rain forest in Ecuador between May 1998 and April 2003

	MC1		MC2		MC3	
	Mean	Range	Mean	Range	Mean	Range
			(mm)			
Incident rainfall[a]	2540	2420–2730	2410	2170–2610	2610	2440–2770
Throughfall	1220	940–1590	1530	1290–1790	1580	1290–1970
Stemflow	25	21–26	25	20–31	23	20–25
Throughfall + stemflow	1250	970–1620	1550	1310–1820	1600	1310–2000
Interception loss	1290	1110–1510	860	670–1300	1010	770–1190
Surface runoff	960	720–1170	1100	1020–1200	1030	880–1140
Transpiration[b]	290	73–450	450	130–780	570	360–860
Evapotranspiration	1580	1420–1760	1310	1090–1470	1580	1400–1700

[a] For the hydrological years 1998/1999 and 1999/2000, missing incident rainfall data were calculated by regression of the 2000–2003 data collected at gaging station 2 on those of gaging stations 1 and 3 (see Fig. 1.2, Chapter 1 in this volume).
[b] Neglecting direct evaporation from the soil surface.

(annual totals differed from those shown in Table 12.1 because of different observation periods); it varied little among the gaging stations. Missing incident rainfall data of gaging stations 1 and 3 for the period April 1998 to May 2000 were reconstructed by regression of the incident rainfall rates at these stations on the values obtained at gaging station 2 for the following 1-year period (Fleischbein 2004). The mean annual interception loss at the five study transects in the forest during the same period varied between 590 mm and 1320 mm, i.e. between 25% and 52% of the incident rainfall. The throughfall at our study transects, accounting for 48–75% of the incident rainfall, is at the lower end or even lower than the range presented by Bruijnzeel (2001) as 55–101% for lower mountain forests. Throughfall is difficult to measure, because of the possible large spatial variability caused by the vegetation (Lloyd and Marques 1988). The low number of collectors at the individual measurement transects might have resulted in an underestimation of the true throughfall with resulting overestimation of the interception loss. The quality of our throughfall measurement has been discussed extensively by Fleischbein et al. (2005). The stemflow only accounted for 1.0–1.1% of net precipitation during the period used for determination of the hydrological canopy properties. An explanation of the high interception losses may be the particularly high canopy density as indicated by the high leaf area index (LAI) ranging from 5.2 to 9.3 above our 40 throughfall collectors, a high epiphyte loading, and the high solar radiation because of the proximity of the equator. The LAI is at the upper end of the range of LAI values for tropical forests in the literature of 4.6–6.1 in an Amazonian lowland rain forest (Roberts et al. 1996) and a range from 2.0 (for stunted elfin cloud forest in Puerto Rico; Weaver et al. 1986) to 7.7 (for mountain rain forest Costa Rica; Köhler 2002) in tropical mountain forests. Epiphytes, mostly bryophytes, covered up to 80% of the trunk and branch surfaces.

Mean S was estimated at 1.9 mm for relatively dry weeks; mean S_t was 0.04 mm, mean p 0.42, and mean p_t 0.003, respectively. The range of S values (0–8.0 mm) for the canopy above the individual throughfall collectors covers the whole range of the storage capacities of similar vegetations and climates given in the literature (0.04–8.3 mm). The range of p values (0.14–0.98) is well within the range of values reported in the literature (0.05–1.0 mm). The ranges of S_t and p_t values are at the lower end of the spectrum reported in the literature (0.01–0.39 mm and 0.003–0.20 mm, respectively; for an overview of S, S_t, p, and p_t values in the literature, see Fleischbein et al. 2005). Thus, S_t contributes little to the total storage capacity of the vegetation, and stemflow also plays a minor role in the water budget of the studied forest.

The fraction of direct throughfall (p) and the LAI correlated significantly with interception loss (r = −0.77 and 0.35 after removal of four outliers, respectively, n = 40; Fig. 12.1). The significant correlation between LAI and interception loss indicates that canopy density is an important control of the interception loss. A similar conclusion has already been drawn by Burghouts et al. (1998) who found a negative correlation between throughfall and litterfall at 30 points under lowland rain forest in Borneo. However, the low coefficient of determination also suggests that it is not possible to reliably estimate the interception loss from the LAI. This may partly also be related with the fact that the area of the LAI covered by our measurement and the catch area of the throughfall samplers had different sizes. The direct throughfall fraction p seems to be a better predictor of the interception loss in our study forest (Fig. 12.1). A relationship between p and LAI, as found here, was also reported by Pitman (1989) for a fern canopy and by Van Dijk and Bruijnzeel (2001) for corn and yucca canopies. In both studies, LAI was an exponential function of p.

Bryophyte and lichen coverage tended to decrease S_t while vascular epiphytes tended to increase it, although there was no significant correlation between epiphyte coverage and interception loss. The negative correlation between epiphyte coverage and S_t in our study is in contrast to findings of Steinhardt (1979) showing that a higher epiphyte loading results in higher S_t. However, a negative effect of lichen and bryophyte coverage on S_t was only observed for relatively dry periods (only one rain event/week). During dry periods, lichens and bryophytes may even repel water because of their hydrophobic surfaces (Crockford and Richardson 2000). Prolonged rainfall or events with a higher intensity resulting in saturation of the bryophytes and lichens may change their influence on the storage capacity of the trunk and branches (Crockford and Richardson 1990). The consequences of the heterogeneous water passage through the canopy are discussed by Homeier and Breckle (Chapter 23 in this volume).

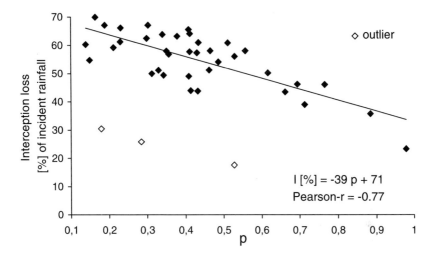

Fig. 12.1 Relationship between Leaf Area Index and the direct throughfall fraction (p, determined with the regression approach) above 40 throughfall collectors. For the calculation of the regression equations outliers were not considered

12.3.2 Flow Regime in Soil

After having passed the canopy with the associated interception loss the remaining water reaches the soil as throughfall or stemflow. The further water redistribution in soil depends on soil water conductivity and soil saturation at the time of rainfall. Both soil water conductivity and soil saturation are highly heterogeneous in the study area, as shown by Bogner et al. (Chapter 30 in this volume), who partly attribute this heterogeneity to the high stoniness of the studied soils. In the study area, the concentrations of some elements such as organic C and N, Al, and trace metals in stream water increased during rainstorms instead of decreasing due to the dilution effect of the electrolyte-poor rain water (Wilcke et al. 2001). Therefore, we hypothesized that rainstorm water reaches the stream via near-surface lateral flow because the topsoil is richer in organic matter and most nutrients than the deeper soil (Wilcke et al. 2001; Goller 2004; Goller et al. 2006). Near-surface lateral fluxes, explaining the observed distribution of nutrients in the landscape, would be favored by abrupt decreases in saturated water conductivity (K^*) with increasing soil depth possibly causing water ponding at the interface between the differently conductive layers such as the organic layer and the mineral soil.

The K^* determined at 20 locations with the Guelph parameter varied in the organic layer between the orders of 10^{-4} m s^{-1} and 10^{-3} m s^{-1} (Fleischbein 2004). In the mineral topsoil, K^* ranged from 10^{-7} m s^{-1} to 10^{-4} m s^{-1}, and in the subsoil between 10^{-9} m s^{-1} and 10^{-4} m s^{-1}. The large difference between the high K^* in the organic layer and the low K^* in the mineral soil, which was also reported by Huwe et al. (Chapter 29 in this volume), possibly favors water ponding on top of the mineral soil and subsequent lateral flow in the organic layer (Elsenbeer 2001).

As a further test of this hypothesis we analyzed the oxygen isotopic signature of the ecosystem water fluxes during an arbitrarily selected rain storm with standard mass spectrometric methods (Goller et al. 2005). Between 16 August 2000 and 15 August 2001, we monitored the $\delta^{18}O$ values, i.e. the difference between the ratio of $^{18}O/^{16}O$ of the sample and the same ratio of the "Vienna Standard Mean Ocean Water (V-SMOW)", in all ecosystem fluxes. The $\delta^{18}O$ values of rainfall varied between −12.6‰ and 2.1‰. There was no correlation of $\delta^{18}O$ values in rainfall with temperature and rainfall volume indicating that the variation in $\delta^{18}O$ values of rainfall was attributable to air masses of different proveniences. The $\delta^{18}O$ values of throughfall and lateral flow were similar to those in rainfall. Variations in $\delta^{18}O$ values of the soil water and the streamflow were smaller (−9.1‰ to −3.0‰ and −5.8‰ to −8.7‰, respectively) than those of rainfall, throughfall, and lateral flow. This dampening of the $\delta^{18}O$ signal between rainfall and stream water is typical of forested ecosystems (Förstel 1996). The differences in the $\delta^{18}O$ values of rainfall and soil water enable to distinguish the sources of water to the stream during rainstorms.

On 19 September 2000, the day of the selected rainstorm, $\delta^{18}O$ values in stream water increased immediately from a value around −7‰ to −6.0‰ (MC1), −6.4‰ (MC2), and −5.9‰ (MC3) in response to the rainstorm (Fig. 12.2). This demonstrated that during the storm a contribution of water from a different source reached the stream. This water source had a $\delta^{18}O$ value similar to those of rainfall and lateral flow indicating that during elevated rainfall the water flows rapidly in the organic layers to the stream channel paralleling the surface. This finding was confirmed by the higher volume of water in the organic layer than in the upper mineral soil during the rainstorm event (Goller et al. 2005). Our results suggest that water flow paths through the ecosystem are dominated by vertical directions through the soil profile to the stream channels during normal wet conditions, interrupted by short-term flow direction changes to lateral pathways, mainly in the organic layers, during rainstorm events.

12.3.3 Water Budget

We set up the water budget of our catchments to determine evapotranspiration rates and as a prerequisite for element budgets. Results of the element budgets are presented in Chapter 13 in this volume. At MC2, some preliminary geophysical assessments of surface sediment layers had been undertaken suggesting that there were potentially dense, surface-parallel layers in the subsoil (S. Hecht, personal communication). Furthermore, our $\delta^{18}O$ study did not indicate water of unknown sources in the stream. We did not observe surface water flow elsewhere as in the streams, except in MC3 where an unknown but visually seemingly small surface water flow could not be completely captured.

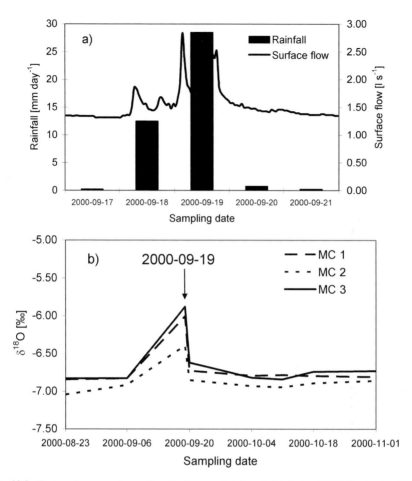

Fig. 12.2 Temporal course of: **a** surface discharge and daily rainfall during 17–21 September 2000, **b** $\delta^{18}O$ values in stream water between 23 August and 1 November 2000. Note that **a** shows a shorter time-span around the rainstorm on 19 September 2000

The mean annual incident rainfall at the three gaging stations between April 1998 and April 2003 ranged between 2410 mm and 2610 mm (Table 12.1). The mean annual throughfall ranged between 1220 mm and 1580 mm and the mean annual stemflow between 23 mm and 25 mm, yielding an interception loss by evaporation from the canopy of 860–1290 mm (Table 12.1). Our measurement of incident precipitation might be low because of the error of the Hellmann-type rain gages and because we did not consider fog, cloud water, and horizontal rain inputs. Furthermore, we only measured the incident precipitation at the lower end of our catchments. The quality of our rainfall measurements was assessed in a comparison study with all other used devices. It was found that all devices yielded similar results within an error of <10% (Rollenbeck et al. 2007). There were only a few weeks in the first monitored year during which the mean throughfall volume of an individual transect was higher than the rainfall volume, indicating some hidden precipitation like fog (Wilcke et al. 2001). Bendix and Rollenbeck (2004) determined horizontal precipitation with a mesh grid according to Schemenauer and Cereceda (1994) at the meteorological station at 1952 m a.s.l. and estimated an additional water input of 12% of the incident precipitation collected with Hellmann-type gages. As mesh grids do not have the same water-catching properties as the forest canopy, this is a rough estimate. The incident rainfall gradient increased with altitude. Measurements taken above an altitude of ca. 2200 m a.s.l. showed a stronger increase in precipitation per 100 m change in altitude at higher areas (18%) than at lower altitudes (2%) where our catchments were located (see Chapter 8 in this volume). Because our catchments were entirely below the first condensation level in the RBSF at ca. 2200 m and because the altitudinal gradient between 1900 m and 2200 m was small, the error of our total water input estimate should also be small.

From the visual comparison of TOPMODEL surface flow rates and surface flow rates based on the weekly manual water level measurements, we judged that our model reflected the real surface flow rates satisfactorily most of the time (Fig. 12.3). Exceptions were some peak surface flow rates, which were underestimated in the model, particularly for MC3 (Fleischbein et al. 2006). Further assessments of the quality of the model fit are given by Fleischbein (2004).

Using an optimized parameterization of TOPMODEL (Fleischbein 2004; Fleischbein et al. 2006) we determined the mean annual surface runoff of between 960 mm and 1100 mm (Table 12.1). During the first four years, we recorded one event of overflow of the weirs in MC1 and two events in each of MC2 and MC3 with our weekly measurements. The overflow times during the hourly measurements of the water levels accounted for 0.2–0.3% of the total measurement period. Thus, during some short peak flow periods our water flow data underestimated the real surface flow. In MC3, we observed several times extremely large surface flow increases within short time periods. We suspect that the catchment received additional water from outside the core catchment during some rainstorms. During these occasions, the assumption of a closed system would have been not maintained for MC3.

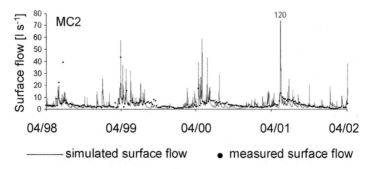

Fig. 12.3 TOPMODEL simulations for the period April 1998 to April 2002 in hourly resolution and manually measured flow rates in weekly resolution (*black dots*) at MC2 (Fleischbein et al. 2006)

If the direct evaporation from the soil was neglected, mean annual transpiration of the three catchments ranged between 290 mm and 570 mm (Table 12.1). The mean annual evapotranspiration ranged between 1310 mm and 1580 mm. The annual mean transpiration accounted for 18–36% of total evapotranspiration. The range of mean total evapotranspiration at our study site is at the upper end of the range of 886–1606 mm year^{-1} reported by Bruijnzeel (2001) for tropical lowland forests. It is much higher than the 310–390 mm year^{-1} reported by Bruijnzeel (2001) for strongly fog-influenced upper mountain forests. However, even for a strongly fog-influenced forest, Hutley et al. (1997) reported an extreme evapotranspiration value of 1260 mm year^{-1}. Motzer (2003) estimated the potential annual evapotranspiration to be only 561 mm for the same study area. His estimate is based on a data set covering less than a year during a relatively dry period (see also Chapter 16 in this volume). Our high estimates of annual evapotranspiration are mainly attributed to high interception losses (Fleischbein et al. 2005). The range of mean annual transpiration (neglecting direct evaporation from the soil) of 290–570 mm (Table 12.1) was lower than that reported for a Puerto Rican rain forest of 632–815 mm (Schellekens et al. 2000) but in the range of the annual transpiration in other tropical mountain forests of 170–845 mm (Bruijnzeel 2001).

12.4 Conclusions

The study catchments show a particularly high interception loss because of the strong insolation, additional advective energy, and the dense canopy. The interception loss is partly controlled by the canopy density, as indicated by the correlations between leaf area index and direct throughfall coefficient and interception loss. Epiphytes do not seem to play an important role for interception loss. The stream flow of the studied catchments responds quickly to rainstorms because of fast near-surface flow. This has been proven by the fact that the $\delta^{18}O$ signal in stream water quickly increased to that of near-surface soil solution during a storm event. As the near-surface water passes by soil regions which are particularly rich in organic matter and nutrients, rainstorms enhance C and nutrient export from the study catchments. The water budget confirmed that the total evapotranspiration ranged at the upper end of that reported in the literature for comparable forests. Total evapotranspiration is mainly driven by the interception loss accounting for 64–82% of the total evapotranspiration (if direct evaporation from the soil surface is neglected).

Acknowledgments We thank Karin Wagner, Melanie Leng, Heinz-Jürgen Tobschall, Klaus Knoblich, Manfred Küppers, and numerous German and Ecuadorian student helpers for their contribution to this work.

Chapter 13
Nutrient Status and Fluxes at the Field and Catchment Scale

W. Wilcke(✉), S. Yasin, K. Fleischbein, R. Goller, J. Boy, J. Knuth, C. Valarezo, and W. Zech

13.1 Introduction

Rainfall, litter fall, and organic matter turnover are the major drivers of nutrient fluxes in forest ecosystems. During its passage through the forest, the quantity and chemical composition of water – which enters the forest via rainfall and clouds – changes (Parker 1983). The chemical quality of throughfall and stemflow is controlled by the intensity of rainfall, dry deposition, and soil fertility (Parker 1983; Schaefer and Reiners 1990). The few studies on litter leachate, i.e. the water percolating through the organic layer, in tropical forests indicate that the concentrations of all nutrients increase compared with rainfall and throughfall (e.g., Steinhardt 1979; Hafkenscheid 2000; Wilcke et al. 2001a). In stream water, nutrient concentrations are lower than in litter leachates, except for elements which are released by weathering (Bruijnzeel et al. 1993; McDowell 1998).

The role of organic matter turnover for nutrient fluxes depends on its storage of nutrients and the nutrient turnover times. The turnover of organic matter is comparatively fast for fine litter, intermediate for coarse woody debris, and slow for living trees (Clark et al. 2002). In tropical forests on strongly weathered or inherently nutrient-poor soils, the soil organic matter contains most plant available nutrients (Cuevas and Medina 1986; Grubb 1995; Kauffman et al. 1998). The plant availability of N, P, and S in organic layers depends largely on the mineralization rates, while base metals are mainly bound in exchangeable form. A common approach when assessing turnover times of organic matter in forests is to relate organic matter storage in the soil organic layer to litterfall as a measure for the proportion of litter decomposed in a given period (Vogt et al. 1986; Proctor 1987).

The catchment approach described by Wilcke et al. in the previous chapter (Chapter 12) not only allows for determining evaporation and transpiration but is also useful to set up element budgets which enable a better understanding of biogeochemical processes (Bruijnzeel 1990; Likens and Bormann 1995; Matzner 2004). Catchment budgets of elements indicate directions of ecosystem development in response to external stresses, such as nutrient and pollutant inputs, climate change or direct anthropogenic interferences. The knowledge of this response is necessary to predict the future development of the Andean landscapes and to develop more sustainable land-use options as are presently in effect.

In order to set up complete catchment budgets, we determined the size and chemical quality of all ecosystem fluxes and budgeted inputs and outputs in small catchments of the RBSF, where the water budget was also determined. Again we focused on three 8–13 ha catchments in the area between 1900 m and 2200 m a.s.l. as described in Chapter 12 (see also Fig. 1.2 in Chapter 1).

13.2 Methods

To determine deposition rates and concentrations and fluxes of elements in rainfall, throughfall and stemflow, we used the same collectors as described in Chapter 12. The dry deposition was estimated with the canopy budget model of Ulrich (1983). Soil solution was collected below the organic layer and at 0.15 m and 0.30 m depth in the mineral soil with three lysimeters at each depth of each measurement transect (see Fig. 1.2 in Chapter 1). The single samples were bulked to a composite sample per rainfall gaging station or measurement transect for each type of collector. Rainfall, throughfall, stemflow, and litter leachate were sampled since April 1998, soil solution since May 2000. To determine nutrient export with surface flow, we collected water samples from the center of the stream. The instrumentation is described by Goller et al. (2006). Samples were collected weekly.

Volume-weighted means (VWM) of element concentrations were calculated for rainfall, throughfall, and stemflow, and flow-weighted means (FWM) for surface flow. Element fluxes were calculated by multiplying VWM or FWM concentrations with annual water fluxes. In soil, no element fluxes were calculated because of the unknown size of water flow. Medians were used for comparing the element concentrations.

Each of the five transects was equipped with three litter collectors. Samples were collected weekly and dried at 40 °C in an oven. The samples were bulked to a monthly sample per transect (Wilcke et al. 2002). To determine the coarse woody debris, we selected 16 plots of 400 m^2 in MC2 and quantified the mass and chemical properties of all fallen and standing dead wood with a diameter >0.1 m (Wilcke et al. 2005). All organic samples were digested with concentrated HNO_3. In all water samples and organic matter digests we determined P, K, Ca, and Mg concentrations with AAS or ICP/OES. In the rainfall, throughfall, and stemflow samples, additionally N and Cl$^-$ concentrations – needed for Ulrich's canopy budget approach – were measured with a continuous-flow analyzer and pH with a glass electrode. The N concentrations in solid phase samples were determined with an elemental analyzer.

13.3 Results and Discussion

13.3.1 Deposition and Canopy Interaction

13.3.1.1 Element Concentrations

The volume-weighted mean (VWM) pH of the rainfall was 5.3, ranging over 5.2–5.7 between the five study years and 5.1–5.3 between the three gaging stations. The pH of rainfall in our study was similar to that in other mountain forests of the northern Andes (4.4–5.6; Steinhardt 1979; Veneklaas 1990). Rainfall was slightly more acid in most years than was expected in equilibrium with atmospheric CO_2, indicating that there were some inputs of acids. Mineral acids might originate from forest fires in the Amazon basin (Fabian et al. 2005; see also Chapter 22 in this volume). Furthermore, organic acids are released from the canopy of the Amazon lowland forest and might reach our study site (Forti and Neal 1992). During the passage through the canopy, the pH was buffered to a VWM of 6.2 in throughfall, with little variation among study years (5.9–6.5) and the five study transects (6.1–6.5). The throughfall at the Venezuelan site of Steinhardt (1979) was more acid. Thus, the canopy of the study forest had a high buffer rate and consumed most deposited H^+.

The VWM concentrations of N, P, K, Ca, and Mg in incident rainfall were similar or in the lower half of the range of concentrations reported by Hafkenscheid (2000) for a range of tall-statured lower mountain rain forests mainly in Central and South America (Table 13.1). The reason for low element concentrations in rainfall was the location far away from anthropogenic emission sources and from the sea. The concentrations of the base metals in rainfall varied markedly between the study years (Table 13.2).

The concentrations of the studied nutrients increased in throughfall relative to rainfall. In throughfall, VWM concentrations of all elements were at the upper end or above the range of concentrations in lower mountain rain forests (Hafkenscheid 2000). The reasons for the "throughfall enhancement" (Parker 1983) include concentration effects because of evaporation of intercepted water, particulate and gaseous dry deposition, hidden depositions with cloud water which does not reach the soil surface because of complete evaporation, and canopy processes. The most important canopy processes are leaching from the plant tissue, N fixation, decomposition of plant debris in the crown, and insect excretions (Parker 1983; Schaefer and Reiners 1990). The particularly strong increase in K concentrations after the passage of rainfall through the forest canopy has frequently been observed. It is attributable to the high leachability of K from leaves (Tukey 1970).

Table 13.1 Ranges of volume-weighted mean concentrations of major nutrients in incident rainfall, throughfall, and stemflow at three gaging stations for incident rainfall near each of three microcatchments (*MC1–MC3*) and at three stemflow and five throughfall collection sites in the catchments in five hydrological years (May–April) between 1998 and 2003

		MC1	MC2/1	MC2/2	MC2/3	MC3	Literature[a]
		(mg l^{-1})					
Incident rainfall	N[b]	0.32–0.48	0.36–0.44	–	–	0.31–0.47	0.17–0.85
	P[b]	0.01–0.05	0.01–0.05	–	–	0.02–0.10	0.01–0.07
	K	0.18–0.37	0.14–0.72	–	–	0.18–0.33	0.09–0.38
	Ca	0.10–0.13	0.09–0.73	–	–	0.10–0.12	0.10–0.79
	Mg	0.05–0.06	0.03–0.31	–	–	0.03–0.05	0.03–0.33
Throughfall	N[b]	1.2–2.2	1.0–1.4	1.0–1.3	1.2–2.3	1.7–2.3	0.33–1.4
	P[b]	0.33–0.61	0.06–0.14	0.01–0.18	0.14–0.38	0.60–0.95	0.11–0.12
	K	8.2–14	3.8–8.1	3.8–7.6	4.8–12	11–20	2.9–5.5
	Ca	1.2–2.4	0.33–1.64	0.36–1.4	1.0–1.6	2.1–3.5	0.55–1.6
	Mg	0.57–1.1	0.21–0.62	0.26–0.5	0.6–1.1	1.5–2.4	0.26–0.58
Stemflow	N[b]	1.2–2.1	1.1–1.6	–	–	1.7–2.2	–
	P[b]	0.36–0.69	0.14–0.29	–	–	0.53–0.82	–
	K	9.7–12	5.4–10	–	–	9.8–18	–
	Ca	1.2–2.3	0.45–1.4	–	–	1.8–3.1	–
	Mg	0.56–0.87	0.24–0.53	–	–	0.92–1.5	–

[a] Compilation for a range of tropical mountain forests (Hafkenscheid 2000).
[b] N and P were not measured in the first hydrological year (1998/1999).

Table 13.2 Ranges of annual deposition rates of major nutrients in incident rainfall, throughfall, and stemflow at three gaging stations for incident rainfall near each of three microcatchments (*MC1–MC3*) and at three stemflow and five throughfall collection sites in the catchments and according to Ulrich (1983) calculated dry deposition in five hydrological years (May–April) between 1998 and 2003

		MC1	MC2/1	MC2/2	MC2/3	MC3	Literature[a]
		(kg ha^{-1} year^{-1})					
Incident rainfall	N[b]	8.6–13	9.3–9.8	–	–	8.2–13	6.5–18
	P[b]	0.29–1.3	0.21–1.4	–	–	0.43–2.6	0.05–1.1
	K	4.8–9.2	3.6–16	–	–	5.0–8.8	2.6–14
	Ca	2.7–3.5	2.4–16	–	–	2.6–3.3	3.6–28
	Mg	1.2–1.6	0.86–6.7	–	–	0.81–1.3	1.3–5.2
Throughfall	N[b]	13–27	16–17	14–20	17–32	29–36	1.0–3.0
	P[b]	3.5–7.3	0.96–2.0	0.26–2.6	1.9–5.3	12–13	0.43–2.5
	K	86–156	65–157	70–102	65–164	148–279	50–87
	Ca	18–25	5.6–24	6.7–19	18–22	29–50	1.3–19
	Mg	9–12	3.6–9.0	4.7–6.8	10–15	23–34	1.9–11
Stemflow	N[b]	0.30–0.47	0.30–0.35	–	–	0.37–0.55	–
	P[b]	0.09–0.18	0.04–0.07	–	–	0.10–0.19	–
	K	2.4–3.2	1.4–2.1	–	–	2.0–4.1	–
	Ca	0.29–0.40	0.13–0.31	–	–	0.40–0.72	–
	Mg	0.14–0.22	0.07–0.12	–	–	0.20–0.34	–
Dry deposition	N[b]	5.4–24	6.7–19	–	–	2.0–18	–
	P[b]	0.39–1.8	0.15–2.7	–	–	0.22–3.6	–
	K	5.5–13	3.0–12	–	–	1.5–12	–
	Ca	1.8–7.4	1.7–12	–	–	0.65–4.3	–
	Mg	0.74–4.1	0.61–5.2	–	–	0.20–1.7	–

[a] Compilation for a range of tropical mountain forests (Hafkenscheid 2000).
[b] N and P were not measured in the first hydrological year (1998/1999).

13.3.1.2 Element Fluxes

The rainfall deposition of N and P at our study site was similar to or above the range of data collected by Hafkenscheid (2000), that of the base metals was consistently at the lower end (Table 13.2). Throughfall deposition for all elements was higher than the range of values given by Hafkenscheid (2000) reflecting their elevated concentrations in throughfall. For all elements, throughfall deposition was higher than rainfall deposition, except for H^+. Protons were buffered during the passage through the canopy, mainly resulting in the release of base metals. Dry deposition for all elements had a size similar to that for bulk deposition with incident rainfall, indicating dust and gas inputs into our study area.

The weekly Ca deposition was highly variable during the monitored period (Fig. 13.1). Rainfall deposition at gaging station 2 in the hydrological years 1998/1999, 2000/2001, 2001/2002, and 2002/2003 was similar (2.4–4.4 kg ha^{-1} year^{-1}) and much lower than in 1999/2000 (16 kg ha^{-1} year^{-1}). The rainfall deposition of Ca in incident rainfall was closely correlated with that of Mg ($r = 0.94$) and K ($r = 0.75$, data from gaging station 2). Rainfall volumes only explained 10% of the variation in weekly Ca deposition rates, indicating that differences in Ca deposition were not attributable to dilution/concentration effects. The variation in annual rainfall deposition rates of Ca (and K and Mg) at our site during the five study years could also not be explained by seasonality. Therefore, we tested whether the interannual variation in deposition rates is related with long-term climatic cycles. As an indication of long-term climatic cycles such as the El Niño Southern Oscillation (ENSO)-phenomenon, we used the Pacific Ocean surface temperature. The results of a

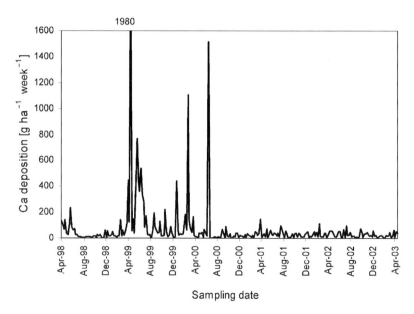

Fig. 13.1 Temporal course of the weekly Ca deposition rates at incident rainfall gaging station 2 between April 1998 and April 2003

Chi-square test suggested a 94–98% similarity of the weekly Ca deposition rates to the changes in the Pacific Ocean surface temperature for March 1998 to December 2000. Although the underlying mechanisms are unclear, this suggests that climatic variations of the ENSO cycle might have an impact on the nutrient inputs at our study site. However, to test this hypothesis, a data set covering several ENSO cycles would be needed.

In three of the five monitored hydrological years, the mean throughfall deposition rates of Ca were markedly higher than the total deposition (bulk deposition with incident rainfall + dry deposition (data not shown) indicating that Ca was leached from the canopy in these years. However, between 1999 and 2001 (the two years with highest Ca input), Ca was retained by the canopy, suggesting a higher Ca requirement of the forest trees or the organisms in the canopy.

13.3.2 Litterfall and Coarse Woody Debris

The annual fine litter fall was similar at all study transects. On average of all five years it ranged between $9.1\,t\,ha^{-1}\,year^{-1}$ and $12\,t\,ha^{-1}\,year^{-1}$ among the five study transects. Furthermore, it did not vary much among the study years (annual means of all five transects were $9.1–11\,t\,ha^{-1}\,year^{-1}$). The same was true for mass-weighted element concentrations and element deposition rates with litterfall (Table 13.3). The annual litterfall was at the upper end of the range for various tropical mountain forests globally (Vitousek 1984; Bruijnzeel and Proctor 1995; Hafkenscheid 2000,

Table 13.3 Ranges of mass, mass-weighted mean nutrient concentrations, and nutrient deposition of fine litterfall at five litter collection sites in the three studied catchments ("microcatchments", *MC1–MC3*) between 1998 and 2003. Note: mass data between May 1999 and April 2000 were lost and substituted as means of the monthly values of the three following years

	MC1	MC2/1	MC2/2	MC2/3	MC3	Literature
Litter fall (t ha^{-1} year^{-1})						
Mass	8.5–13	9.7–12	7.9–11	8.6–13	9.2–14	0.5–12[a,b,c]
Concentration (mg kg^{-1})						
N	16–19	16–20	15–20	17–21	18–22	6.0–15[c]
P	1.4–1.9	0.95–1.2	0.85–1.1	1.2–1.5	1.5–1.6	0.17–0.95[c]
K	9.1–11	6.7–9.2	6.1–7.2	8.0–10	3.7–9.1	0.90–8.9[c]
Ca	18–20	10–12	10–13	13–17	16–17	1.9–22[c]
Mg	4.0–5.8	3.0–3.8	3.2–4.5	4.2–5.6	4.6–5.7	1.8–4.4[c]
Deposition (kg ha^{-1} year^{-1})						
N	165–204	173–225	117–185	152–218	182–280	29–101[c]
P	12–20	10–14	6.8–9.7	11–16	14–23	0.70–7.7[c]
K	77–131	65–112	49–76	15–108	76–130	4.6–59[c]
Ca	154–250	103–126	81–118	117–188	149–224	6.7–119[c]
Mg	49–57	31–43	25–42	37–55	48–65	6.3–25[c]

[a] Vitousek (1984).
[b] Bruijnzeel and Proctor (1995).
[c] Compilation for a range of tropical mountain forests (Hafkenscheid 2000).

Table 13.4). The forest produced as much biomass as found in tropical lowland forests (7.5–13.3 t ha^{-1} year^{-1}; Proctor 1987).

The mass-weighted mean concentrations of all five major nutrients in fine litterfall were at the upper end or greater than the range of concentrations reported by Hafkenscheid (2000) for other tropical mountain forests (Table 13.3). The mass-weighted mean concentrations of all elements in fine litterfall differed considerably between transects. The concentrations of K, Mg, and P were higher on transects MC1, MC2/3, and MC3 than on transects MC2/1 and MC2/2. The concentration of Ca was higher on transects MC1 and MC3 than on all other transects. The pH of the organic layer was markedly different among the five study transects. The organic layers at the transects in MC2 were more acid (pH in H$_2$O: 4.4–4.7) than in MC1 and MC3 (6.2–6.3). There were significant correlations between the concentrations of Ca ($r = 0.99$), Mg ($r = 0.87$), P ($r = 0.97$), and Zn ($r = 0.85$) in the organic layer and their respective mean concentrations in the litterfall of the first monitored year. The finding that the concentrations of P, Ca, and Mg in litterfall were closely correlated with their respective concentrations in the organic layer indicates that the uptake of these elements by the vegetation is affected by the soil values. Furthermore, the uptake of these elements is greater and the cycling is faster on soils with higher pH in the organic layer.

Fine litterfall was the most important source of flux to soil for all elements except for K. In the first monitored year, 16–29 kg ha^{-1} N, 4.0–8.6 kg ha^{-1} P, 77–168 kg ha^{-1} K, 16–29 kg ha^{-1} Ca, and 7.2–21 kg ha^{-1} Mg reached the soil with throughfall + stemflow (Wilcke et al. 2001a). In the same year, the contribution of litterfall to the total flux (= throughfall + stemflow + litterfall) to soil increased in the order, K (36) < P (67) < Mg (81) < Ca (87) < N (90).

The average mass of coarse woody debris was 9.1 t ha^{-1}; 40% of the CWD mass did not have soil contact. The individual coarse woody debris masses

Table 13.4 Quotients of the storages of mass and nutrients in the Oi horizon to the annual litterfall (K_{Oi}) and of the storages of mass and nutrients in the whole organic layer (Oi+Oe+Oa horizons, K_{OL}) to the annual litterfall at the five sampling locations

Location	Mass	N	P	K	Ca	Mg
		K_{Oi} (years)				
MC1	0.94	0.95	0.72	0.47	0.89	0.68
MC2/1	0.96	0.89	0.54	0.45	0.85	0.80
MC2/2	1.1	1.1	0.95	0.49	0.95	1.2
MC2/3	1.5	1.4	0.72	0.41	1.9	1.5
MC3	1.0	0.89	0.70	0.52	0.95	1.1
		K_{OL} (years)				
MC1	8.8	12	7.4	3.8	9.3	6.2
MC2/1	13	15	13	6.6	5.2	4.5
MC2/2	16	15	16	20	5.7	7.7
MC2/3	15	18	12	7.0	10	7.8
MC3	9.9	11	7.9	5.3	9.0	9.0

were highly variable among the nine plots (0.4–23 t ha^{-1}). We did not detect any significant relationship between topographic position or stand properties and mass and C and nutrient storage of coarse woody debris. The coarse woody debris contributed <1.8% to the nutrient storage in aboveground dead biomass (i.e. coarse woody debris + organic layer including Oi, Oe, Oa horizons; Wilcke et al. 2005).

13.3.3 Organic Matter Turnover

We calculated the quotient of the storage of mass or nutrient in the Oi horizon to the annual flux of mass or nutrient by litterfall of the first monitored year (K_{Oi}) being most likely similar to the inverse of the K_L value used in the literature (e.g. Edwards 1982; Heaney and Proctor 1989; Smith et al. 1998). We also calculated the quotient of storage of a nutrient in the whole organic layer to the annual flux of the same nutrient by litterfall (again only first monitored year, K_{OL}, Table 13.4). The K_{Oi} and K_{OL} values represent the mean residence time of organic matter or an element in the Oi horizon or the whole organic layer. The residence time integrates all dissipation processes (i.e., release of CO_2 and other gases, leaching, uptake by plants and animals) and may be biased by retention of element reaching the soil via throughfall and stemflow and because root litter input is not considered. The K_{Oi} values of the organic matter mass in our study (0.9–1.5 years; Table 13.4) are at the upper end or greater than the range of 0.6–1.2 reported for tropical mountain forests elsewhere (Edwards 1982; Heaney and Proctor 1989; Bruijnzeel et al. 1993). The mean K_{OL} value of the mass is 11 times greater than the K_{Oi} value. This indicates that in our study forest the turnover of organic matter in the organic layer is slower than in the other tropical mountain forests.

The estimated mean decomposition constant of coarse woody debris was 0.09 year^{-1}. The annual nutrient release from coarse nutrient debris contributed at most 1.5% to the totally plant-available nutrients in the forest soil (nutrient input with throughfall and stemflow + release from organic layer + release from coarse woody debris). Our results demonstrate that in the studied mountain forest the coarse woody debris contribution to nutrient release is small. This might be specific for forests with a thick organic layer.

13.3.4 Soil Solution

The average median of the five hydrological years (1998–2003) of the pH in litter leachates on transects MC2/1 (4.7) was lower than the VWM pH in throughfall (6.3) and those at MC2/2 (5.7), MC2/3 (5.9), and MC1 (6.0) were slightly lower than the VWM pH in throughfall (6.2, 6.1, and 6.5, respectively), indicating that additional acids [i.e. acidic components of dissolved organic matter (DOC)] were leached from the organic layer. On transect MC3, the average median pH was even higher in litter leachate (6.5) than in throughfall (6.1). Thus, the pH of the litter leachate reflects the pH of the organic layer illustrating that the organic layer of the catchments MC1 and MC3 are better buffered than in the more acid MC2.

The average medians of all major nutrient concentrations in litter leachate were higher than the range of concentrations in two Jamaican upper montane cloud forest types except for P at all study transects and the base metals at transects MC2/1 and MC2/2 (Hafkenscheid 2000; Table 13.5). Nutrient concentrations were highest at the least acid transect in MC3 and decreased with increasing acidity to the most acid transect MC2/1, reflecting the different nutrient concentrations in the organic layer and decreasing biological turnover with increasing acidity. The litter leachate had the highest concentrations of N, Ca, and Mg of all ecosystem fluxes, while the highest concentrations of P and K occurred in throughfall. These results demonstrate that there was a considerable variation in nutrient availability at the small scale.

Nutrient concentrations in the mineral soil solution at MC3 were at the higher end or higher than the comparison values from Jamaica in Hafkenscheid (2000), while they were at the lower end or lower at MC2.1, again illustrating the marked differences in nutrient availability in relation to pH. In the mineral soil solutions, the concentrations of all nutrients decreased from a depth of 0.15 m to 0.30 m at most study sites, indicating that they were retained in the mineral soil or taken up

Table 13.5 Ranges of the medians of the concentrations of major nutrients in litter leachate and mineral soil solution at 0.15 m and 0.30 m depth at five collection sites (*MC*) in the catchments in five hydrological years (May–April) for litter leachate (1998–2003) and three hydrological years for soil solutions (2000–2003). *n.d.* Not detected

		MC1	MC2/1	MC2/2	MC2/3	MC3	Literature[a]
		(mg l^{-1})					
Litter leachate	N	2.3–4.8	2.0–2.3	1.7–4.2	2.6–6.6	3.8–7.7	0.81–1.1
	P	n.d.–0.21	–	–	n.d.–0.13	0.22–0.55	<0.04–0.53
	K	3.2–12	2.6–4.0	4.2–13	9.9–18	6.4–30	2.9–4.1
	Ca	4.1–6.1	0.92–2.9	0.82–2.2	2.8–5.3	6.4–9.2	1.7–2.5
	Mg	1.3–2.2	0.69–1.8	0.67–1.6	1.3–2.9	4.1–5.1	0.93–1.1
Soil solution 0.15 m depth	N	1.4–1.8	0.90–1.0	1.3–2.0	0.98–1.4	1.7–3.7	0.70–0.97
	P	–	–	–	–	–	0.03–0.05
	K	0.44–1.3	0.09–0.54	0.31–1.2	0.10–0.22	0.85–1.5	0.85–0.97
	Ca	1.3–2.2	0.12–0.21	0.16–0.19	0.99–1.4	2.3–3.7	0.46–0.57
	Mg	0.37–0.77	0.07–0.12	0.43–0.68	0.29–0.34	1.4–2.8	0.54–0.84
Soil solution 0.30 m depth	N	0.98–1.5	0.58–0.62	1.0–1.7	0.94–1.2	1.4–2.9	0.38–0.39
	P	–	–	–	–	–	0.01–0.02
	K	0.68–0.85	0.08–0.10	0.21–1.5	0.14–0.37	0.35–0.57	0.37–0.45
	Ca	0.75–1.0	0.08–0.09	0.19–0.33	0.90–1.2	1.5–2.6	0.11–0.38
	Mg	0.27–0.39	0.04–0.08	0.28–0.65	0.24–0.30	1.5–2.5	0.41–0.42

[a] Data for two upper montane cloud forests in Jamaica (Hafkenscheid 2000). Our soil solution values at 0.15 m depth correspond to soil water in the A horizon and those at 0.30 m depth to drainage water from the B horizon in Jamaica.

by (relatively few) roots and associated mycorrhiza. For N, it is also possible that denitrification favored by frequent water-logging decreased concentrations in soil solution. However, we did not measure gaseous N fluxes.

13.3.5 Surface Flow

Although the mineral soils were strongly acid (pH in H_2O of 3.9–5.3 in the A horizons of the five study transects; Wilcke et al. 2001a), the flow-weighted mean (FWM) pH of the stream water was 6.1 (for all three catchments and all five years). On average for all five years, the stream water of MC3 had the lowest pH (5.7) while that of MC1 and MC2 had a similar and higher pH (6.6–6.7). In these two catchments, there were some sampling dates with a pH in stream water below 6.0 related with storm events. The higher pH of the stream water than of the mineral soil indicated that H^+ were buffered in the subsoil by mineral weathering and possibly also consumed by chemical reduction processes. While the stream water of MC3 had a pH which is typical of the equilibrium with atmospheric CO_2, the pH of stream water of MC1 and MC2 was above this value, suggesting that stronger chemical reduction processes producing alkalinity occurred in these two catchments compared with MC3. During storm events, water traveling rapidly in macropores (which is more acid than matrix flow because of less equilibration time with the mineral soil) reached the stream (Cozzarelli et al. 1987; Neal et al. 1989; McDowell and Asbury 1994; see Chapter 12 in this volume). A rain forest in Puerto Rico showed a similar course of pH from rainfall to stream water (McDowell 1998).

In the stream water, the FWM N and P concentrations were higher than the range of values in the literature used for comparison, FWM K concentrations were similar, and FWM Ca, and Mg concentrations were at the lower end (Table 13.6). The higher N and P concentrations in stream water of our study site compared with the literature might indicate that the demand of the vegetation for N and P is smaller. Much lower Ca and Mg concentrations at our study site than in the Puerto Rican stream water of McDowell and Asbury (1994) are explained by the lower metal release by weathering of the bedrock at our geologically old site than at the geologically younger, magmatic catchment in Puerto Rico. Furthermore, the nutrient deposition at the Puerto Rican site was greater than at our site because of the proximity of the sea. Elevated K concentrations in stream water were the consequence of the weathering of mica and illites in parent rock (e.g., phyllite) and soil (Schrumpf et al. 2001).

There was considerable variation in nutrient concentrations among the five study years except for N. The nutrient export paralleled the variations in total deposition as indicated by close significant correlations between annual total deposition (measured bulk, calculated dry deposition) and nutrient export with surface runoff ($r > 0.6$ for all nutrients). Thus, changing inputs from the atmosphere have an impact on nutrient export with surface runoff.

High-flow events were associated with increased total N and P concentrations in stream water (Fig. 13.2, P not shown). The concentrations of the base metals in

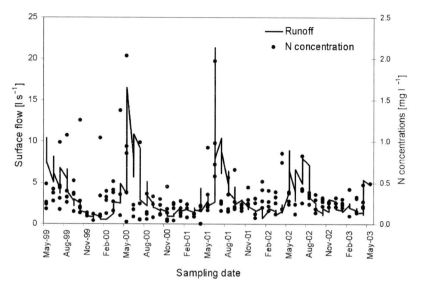

Fig. 13.2 Temporal course of the mean runoff and total N concentrations in stream water of the three microcatchments

contrast did not change irrespectively of flow condition (data not shown). Goller et al. (2006) also reported increases in dissolved organic C and N concentrations and decreased pH during high-flow conditions. The finding that there was no dilution of the N concentrations with increasing electrolyte-poor rainfall suggests that the fast percolating near-surface water after strong rainstorms on already presaturated soils carries organic matter and organically bound nutrients to the stream (Goller et al. 2005; see also Chapter 12 in this volume). However, this was not the case for the base metals, suggesting that the higher leaching during rainstorms of these metals which accumulate in the organic layer and topsoil (Wilcke et al. 2002) was compensated by other sources during baseflow conditions, such as mineral weathering and deep leaching.

13.3.6 Catchment Budget

The catchment budget of elements was calculated as the difference between inputs [measured rainfall plus (according to the model of Ulrich 1983) calculated dry deposition] and outputs (export with surface runoff). Our N budget is incomplete because we did not measure gaseous N losses (Wilcke et al. 2001b).

The mean net nutrient budget of the three studied catchments was positive for most studied elements in most years (Fig. 13.3). There was considerable variation in the net budgets among hydrological years. The apparent accumulation of N is probably at least partly compensation by denitrification losses. The accumulation of P is attributable to the strong sorption of P to Fe oxides in soil and the precipitation of Al phosphates. The largest variation occurred for the base metals. Calcium

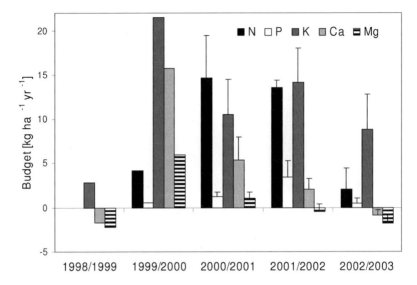

Fig. 13.3 Mean budget of N (excluding gaseous N fluxes), P, and base metals (K, Ca, Mg) of the three microcatchments between 1998 and 2003. N and P were not determined in 1998/1999 and only in MC2 in 1999/2000. Error bars show standard deviations. In 1998/1999 and 1999/2000 nutrient budgets could only be calculated for MC2. Negative values indicate net depletion and positive values net accumulation

and Mg were depleted in the hydrological years 1998/1999 and 2002/2003 and Mg also in 2001/2002. In these years, our study catchments lost base metals presumably because of higher weathering than deposition rates. However, the net base metal losses – similar to the base metal fluxes with surface runoff (Table 13.6) – were lower than in the geologically younger Puerto Rican catchment studied by McDowell and Asbury (1994), indicating that the release of base metals by weathering is probably lower at our study sites.

Interestingly, the most pronounced accumulation of base metals was observed in the hydrological year 1999/2000, which is at the same time the year of the highest base metal input (Fig. 13.3). As there was no marked increase in base metal concentrations in the litter leachate during this year, this suggests that the deposited K, Ca, and Mg were retained in the vegetation. We hypothesize that the base metals were taken up by the vegetation because of increased growth during these years.

13.4 Conclusions

The studied catchments received similar high element depositions via bulk and dry deposition. We observed strong interannual variation in deposition rates of nutrients. This nutrient input has an effect on the whole catchment budget. Thus, changing

Table 13.6 Ranges of surface flow, flow-weighted mean concentrations in stream water, and element export with surface runoff in the three catchments (*MC1–MC3*) between May 1998 and April 2003

		MC1	MC2	MC3	Literature
		(mg l^{-1})			
Flow-weighted mean concentrations	N	0.34–0.47	0.26–0.41	0.29–0.47	0.22–0.24[a]
	P	n.d.–0.06	n.d.–0.13	0.01–0.08	0.002[a]
	K	0.26–1.2	0.21–0.56	0.22–1.8	0.03–1.4[a,b]
	Ca	0.50–1.6	0.36–1.2	0.33–1.2	n.d.–19[a,b]
	Mg	0.32–0.60	0.35–0.55	0.22–0.59	n.d.–5.1[a,b]
		(kg ha^{-1} $year^{-1}$)			
Export	N	3.0–5.1	3.0–4.9	2.8–4.9	4.3–9.4[a]
	P	0.04–0.68	0.02–1.6	0.10–0.87	0.03–0.08[a]
	K	2.5–12	2.2–6.1	2.5–18	4.9–17[a]
	Ca	5.8–16	3.7–12	3.5–13	44–96[a]
	Mg	2.7–6.0	4.2–6.0	2.5–6.2	28–63[a]

[a] Nutrient export from a geologically young magmatic tropical rain forest catchment in Puerto Rico (McDowell and Asbury 1994).
[b] Surface waters of tropical forests (Forti and Neal 1992).

deposition because of climatic changes or increased anthropogenic activity likely results in a drift of the studied tropical mountain forest.

The study forest has similar litter productivity as tropical lowland forests. The annual litter is completely turned over within less than 1.5 years and the whole organic layer at the scale of 2–3 decades. The total coarse woody debris has a similar size as the annual litterfall and is turned over at a similar rate as the whole organic layer. Coarse woody debris therefore contributes little to nutrient supply via mineralization.

There was a considerable variation in the quality of ecosystem solutions such as throughfall, litter leachate, and soil solution among the five study transects, coinciding with the variation in soil solid phase properties described in Chapter 9 in this volume. This corroborates the previous conclusion that the studied mountain forests provide a multitude of different ecological niches at the small scale.

Ecosystem losses of N and P were influenced by fast near-surface flow in response to few rainstorms.

Acknowledgments We thank Uwe Abramowski, Christoph Bengel, Timo Hess, and numerous Ecuadorian and German student helpers for their support.

Chapter 14
Biotic Soil Activities

S. Iost, F. Makeschin(✉), M. Abiy, and F. Haubrich

14.1 Introduction

There are few studies investigating the different carbon pools and fluxes of tropical soils. Even more restricted is our knowledge of tropical mountain forest soils that cover about 11% of the total tropical forest area (Doumenge et al. 1995). In these ecosystems, decomposition of root and leaf litter is reduced in comparison with lowland tropical forest (Coûteaux et al. 2002). Thus, tropical mountain forests contribute substantially to carbon sequestration in soils. An important component of ecosystem carbon cycling is soil CO_2 efflux from the soil to the atmosphere, which results from the activity of soil microbes and root respiration. Globally CO_2 efflux amounts to approximately 80 Pg C year^{-1}. Tropical and subtropical evergreen broadleaved forests contribute the largest parts of approx. 22 Pg C year^{-1} (Raich et al. 2002). Through human activity, especially burning of fossil fuels and conversion of forests into other land use forms, about 7 Pg C enter the atmosphere additionally per year as carbon dioxide. It is assumed that increasing CO_2 in the atmosphere are a main cause for higher temperatures observed (Rustad 2001), but as long as it remains unclear how carbon pools and fluxes react on this rise of the temperatures, it is also remains unclear how the global carbon cycle will be altered by climate change in the long term. The objective of this study was to quantify total carbon and nitrogen contents, microbial biomass and total, heterotrophic and root respiration between 1050 m and 3060 m along an altitudinal gradient in natural forests of the research area. Details of the study sites are given by Moser et al. (Chapter 15 in this volume).

14.2 Methods

Total soil respiration (TSR) was measured fortnightly at five sites along the altitudinal gradient between July 2003 and September 2004, and at 1050 m, 1890 m and 3060 m measurements continued until August 2005. For in situ measurements a portable closed chamber system connected to an infra-red gas analyser (EGM-4, SRC-1; PP Systems) was used by 16 continuous points per plot. The contribution of roots to total soil respiration was determined using the trenching method at 1050 m, 1890 m and 3060 m (September 2004 until August 2005). At each plot eight 50×50 cm subplots were established by trenching the soil along the sides of the subplots to a depth of 50 cm. At each subplot two collars were installed accordingly to TSR measurements. As control TSR measurements at the same altitudes were used. Partitioned soil CO_2 efflux (R_H) was measured at the same dates and within two hours of the TSR measurements at the respective plots. Root contribution (RC%) was calculated as the difference between TSR and R_H after correction of the latter by the amount of carbon that resulted from the decomposition of the trenched roots. Different authors have discussed the constraints of the trenching method and the interested reader is referred to these (Hanson et al. 2000; Rey et al. 2002; Lavigne et al. 2003; Lee et al. 2003). Total soil carbon (SOC) and nitrogen (TN) were determined by the combustion method (Vario El; Heraeus). Microbial biomass C and N were determined by the chloroform fumigation–extraction method (Vance et al. 1987) and calculated using a k_{EC} factor of 0.43 (Martens 1995) and a k_{EN} factor of 0.45 (Jenkinson et al. 2004). Microbial community structure was determined by analysing phospholipid fatty acid patterns in the soil (Zelles 1999).

14.3 Results and Discussion

14.3.1 Soil CO_2 Efflux

TSR was inversely related to altitude (Fig. 14.1; $r_s = -0.83$). Spatial and temporal variability of TSR decreased with increasing altitude, suggesting that TSR at lower altitudes is more influenced by factors like fluctuating litter fall and precipitation (see Chapter 8 in this volume). Furthermore, at high altitudes TSR was more evenly distributed over the whole measurement period. Soil CO_2 efflux was slightly variable during the day but did not show any systematic diurnal patterns at any site (data not shown). The annual fluxes calculated on the basis of the collected data (Fig. 14.1) ranged amongst the highest worldwide at 1540 m, while fluxes at higher altitudes were similar to those reported for boreal and cold temperate biomes (Raich 2003) and were comparable with results summarized by Bond-Lamberty et al. (2004).

Over the two years of measurement TSR at 1050 m and 1890 m was significantly lower during June to October 2004 and from May to the end of the measurement period in August 2005 (Fig. 14.2). In contrast, soil respiration at 3060 m showed a smaller variation throughout the whole measurement period, missing systematic seasonal tendencies. The development of TSR at 1890 m clearly coincides with the distribution of rainfall over the year (see Chapter 8 in this volume). During periods

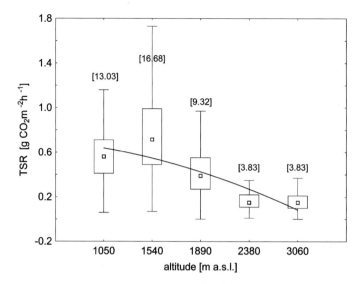

Fig. 14.1 Average total soil respiration (TSR) along the altitudinal gradient and calculated annual C fluxes. Respiration values based on 108 weeks at 1050 m, 1890 m and 3060 m and 56 weeks at 1540 m and 2380 m a.s.l. *Small boxes* indicate the median, *large boxes* 25–75% of the data and *error bars* minimum and maximum. Average annual C fluxes are displayed *in square brackets* as the sum of TSR in Mg C ha^{-1} year^{-1} referring to each measurement period, respectively

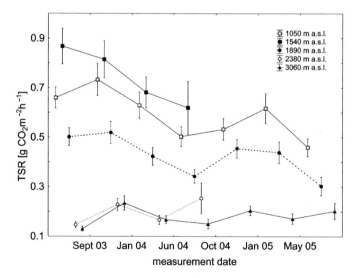

Fig. 14.2 Total soil respiration during two years (June 2003 to August 2005) along the altitudinal gradient. Values are given as average respiration based on ten measurement dates with 16 measurements each. *Error bars* indicate confidence intervals ($P = 0.95$; $n = 160$)

of lower rainfall TSR is enhanced, and suppressed during months with more than 200 mm precipitation. This data also indicates that drying and rewetting of the soil causes flushes of CO_2 efflux and that TSR response on precipitation is delayed.

These results are supported by soil moisture data of Moser et al. (Chapter 15 in this volume). Over the whole altitudinal gradient TSR was inversely correlated with water content in the organic layer ($r_s = -0.72$; $P \leq 0.05$) and in the mineral soil ($r_s = -0.76$; $P \leq 0.05$) while positively linked to temperature in the organic layer ($r_s = 0.76$; $P \leq 0.05$) and the mineral soil ($r_s = 0.78$; $P \leq 0.05$). Multiple regression analysis showed, that these two abiotic factors alone could explain 58% of the variation of TSR. Our findings are well supported by the data from other studies in Brazil, Hawaii and Costa Rica (Raich and Tufekcioglu 2000; Schwendenmann et al. 2003; Doff-Sotta et al. 2004).

TSR was positively correlated with above ground leaf litter fall ($r_s = 0.975$; $P < 0.05$), which suggests that heterotrophic soil respiration is mainly driven by the availability of easily decomposable organic compounds and also nitrogen supplied by senescent leaves (Raich and Tufekcioglu 2000). Furthermore, TSR increased consistently with increasing leaf area index ($r_s = 0.872$), which also indicates that recently assimilated photosynthates regulated TSR via root respiration (Lavigne et al. 2003).

Root contribution to TSR was inversely related to altitude. It decreased from 41.15% at 1050 m to 7.78% at 3060 m. Figure 14.3a shows that R_H at 1050 m was much less variable in time than TSR throughout the investigation period. From the fine root litter decomposition data of Moser (personal communication), we assume that during this time more than 50% of the cut fine roots were decomposed. As R_H shows a general decline and little fluctuation we consider root decomposition to be more or less even. At 1890 m periods of enhanced and strongly fluctuating R_H occurred right after the trenching and after more than one year (Fig. 14.3b). At this altitude about 60% of the cut fine roots were decomposed during the course of the experiment but this decomposition apparently fluctuated more strongly than at 1050 m, causing peaks in the soil CO_2 efflux. At 3060 m only 5% of the cut roots were decomposed within one year. Enhanced R_H at the end of the measurements (Fig. 14.3c) may indicate that decomposition is accelerating after approximately one year. For evaluation of the root exclusion method remaining root mass in soil samples of the trenched plots were determined after the conclusion of the experiment. At none of the plots did any living roots remain, but root necromass at 3060 m was significantly higher than at lower altitudes. These findings match the enhanced R_H at the end of the experiment and can be explained by either suppressed dead root decomposition or higher root biomasses at higher altitudes (see also Chapter 15 in this volume).

There are very few comparable studies on root contribution in tropical forests, especially mountain forests. The few studies that applied trenching had results comparable with our plot at 1050 m. Li et al. (2004) determined root contribution in a pine plantation and a secondary forest at 400 m a.s.l. in Costa Rica as 56.2% and 69.5%, respectively. In a tropical lowland forest in Brazil root contribution was

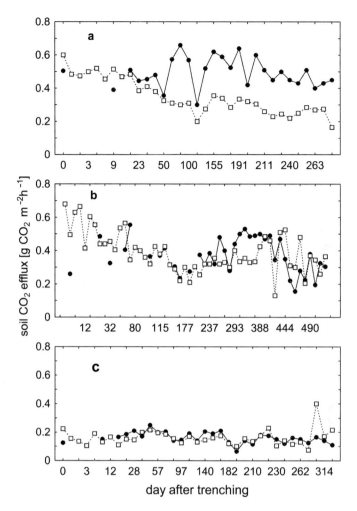

Fig. 14.3 Total soil respiration (*filled boxes*: TSR) compared with soil CO_2 efflux from root exclusion plots (*empty boxes*: R_H) at 1050 m (**a**), 1890 m (**b**) and 3060 m (**c**) determined from September 2004 to August 2005 (**a, c**) and from February 2004 to August 2005 (**b**)

between 24% and 35% in a clay soil and about 35% in a sandy soil (Silver et al. 2005). Root contribution at 1890 m ranges within the span of 10–90% given by Hanson et al. (2000) and the results at 3060 m are considered quite low. Lavigne et al. (2003) found root contribution to total soil CO_2 efflux to be positively correlated with soil temperature along a gradient of average soil temperatures between 1 °C and 7 °C. Even though soil temperatures at our plots ranged from 9.6 °C to 19.4 °C we assume this relation to be of significant influence along the investigated gradient and therefore our results consistent.

14.3.2 Soil Organic Carbon and Nitrogen

At altitudes >2380 m SOC was significantly higher in the L layer than at the lower sites ($r_s = 0.45$, $P \leq 0.05$; Fig. 14.4a). No differences in SOC content along the gradient could be detected in the densely rooted OeOa horizon. In the top mineral soil (0–10 cm) SOC at 3060 m was considerably high, already indicating peaty characteristics in these gleyic podzols ($r_s = 0.55$, $P \leq 0.05$). TN in the L layer decreased significantly along the gradient ($r_s = -0.86$, $P \leq 0.05$; Fig. 14.4b). Similar to SOC, in the OeOa no significant differences of total nitrogen were detected except for

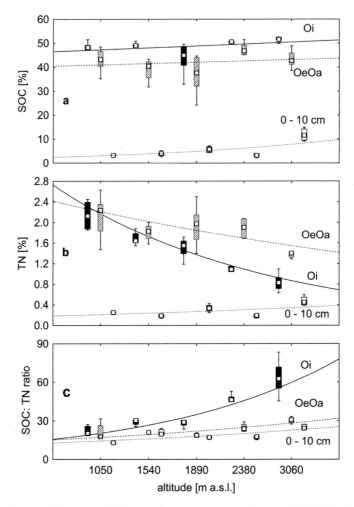

Fig. 14.4 Total organic carbon (*SOC*; **a**, total nitrogen contents *TN*; **b**) and SOC/TN ratio **c** along the altitudinal gradient

3060 m; there TN was very low. In contrast, TN in the mineral soil (0–10 cm) increased along the elevation gradient. Consequently, SOC/TN ratios increased with increasing altitude in all top soil horizons, especially in the Oi layer (Fig. 14.4c). At all sites the values decreased along the vertical gradient with depth. Extremely high SOC/TN ratios of >60 appeared in the Oi horizon of site 3060 m indicating a reduced decomposability of the leaf litter. Along the altitudinal gradient total SOC stocks in top soils increased from 44 t ha^{-1} to 128 t ha^{-1} (data not shown).

SOC contents and stocks in top soil horizons indicate the status of carbon gains and losses as influenced by litter input and quality on the one hand, and heterotrophic respiration and leaching of dissolved organic carbon on the other. Besides litter production and litter quality, soil reaction with associated base saturation and/or nutrient supply and the hydromorphic conditions may also alter the input–output processes for soil carbon. Grieve et al. (1990) and Schrumpf et al. (2001) report on increasing SOC concentrations along altitudinal gradients in Costa Rica and along the altitudinal gradient of our study region between 1850 m and 2400 m, respectively. Our results fit well with data reported from other studies (Sarmiento and Bottner 2002; Schwendenmann et al. 2002; Zou et al. 2005). Schuur et al. (2001) found SOC to increase with increasing mean annual precipitation in mesic to wet ecosystems. According to the data of Moser et al. (Chapter 15 in this volume) there is a clear, positive gradient of top soil moisture by elevation while top soil temperature is negatively correlated. Also root necromass increased by elevation showing higher C/N ratios and contents of inhibitory substances. It seems to be probable that all these factors generally reduce biomass production, litter quality and organic matter turnover by elevation from 1000 m to about 3000 m in our investigation area. According to the general assumption that heterotrophic microorganisms are N-limited above the critical C/N ratio of 30 for detritus (Merilä et al. 2002), we assume that decomposition of above- and belowground litter at altitudes above 2380 m is strongly N-limited. At the upper elevations the very high moisture content in the organic layer and top mineral soils up to maximum values of 60% as reported by Moser et al. (Chapter 15) strengthen this effect, resulting in low soil respiration values and high carbon accumulation in organic layers.

14.3.3 Microbial Biomass

In our study microbial C (Cmic) in the Oi and OeOa layers ranged from 4.20 mg g^{-1} to 8.92 mg g^{-1}, and in the top mineral soil between 0.54 mg g^{-1} and 1.16 mg g^{-1} (Fig. 14.5a). Microbial N (Nmic) amounted to 0.69–1.32 mg g^{-1} in the Oi and OeOa horizon and 0.15 mg g^{-1} in the top mineral soil. Both parameters decreased significantly with altitude in the Oi horizon [r_s(Cmic) = −0.31, $P < 0.05$; r_s(Nmic) = −0.47, $P < 0.05$] but not in the OeOa layer or top mineral soil. At the lower altitudes Cmic and Nmic contents clearly decreased with soil depth showing lowest values in the top mineral soil (Fig. 14.5). The two upper sites (2380 m, 3060 m a.s.l.) had the highest Cmic and Nmic contents in the densely rooted OeOa horizon; this indicates

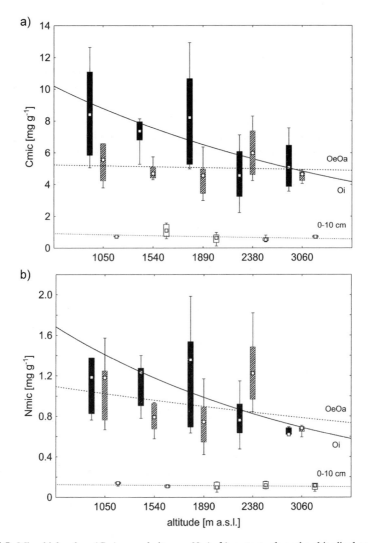

Fig. 14.5 Microbial carbon (*Cmic*; **a** and nitrogen *Nmic*; **b**) contents along the altitudinal gradient

that microbial biomass at these altitudes was rather connected to roots and that microbes utilize root-derived organic matter as the primary energy source (Kelting et al. 1998; Kuzyakov 2002). We attribute the lower microbial biomass in the Oi horizon of the two upper plots mainly to very high C/N ratios of the leaf litter, low soil temperatures and high water contents of both organic layer and mineral soil. Median Cmic/Nmic ratios of the Oi layer of the investigated sites were significantly correlated with altitude and ranged from 6.15 at 1540 m to 7.5 at 3060 m ($r_s = 0.26$; $P < 0.05$).

Since the Cmic/Nmic ratio of soil bacteria generally ranges between 5 and 8 and that of soil fungi between 10 and 15, a structural shift in microbial community composition in the Oi horizon towards a higher dominance of fungi at the higher and consistently wetter plots could be hypothesized. This assumption is supported by Ruan et al. (2004) who report a positive correlation between fungal biomass and soil moisture. In the OeOa horizon Cmic/Nmic ratios ranged between 4 and 8 and did not change significantly along the gradient. In the top mineral soil Cmic/Nmic differed between the respective sites and ranged from an average of 5 at 1050 m to 10.3 at 1540 m and was not correlated with altitude.

As an indicator for microbial activity and carbon turnover in soils the proportion of microbial carbon of SOC (Cmic/SOC) was calculated. In the tropics Cmic/SOC usually ranges between 1% and 5% (Priess and Fölster 2001). Ratios of the current study ranged between 0.66 and 3.07, thus well within the cited span. In the Venezuelan Andes, Sarmiento and Bottner (2002) found Cmic/SOC between 0.3 and 0.6, which is low compared with our results at 3060 m. The proportion of microbial active carbon of SOC decreased significantly with altitude in the Oi ($r_s = -0.50$; $P < 0.05$) and in the top mineral soil ($r_s = -0.64$, $P < 0.05$). The lacking of such a gradient in the OeOa layer may be explained by high mycorrhizal activity in this densely rooted part of the organic layer along the altitudinal gradient. In the Oi layer at 2380 m and 3060 m, as well as in OeOa and 0–10 cm mineral soil at 3060 m, Cmic/SOC are below or close to 1%, which indicates reduced turnover of carbon in these soils (Joergensen et al. 1994).

Total amount of phospholipids fatty acids (totPLFA) in the organic layer horizons decreased from 2084.62 nmol g^{-1} in OeOa at 1050 m to 536.44 nmol g^{-1} in Oi at 3060 m (data not shown). In the mineral soil totPLFA concentrations increased from 42.19 nmol g^{-1} (1050 m) to 276.19 nmol g^{-1} (3060 m). The amount of PLFA extracted from the samples was in the range given in the literature for soils in Mediterranean climate (Steenwerth et al. 2002; Fierer et al. 2003), and higher than reported for temperate soils (Peacock et al. 2001; Ponder Jr and Tadros, 2002). In our study, total PLFA was significantly correlated with Cmic ($r_s = 0.73$, $P < 0.05$) and Nmic ($r_s = 0.87$, $P < 0.05$). Significant correlations of totPLFA and microbial biomass carbon were also reported by (Zelles 1999; Yao et al. 2000; Bailey et al. 2002; Bååth and Anderson 2003). In total, 30 fatty acids were identified and 16 of these were known as specific biomarkers. Gram positive bacteria were represented by the fatty acids i15:0, a15:0, i16:0 and i17:0. The amount of Gram negative bacteria was calculated by summing amounts of 16:1n7c, 17:0cy9,10, 17:1n7c, 18:1n9c, 18:1n7 and 19:cy9,10. Actinomycetes were represented by 16:0 10Me and 18:0 10Me, and protozoa by 20:2n6c and 20:4n6. Furthermore, 18:2n6 and 18:3n6 were designated as fungal indicators.

Table 14.1 gives an overview over the summed proportions, i.e. abundances of these five taxonomic groups in the respective soil samples. Generally, these proportions were not continuously correlated with altitude or profile, even though there were some distinct trends within the vertical soil gradient. At all elevations the Gram negative bacteria and actinomycetes increased from Oi to the top mineral

Table 14.1 Relative abundances (in Mol%) of large microbial taxonomic groups

Elevation (m)	Horizon	Gram+ (%)	Gram- (%)	Actinomyceta (%)	Protozoa (%)	Fungi (%)
1050	Oi	4.9	22.6	1.4	0.7	37.1
1890	Oi	6.6	28.8	1.7	1.2	33.8
2000	Oi	10.0	31.7	2.6	0.9	13.6
3060	Oi	3.4	6.5	1.4	1.2	37.8
1050	OeOa	10.1	31.3	3.3	1.3	24.2
1890	OeOa	14.7	33.9	5.5	1.1	20.6
2000	OeOa	13.2	34.0	5.1	1.3	19.9
3060	OeOa	6.7	35.9	3.4	0.0	30.9
1050	0–10 cm	22.2	35.2	6.3	0.0	7.1
1890	0–10 cm	6.1	62.4	5.4	0.5	25.6
2000	0–10 cm	7.3	60.0	4.9	0.6	9.8
3060	0–10 cm	10.4	51.6	4.8	0.2	13.3

soil. In contrast, fungi decreased at 1050 m and 3060 m while showing no clear differences in the middle elevation zones. Principal component analysis (PCA) revealed, that the PLFA pattern in the mineral soil of 1050 m was dominated by actinomycetes and Gram negative bacteria. Organic layer samples from the different altitudes had similar abundances of 15:0, 17:0, 18:0Me and 19:0 cy9,10; and OeOa and 0–10 cm at 3060 m and mineral soil at 1890 m formed two intermediate groups. With respect to three indicators of Gram negative bacteria (17:1, 18:1n9c, 18:1n7c) differences between Oi layers were small along the altitudinal gradient and also small along the depth gradient. Consequently, the microbial community structure along the altitudinal gradient was similar in the Oi layer according to PLFA analysis which contradicts the assumption that increasing Cmic/Nmic ratios in this layer indicate a shift in the microbial community structure especially towards fungal dominance. This finding is supported by Salamanca et al. (2002), who were not able to link high Cmic/Nmic ratios to high fungal biomass. Thus, the variation of Cmic/Nmic might indicate a trend towards higher fungal biomasses at higher altitudes.

14.4 Conclusions

Along the investigated altitudinal gradient from 1050 m to 3060 m from lower to upper mountain rain forests in Southern Ecuador decreasing mineralization of carbon and increasing soil organic carbon stocks in the soils are reflected by increasing organic layer thickness. Since soil chemical parameters varied little with stable cation exchange capacities and slightly decreasing pH over the gradient, we therefore assume a priori comparability of the study sites. In the current study the main indicator for declining mineralization of carbon and therefore microbial activity was TSR that decreased substantially within the reported range. As the most decisive abiotic factors of TSR, volumetric water content and soil temperature were determined. Furthermore, TSR was significantly linked to LAI and litter fall, i.e. TSR was also strongly influenced by the vegetation. The microbial biomass, whose activity contributed between 50% and 90% to TSR and therefore exerted a strong influence on TSR, declined slightly but not significantly with altitude and was not significantly correlated to TSR. Between Cmic/SOC and TSR a stronger, yet not significant relation was detected. These findings underline the hypothesis that the mere size of microbial biomass as determined by the CFE method does not yield information on its activity as microbial communities always contain varying percentages of inactive organisms, including dormant forms (Mamilov and Dilly 2002). A further constraint for microbial activity and mineralization at the investigated sites were very high C/N ratios of leaf and root litter, which leads to the assumption that organic matter decomposition in tropical mountain rain forests is strongly N-limited.

Chapter 15
Altitudinal Changes in Stand Structure and Biomass Allocation of Tropical Mountain Forests in Relation to Microclimate and Soil Chemistry

G. Moser, M. Röderstein, N. Soethe, D. Hertel, and C. Leuschner(✉)

15.1 Introduction

In tropical montane forests, the decline of tree size with increasing elevation is a well recognized phenomenon (Lieberman et al. 1996; Raich et al. 1997). The decrease aligns with a continuous species shift from lowland forests, to lower, middle and upper montane forests (Gentry et al. 1995). Leaf area index (LAI) also decreases with elevation from lowland to upper montane forest (Kitayama and Aiba 2002).

With respect to other structural and functional parameters such as plant biomass and productivity, however, only very limited data exist from tropical montane forests. Altitudinal changes in aboveground biomass and productivity were studied in transects in Malaysia (Kitayama and Aiba 2002), Hawaii (Raich et al. 1997), Puerto Rico (Weaver and Murphy 1990) and Jamaica (Tanner 1980), some of them covering only a few hundred meters of altitudinal distance. The data base is even more limited if belowground biomass is considered: for example, a combined assessment of above- and belowground biomass in neotropical montane forests has been conducted in not more than 16 different stands so far, and only exceptionally included altitudinal transects.

A better understanding of the causes of tree size reduction with elevation in tropical mountains is closely linked to information on altitudinal changes in biomass, carbon allocation and productivity of montane forests. Although numerous hypotheses focusing on climatic or edaphic constraints of tree growth have been formulated in order to explain this phenomenon (e.g. Bruijnzeel and Proctor 1995; Flenley 1995), all of them are eventually linked to carbon gain and allocation of the trees and their control by the environment. Thus, tree biomass and productivity data (see Chapter 17 in this volume) are of paramount importance.

In this chapter, we present detailed above- and belowground biomass data of an altitudinal transect study in the Ecuadorian Andes. Study aim was to analyze altitudinal changes in forest biomass and tree root/shoot ratio, and to relate them to possible underlying climatic and edaphic factors.

15.2 Study Sites

This analysis is based on an in-depth study in five forest stands along a 2000-m altitudinal gradient in South Ecuador between 1050 m and 3060 m a.s.l., at a maximum distance of the stands to each other of 30 km (Fig. 1.3, in Chapter 1). The lowermost stands 1 and 2 are within the Podocarpus National Park close to the entrance in Bombuscaro, south of the province capital Zamora. Stands 3 and 4 are located in the RBSF. Stand 5 is an elfin forest located in the Cajanuma area at the northwestern gate of the National Park. Patches of alpine Paramo are found about 200 m upslope of this stand.

The sites are situated on moderately steep slopes facing northeast to northwest, except for site 2 (Table 15.1). The study plots of 20×20 m were selected in parts of the forests that were representative for the respective forest types and elevations characterized by Balslev and Øllgaard (2002). All stands were selected in areas with no or only minimal signs of human influence as indicated by a more or less homogenous canopy structure. All stands had closed canopies with no larger gaps in direct proximity to the study plots.

Precipitation principally increases with elevation from 1050 m to 3060 m (Table 15.1; P. Emck and M. Richter, unpublished data; see Chapter 8). The soil types change along the gradient due to changes of bedrock and hydrology (see Fig. 1.3 in Chapter 1). The mineral topsoil is generally acid and pH ($CaCl_2$) decreases with elevation. There is a general, but not continuous, increase in depth of the ectorganic layer with elevation; in parallel, the C/N ratio of the organic layer increases as well (more details on soil conditions are given in Chapter 14; see also Chapter 9).

Canopy trees belonging to the Melastomataceae occurred in all elevations while five additional plant families were present at least in four of the five plots. Most other families showed clear preferences in their altitudinal distribution.

15.3 Methods

15.3.1 Microclimate and Soil Chemical and Soil Moisture Measurements

In all five stands, air temperature and relative air humidity (1.5 m above ground), as well as soil temperature and soil moisture (in the organic layer and in 10 cm of mineral soil) were measured continuously in the study period (April 2003 to April 2004). Air temperature and air humidity were measured once per hour. Soil temperature was measured in the middle of the organic layer and at a depth of 10 cm in the mineral soil and was recorded once per hour. Soil moisture was determined as volumetric water content by TDR sensors in the same depths as soil temperature and was read by data loggers every 6 h. Given are means, maximum and minimum values for the period April 2003 to March 2004.

Table 15.1 Location and characteristics of the five study plots in South Ecuador. pH (CaCl$_2$) of the mineral topsoil (0–30 cm), C/N ratio of the organic layer (L/Of1) and soil classification (FAO system; after S. Iost, unpublished data). Rainfall data are extrapolated from measurements in gaps at 1050 m (authors' own measurements) and at 1950, 2680 and 3170 m done by P. Emck (unpublished data)

Plot	Coordinates	Elevation (m a.s.l.)	Inclination (degrees)	pH (CaCl$_2$)	C/N (L/Of1)	Soil type	Organic layer thickness (mm)	Rainfall (mm year^{-1})
1	04° 06' 54" S, 78° 58' 02" W	1050	26	3.94	22	Alumic acrisol	48	ca. 2230
2	04° 06' 42" S, 78° 58' 20" W	1540	10	3.90	29	Alumic acrisol	243	ca. 2300
3	03° 58' 345" S, 79° 04' 648" W	1890	31	3.52	28	Gleyic cambisol	305	ca. 1950
4	03° 59' 19" S, 79° 04' 55" W	2380	28	3.26	46	Gleyic cambisol	214	ca. 5000
5	04° 06' 71" S, 79° 10' 581" W	3060	27	2.86	63	Podzol	435	ca. 4500

Soil pH (CaCl$_2$) and C/N ratio were measured in the same horizons by S. Iost (see Chapter 14).

15.3.2 Analysis of Stand Structure

Tree biometric data were investigated in populations of 80 trees each per stand that covered, respectively, 827 m^2, 360 m^2, 343 m^2, 290 m^2 and 96 m^2 (in horizontal projection) of stands 1, 2, 3, 4 and 5. Thus, the size of the inventory plots decreased upslope due to increasing stem density. All trees reaching the canopy were investigated for stem length and diameter at breast height (DBH, at 1.3 m) using a Vertex III Forestor device (Haglöf, Sweden) and a mesh tape. Therefore minimum DBH at plot 1 was 5 cm and at the uppermost plot 3 cm. Stem length and tree height were determined independently because many trees did not grow in erect position, therefore stem inclination was measured additionally (see also Chapter 10.3). After projecting the 80-tree inventory plots on the horizontal basis, we calculated stem density and stand basal area (sum of the cross-sectional areas of all trees).

Stem wood density (dry mass per fresh wood volume) was measured from stem wood samples (from the bark to the bole centre) extracted with a stem corer at 1.5 m height ($n=20$ per plot).

Leaf area index (LAI) was estimated by each ten measurements per plot at random positions with a LAI-2000 system (Licor, USA).

15.3.3 Estimation of Aboveground Biomass

Aboveground tree biomass was estimated with allometric equations for the 80 canopy trees per plot based on the measured DBH, tree length, and stem wood density data. We ignored understorey trees and shrubs and standing or lying dead trunks (see Wilcke et al. 2005) since understorey biomass in mature moist tropical forests may comprise less than 3% of the aboveground biomass (Brown 1997).

We screened the literature for allometric equations available for the humid tropical moist forests. In the 29 relevant studies we found 129 different allometric equations for total aboveground biomass or different fractions of it. The only allometric equation that seems to exist for Andean mountain forests was developed for a cloud forest in Venezuela (Brun 1976) with a high specifity to the local conditions in that forest. Since specific equations applicable for lower to upper tropical montane forests have only been established for Malaysia (Yamakura et al. 1986), Jamaica (Tanner 1980) and Hawaii (Raich et al. 1997), we tested the pan-tropical equations established by Brown and Iverson (1992) and Chave et al. (2005) for tropical wet conditions for estimating aboveground tree biomass only. These pan-tropical equations may be applicable to the whole range of Andean mountain forest types from the lower to the upper montane vegetation belt. The equations were developed for estimating the total aboveground biomass including leaves, twigs, branches, bark and boles of trees, based on data of 169 (Brown and Iverson 1992) or 2410 (Chave et al. 2005) harvested trees from all over the tropics.

The applicability of the allometric equations to our stands was assessed with three recently wind-thrown tree individuals per study site, which were analyzed for stem length, DBH, wood volume and specific wood gravity. Due to best fit to these empirical data we selected the allometric equation of Chave et al. (2005):

$$B = \exp[-2.557 + 0.940 \ln(\sigma D^2 H)]$$

where B is tree aboveground biomass (kg per tree), D is DBH (cm), H is stem height (m) and σ is wood density (g cm^{-3}). The stand aboveground biomass total (AGB) was obtained by summing up the calculated masses of the 80 trees per plot.

15.3.4 Root Sampling and Analysis

Coarse root biomass for each stand was determined at 12–16 soil pits (40×40 cm) that were dug to 60 cm soil depth. Biomass (live roots) and necromass (dead roots) of all roots with a diameter >2 mm were excavated in steps of 10 cm horizons in the organic layer and the mineral soil. In the laboratory, all roots were washed and dried at 70 °C to constant dry mass.

For analyzing bio- and necromass of fine roots (diameter <2 mm), soil coring was conducted from March to May 2003 in soil profiles of 30 cm depth (organic layer and the mineral soil) under the five stands. Preliminary investigations of a lower number of soil cores to 60–80 cm depth revealed that the organic layer and the mineral soil to 30 cm depth must contain about 75% or more of the profile total of tree fine root biomass since fine root densities in the subsoil were very low. Root sampling was conducted with a steel corer (33 mm in diameter, $n=20$ per plot). The soil material was stored at 4 °C in the laboratory at Estación Científica San Francisco where processing took place within 30 days. Fine root biomass and necromass were separated under a microscope according to the procedure described by Leuschner et al. (2001).

15.3.5 Statistical Analysis

Differences of stem length, DBH, wood density, LAI and leaf biomass among the five Ecuadorian stands were analyzed with a non-parametric analysis of variance (Kruskal–Wallis test) and a Mann–Whitney two-sample test (U test) using the SAS ver. 8.2 package (SAS Institute, USA).

Linear and simple non-linear regression analyses were applied to identify significant effects of elevation, mean air temperature, vpd, mean annual precipitation, mineral soil moisture, mineral soil proton concentration and C/N ratio of the organic layer on tree height, DBH, basal area, stem density, aboveground biomass, belowground biomass, total biomass and root/shoot biomass ratio. All calculations were done using Xact ver. 8.0 software (SciLab, Germany).

15.4 Results and Discussion

15.4.1 Altitudinal Change in Microclimate and Soil Conditions

The annual mean air temperature inside the forest stands at 1.5 m above the forest floor decreased between 1050 m and 3060 m with a temperature lapse rate of 5 K km^{-1} (Table 15.2). The annual mean of relative air humidity inside the stands increased slightly along the transect revealing a significant positive correlation with elevation ($y = 85.8 + 0.003x$, r^2 adj = 0.85, $P = 0.008$). The corresponding vapor pressure deficit decreased from 1050 m to 3060 m ($y = 3.9 - 0.009x$, r^2 adj = 0.93, $P = 0.002$). vpd extremes exceeded 3 kPa in stands 1 and 3, but were lower in stands 2, 4 and 5.

Annual mean temperature of the organic layer was between 0.3 °C and 1.7 °C higher than mean air temperature in all stands (Table 15.2); that of the mineral topsoil was very close to air temperature.

Table 15.2 Microclimatic and soil moisture conditions of the five study plots in South Ecuador. Given are means, maximum and minimum values for the period April 2003 to March 2004. Air temperature and air humidity were measured at a height of 1.5 m inside the forest stands. Soil temperature was measured in the middle of the organic layer and at a depth of 10 cm in the mineral soil. Soil moisture was determined every 6 h by TDR sensors in the same depths as soil temperature

Plot	Elevation (m a.s.l.)	Air temperature (°C)			Air humidity (%)			Vapour pressure deficit (hPa)			Organic layer temperature (°C)			Mineral soil temperature (°C)			Organic layer moisture content (vol.%)			Mineral soil moisture content (vol.%)		
		Mean	Max	Min	Mean	Max	Min	Mean	Max	Min	Mean	Max	Min	Mean	Max	Min	Mean	Max	Min	Mean	Max	Min
1	1050	19.4	30.2	11.5	88.7	100	15.5	2.7	31.1	0	20.0	28.7	14.4	19.4	27.4	14.4	9.9	16.0	4.4	29.7	38.5	15.3
2	1540	17.5	26.7	11.2	88.9	100	31.7	2.5	22.2	0	18.9	20.6	16.1	18.5	19.2	17.3	12.9	23.9	3.4	30.3	43.5	20.4
3	1890	15.7	29.4	7.9	90.7	100	15.7	2.0	31.5	0	16.0	18.8	11.7	16.4	18.2	15.3	11.6	22.3	3.6	35.4	44.7	27.4
4	2380	13.2	25.1	7.0	93.2	100	34.0	1.2	20.4	0	14.9	16.6	13.3	13.0	13.9	11.6	34.0	39.8	23.8	44.7	48.7	35.7
5	3060	9.4	18.8	3.1	93.5	100	28.6	0.9	14.5	0	9.7	10.8	8.1	9.8	10.7	8.7	45.3	61.7	30.5	49.1	59.5	39.5

Average soil moisture of the organic layer and the mineral topsoil increased greatly between plots 1 and 5; the seasonal minima in the organic layer of stands 1–3 were quite low due to the small depth of this horizon. The seasonal minima in the mineral soil were not as low as in the organic layer and water shortage did not occur (Table 15.2).

15.4.2 Altitudinal Change in Stand Structure

Canopy height declined 3.5-fold and mean stem length 3.0-fold between 1050 m and 3060 m (Table 15.3). In the uppermost stand (stand 5) in the elfin forest, strongly inclined trunks with low canopy height were responsible for the relatively high maximum stem length (Table 15.3). Tree height showed a very close correlation to both elevation and temperature; however, vpd, soil moisture, soil proton concentration and C/N ratio of the organic layer had significant effects on tree height as well (Table 15.4).

In the large-scale altitudinal transects in Malaysia (Aiba and Kitayama 1999), Costa Rica (Lieberman et al. 1996) and Hawaii (Raich et al. 1997), tree height showed a more or less continuous decrease with altitude as it did in our Ecuadorian transect. Similarly, a continuous decrease of tree height from the lowlands to the timberline was also reported for undisturbed temperate mountain forests, e.g. in the Southern Alps of Italy (Reisigl and Keller 1999) and in Tierra del Fuego (Pollmann and Hildebrand 2005). However, in certain mountain transects as on Tenerife, tree height seems to decrease upslope only slightly; the timberline may then consist of tall trees (Srutek et al. 2002). We speculate that a more or less continuous decrease in tree height occurs on mountain slopes where the altitudinal temperature decrease is the main environmental factor that controls tree growth and microbial activity in the soil. However, in all those mountains, where additional environmental constraints such as water logging, drought or strong winds are influencing tree growth, these factors may overlay the temperature effect causing a more or less abrupt transition between tall high-elevation trees and low-statured krummholz or alpine non-forest vegetation higher upslope. An important second cause of abrupt timberlines is human impact that has lowered timberline elevation in many mountains of the tropics.

Altitudinal comparisons of mean DBH are often problematic because authors tend to define lower stem diameter limits at higher elevation where thinner stems prevail than in low-elevation forests with thicker stems. If the lower DBH limit varies along the slope, a larger altitudinal decrease in mean DBH would be detected than exists in reality. Lieberman et al. (1996) fixed a DBH minimum in their Costa Rican elevation gradient between 100 m and 2600 m of 10 cm and did not find altitudinal dependence but found highest means at high elevations. To avoid such a bias in the Ecuadorian transect we investigated all stems that reach the canopy, irrespective of diameter. We found a 8.6-fold decrease in maximum DBH and in mean DBH by a factor of 2.5 between stand 1 and stand 5 which is roughly proportional to the 3.0-fold reduction in mean stem length, but contrasts with an 8.6-fold increase of stem

Table 15.3 Aboveground stand structural characteristics of the five study plots. Canopy height, mean stem length, diameter at breast height, stem density, basal area, stem wood density, and LAI of the five study plots in South Ecuador (80 trees per plot). Stem density is given as plot total and separately for five different DBH classes. Stem wood density was measured at 20 tree individuals per plot. LAI was measured at ten points per plot. *Different letters* indicate significant differences between the plots ($P < 0.05$)

Plot	Elevation (m a.s.l.)	Canopy height (m)	Stem length (m) Mean ± SE	Stem length (m) Max.	Diameter at breast height (cm) Mean ± SE	Diameter at breast height (cm) Max.	Stem density (n ha^{-1}) 3–5 cm	5–10 cm	10–20 cm	20–30 cm	30–70 cm	Total	Basal area (m^2 ha^{-1})	Wood density (g cm^{-3}) Mean ± SE	LAI (m^2 m^{-2}) Mean ± SE
1	1050	31.8	15.6 ± 0.7[a]	39.7	17.3 ± 1.3[a]	69.2		315	399	133	121	968	33.6	0.64 ± 0.03[a]	5.1 ± 0.1[a]
2	1540	21.7	12.1 ± 0.5[b]	27.2	11.5 ± 0.6[b]	27.3	55	1028	1000	139		2222	27.5	0.65 ± 0.03[a]	4.6 ± 0.1[b]
3	1890	18.9	10.1 ± 0.4[c]	24.8	12.2 ± 0.8[b]	27.3	88	1283	904	58		2333	36.9	0.60 ± 0.04[a]	3.9 ± 0.2[c]
4	2380	12.0	7.4 ± 0.3[d]	13.4	9.8 ± 0.6[c]	32.5	241	1583	826	103		2753	27.2	0.61 ± 0.04[a]	3.6 ± 0.1[c]
5	3060	9.0	5.2 ± 0.3[e]	19.2	7.2 ± 0.4[d]	16.5	2703	4158	1455			8317	42.2	0.69 ± 0.03[a]	2.9 ± 0.3[d]

Table 15.4 Regression analysis of stand biometric data as dependent on abiotic site factors. Regression analysis between forest stand structural parameters, biomass fractions and the root/shoot ratio as dependent variables and topographic, climatic and edaphic factors as source parameters in the elevation transect in South Ecuador. Data on annual precipitation of stands 3, 4 and 5 from P. Emck and M. Richter (unpublished data). Data on proton concentration of the mineral topsoil and C/N ratio of the organic layer (L/Of1) from S. Iost (unpublished data). r^2 and adjusted r^2 (*adj*) refer to linear or simple non-linear regression models. Relationships with $P < 0.05$ are printed in italics, exponential functions are marked by E

Dependent variable	Elevation			Temperature			VPD			Precipitation			Soil moisture			Soil proton concentration			C/N ratio (L/Of$_1$)		
	r^2	r^2 adj	P	r^2	r^2 adj	P	r^2	r^2 adj	P	r^2	r^2 adj	P	r^2	r^2 adj	P	r^2	r^2 adj	P	r^2	r^2 adj	P
Tree height	0.97	0.97	*0.001*	0.97	0.96	*0.005*	0.96	0.91	*0.019 E*	0.63	0.50	0.055	0.90	0.86	*0.006*	0.92	0.85	*0.036 E*	0.94	0.88	*0.026 E*
DBH	0.87	0.82	*0.009*	0.89	0.80	*0.050 E*	0.75	0.67	*0.028*	0.50	0.34	0.090	0.70	0.60	*0.038*	0.63	0.51	0.054	0.78	0.70	*0.023*
Stem density	0.98	0.95	*0.010 E*	0.98	0.95	*0.009 E*	0.97	0.93	*0.014 E*	0.37	0.16	0.140	0.97	0.93	*0.014 E*	0.95	0.93	*0.002*	0.97	0.94	*0.012 E*
Basal area	0.23	−0.04	0.209	0.25	0.001	0.196	0.14	−0.14	0.236	0.001	0.33	0.479	0.14	−0.14	0.268	0.44	0.25	0.110	0.2	−0.07	0.226
AGB	0.92	0.85	*0.036 E*	0.93	0.82	*0.034 E*	0.73	0.63	*0.033*	0.56	0.41	0.073	0.65	0.53	*0.050*	0.41	0.21	0.124	0.99	0.99	*0.002 E*
BGB	0.69	0.58	0.040	0.73	0.64	*0.032*	0.92	0.83	*0.041 E*	0.68	0.57	*0.043*	0.91	0.82	*0.043 E*	0.84	0.78	*0.014*	0.90	0.87	*0.006*
Total tree biomass	0.94	0.88	*0.025 E*	0.93	0.88	*0.027 E*	0.84	0.69	*0.080 E*	0.43	0.24	0.114	0.52	0.36	0.085	0.91	0.83	*0.040 E*	0.97	0.93	*0.014 E*
Root/shoot ratio	0.86	0.81	*0.011*	0.88	0.83	*0.009*	0.93	0.87	*0.031 E*	0.76	0.68	0.26	0.93	0.86	*0.033 E*	0.86	0.81	0.114	0.98	0.97	*0.001*

density along the 2000 m elevation transect. Mean DBH correlated significantly with elevation, air temperature and C/N ratio of the organic layer (Table 15.4). The lowermost plot (plot 1) was the only stand where trees with a DBH >30 cm occurred, but the most frequent DBH class was the 10–20 cm class. More upslope, the DBH class of 5–10 cm included the highest number of canopy trees. At the two uppermost sites (stands 4, 5), a higher number of trunks belonged to the 3–5 cm DBH class than to the 10–20 cm class (Table 15.3).

Smaller trees with thinner stems and less extended crowns allow for higher tree densities per ground area when moving upslope in mountains. Altitudinal increases in tree density were not only reported for South Ecuador, but also for a transect on Mt. Kinabalu in Malaysia (Takyu et al. 2002). In contrast, Heaney and Proctor (1990) found only minor changes in stem density between 100 m and 2600 m on Volcan Barva, Costa Rica, probably because they used also a minimum DBH of 10 cm for their inventories at all elevations. Similarly, upper-montane *Quercus* forests in the Sierra de Talamanca (Costa Rica) at 2900 m a.s.l. had exceptionally low stem densities when only stems with a DBH > 10 cm were considered (390 ha^{-1}). These values are not higher than in many tropical lowland forests. However, if all stems >3 cm DBH are included, stem density increases by a factor of nearly ten (3460 ha^{-1}; Köhler 2002). On subtropical Mt. Teide, Tenerife, the density of *Pinus canariensis* trees remains constant with elevation or decreases (Srutek et al. 2002). We conclude that changes in tree density along tropical mountain slopes seem to be highly dependent on the floristic composition and the stand dynamics of the respective forest communities. In addition, tree density in high-elevation forests may also depend on local edaphic and climatic conditions such as the occurrence of waterlogging or exposure to strong winds. Temporarily waterlogged soils favor woody plants that are able to resprout and to form multiple stems giving them a high morphological plasticity. For example, the widespread species *Weinmannia loxensis* (Cunoniaceae) in the uppermost stand (stand 5 at 3060 m) of the Ecuador transect forms creeping belowground stems which connect up to three shoots (Soethe et al. 2006b). The ability of many tree species to grow multiple stems on temporarily waterlogged soils partly explains the very high stem densities that were counted in stand 5.

The values of Madsen and Øllgard (1994) from upper montane forests of this region (2700 m, 2900 m elevation) are lower than this value but they also indicate high tree densities close to the timberline.

Stand basal area is a function of mean DBH and stem density which both showed opposite trends with elevation in South Ecuador. In this transect, we found no clear altitudinal dependence of basal area although the highest value was measured at the uppermost stand (Table 15.3). It correlated neither with elevation nor temperature, and was not influenced by any of the soil parameters (Table 15.4).

This is in line with results obtained by Lieberman et al. (1996) in Costa Rica. These results together with literature data indicate that basal area of tropical mountain forests seems to be only weakly dependent on climatic, edaphic or stand structural parameters such as maximum tree height (Aiba and Kitayama 1999).

Stem wood specific gravity did not differ significantly between the five studied stands and did not correlate with elevation (Table 15.3).

In Ecuador, leaf area index decreased continuously with elevation (Table 15.3), as was found in the transect on Mt. Kinabalu (Kitayama and Aiba 2002). This implies a decrease in canopy carbon gain with increasing elevation in the Ecuador transect, because neither leaf area-related foliar N content nor solar radiation changed significantly between 1000 m and 3000 m elevation (M. Unger, personal communication). Carbon gain of forest canopies is mainly controlled by three factors, leaf photosynthetic capacity, which is linked to foliar N content, photosynthetic photon flux densities and leaf area index. As a consequence, the stand assimilation rate must decrease upslope along the transect, thus being one of the factors reducing tree height and aboveground biomass of high-elevation forests in South Ecuador.

15.4.3 Altitudinal Change in Tree Biomass

Total aboveground tree biomass (AGB) declined 2.5-fold from plot 1 to plot 5 and was significantly dependent on elevation (Table 15.5). In stand 1, trees with a DBH >30 cm accounted for more than two-thirds of the total aboveground tree biomass. In contrast, in all other stands, trees with a DBH of 10–20 cm were responsible for the largest portion of AGB. Trees with a DBH <5 cm contributed only at plot 5 more than 1% of the aboveground tree biomass. AGB was highly dependent on the C/N ratio of the organic layer, but was also related to mean air temperature and soil moisture (Table 15.4).

A striking result of this study is the high fine and coarse root biomass of the high-elevation forests in Ecuador (Table 15.5). Belowground biomass (BGB) of the

Table 15.5 Aboveground, belowground, and total biomass of the five stands in South Ecuador. Estimates of total aboveground tree biomass (AGB) are based on an allometric equation given by Chave et al. (2005), which relates biomass to tree height, DBH and wood density. Given are AGB values for five different DBH classes. Belowground biomass (BGB) is given as the sum of coarse root biomass (diameter > 2 mm) and fine roots (diameter <2 mm). Total biomass is the sum of AGB and BGB

Plot	Elevation (m a.s.l.)	AGB (Mg ha^{-1})						BGB (Mg ha^{-1})	Total biomass (Mg ha^{-1})
		3–5 cm	5–10 cm	10–20 cm	20–30 cm	30–70 cm	Total		
1	1050		7.7	40.9	41.1	195.3	285.1	32.1	317.2
2	1540	0.4	19.8	97.3	50.0		167.5	36.3	203.8
3	1890	0.4	22.4	119.4	30.8		173.0	25.8	198.8
4	2380	0.8	19.1	49.8	7.2	22.9	99.8	39.2	139.0
5	3060	8.8	51.1	52.3			112.2	62.7	174.9

uppermost stand was about two-fold greater than that of stands 1–3 (1050–1890 m) and showed significant relations to all seven tested parameters; strongest correlation existed with the C/N ratio of the organic layer and the proton concentration of the mineral topsoil (Table 15.4). Corresponding trends in fine root biomass with elevation have been reported by Kitayama and Aiba (2002) on Mt. Kinabalu and in a meta-analysis of fine root data of paleo- and neotropical forests by Hertel and Leuschner (2007). In the Ecuador transect, the high fine root biomass of stand 5 was not a consequence of a longer root lifespan, but must have resulted from intensive fine root growth (S. Graefe, personal communication). Thus, a direct effect of lowered temperature on aboveground meristematic activity and tree growth is unlikely because air and soil temperature differed by not more than 1 °C from each other, and branch and root meristems should have responded similarly to a reduction in temperature along the slope. Rather, a direct or indirect effect of lowered temperature or reduced soil resources on carbon allocation patterns of the trees must play a prominent role for the altitudinal reduction in AGB. Figure 15.1 shows the contrasting trends of above- and belowground biomass with increasing altitude. The two trend lines meet at an elevation of about 3400 m a.s.l. where tree above- and belowground biomass are expected to reach equal size and the alpine tree line in South Ecuador occurs.

Total aboveground tree biomass significantly decreased with altitude in the Ecuadorian transects (Table 15.5). The portion of belowground biomass in total tree biomass increased from 10.1% to 35.8% between the lowermost and the uppermost stand. A five-fold increase in the root/shoot ratio of the trees between 1050 m and 3060 m in the Ecuador transect underlines the large carbon allocation shift from aboveground to belowground tree organs that takes place along this slope (Fig. 15.2).

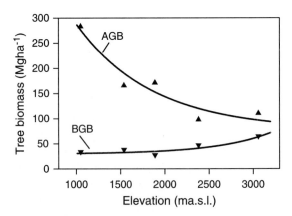

Fig. 15.1 Stand above- and belowground biomass along the elevation transect in Ecuador. Aboveground biomass (*AGB*) was derived from tree height, DBH (each $n=80$) and wood density data ($n=20$) using the allometric equation of Chave et al. (2005; see text). Belowground biomass (*BGB*) is the sum of coarse root biomass (diameter>2 mm, $n=12/16$) and fine root biomass (diameter <2 mm, $n=15$). For details on regression analysis see Table 15.5

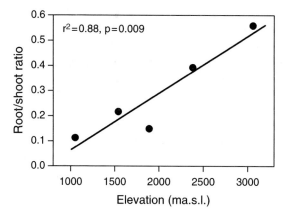

Fig. 15.2 Linear regression analysis between the root/shoot biomass ratio, calculated as the quotient of BGB/AGB and elevation

Root/shoot ratio was correlated to all tested site factors, with closest relationship to the C/N ratio of the organic layer, air temperature and elevation (Table 15.4).

A causal explanation for the carbon allocation shift must remain speculative since elevation, temperature, soil C/N ratio and also soil pH are closely related to each other in the Ecuadorian transect. Specific hypotheses require experimental testing.

15.5 Conclusions

According to the resource balance hypothesis of Bloom et al. (1985) the impressive change in allocation patterns is best explained by a growing importance of limiting soil resources over limiting light with increasing elevation. Increasing carbon and nutrient allocation to roots on the cost of aboveground biomass would then represent a compensatory response of the trees to cope with an increasing limitation by nutrient and/or water shortage at high elevations.

Water shortage is unlikely to occur at the high-elevation sites in South Ecuador; rather, water logging may have contributed to a slowing down of decomposition and thus a poor nutrient supply in the uppermost stands (Schuur 2001). There are several possible pathways by which low temperatures may have resulted in impaired nutrient supply or uptake, among them: (a) a reduction in decomposition rate, (b) a low activity of mycorrhizal fungi and (c) a reduction in membrane transporter activity and thus a lowered nutrient uptake rate.

Unfavorable soil chemical conditions as a consequence of low nutrient contents in plant litter, or high concentration of free aluminum and elevated Al/Ca ratios in the soil solution (Hafkenscheid 2000) may also negatively affect root vitality and root growth, additionally impairing nutrient uptake. Finally, the observed large stocks of coarse root biomass in the uppermost stands, which mainly serve for tree anchoring, point at the growing need for tree stabilization on steep and wet slopes at high elevations (Soethe et al. 2006b).

While it is unlikely that a single factor is responsible for the observed remarkable carbon allocation shift with increasing elevation, there is an urgent need of well designed field experiments aimed at disentangling the possible influencing factors that may limit nutrient supply and nutrient uptake in these high-elevation tropical forests.

Acknowledgements We are grateful to J. Homeier who provided floristic inventory data for the five Ecuadorian stands. We also thank S. Iost and F. Makeschin (University of Dresden) for contributing the pH and C/N ratio data and information on soil types, P. Emck and M. Richter (University of Erlangen) for supplying the rainfall data, and M. Unger (University of Göttingen) for leaf chemical and radiation data. For technical support with the LAI-2000 measurements we thank M. Küppers (University of Hohenheim).

Chapter 16
Stand Structure, Transpiration Responses in Trees and Vines and Stand Transpiration of Different Forest Types Within the Mountain Rainforest

M. Küppers(✉), T. Motzer, D. Schmitt, C. Ohlemacher, R. Zimmermann, V. Horna, B.I.L. Küppers, and T. Mette

16.1 Introduction

Lösch (2001) concluded that information on the ecophysiology of tropical mountain trees – key components for the understanding of the functional role of vegetation in these ecosystems – is entirely missing; therefore we concentrate on transpiration. Available knowledge on stands is almost exclusively from catchment studies (e.g. Bruijnzeel and Proctor 1995; see Chapter 12 in this volume), but types of "gully", "slope" and "stunted" forests are found everywhere. To evaluate their contribution to water consumption and buffering in the landscape it is essential to characterize these forest types individually, here via up-scaling from sap flow measurements on individual trees to plot-scale. In this context it is important to know whether sap flow varies species-specifically (then many species have to be studied) or whether the size of individual plants, their life-forms, and associations with functional groups are more important (then only a few representative individuals need to be investigated).

16.2 Methods and Sites

16.2.1 Methods Used

A detailed description of measurements of environmental parameters, sap flow, tree diameter at breast height (DBH), basal area (A_b), and sap wood area (A_s) is given by Motzer et al. 2005. Meteorological parameters were measured within and above stands (air temperature and relative humidity in three heights, photosynthetic active radiation (PAR) in four heights, wind velocity in two heights (all within stand), global radiation (G) and net radiation balance (Q), PAR, wind, temperature, and humidity also 2 m above the stand), soil temperature and moisture in four depths using standard meteorological equipment (see Motzer 2003, p. 29). Measurements of leaf area index (LAI) were made on overcast days with LiCor 2000 canopy analysers (LiCor, Nebraska, USA) on a $2\times 2\,m^2$ plot grid using detector rings 1+2 for best resolution at small azimuth angles; however, these measurements were not corrected for an unknown but small contribution from woody plant parts. Leaf conductances ($n\geq 4$ leaves per plant) were determined at ambient microclimatic conditions (Motzer 2003) using either the LiCor 1600 H_2O porometer or the type AP4 (ΔT-Devices, Cambridge, UK).

16.2.2 Sites and Altitudinal Gradient

Studies were performed within the closed mountain forest area opposite of the Estación Científica San Francisco east of Loja and at an elfin forest/Páramo ecotone at Cajanuma, all inside or adjacent to the Podocarpus NP. Six plots (between $12\times 12\,m^2$ and $20\times 24\,m^2$) were chosen (Table 16.1) representing the three major types "slope forest" (at 1975 m and 2125 m a.s.l.), "gully forest" (one in transition to slope forest at 1950 m, the other in the typical gully at 2050 m a.s.l.) and "elfin forest" (at 2240 m and 3060 m a.s.l.). In some cases results are included from a lower montane forest of the same mountain range at Cerro Tambo (North Peru, 1400 m a.s.l., 5° 42.81′ S, 77° 15.93′ W; mean annual temperature 19.4 °C, annual precipitation 2100–2200 mm).

At each site (co-)dominant trees of upper canopy were investigated. They exhibited the largest crown areas and were expected to contribute a major part to stand transpiration. Otherwise, representative plants of each canopy layer were studied individually, where possible identified to their species or genus, so that we were later capable of distinguishing results species-wise or in relation to "functional types" of species.

16.3 Results and Discussion

16.3.1 Stand Structure

In the slope forest tree heights reach 10–14 m, in the moister gully 40 m, and in transition forest 18–20 m (Table 16.1), with plant crowns arranged in three or four layers (overstorey, mid- and sub-canopy, understorey; Motzer 2003). In the elfin forests trees are significantly lower (6–9 m, two leafy layers: canopy, understorey).

16.3.1.1 Leaf Area Index: Magnitude and Variation over Time

For interpretation of plot transpiration total plant surface, the magnitude and variation over time has been determined as LAI (including an unknown but small contribution from the projection of woody plant parts). It is highest in the gully (mean: 9.8; min.

Table 16.1 Site-specific parameters of stand structure for different forest types at the ECSF environment (Cerro Tambo for comparison), separated for trees and vines where applicable. The gully-forest resembles the vertical structure of lowland forests

		Tall mountain forests					Stunted forests	
		Gully forest, ECSF	Transition gully to slope forest, ECSF	Slope forest, ECSF	Slope forest, ECSF	Slope forest, Cerro Tambo	Heath forest, Cerro Tambo	Elfin forest, ECSF	Elfin forest, at treeline, Cajanuma[a]
	Altitude (m a.s.l.)	2050	1950	1975	2125	1400	1400	2240	3060
Trees	Max. tree height (m)	40	18–20	13–14	10–12	29	15	6–8	9
	Max. basal diameter (mm)	1085	430	600	236	560	197	216	165
	Mean basal diameter (mm)	62	70	71	17	–	–	47	7.3
	Max. crown projected area (individual; m^2)	149	75	100	27	112	12	13	n.d.
	Rel. basal stand area (%)	3	0.31	0.53	0.3	0.50	0.17	0.3	0.43
	Mean LAI (m^2/m^2)	9.8	6.7	6.4	5.9	4.4	4.4	3.0[b]	2.9
Vines	Max. basal diam. (mm)	36	30	25	20	–	–	–	–
	Mean basal diam. (mm)	15.4	7	6.5	7	–	–	–	–
	Rel. basal stand area (%)	0.090	0.015	0.01	0.075	–	–	–	–

[a] G. Moser, M. Röderstein, J. Homeier, unpublished data
[b] M. Oesker, unpublished data

7.3, max. 13.1) and transition forest, decreases slightly from 6.7 (min. 3.7, max. 9.8) at 1950 m to 5.9 (min. 2.8, max. 9.3) at 2125 m a.s.l. (Table 16.1) and is lowest in the elfin forests. Even in the closed forest local spatial heterogeneity reaches a ΔLAI_{space} = 6.1. Spatial clumping of leaves increases with altitude. After a month of little rain (January 2000) the mean LAI decreased by 0.3. Maximum shedding occurred from spots of highest and lowest from spots of lowest LAI, so that mean values varied little. Within-plot spatial variability permanently changed its pattern in periods as short as three months (Fig. 16.1). The mean temporal ΔLAI_{time} range from −0.92 to +0.20, while simultaneous plot-specific decreases and increments in LAI may reach:

$$\Delta LAI_{time} = |+4.12| + |-3.44| = 7.5.$$

16.3.1.2 Tree and Woody Vine Structural Parameters

Tree height, trunk size (basal diameter), and maximum crown projected area in the ECSF forests decreased significantly with altitude ($P<0.003$; but not in vines; Table 16.1). Significantly thicker vines were only found in the gully. Cerro Tambo slope forest had almost no vines. Tree age there reached more than 140 years (Zimmermann et al. 2002).

Neither exact lengths of main shoots nor crown projected areas could be determined in vines. However, in both woody life forms basal stand areas (% of plot area) declined significantly with altitude ($P<0.025$; Table 16.1). Vines add little to the forest biomass (stand basal area <0.1%).

Allometric studies yield useful regressions for tree-to-stand up-scaling of transpiration. We observed highly significant ($P<0.0001$) exponential relationships between tree height h and basal A_b as well as crown projection areas A_c (Motzer 2003, p. 52). Irrespective of h, A_c increases linearly with A_b and sapwood area A_s ($r^2 > 0.85$). In the elfin forest A_c values are generally smaller for a given sapwood area A_s (data not shown).

Several studies indicate that basal area (A_b) and sapwood area (A_s) are strongly correlated with a plant's total leaf area (e.g. McDowell et al. 2001; Meinzer et al. 2001). In our study total plant leaf area could not be determined unambiguously and was approximated by A_c, yielding excellent linear relations to A_s ($r^2 = 0.80$–0.91). Moreover, as observed in other tropical trees (Goldstein et al. 1998; Phillips et al. 2001) sapwood area mostly contributes to 80–90% of the basal area A_b (see A_s/A_b in Table 16.3).

16.3.2 Transpiration of Individual Trees and Vines

16.3.2.1 Leaf Level: Conductances for Water Vapour and Transpiration Rates

Leaf conductance (g) and leaf transpiration (E) of canopy trees (in mid-sun crown), of vines on these trees, and of sub-canopy and understorey plants have been measured

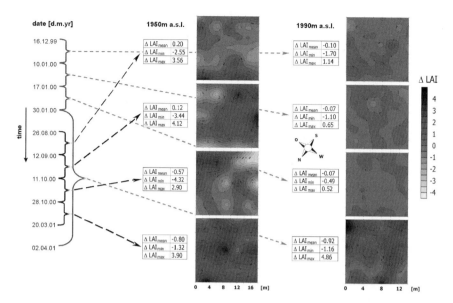

Fig. 16.1 Changes in leaf area index over time (ΔLAI_{time}; m²/m²) as observed for the plots in the transition (1950 m a.s.l.) and slope forest (1975 m a.s.l.). ΔLAI is given as the difference between two successive measurements in the form of iso-lines (steps of 0.5 m²/m²) and steps according to the plate. Measurement dates: column on the left (*d* day, *m* month, *yr* year). Mean over the plot, plus minimum and maximum observed changes (at one measurement point) are indicated

on dry leaf surfaces (Table 16.2). Highest *g* and *E* were observed in sun leaves of trees. Vine leaves and tree shade leaves behaved similarly. On dry days partial stomatal closure reduced transpiration in both life forms (data not shown), but much more effectively in liana leaves.

16.3.2.2 Plant Level: Sap Flow Densities

Sap flow density (J_s) is a measure of how much water flows within a given time span through a certain area, here the hydro-active part of the xylem. It is a kind of "conductance" of the sapwood but is simultaneously affected by microclimate in the canopy as the driving "force" of transpiration. Highest $J_{s(max)}$ values were found in lianas and in the pioneer tree *Cecropia montana* (Table 16.3) whereas no clear differences are indicated between canopy and trees of the forest interior. Lowest $J_{s(max)}$ values were observed for humid days, again with higher $J_{s(max)}$ values in lianas. Altitudes ranging from 1400 m to 1975 m a.s.l. showed no effects.

Higher J_s vlaues in vines compared with trees are well known from the literature (e.g. Fichtner and Schulze 1990; Larcher 2003; Phillips et al. 1999). They result from the fact that slim liana stems support water flow into a large canopy leaf area. Since vines prefer moist sites (Table 16.1) their comparatively low leaf conductances, $g_{max(obs)}$, indicate the need to avoid excessive transpiration rates.

Table 16.2 Maximum observed leaf conductances ($g_{max(obs)}$), maximum observed ($E_{max(obs)}$) and mean leaf transpiration rates (E_{mean}) at ambient conditions for plants of different woody life-forms growing in the overstorey (canopy) or forest interior. Photosynthetic active radiation (PAR), leaf temperature (LT) and leaf-to-air water vapour concentration difference (Δw) were measured in parallel to $g_{max(obs)}$. n.d. not determined

Species	$g_{max(obs)}$ (mmol m^{-2} s^{-1})	PAR (µmol m^{-2} s^{-1})	LT (°C)	Δw (Pa kPa^{-1})	$E_{max(obs)}$ (mmol m^{-2} s^{-1})	E_{mean} (mmol m^{-2} s^{-1})
Trees and shrubs						
Canopy						
Trichilia guianensis Klotzsch ex C. DC.	>450	n.d.	n.d.	n.d.	>6.5	2.2
Hedyosmum anisodorum Todzia	>360	n.d.	n.d.	n.d.	4.3	1.7
Alzatea verticillata Ruiz & Pav.	~300	100–200	17–20	10–15	4.5	~1.8
Naucleopsis sp.	>240	n.d.	n.d.	n.d.	2.7	1.1
Ruagea cf. *pubescens* H. Karst	170	~100	15	5	~2.2	0.84
Sub-canopy and understorey						
Palicourea sp.	225	20–40	~15	~10	1.4	0.62
Guarea sp.	200	20–40	20	9	1.8	0.66
Piper sp.	170	20–40	~15	~10	1.7	0.47
Psychotria brachiata Ruiz & Pav.	150	40	15	7	~1.1	0.52
Vines						
Mikania matezki H. Rob. &W.C. Holmes	~250	200	20	10	3	~0.5
Paullinia sp.	>185	20–40	~15	~10	1.3	~0.6
Mikania sp.	>100	100	16	6	1.5	~0.5

Table 16.3 Maximum observed sap flux densities ($J_{s(max)}$), photosynthetic active radiation (PAR), and vapour pressure deficit (D) at time of observation for plants of different woody life-forms growing in the overstorey (*Canopy*) or forest interior, either at 1950m and 1975m a.s.l. (ECSF) or at 1400m a.s.l. (Cerro Tambo). *DBH* diameter at breast height, *h* height of plant, A_s/A_b fraction of sapwood area in basal area, *n.d.* not dermined

	Max. observed sap flux density $J_{s(max)}$	PAR (at upper crown surface)	D (at upper crown surface)	DBH	Height (h)	Sapwood area (A_s)	A_s/A_b
	mmol m^{-2} s^{-1}	µmol m^{-2} s^{-1}	hPa	mm	m	cm^2	
Trees and shrubs							
Canopy							
Cecropia montana Warb. ex Snethl.[a]	6822	1800	19.0	283	17.0	640.4	1.00
Ruagea cf. *pubescens* H. Karst	3363	1800	19.0	382	18.0	1036.7	0.90
Unidentified species (Anacardiaceae)	3223	1660	11.3	289	~14	423.1	0.65
Trichilia guianensis Klotzsch ex C. DC.	3032	1800	19.0	404	18.0	1166.3	0.91
Guarea sp.	2558	1800	19.0	420	18.0	1228.9	0.89
Aniba cf. *muca* Ruiz & Pav.	2243	1800	19.0	127	12.0	126.9	1.00
Vismia tomentosa auct. non Ruiz & Pav.	1814	1660	11.3	175	12.0	205.1	0.85
Tapirira guianensis Aubl.	1756	1660	11.3	311	13.0	423.1	0.65
Alzatea verticillata Ruiz & Pav.	1109	1660	11.3	471	13.0	871.8	0.50
Cerro Tambo, N. Peru							
Pouteria sp.	3000	1650	10.8	172	22.5	85	0.45
cf. *Clarisia biflora*	1667	1140	3.2	274	27.0	152	0.28
cf. *Clarisia biflora*	1333	1710	10.4	360	22.9	192	0.23
cf. *Clarisia biflora*	1167	1575	13.5	261	29.0	142	0.30
cf. *Clarisia biflora*	556	1560	13.2	458	28.5	261	0.18
Mid- to sub-canopy							
Piper obtusifolium L.	3000	~90–100	<19	111	12.0	96.4	1.00
Ocotea sp.	2243	~90–100	<19	127	13.0	106.2	0.84
Aniba sp.	1737	~90–100	<19	159	14.0	132.7	0.67
Graffenrieda emarginata Ruiz & Pav.	1709	~90–100	<19	178	11.0	236.9	0.95

(continued)

Table 16.3 (continued)

	Max. observed sap flux density $J_{s(max)}$ mmol m^{-2} s^{-1}	PAR (at upper crown surface) μmol m^{-2} s^{-1}	D (at upper crown surface) hPa	DBH mm	Height (h) m	Sapwood area (A_s) cm^2	A_s/A_b
Psychotria brachiata Sw.	1406	~90–100	<19	61	6.0	20.7	0.71
Persea cf. *caerulea* Ruiz & Pav.	1080	~350	7[b]	13	n.d.	122.7	0.92
Cerro Tambo, N. Peru							
Unidentified species	5111	1125	13.1	127	17.8	60	0.57
cf. *Nectandra* sp.	3278	1125	13.1	172	19.0	77	0.48
Pera officinalis	2055	315	13.5	124	11.2	60	0.57
Unidentified species	1444	228	3.2	127	13.5	62	0.56
Vines							
Paullinia sp.	24774	~1800	17.0	22.85	n.d.	2.79	0.56
Liabum sp.	8144	~1800	19.0	19.15	n.d.	1.36	0.66
Mikania matezkii H.Rob. & W.C.Holmes	5539	900	15.0	31.50	n.d.	4.35	0.56
Mikania sp.	3159	900	15.0	25.53	n.d	3.58	0.70
Mikania sp.	1466	~350	7[b]	2.60	n.d.	5.31	1.00

[a] Pioneer
[b] Humid day

In the diurnal course J_s is driven by radiation (PAR) and the vapour concentration difference between leaf and air (Δw, approximated by vapour pressure deficit, D). Taking the daily sum of J_s as a linear function of the daily sum of PAR we observed $r^2 > 0.81$ over 85 days (Motzer 2003). Correlating the simultaneous sap flows of all individuals with each other yielded highly significant linear relationships ($r^2 > 0.88$; $P < 0.01$ two-tailed). These results form an excellent base for scaling-up water consumption from plant to stand and from days to years.

16.3.2.3 Plant Level: Sap Flow of Individuals from Different Forest Types

On a day-by-day base sap flow is an excellent measure of total tree transpiration, F (for sapwood hydro-activity, affecting the calculation of F from J_s, see below). In order to be able to compare individual plant water losses measured non-simultaneously and at different sites, mean daily water flows, F_{mean}, have been calculated from 80 to 170 days of typical weather conditions. We found a significant linear increase in F_{mean} with A_b at all studied sites (Fig. 16.2a, c, d) with clear differences between the forest types: For a given stem size (A_b) most water was consumed at the transition, intermediate at the slope, and minimal at the elfin forest (Fig. 16.2a). Trees and shrubs from ECSF elfin and slope forests follow the relationship (F_{mean} = 3.155+227.84*A_b; $r^2 = 0.81$). Moreover, although both stunted forests grow at different altitudes and are ca. 250 km apart from each other, their plants behaved in the same way (Fig. 16.2c). Clearly physiognomy and plant size (here for simplicity of measurement expressed as A_b which was found to be linear to A_c and in a saturation function to plant height; Motzer 2003, p. 52f.) have an impact on water consumption beside weather conditions, irrespective of the causes for the physiognomic similarities: At the ECSF's sub-páramo it is for the cold, windy climate with large amounts of rain while at Cerro Tambo it is for extremely nutrient poor soils and a more seasonal climate. At Cerro Tambo F_{mean} is low at LAI $\leq 4.4\,m^2\,m^{-2}$ (Table 16.1, Fig. 16.2d).

In Fig. 16.2b F_{mean} has been normalized for differences in stem size of plants. Irrespective of A_b and site, almost all plants transpire below 750–1000 kg day^{-1} ($m_{basal\,area}$)$^{-2}$ while the typical pioneer tree *Cecropia montana* consumes more than twice as much water per basal area. Despite the higher $J_{s(max)}$ in vines (Table 16.3) their F_{mean} values (*Paullinia* 3.06 kg day^{-1}, *Mikania* 0.45 kg day^{-1}, *M. matezkii* 0.80 kg day^{-1}, *Liabum* 0.44 kg day^{-1}) were as low as or even lower than in the elfin forest.

16.3.3 Scaling Up from Individual Plant to Transpiration on the Stand Level

Extrapolation from sap flow of individuals to plot scale is a function of the sum of sap flows ΣF of measured trees, of structural parameters such as sapwood A_s, crown projection A_c and basal area A_b, LAI, plant height h, plant-, species- or life form-specific parameters and soil characters. Up-scaling must rely on those parameters which can

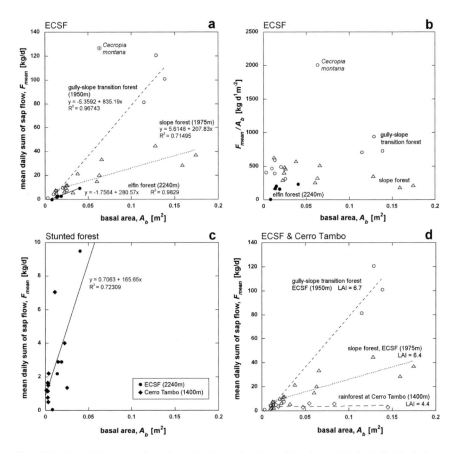

Fig. 16.2 Mean daily sum of sap flow (F_{mean}) as a function of basal area (A_b) for individual plants from different forest types (maximum standard deviation observed – resulting from different day-by-day weather conditions – did not exceed 20% of the absolute F_{mean} value). **a** Transition-, slope- and elfin forests at 1950–2240 m a.s.l.. Note the pioneer species *Cecropia montana*. **b** For the same forests and data as in **a** but with mean daily sap flows normalized for basal areas as F_{mean}/A_b. The special role of the pioneer species becomes evident. **c** Data for ECSF and Cerro Tambo stunted forests obey the same linear regression. **d** Transition- and slope-forests as in **a** but here in comparison with the rainforest at Cerro Tambo. *LAI*: Leaf area index

be determined with an adequate reliability in the specific situation of the plot. In a generalized form the resultant stand transpiration (E_c) is the product of mean A_s-, A_c-, or A_b-related sap flow (F_x) of measured individuals (x) and the corresponding structural parameter (SP), combined as the mean value ($\overline{F_{SP,x}}$) in the form:

$$E_c = \overline{F_{SP,x}} \cdot \sum_{i=1}^{n} SP_i.$$

In most approaches either A_s, A_b (partly also A_c), or a combination are taken (e.g. Aboal et al. 1999; Čermák and Kučera 1990; Köstner et al. 1998b; Santiago et al. 2000). Up-scaling was performed for the gully-slope transition forest at 1950 m a.s.l.. Here we distinguished between plants forming the overstorey (OV) above and the mid- to sub-canopy (SC) below 14 m stand height. Then stand transpiration follows from:

$$E_c = \frac{(A_{b,OV} \sum \overline{F_{b,OV}}) + (A_{b,SC} \sum \overline{F_{b,SC}})}{A_{plot}}$$

with the mean sap flow $\overline{F_b}$ (basal area related) of measured trees and total basal area of all trees on the plot, $A_{plot} = A_{b,OV} + A_{b,SC}$, but calculated separately for over-storey and mid- to sub-canopy trees. This term contains uncertainties from measurements on structural parameters, fluctuations of daily microclimate affecting sap fluxes and from distribution of cross-sectional xylem activity (see below; and for details, see Motzer 2003).

For plants in the understorey, transpiration was approximated from porometry (Motzer et al. 2005, 2007). Although an exact determination was impossible for the highly dynamic light and its consequences on stomatal behaviour (e.g. Küppers et al. 1997, 1999; Timm et al. 2004) its amount of 2–5% of E_c is so small that errors do not affect the overall result. Assuming homogenous xylem hydro-activity over each sapwood area (A_s) resulted in 20–24% overestimation of E_c on the stand level, whereas only a minor difference of 3.8% became evident assuming either a linear or an exponential decline of the hydro-activity from outer to inner xylem (Motzer 2003, pp 152–153). Measurements indicate this decline, resulting in stand transpiration from 1.0 mm day^{-1} to 2.4 mm day^{-1} (October 2000 to March 2001). Vines – here at their altitudinal limit of existence – contribute little to transpiration in transition and slope forests.

16.3.4 Energy Balance, Evapotranspiration and Transpiration of the Stand

Measured (2 m above the stand) annual global radiation input (G_{ann}) and annual net radiation balance (Q_{ann}) are given in Table 16.4. Knowing stand transpiration E_c, microclimatic conditions, and Q, it is possible to determine the contribution of E_c to actual stand evapotranspiration (E_a) from the term $Q = H + \lambda E + B + \mu A$. Here H indicates long-wave heat radiation (e.g. in sun from the warmer canopy to cooler ambient air), B a sensible heat flux into the stand, μA the amount of energy exchanged by metabolic processes, and λE ($=\lambda E_a$) radiation energy consumed for vaporization of water. μA is negligible (1–3% of Q). B is relevant only in the short-term when energy mainly from radiation is stored as heat in the stand.

Table 16.4 Amounts and percentages of evapotranspiration at the gully-slope transition forest (1975 m a.s.l.) for the observation period October 2000 to September 2001. P_{ann} = annual precipitation (mm), $E_{a(ann)}$ = annual actual evapotranspiration (mm), $E_{0(ann)}$ = annual potential evapotranspiration (mm), Q_{ann} = annual net radiation balance, here expressed as equivalent of vaporization ($Q_{E\text{-equivalent}}$; mm)

	$E_{a(ann)}$	$E_{0(ann)}$	$Q_{E\text{-equivalent}}$	P_{ann}
P_{ann}	26.4%	30.9%	37.9%	–
Q_{ann}	69.9%	81.6%	–	–
$E_{0(ann)}$	85.6%	–	–	–
Σ (mm)	560.7	654.2	802.5	2120

Annual radiation input (measured) G_{ann} = 3.94 GJ m^{-2}
Annual net radiation balance (measured) Q_{ann} = 1.996 GJ m^{-2}
Annual stand transpiration (measured) $E_{c(ann)}$ = 477–533 mm
Annual interception (calculated) $E_{i(ann)}$ = 28–84 mm
Mean annual vapour pressure deficit (measured) D_{ann} = 1.5–2.0 hPa
Mean annual air temperature (measured) t_{ann} = 15.7 °C (min. 12.5 °C; max. 19.0 °C)

λE was determined by modification of the Penman–Monteith equation (Monteith 1965) in analogy to stomatal conductance models (e.g. Jarvis 1995) including aerodynamic boundary (g_a) and canopy conductances (g_c; for details, see Motzer 2003, p. 165). Knowing g_c, the fraction of energy consumed for "imposed evaporation rate" $\lambda E_{(imp)}$ – driven by D – could be separated from $\lambda E_{(eq)}$, driven by the available energy, $Q - B$. For most days atmospheric coupling of the canopy was good [decoupling coefficient Ω (McNaughton and Jarvis 1983) decreased to 0.2 (Motzer et al. 2005)], indicating a generally high g_a (for further details, see Motzer 2003, pp 155–158, 173, 174).

The driving variables for λE are $Q - B$, D, g_c, and g_a. Knowing those it becomes possible to separate leaf-controlled stand transpiration E_c from passive evaporation $E_i = E_a - E_c$ (e.g. from interception). Measurements of diurnal courses show that on (rare) cloudless days E_a may reach water loss rates of 0.4 mm h^{-1}, with E_c only slightly below it, demonstrating the dominant role of stand transpiration in the overall evapotranspiration. For the mid- and sub-canopy the daily period of transpiration (E_{SC}) is clearly shorter than for E_c. In the long term E_{SC} contributes to 20–25% of the daily E_c, but on cloudless days it may reach 50% over short periods, especially when turbulences reach the forest interior and generally reduce Ω. On still, covered and rainy days sub-canopy transpiration is suppressed by low D, moisture on leaf surfaces and increased Ω.

Mean values of daily sums of transpiration amounted to 1.8 mm day^{-1} (dryer part of season, October 2000 to March 2001), on clear days up to a maximum of 2.8 mm day^{-1} when highest hourly rates of 0.38 mm h^{-1} were observed. But even for rainy days with long lasting precipitation daily transpiration amounted to 0.7–1.0 mm day^{-1} (see Motzer 2003, pp 175ff).

16.3.5 Annual Precipitation, Annual Potential and Actual Evapotranspiration and Annual Transpiration on the Stand Level

Figure 16.3 shows annual courses of monthly based precipitation (P), net radiation balance (Q), potential (E_0), and actual evapotranspiration (E_a) determined from day-by-day measurements. E_0 is calculated solely from microclimatic parameters (Penman 1948), not including vegetation effects. The driest months coincided with highest Q, the highest P was observed in June. Lowest Q was measured in February during long lasting cloud cover and modest rain. During the rainy season E_a was almost as high as E_0, but in the dryer part of the year E_a was much lower. Since 85% to >90% of E_a are contributed by stand transpiration E_c, the difference between E_0 and E_a indicates at what time of the year vegetation (mainly through stomata) exerted controlling effects on E_a: this was the case during the dryer period and only to a little extend during the rainy season.

The measured annual net radiation balance (Q_{ann}) would have allowed for vaporizing of an equivalent ($Q_{E\text{-equivalent}}$) of 37.9% of P_{ann} (Table 16.4), a fact demonstrated by the low means of daily potential (1.8 mm day^{-1}) and actual (1.52 mm day^{-1}) evapotranspiration rates. The measured annual stand transpiration [$E_{c(ann)}$] consumed 59–66% of the total available energy (Q_{ann}) but only 22–25% of annual precipitation. Some 1.3–4.0% of P_{ann} were vaporized as annually intercepted water [$E_{i(ann)} = E_{a(ann)} - E_{c(ann)}$]; soil evaporation was negligible due to stand structure.

One may ask the question of whether species diversity affects these parameters of transpiration on different scales. This depends on the definition of "diversity" used: If it considers number of species only, then "diversity" has no effect as long as all species belong to the same "functional type" and are of similar physiognomy. However, if "diversity" contains a component of frequency of individuals of a

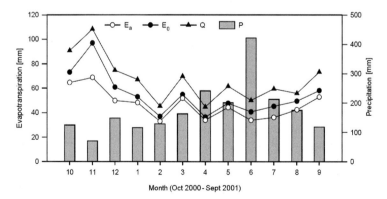

Fig. 16.3 Monthly sums of evapotranspiration, precipitation (P; mm) and monthly net radiation balance (Q; mm, as equivalent to vaporization) for the observation period October 2000 to September 2001. E_a = monthly actual, E_0 = monthly potential evapotranspiration

species, than it does have an effect, for large numbers of individuals enforce smaller individual tree sizes. "Functional type diversity" (here we have three different functional types) may have an effect, but only when all functional types are represented by a sufficiently high number of individuals.

16.4 General Discussion and Conclusions

All studies have been performed in forests of high species diversity (Homeier et al. 2002). This raises the question of whether species-specific characteristics are decisive for sap flow besides the environmental parameters. But in all cases they were much less important in determining sap flow than the "social" position of the investigated tree (indicated by position (overstorey to sub-canopy) or size (Fig. 16.2): All individuals follow the same linear relationship of daily sap flow (F_{mean}) and basal area (A_b, a measure of "social position" for its relationship to crown extension A_c and tree height h). Only when a species belongs to another functional group (like *Cecropia montana*, pioneer; Fig. 16.2a, b) or when the forest type changes (Fig. 16.2a, c, d) did the relationship vary. Obviously, sap flux can be measured across tree species to yield a stand-specific pattern, so that species are, with respect to this particular question, not important.

Tree based daily transpiration rates given in Fig. 16.2a are in the lower third of those from tropical lowland rainforests (e.g. Anhuf et al. 1997; Dünisch and Morais 2000; Granier et al. 1992, 1996; James et al. 2002; Oren et al. 1996). Data for trees of mountain forests are rare and contradictory; some are of similar (Zotz et al. 1998) others of much less (Santiago et al. 2000: 35%) or much higher magnitude (e.g. James et al. 2002; Jordan and Kline 1977: up to 3-fold). Presumably this is due to the large physiognomic, climatic, and ecosystemic variability of mountain forests as indicated for the two adjacent ECSF forests in Fig. 16.2a and the taller, dryer and more open rainforest at Cerro Tambo (Fig. 16.2d). In contrast, such physiognomic different stands like stunted and slope forests show great similarities in the relationship between daily tree transpiration (F_{mean}) and basal area (A_b; Fig. 16.2a, c). However, lower daily transpiration rates as compared with lowland forest trees are not a consequence of differences in hydraulic properties since short-term maximum rates of up to 20kg h^{-1} tree^{-1} and maximum sap flux densities fully coincide with those of lowland trees (Oren et al. 1998). The lower rates result from long lasting cloud covers.

Irrespective of large tree transpiration rates, stand transpiration E_c was low for the low abundance of huge trees. Daily rates of E_c = 1.8mm day^{-1} are typical and are even 20–30% lower in the rainy season.

The annual actual stand evapotranspiration $E_{a(ann)}$ of tropical lowland rainforests ranges over 1000–2000mm year^{-1} (40–75% of annual precipitation P_{ann}) and the stand transpiration [$E_{c(ann)}$] ranges over 800–1500mm year^{-1} (30–56% of P_{ann}; e.g. Hölscher et al. 1997; Jordan and Heuveldop 1981; Leopoldo et al. 1995; Lesack 1993; Roberts et al. 1993). This is 3.6-fold more than the $E_{a(ann)}$ [3.1-fold more than the $E_{c(ann)}$] of the transition forest.

In an independent catchment study, Wilcke et al. (Chapter 12 in this volume) and Fleischbein et al. (2006) report $E_{a(ann)}$ ranging over 954–1570 mm year^{-1} (April 1998 to April 2002) for three micro-catchments in the vicinity of the transition forest plot at stand transpiration rates $E_{a(ann)}$ of 471–568 mm year^{-1}. Their $E_{a(ann)}$ are up to 2.8-fold higher than our estimates (Table 16.4). An 18% higher precipitation in their study cannot account for this discrepancy. In contrast to the statement of Fleischbein et al. (2006) and Wilcke et al. (Chapter 12) that Motzer's (2003) "estimate is based on a data set covering less than a year during a relatively dry period", Motzer did continuously monitor solar radiation input (G) and net global radiation balance (Q) 2 m above the canopy–atmosphere interface for *more* than a year at the transition forest plot, which is typically exposed in comparison to the catchment forests. Motzer's measured annual global radiation input of G_{ann} = 3.94 GJ m^{-2} year^{-1} (4.81 GJ m^{-2} year^{-1} at the nearby reference station over an open field) and measured annual net radiation balance Q_{ann} = 1.996 GJ m^{-2} year^{-1} at the plot are reasonable estimates for what is known for such mountain environments of frequent fog and cloud cover and typically overcast days (compare e.g. Larcher 2003; Monteith and Unsworth 1990; Schulze et al. 2005). Since it was obtained in a year with a relatively dry period, the value is likely to be above the long-term mean. It allows for a theoretical maximum evaporation of $Q_{E\text{-equivalent}}$ = 802.5 mm year^{-1} if all available radiation energy is consumed for vaporization of water (Larcher 1975 gives 2.3 mm day^{-1} = 839.5 mm year^{-1} for a similar environment), a value clearly lower than any deduced $E_{a(ann)}$ from the catchment studies of Wilcke et al. (Chapter 12) and Fleischbein et al. (2006). The latter give an annual interception of $E_{i(ann)}$ = 713–995 mm year^{-1} (which is equivalent to the full vaporization of 100–200 mean central European thunderstorm events!) in a normally overcast, foggy mountain environment where temperatures are always too low for massive vaporization [observed min. 12.5 °C, max. 19.0 °C, annual mean 15.7 °C at a mean annual vapour pressure deficit (D) of 1.5–2 hPa (maximum 11 hPa in one event), mean annual absolute humidity of 12.5 g kg^{-1} at very little variation, all above canopy; Motzer 2001, p. 71–72]. Even for their $E_{i(ann)}$, the measured Q_{ann} (Table 16.4) is too low. Catchments integrate over several forest types without being able to differentiate, and uncertainties concerning appropriate measurements of P_{ann}, canopy throughfall (see Chapter 18; also see Rollenbeck et al. 2007, Fig. 8) and leakages are unavoidable. However one must take care when scaling-up from plants to plots and stands, especially for uncertainties in the hydro-activity of sap wood areas in trees.

The few mostly catchment based studies on mountain forests give $E_{a(ann)}$ of 300–1600 mm year^{-1} (e.g. Cavelier et al. 1997; Fleischbein et al. 2006; Santiago et al. 2000). According to Bruijnzeel (2001), high mountain forests under strong effects of fog exhibit an $E_{a(ann)}$ of 310–390 mm year^{-1}, a value gradually increasing to 1155–1380 mm year^{-1} in lower montane forests at decreasing cloud effects, then becoming similar to that of lowland rainforests. Following this differentiation and in excellent agreement with the results of Bruijnzeel (2001), the transition forest studied here [$E_{a(ann)}$ = 560.7 mm year^{-1}; Table 16.4] belongs to the hydrological type of mountain forests under moderate influence of fog, mainly through reduction of radiation rather than by water coming from fog.

It is concluded that:

1. Species-specific characters are generally less important than life-forms (vine, tree), any association with a functional group (pioneer to late-successional), and the social position of a plant.
2. In all forest types total tree water consumption is directly related to plant size (linearly to basal diameter) but is affected by vegetation type.
3. A detailed survey of xylem activity in tropical trees is needed.
4. Measured annual stand evapotranspiration in the slope forest is low for the high (relative) air humidity and permanent cloud cover: it consumes 70% of the available energy, 26% of precipitation. Interception is low for the same reasons: 15% of evapotranspiration, 10% of available energy. Consequently, most precipitated water leaves the ecosystem via run-off or as sub-surface flow.

Chapter 17
Plant Growth Along the Altitudinal Gradient – Role of Plant Nutritional Status, Fine Root Activity, and Soil Properties

N. Soethe(✉), W. Wilcke, J. Homeier, J. Lehmann, and C. Engels

17.1 Introduction

In tropical montane forests, aboveground net primary productivity (ANPP) usually decreases with increasing altitude. Besides low photosynthesis (Kitayama and Aiba 2002) and direct impact of low temperatures on plant growth (Hoch and Körner 2003), low ANPP at high altitudes has often been attributed to nutrient limitation (Bruijnzeel et al. 1993; Bruijnzeel and Veneklaas 1998; Tanner et al. 1998).

Plant growth is often correlated with nutrient availability in tropical montane forests. For example, the exceptionally high tree stature in a montane forest stand in Papua New Guinea was attributed to its nutrient rich soil parent material (Edwards and Grubb 1977). In montane forests of Jamaica (Tanner et al. 1990), Hawaii (Vitousek and Farrington 1997; Vitousek et al. 1993), and Venezuela (Tanner et al. 1992), trunk diameter growth and leaf production of several native tree species were enhanced by addition of N or P.

The nutritional status of plants is governed by the amounts of chemically available nutrients in soil and the ability of fine roots for nutrient acquisition. The ability for nutrient acquisition comprises the spatial exploitation of the soil by roots and nutrient uptake activity. Chemical nutrient availability in tropical montane forests may be affected by parental substrate, weathering intensity, cation exchange capacity, the rates of litter decomposition, or extracellular phosphatase activity (Treseder and Vitousek 2001; Kitayama and Aiba 2002; Wilcke et al. 2007). Spatial nutrient availability is dependent on the exploitation of soil by roots or mycorrhizal hyphae and the mobility of the respective nutrient in soil. High abundance of mycorrhizal fungi contributes to high spatial availability of nutrients in the organic surface layer (Treseder and Vitousek 2001; Haug et al. 2004; Kottke et al. 2004). Also fine root abundance in the organic layers is generally very high (Hertel et al. 2003). Unfavourable soil conditions such as shallow mineral soils (Ostertag 2001), oxygen deficiency (Santiago 2000), low nutrient concentrations (Cavalier 1992), and low pH (Godbold et al. 2003) may cause a superficial distribution of fine roots, and may impair the physiologically based ability of roots for nutrient uptake in deeper soil layers.

We present two different approaches that elucidate the impact of nutrient supply on plant growth along the altitudinal gradient. In a soil-based approach soil nutrient concentrations were correlated with tree basal area growth rates. In a plant-based approach, the nutritional status of trees was assessed by foliar analysis, and the ability of plants for nutrient acquisition was estimated from measurements of the vertical pattern of rooting and nutrient uptake. It was hypothesized: (a) that plant growth at high altitudes is limited by nutrient deficiency, and (b) that nutrient deficiency is related both to decreased nutrient contents in soil and impaired rooting in mineral soil.

17.2 Relationship Between Tree Growth Along the Altitudinal Gradient and Soil Properties

17.2.1 Tree Growth Along the Altitudinal Gradient

Tree diameter growth rates (for methods, see Homeier 2004) were measured at eight permanent 400 m^2 plots between 1960m and 2450m a.s.l. within the Reserva Biologica San Francisco.

The tallest trees were located on slopes at the lower elevations of the study area (forest type I, see Chapter 10.1). At the lower end of our transect Homeier (2004) found a larger variation in tree growth rates than at higher altitudes. Tree basal area growth rates decreased significantly with increasing altitude (see Chapter 21).

17.2.2 Relationship Between Soil Properties and Tree Area Growth Rate

The concentrations of all macronutrients (N, S, P, K, Ca and Mg) in the organic layer decreased with increasing altitude (see Chapter 9). Tree basal area growth rates were most tightly correlated with P concentration (Fig. 17.1).

There were further positive correlations between basal area growth rates and concentrations of N, K, and Mg in the organic layer (Table 17.1). The high variability of growth rates at the two lowermost study sites reduced the quality of the fit. This was particularly true for K and Mg. Total Ca and S concentrations in the soil organic layer were not related to basal area growth rates. In the organic layer most K, Ca, and Mg is bound in exchangeable form, and thus, is readily available to plants, while most N, P, and S is incorporated into soil organic matter, and their availability is governed by mineralization rates (Wilcke et al. 2002). At lower altitudes of the study area average turnover time for N and P in the organic layer is 14 years and 11 years, respectively, calculated as the quotient of storage in the organic layer to flux by litterfall (Wilcke et al. 2002). Taking only aboveground fine litterfall into account results in an underestimation of turnover rates because the belowground litter input is likely to be as large as the aboveground litter input (Röderstein

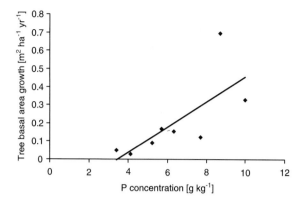

Fig. 17.1 Relationship between total P concentrations in the organic layer and tree basal area growth

Table 17.1 Regression equations (R.E.) and correlation coefficient (r) of tree area growth rates (Y; m² ha⁻¹) against nutrient concentrations (g kg⁻¹) and C to nutrient ratios (x) in the organic layer

Nutrient	R.E.	r
N	$Y = 0.05x - 0.60$	0.67
P	$Y = 0.06x - 0.21$	0.70
S	$Y = 0.28x - 0.24$	0.51
K	$Y = 0.08x + 0.03$	0.64
Ca	$Y = 0.06x + 0.09$	0.60
Mg	$Y = 0.29x - 0.08$	0.67
C:N	$Y = -0.03x + 0.97$	-0.73
C:P	$Y = -0.01x + 0.66$	-0.74

et al. 2005). Gross mineralization rates depend on the N and P stocks in soil organic matter – mainly in the organic layer. At 47 study sites between 1900 m and 2200 m, the organic layer had a thickness of 2–43 cm, a mass of 30–713 t ha⁻¹, and a storage of 0.87–21.0 t N ha⁻¹ (mean 5.5 t ha⁻¹) and 0.03–0.70 t P ha⁻¹ (mean 0.21 t ha⁻¹). Thus, between 1900 m and 2200 m several hundreds of kilograms of N and several tens of kilograms of P per hectare would be released annually from the organic layer by gross mineralization.

At higher altitudes, nutrient supply in soil may become deficient to plants not only because of lower nutrient concentrations but also because of lower mineralization rates. Accordingly, tree basal area growth rates were significantly negatively correlated with C:N and C:P ratios of the organic layers (Table 17.1). High C:nutrient ratios indicate low mineralization rates of nutrients in the organic layer. Mineralization rates are further decreased at higher altitudes because of lower soil pH, lower temperatures and less oxygen availability due to waterlogging (Schrumpf et al. 2001; Wilcke et al. 2002; Wegner et al. 2003; McGroddy et al. 2004; Soethe et al. 2007). In conclusion, our data indicate that reduced tree growth rates at higher altitudes correlate with lower nutrient supply from the organic layer.

17.3 Foliar Nutrient Concentrations and Ability for Nutrient Acquisition Along the Altitudinal Gradient

17.3.1 Nutritional Status of Plants at Different Altitudes

Nutritional status of plants at different altitudes was determined by foliar analysis. Therefore, nutrient concentrations (N, S, P, K, Ca, Mg) in mixed leaf samples were assessed separately for the plant groups "trees" (lignified and higher than 3.0 m), "shrubs" (lignified and with a height of 0.5–3.0 m) and "herbs" (not lignified).

At 1900 m, foliar nutrient concentrations of trees were within the ranges of nutrient sufficiency given in textbooks for temperate broadleaved tree species (Table 17.2; Bergmann 1993). At this altitude, foliar concentrations of all nutrients were higher than the ranges of nutrient deficiency found in other studies for trees of tropical montane forests (Vitousek et al. 1995; Vitousek and Farrington 1997) and tropical lowland forests (Drechsel and Zech 1991). Also foliar micro-nutrient concentration indicates that at 1900 m growth was not limited by nutrient deficiency. This is in contrast to other studies where growth limitation of tropical forestes by N or P deficiency has been suggested (Tanner et al. 1990; Vitousek et al. 1993).

In comparison with lower altitudes, forest growth at 1900 m was reduced also in South Ecuadorian montane forests. Possible causes for reduced growth in these forests include increased cloudiness (Bruijnzeel and Veneklaas 1998) and direct growth reduction by low temperatures (Hoch and Körner 2003).

The effect of increasing altitude on foliar nutrient concentrations of trees varied depending on the specific nutrient (Table 17.2). While concentrations of Ca and Mg were not affected by altitude, the concentrations of all other nutrients were lower at 2400 m and 3000 m than at 1900 m. The decrease of the concentrations of N, P, S, and K at higher altitudes was similar for N, P, S, and K, ranging from 30% to 48%.

Edwards and Grubb (1982) suggested that decreasing foliar N concentrations at high altitudes of tropical montane forests reflect the development of thicker cell walls, which may not be related to N shortage but e.g. aid in minimizing the infestation of fungi under cool and moist conditions. It was found that foliar N and P concentrations increased with increasing altitude when expressed on a leaf area basis, whereas concentrations decreased when based on leaf dry matter (Vitousek et al. 1992; Kitayama and Aiba 2002). In the present study foliar N and P concentrations decreased with increasing altitude, regardless of whether concentrations were based on leaf dry matter or leaf area (Soethe et al., unpublished data).

This shows that at higher altitudes nutrient acquisition of plants was even more affected than biomass growth leading to dilution of nutrients in leaf dry matter. At 2400 m and 3000 m, foliar concentrations of N, S, and K were below the ranges of

Table 17.2 Effect of altitude on nutrient concentrations in youngest fully developed leaves of trees, shrubs and herbs. Different upper-case *letters* indicate significant differences between altitudes (Tukey test; $P=0.05$); data in parentheses show standard errors

	Location	N	P	S	K	Ca	Mg
		Nutrient concentrations (mg g^{-1})					
Trees	This study						
	1900 m	21.7 (1.3)^A	2.2 (0.2)^A	1.9 (0.1)^A	10.0 (1.2)^A	6.2 (1.7)^A	3.4 (0.6)^A
	2400 m	13.5 (1.2)^B	1.2 (0.1)^B	1.1 (0.1)^B	7.0 (0.4)^{AB}	7.4 (1.0)^A	2.7 (0.2)^A
	3000 m	11.3 (0.3)^B	1.4 (0.1)^B	1.2 (0.1)^B	5.3 (0.4)^B	4.9 (0.3)^A	3.2 (0.2)^A
Ranges of nutrient sufficiency							
Temperate zones[a]		17–40	1.5–3.0	–	10–18	2–18	1.5–4.0
Tropical zones[a]		14–30	1.0–2.5	1.5–2.5	10–23	9–40	2.5–8.0
Ranges of nutrient deficiency							
Tropical montane forest[b]		8.7–14.2	0.6–1.0	–	–	–	–
Tropical zones[c]		6–20	0.3–1.1	0.3–0.5	1.0–5.7	–	0.7–0.9
Shrubs	This study						
	1900 m	22.3 (2.6)^A	2.0 (0.4)^A	1.8 (0.2)^A	10.2 (1.2)^A	9.2 (2.4)^A	3.6 (0.6)^A
	2400 m	14.9 (0.9)^B	1.3 (0.1)^A	1.3 (0.2)^A	7.7 (0.8)^A	9.7 (1.4)^A	2.8 (0.3)^A
	3000 m	13.4 (0.7)^B	1.7 (0.1)^A	1.2 (0.1)^A	9.4 (0.6)^A	6.4 (0.9)^A	3.5 (0.5)^A
Herbs	This study						
	1900 m	23.7 (1.9)^A	3.3 (0.4)^A	2.2 (0.2)^A	22.4 (1.6)^A	10.0 (1.5)^A	4.5 (0.5)^A
	2400 m	15.2 (1.2)^B	1.5 (0.1)^B	1.4 (0.4)^B	15.4 (1.3)^B	5.9 (0.6)^{AB}	2.6 (0.1)^B
	3000 m	12.3 (0.7)^B	1.5 (0.1)^B	1.1 (0.1)^B	11.8 (1.4)^B	5.6 (0.8)^B	3.5 (0.4)^{AB}

[a] Ranges of foliar nutrient concentrations where neither deficiency nor toxicity occurs on temperate or tropical broadleaved tree species, respectively (Bergmann 1993).
[b] Ranges of foliar nutrient concentrations of the tree species *Metrosideros polymorpha* growing in Hawaiian tropical montane forests on sites where the respective nutrient was limiting plant growth (Vitousek et al. 1995; Vitousek and Farrington 1997).
[c] Ranges of foliar nutrient concentrations where deficiency symptoms occurred at several tropical broadleaved tree species (Drechsel and Zech 1991).

sufficiency according to all reference data given by Bergmann (1993). Foliar concentrations of N, P, and K were within or only slightly higher than the ranges indicating deficiency in tropical forests (Drechsel and Zech 1991; Vitousek et al. 1995; Vitousek and Farrington 1997).

Foliar nutrient concentrations at higher altitudes did not indicate Mg deficiency. This is in apparent contrast to the close correlation between tree area growth rates and Mg concentrations of the organic layer (Table 17.1). At our study sites, the supply of several nutrients from the organic layer decreased at higher altitudes. Reduced nutrient supply was associated with lower biomass growth. For Mg supply from soil and plant uptake were in balance with growth even at high altitudes, and therefore, foliar concentrations were not reduced to the range of deficiency despite the lower Mg supply from soil.

In other tropical montane forests, tree growth was limited by N and P but not by base cations (Vitousek and Farrington 1997). The low supply of K and Mg in the present study is also in line with previous findings of Wilcke et al. (2002) that, at acid micro-sites within the study area, base metals were immobilized during incubation of the organic layer by micro-organisms.

It is well known that foliar nutrient concentrations vary between plant species (Tanner et al. 1990; Vitousek et al. 1995). For example, foliar N concentrations are higher in leguminous species than in non-leguminous species (Drechsel and Zech 1991). Slow-growing species adapted to nutrient-poor habitats usually show higher foliar nutrient concentrations than fast-growing species when growing under nutrient-poor conditions (Chapin 1980). The species composition at our forest sites completely changed with increasing altitude (Homeier 2004). However, compared with the effect of altitude, the effect of plant group (trees, shrubs, herbs) on foliar nutrient concentrations was negligible (with the exception of increased K concentrations in herbs at all altitudes and increased P concentrations of herbs at 1900m; Table 17.2). This indicates that the differences of foliar nutrient concentrations at different altitudes were not due to the occurrence of specific species, and supports the assumption that the decrease of nutrient concentrations at 2400m and 3000m in comparison with 1900m was induced by environmental effects on nutrient availability in soil or nutrient acquisition of plants.

17.4 Plant Ability to Acquire Nutrients at Different Altitudes

The availability of nutrients to plants depends on the exploitation of soil by roots, nutrient uptake by roots, and nutrient supply in soil. As an estimate of the exploitation of soil by roots, the vertical distribution of root length densities (RLD) of fine roots (<2mm diam.) was assessed in three permanent 400 m² plots at 1900m and 2400m in the RSF as well as at 3000m in Cajanuma near to the north-eastern entrance of the Podocarpus National Park. Root length densities were highest in the organic layer (Fig. 17.2), where also high amounts of nutrients were stored (see Chapter 9). This has also been reported for an old tropical montane forest (Hertel et al. 2003) and a lower montane forest (Vance and Nadkarni 1992) in Costa Rica and is in accordance with Ostertag (2001), who found highest root lengths in the upper soil layers of a Hawaiian tropical montane forest. Taking 1.8 cm fine roots cm^{-3} soil (Claassen and Steingrobe 1999; Yanai et al. 2003) as a proxy where most N_{min} in soil is spatially available, RLD suggest that all N_{min} in the organic layer becomes spatially available to plants in the organic layer. For immobile nutrients

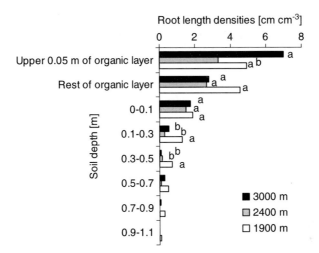

Fig. 17.2 Vertical distribution of RLD at different altitudes and seasons. Lower-case *letters* indicate significant differences of RLD between altitudes (Dunn test; $P < 0.05$; $n = 8$–20). Redrawn from Soethe et al. (2006a) with kind permission of Springer Science and Business Media

such as P considerably higher RLD are necessary to afford access to all chemically available stocks (Claassen and Steingrobe 1999). However it can be expected that, especially in the organic layers, the spatial availability of P is increased by symbiosis with mycorrhizal fungi (Kottke et al. 2004). Thus, nutrient acquisition from the organic layer is presumably governed by the chemical rather than by spatial availability.

With increasing depth in mineral soil, RLD decreased more sharply at 3000 m and 2400 m than at 1900 m (Fig. 17.2). At 2400 m and 3000 m, RLD decreased to values lower than $0.5\,\mathrm{cm\,cm^{-3}}$ at soil depths below 0.1 m and 0.3 m, respectively, while at 1900 m RLD fell below $0.5\,\mathrm{cm\,cm^{-3}}$ only at a soil depth of 0.7–0.9 m. At a depth of 0.1–0.5 m, RLD were significantly higher at 1900 m than at the upper two sites. Below these depths low numbers of replicates did not allow statistical analysis.

In accordance with the impaired exploitation of mineral soil by fine roots at higher altitudes, N uptake activity decreased more sharply with increasing soil depth at 3000 m than at 1900 m (Soethe et al. 2006a). Averaging of all plant groups (trees, shrubs, herbs, saplings), 43% of the N obtained from the organic layer was acquired from 0.05 m soil depth at 1900 m, but only 19% at 3000 m. From 0.4 m soil depth, 32% of the N obtained from organic layer was acquired at 1900 m, in comparison with 2% at 3000 m. The pattern of N uptake activity of trees was very similar to the relative distribution of RLD, indicating that the physiologically based ability for nutrient acquisition of fine roots was not affected by unfavourable soil conditions in deeper layers. However, the pattern of both RLD and N uptake activity indicated that the spatial availability of nutrients in mineral soil was smaller at higher than at lower altitudes, possibly as a result of oxygen deficiency in deeper soil layers and shallower mineral soils (Soethe et al. 2006a). Spatial availability of nutrients in deeper soil layers is further reduced by low abundance of mycorrhizal hyphae (Powers et al. 2005). At higher altitudes, foliar P concentrations indicated P deficiency even though the amount of plant available P, as determined by Mehlich III extraction (Soethe et al. 2007), was high in comparison with the annual requirements of plants (Lodhiyal and Lodhiyal 2003). Thus, at high altitudes, unfavourable soil conditions for root growth in deeper soil layers impaired the plant nutritional status, despite the high chemical availability of nutrients in the mineral soil.

17.5 Conclusions

Foliar concentrations of N, P, S, and K were significantly lower at 2400 m and 3000 m than at 1900 m, showing that at high altitudes uptake of these nutrients was even more reduced than biomass growth. Comparison of our data with critical foliar concentrations given in the literature indicates growth limitation by nutrient deficiency at high altitudes. This suggestion is supported by significant correlations between tree basal area growth rates and nutrient concentrations in the organic layer. Our data on rooting pattern indicate that low spatial exploitation of the mineral soil by plant roots is an important factor that contributes to low nutrient acquisition of plants and nutrient deficiency at high altitudes.

Acknowledgements We thank INEFAN for granting the research permit and the Fundacíon Científica San Francisco for logistic support at the Estacíon Científica San Francisco.

Chapter 18
Spatial Heterogeneity Patterns – a Comparison Between Gorges and Ridges in the Upper Part of an Evergreen Lower Montane Forest

M. Oesker(✉), H. Dalitz, S. Günter, J. Homeier, and S. Matezki

18.1 Introduction

In the RBSF, as in many tropical mountain regions, the topography consists of a complex system of crests, ridges, steep slopes, valleys and gorges. These different physical features have an influence on the vegetation composition, stand structure and further processes, which also change with increasing elevation (Lieberman et al. 1996; see Part III 1 in this volume) as well as at the same elevation in line with the topography along a horizontal gradient (Takyu et al. 2002; Tanner 1977). 'The topography, including the elevation, is a major physical factor, which affects the composition, growth, and distribution of tropical forests' (cf. Basnet 1992). Climate parameters such as wind speed and light condition or transpiration (see Chapter 16 in this volume) may be different at ridges and gorges. Because of erosion of soils, soil contents are expected to accumulate on lower slopes and in valleys and gorges (Kubota et al. 2004). Therefore, the hypothesis for the study was that changes in abiotic parameters alter the vegetation structure and composition along the vertical gradient as well as along the horizontal gradient formed by the topographical structure in the lower part of the RBSF (see Chapter 10.3 in this volume).

18.2 Data and Methods

To quantify the impact of topography in the study area, six permanent plots (20×20 m) were established between 1960 m and 2070 m a.s.l., three on the ridge (T2) and three in the adjacent gorge (Q2) at the same elevation. These areas are classified by Homeier et al. (Chapter 10.1) as forest type I (Q2) and type II (T2). In all six plots the data of canopy structure, throughfall (TF; description of collectors given by Rollenbeck et al. 2007) and soil contents were recorded. Canopy structure, defined through its parameters LAI, leaf angle, canopy openness and values of the canopy permeability to radiation, was investigated using hemispherical images and analysed using the software package HemiView (ver. 2.1; DeltaT). Electrical conductivity (EC), pH and potassium, calcium and magnesium concentrations of the

TF were measured for a one-year period between November 2001 and November 2002 at weekly intervals in 31 collectors in the permanent plots on the ridge and in the gorge, respectively. TF samples, taken in October 2002, were transported in a frozen state and in addition components such as nitrate, ammonium, phosphate, chloride, organic N and total C, Li, Al, total P, Mn and Rb were measured. Soil samples were taken in October 2004 separately at a depth of 0–10 cm and at 10–20 cm. Ammonium acetate extractions were done with a 5 g sample in 100 ml solution (1 M). These extracts were used to provide evidence for the plant available fraction of the following elements in the soil: Al, Ca, K, Mg, Mn, Na and Sr. EC and pH were measured in a 10 g soil sample in 100 ml H_2O.

Air temperature and humidity inside the forest were measured using four HOBO H8 sensors, two at 1960 m a.s.l. in the gorge and ridge and two at 2070 m a.s.l., installed in 2 m height. The data was collected over a period of one year for the ridge and gorge sites and the annual mean was calculated.

Composition and species richness of trees and climbers were studied within all plots. All trees with a diameter at breast height (dbh) of ≥5 cm were inventoried. Tree diameter growth rates with a dbh of ≥10 cm were measured with dendrometer bands for the years 2001 and 2002 (Homeier 2004). To describe climber diversity, all woody climbers (dbh ≥0.5 cm) which were rooted within the plots were recorded. All tree and climber individuals in the permanent plots were identified to species where possible, otherwise to morpho-species. Images are available at www.visualplants.de (Dalitz and Homeier 2004).

In order to verify the horizontal heterogeneity, a diagnostic sampling was carried out on four transects along ridges and four parallel transects along gorges over a total sampling area of 4.6 ha (for details, see Günter and Mosandl 2003). The altitudinal changes in structure were demonstrated by two additional transects between 1850 m and 2400 m a.s.l. along a gorge and the neighbouring ridges. The height, abundance and basal area of the trees were measured using the Bitterlich method (Bitterlich 1948; data from Hilpmann 2003).

For statistical analysis, the Mann–Whitney U-test was chosen to compare the plant composition data and a t-test for chemical and stand structure data. Statistical analysis was performed using STATISTICA software (ver. 7.0; StatSoft).

18.3 Results and Discussion

In general no clear pattern of enrichment of elements in the investigated gorge emerged, however results of the soil extractions with ammonium acetate showed significant differences between ridge and gorge (Table 18.1). EC and pH values differ significantly. The EC values were low at both sites, with higher values on the ridge (about 100 µS cm^{-1}) as compared with 40–60 µS cm^{-1} in the gorge. The pH values were (pH 3.8) significantly lower on the ridge than in the gorge, however the highest values in the gorge were still low (~ pH 4.5). Lower pH values at ridges as

Table 18.1 Comparison of soil data between ridge and gorge. The pH and EC data were measured in soil solutions ($n = 27$). Element concentrations were measured after extraction with ammonium acetate ($n = 31$). Data are given for 0–10cm and 10–20cm depths, as mean with SD. Statistics were done with the Student t-test (* $P \leq 0.05$, ** $P \leq 0.01$, *** $P \leq 0.001$)

Depth	Parameter		Ridge Mean	SD	Gorge Mean	SD	P	Significance
0–10cm	EC	(µS cm⁻¹)	97.80	24.48	56.87	42.19	0.00000	***
	pH	–	3.81	0.25	4.75	0.50	0.00000	***
	Al	(mg kg⁻¹)	11.95	9.04	25.10	12.69	0.00000	***
	Ca	(mg kg⁻¹)	155.21	88.41	1443.52	1358.98	0.00000	***
	K	(mg kg⁻¹)	521.06	145.16	275.42	235.42	0.00000	***
	Mg	(mg kg⁻¹)	422.12	194.69	337.40	273.01	0.04051	*
	Mn	(mg kg⁻¹)	5.82	5.08	39.98	52.18	0.00000	***
	Na	(mg kg⁻¹)	115.24	64.71	136.82	22.24	0.00903	**
	Sr	(mg kg⁻¹)	2.92	1.76	19.96	19.29	0.00000	***
10–20cm	EC	(µS cm⁻¹)	104.79	32.02	38.45	15.59	0.00000	***
	pH	–	3.64	0.18	4.50	0.29	0.00000	***
	Al	(mg kg⁻¹)	10.89	11.80	17.02	8.22	0.00074	***
	Ca	(mg kg⁻¹)	130.70	23.68	650.11	846.82	0.00000	***
	K	(mg kg⁻¹)	369.75	192.62	93.87	76.11	0.00000	***
	Mg	(mg kg⁻¹)	245.06	194.66	145.15	183.68	0.00183	**
	Mn	(mg kg⁻¹)	1.44	1.22	10.24	12.83	0.00000	***
	Na	(mg kg⁻¹)	150.38	18.01	132.64	31.71	0.00375	**
	Sr	(mg kg⁻¹)	1.91	0.79	9.02	11.56	0.00000	***

compared to valleys agreed with the results of Kubota et al. (2004) for southern Japan and Tanner (1977) for Jamaica.

The concentrations of Al, Ca, Mn and Sr in the soil extractions were higher in the gorge, whereas K and Mg concentrations in the ammonium acetate extraction were higher on the ridge. However, total amounts for K were four to eight times higher and more than double that of Mg in the gorge (data not shown here). These results follow Tanner's (1977) findings with higher concentrations of exchangeable K and Mg (exchangeable fraction measured using a 1 N ammonium acetate solution) but a lower concentration of Ca in the top 10cm soil layer at the ridges. A tendency towards similar ratios of exchangeable cations emerged in both investigations. Except for Na, all element concentrations were higher in the top 10cm than at a depth of 10–20cm. This could be a result of nutrient input via TF and leaf litter decomposition.

Furthermore, the amounts of TF differed significantly between ridge and gorge (Table 18.2). Expressed as percentages of incident precipitation (IP; measured at

Table 18.2 Comparison of throughfall data between ridge and gorge. Samples were collected between 1960 m and 2070 m a.s.l. in three plots at a ridge (T2) and three in a gorge (Q2) over a one-year period between November 2001 and November 2002 within a weekly rhythm for pH, EC, K, Ca and Mg. Other concentrations were measured from samples collected in October 2002. Data are given as mean with SD ($n = 31$). Statistical results were calculated with the Student t-test (n.s. not significant, * $P \leq 0.05$, ** $P \leq 0.01$, *** $P \leq 0.001$)

Parameter		Ridge Mean	SD	Gorge Mean	SD	P	Significance
Throughfall	(mm)	1766.1	453.5	1505.3	442.8	0.03719	*
% IP	(%)	85.30	25.12	70.80	22.34	0.03810	*
% SD	(%)	28.00	8.00	26.86	9.88	0.40930	n.s.
pH	–	5.14	0.29	5.19	0.52	0.81527	n.s.
EC	($\mu S\ cm^{-1}\ year^{-1}$)	940.2	609.3	2197.5	1217.3	0.00000	***
K	($kg\ ha^{-1}\ year^{-1}$)	40.74	23.63	95.63	46.62	0.00000	***
Ca	($kg\ ha^{-1}\ year^{-1}$)	6.39	4.14	15.30	13.44	0.00000	***
Mg	($kg\ ha^{-1}\ year^{-1}$)	3.78	2.83	7.63	9.55	0.00018	***
Concentration in throughfall collection in October 2002							
Nitrate	($mg\ l^{-1}$)	0.082	0.093	0.335	0.580	0.00069	***
Ammonium	($mg\ l^{-1}$)	0.090	0.175	0.194	0.202	0.01588	*
Chloride	($mg\ l^{-1}$)	1.71	1.39	2.83	2.20	0.00267	**
Phosphate	($mg\ l^{-1}$)	40.8	120.6	118.4	135.6	0.00000	***
Organic N	($mg\ l^{-1}$)	0.29	0.35	0.53	0.32	0.00147	**
Total C	($mg\ l^{-1}$)	15.91	15.02	17.87	7.13	0.02671	*
Li	($\mu g\ l^{-1}$)	0.16	0.09	0.26	0.16	0.00236	**
Al	($\mu g\ l^{-1}$)	20.44	24.16	34.74	26.89	0.00251	**
Total P	($\mu g\ l^{-1}$)	55.6	123.0	138.2	142.0	0.00000	***
Mn	($\mu g\ l^{-1}$)	14.72	44.93	3.16	7.64	0.26818	n.s.
Rb	($\mu g\ l^{-1}$)	8.51	7.71	10.58	5.93	0.02694	*

1950 m a.s.l.) they reached 85% on the ridge and only 70% in the gorge, with a higher relative standard deviation in the gorge. The pH values of TF did not differ significantly but varied slightly more in the gorge. The annual amounts of K, Ca and Mg input with the canopy TF were 1.5 times higher in the gorge than on the ridge, as was also the EC. Similar tendencies in the results were reported by Bernhard-Reversat (1975) from the Ivory Coast, who found three times more K in the TF in a gorge and 1.3 times more Mg and Ca in a collection of five samples. The concentrations of nitrate, ammonium, chloride, phosphate, organic nitrogen, total carbon, Li, Al, total P and Rb were statistically higher in the TF collected in the gorge. Only the Mn concentration did not differ significantly. The higher element input with TF can be explained by a higher element soil content for most elements, higher transpiration rates in the gorge (see Chapter 16) and a different species composition (see Chapter 10.1).

Unexpectedly, the mean annual temperature did not differ between ridge and gorge (both 14.9 °C). No significant differences within the mean annual humidity could be found.

The canopy structure differed significantly between ridge and gorge in almost all parameters, with the exception of the leaf area index (LAI), see Table 18.3. With 7.5% canopy openness and with 14.2% of the radiation above the canopy the forest

Table 18.3 Comparison of canopy structure data between ridge and gorge calculated with HemiView (ver. 2.1 SR1; DeltaT) from hemispherical images. Data are given as mean with SD ($n = 31$). Statistics were calculated with a t-test (n.s. not significant, *** significant at $P \leq 0.001$). HemiView calculated the radiation above the canopy for this site with 11185.3 MJ m^{-2} year^{-1}

Parameter		Ridge Mean	SD	Gorge Mean	SD	P	Significance
Canopy openness	(%)	7.477	0.941	6.095	1.291	0.00001	***
Indirect site factor	–	0.109	0.012	0.083	0.018	0.00000	***
Direct site factor	–	0.146	0.025	0.099	0.024	0.00000	***
Global site factor	–	0.142	0.023	0.097	0.023	0.00000	***
LAI	–	5.745	1.426	5.514	1.417	0.37908	n.s.
Mean leaf angle	(°)	52.54	19.00	34.90	17.86	0.00032	***
Ground cover	–	0.75	0.17	0.86	0.08	0.00081	***
Diffuse radiation	(%)	1.05	0.11	0.80	0.18	0.00000	***
Direct radiation	(%)	13.16	2.22	8.90	2.14	0.00000	***
Total radiation	(%)	14.16	2.28	9.70	2.27	0.00000	***

at the ridge was significantly more open and lighter than in the gorge with only 6.1% canopy openness and 9.7% of the radiation outside the forest. Even the mean leaf angle was higher at the ridge (52.5% compared with 34.9%), which can be explained by the reduction in sunlight hours in the gorges as a result of the shadows cast by the slopes. On the one hand, the calculated values for ground cover were higher in the gorge than on the ridge, on the other hand the LAI was little lower in the gorge (5.51 vs 5.75). Using a different measurement technique (LAI 2000; Li-Cor), Motzer (2003) determined a LAI of 6.4 in the same gorge (see Chapter 16) and Bengel (2003) determined 5.8 for the inner part of the gorge. This discrepancy in the LAI values might be a result of the different architecture of the stands. In the gorge most leaves are concentrated in the upper storey, whereas on the ridge leaves are distributed more equally along the stand height (M. Oesker et al., personal observation). This irregular distribution might influence the technique of using hemispherical images (Chen and Black 1991).

Tree species diversity (rarefied species numbers) differed significantly between gorge and ridge (Table 18.4). Calculated as the Shannon–Wiener index (N1), values in the gorge (mean = 32.3) were also higher than on the ridge (mean = 23.2; Homeier 2004). While Melastomataceae and Lauraceae were the most abundant tree families in the entire forest, Euphorbiaceae were amongst the most abundant tree families on the ridge and Mimosaceae in the gorge (see also Chapter 10.1).

Tree regeneration followed distinct pathways at ridges and in gorges with differences in seedling dynamics as shown by Homeier and Breckle (Chapter 23).

The species number of woody climbers as a mean of three plots was slightly higher in the gorge (16.7) than on the ridge (14.3) but the difference was not significant. This is confirmed by Takyu et al. (2002), who referred this pattern to more P and N downslope. In contrast, Kubota et al. (2004) found increasing species richness upslope, but they included understorey plants, which are more abundant on ridges. They explained this with higher habitat preferences of the species rather than soil fertility. Oesker et al. (2007) attributed higher diversity to a higher

Table 18.4 Comparison of tree and climber diversity and structural parameters between ridge and gorge. The Mann–Whitney U-test was used to calculate the differences between ridge and gorge. Data are given with mean and standard error (*SE*). Significance level (*Sig.*) was $P \leq 0.05$. * Significant, *n.s.* not significant

		Ridge Mean	SE	Gorge Mean	SE	Sig.
Tree diversity (rarefied species number for $n = 43$)	(400m^{-2})	22.0	1.8	29.3	1.6	*
Tree stem number	(ha^{-1})	899	27	781	30	n.s.
Tree basal area	(m^2 ha^{-1})	35.6	4.1	42.4	4.0	n.s.
Tree basal area growth	(% year^{-1})	0.63	0.06	1.55	0.40	n.s.
Woody climber species richness	(400 m^{-2})	14.3	2.0	16.7	1.5	n.s.
Woody climber density	(ha^{-1})	1875	318	950	50	*
Woody climber basal area	(m^2 ha^{-1})	0.34	0.06	0.51	0.22	n.s.

variability of abiotic factors, such as canopy TF and its element concentrations, and with this to a higher niches variability.

In the plot based study in the gorge Q2 a smaller number of trees (781 ha^{-1} vs 899 ha^{-1} on the ridge) and climbers (950 ha^{-1} vs 1880 ha^{-1}) were found, however the basal area of trees (42.4 m^2 ha^{-1} vs 35.6 m^2 ha^{-1}) and that of climbers (0.51 m^2 ha^{-1} vs 0.34 m^2 ha^{-1}) were higher in this gorge, which was also confirmed by Kubota et al. (2004), Takyu et al. (2002) and Tanner (1977). This could be observed although the differences in gorge Q2 were not very pronounced, as the diagnostic sampling study on 4.6 ha showed (Fig. 18.1). The differences in tree basal area between the ridges and gorges of the catchments were not statistically different, but those of catchments Q3 and Q4 were.

The annual growth rate is higher in the gorge (1.55% year^{-1} or 0.66 m^2 ha^{-1} year^{-1}) than on the ridge (0.63% year^{-1} or 0.22 m^2 ha^{-1} year^{-1}). Changes in plant growth rate along the altitudinal gradient are discussed by Bräuning et al. (Chapter 21).

If all these differences between ridges and gorges measured between 1960 m and 2070 m a.s.l., and described in this chapter, were independent of their elevation, two parallel altitudinal gradients (one for ridges, another one for gorges) would describe the ecosystem. With increasing elevation starting at 1850 m up to 2400 m a.s.l. the height of the 100 dominant trees per hectare does not decrease parallel to the ridges and in the gorges (Fig. 18.2a), but the decrease in the ridges is much higher than in the gorges. Correspondingly the crown cover decreases more strongly on the ridges than in the gorges (Fig. 18.2b). This shows that the horizontal gradient caused by topographical changes is strongly dependent on elevation.

A comparison was made of the horizontal differences at the same elevation of tree basal area and abundance from the plot based study with the altitudinal transect study of Hilpmann (2003). It resulted arithmetically in an altitudinal difference (starting from 1850 m a.s.l.) of around 280 m for both parameters.

In addition to human influences, such as selective logging, cutting and burning in the lower parts of the RBSF, Paulsch et al. (Chapter 10.3) mentioned that, in the mid-part of the RBSF, the distribution of forest types is caused by the relief. Ecological parameters alter with the vertical gradient (elevation; described in Part III.1) and the horizontal gradient (shown in this chapter) in the RBSF. All this

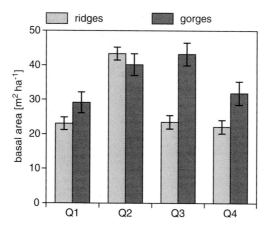

Fig. 18.1 Basal area (means and standard errors) at ridges and in gorges for four catchments in the RBSF (sampling area 4.6 ha). Statistical differences between ridges and gorges: $P = 0.1244$ in Q1; $P = 0.697$ in Q2; $P = 0.0016$ in Q3; $P = 0.0215$ in Q4. Data from Günter and Mosandl (2003)

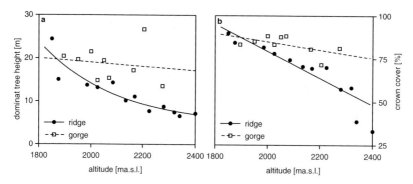

Fig. 18.2 Dominant tree height (**a**) and crown cover (**b**) of the 100 dominant trees per hectare of gorges and ridges along the altitudinal gradient. Data derived from Hilpmann (2003)

intensifies the heterogeneity of the ecological system and generates a complex mosaic of ecological niches, which finally is one pre-requisite for a hotspot in biodiversity (Dalitz et al. 2004; see Chapter 2 in this volume).

18.4 Conclusions

Clear differences between forest types at the same elevation and different topographical positions were found for the forest between 1850 m and 2070 m a.s.l. at RBSF.

The element concentrations of TF were higher in the gorge, as well as the soil concentration, with the exception of K and Mg. The forest canopy was lighter at the ridges. Tree and climber species richness was higher in the gorge. The densities of trees and climbers were higher on the ridge whereas the basal area and growth rates were higher in the gorge resulting in a forest on the ridge with more individuals of smaller sizes. Therefore the changes in vegetation types and species composition went along with changes in abiotic parameter and nutrient fluxes.

In addition to the altitudinal gradient (described in Part III.1) and the vertical gradient described here with the transect study a horizontal gradient caused by topographical complexity could be observed in the plot-based study in this chapter.

However, the differences between many of the parameters studied along the horizontal gradient as they were observed in the lower parts of the RBSF are not transferable to other altitudes. The combination of the gradients in the RBSF increases the heterogeneity of the vegetation and its ecological niches, which finally increases biodiversity (Dalitz et al. 2004; Oesker et al. 2007; see Chapter 2 in this volume).

Chapter 19
The Unique *Purdiaea nutans* Forest of Southern Ecuador – Abiotic Characteristics and Cryptogamic Diversity

N. Mandl(✉), M. Lehnert, S.R. Gradstein, M. Kessler, M. Abiy, and M. Richter

19.1 Introduction

The genus *Purdiaea* is a Neotropical genus of bushes and small trees belonging to the Ericales or heath alliance (Anderberg and Zhang 2002; Ståhl 2004). *Purdiaea nutans* Planch. is the only species of the genus in continental South America, occurring sporadically on nutrient-poor ridges or highly exposed areas in the Venezuelan Guyana Highlands and in the northern Andes from Colombia to northern Peru (Homeier 2005b; Figs. 19.1, 19.2). In all known localities it appears as isolated tree or shrub within mixed forest stands, except in the Rio San Francisco valley, southern Ecuador (Prov. Zamora–Chinchipe), where the only known forests dominated by *P. nutans* occur (Bussmann 2001, 2002; Paulsch 2002; Chapter 10.1 in this volume). These *Purdiaea* forests (*Purdiaeaetum nutantis* Bussmann 2002) cover about 200 ha of upper mountain forest, more than half of which are located within the RBSF (Mandl et al., unpublished data).

Reasons for the occurrence of this forest type in the Río San Francisco valley are unclear. Our hypothesis is that the nutrient-poor soils of the upper montane area of the RBSF might foster the development of the unique *Purdiaea* forest. Phytogeographically, the study area is part of the Amotape–Huancabamba Zone (AHZ) between the Río Jubones system in Ecuador and the Río Chamaya system in Peru, characterized by predominance of crystalline rock and relatively low mountain peaks (Figs. 19.2, 19.3). This zone has a high level of biodiversity and is home to numerous endemic species (Weigend 2002).

The purpose of our study was to characterize the abiotic environment (climate, soil) of the *P. nutans* forest and analyze its cryptogamic plant diversity. Cryptogams (ferns, bryophytes, lichens) do not interact with animals for fertilization and diaspore dispersal and suffer only minor damage by herbivores compared with seed plants. Therefore, their occurrence may reflect prevalent local abiotic factors more directly than do seed plants. If the closed *P. nutans* stands at the RBSF were caused by unusual soil or climate, one would expect a significant difference in the abiotic conditions as well as in the cryptogamic community composition and abundance compared with nearby forest stands where *P. nutans* is scarce or lacking. For the sake of comparison, upper montane forest stands in two other localities, El Tiro and

Fig. 19.1 *Purdiaea nutans*. **a** Habit. **b** Detail of flower. Images by N. Mandl

Tapichalaca, were included in this study. In all three localities, we analyzed 400 m² plots in ridge and slope habitats.

19.2 Results and Discussion

19.2.1 Abiotic Characteristics

The recorded macroclimate of all three study areas is permanently cool and perhumid, with a relative drier period from September until December. Mean daily courses of temperature and air humidity taken over a period of one year show similar curves at all locations. In the investigated ridge and slope forest stands temperatures rarely rise above 17 °C (RBSF) and relative humidity hardly drops below 82%. However, daily courses of temperature and air humidity fluctuate more strongly on the rather open and heterogenic ridges than in the dense slope stands. The *Purdiaea* forest at RBSF receives up to 5000 mm annual precipitation, which is about 2000 mm more than at El Tiro and 1000 mm more than at Tapichalaca. This difference might cause stronger leaching of minerals at RBSF than at the other study sites and therefore poorer soils in terms of available nutrients were expected.

A comparison of soils in the three study sites, however, did not reveal significant differences (Mandl et al., unpublished data). At all three sites soils were acidic, dystric cambisols and gleysols, with organic layers made up of moor-like wet humus followed by thin Ai and thick Ae and Aa horizons. Organic carbon contents in the organic layer were similar in all sites, nitrogen contents increased slightly from El Tiro to Tapichalaca, and total N was higher on slopes than on ridges. In general, soils of Tapichalaca slopes were slightly richer in nutrients, with higher K,

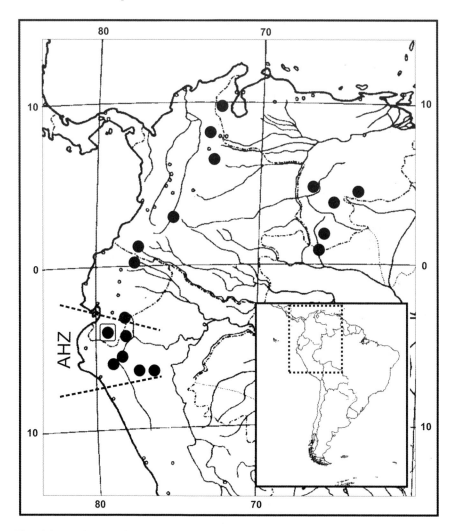

Fig. 19.2 Distribution of *Purdiaea nutans* in South America, with delimitation of the Amotape–Huancabamba zone (*AHZ*). The RBSF study area is marked with a *frame*

Ca and Mg values and higher C, N and P turnover rates. Ridge soils at Tapichalaca, however, were not different from those at the other sites.

The low and open canopy is a striking structural feature of ridge-top forests in this part of southern Ecuador. Differences in canopy height between ridge and slope forests were less pronounced at RBSF than at the other sites. A crucial finding of our investigation was that *P. nutans* occurred mainly on ridges. *Purdiaea*-dominated stands with 80–100% prevalence among trees with ≥10 cm dbh were found only along crests and broad ridge plateaus of the Río San Francisco valley. Downhill, the share of *Purdiaea* diminished to 40–60% and about 100 m below the crest only scattered, isolated mature trees occurred. Studies of seedlings in the RBSF have

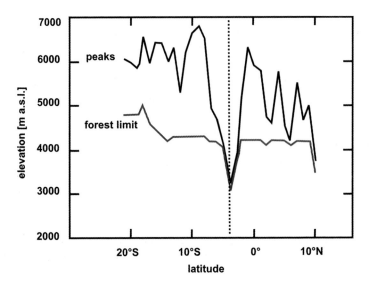

Fig. 19.3 Maximum elevation of mountain peaks and upper forest limit along the latitudinal gradient in the tropical Andes. Note the deep dent at ca. 4 °S (study area)

shown that *Purdiaea* seeds may germinate on lower slopes but soon die in these habitats (Knörr unpupl. data) where the substrate is enriched by the downhill flow of nutrients (Silver et al. 1994). Apparently the species is competitively inferior to others for light due to its slow growth (Homeier 2005b).

All these abiotic features cannot explain the dominance of *P. nutans* at the RBSF. However, there is evidence for fire events at the RBSF dating back to about 850 years ago (see Chapter 10.2 in this volume). Records of maize (*Zea*) pollen following those fires lead us to suspect that fires may have been related to human agricultural practices. Personal field observations suggest that *Purdiaea* is highly fire-resistant and even indicate that germination may be triggered by fire. Consequently, it seems likely that *Purdiaea* may have survived pre-Colombian fire events better than other tree species and may have been able to take advantage of the resulting open vegetation structure to achieve extensive regeneration. Slow growth rates of *Purdiaea* and a dense terrestrial herbaceous vegetation cover may limit the establishment of new tree seedlings and hereby preserve the dominance of *P. nutans*.

19.2.2 Cryptogamic Diversity

Almost 250 species of pteridophytes were recorded in the RBSF (Lehnert et al. 2007) which is one of the highest numbers of fern species recorded from such a small area (approx. 11.2 km^2) in the Andes. About 45 occur on the ridges dominated by *P. nutans*. Although endemic species are lacking, the fern flora of the *Purdiaea* forest stands out by the common occurrence of several rare species such as *Pterozonium brevifrons* (A.C. Sm.) Lellinger, with a range similar to that of *Purdiacaea nutans* (Guyana Highlands and scattered occurrences in the northern Andes, Fig. 19.2; Tryon and Tryon 1982), the northern Andean species *Blechnum schomburgkii* (Klotzsch) C. Chr. and *Cyathea peladensis* Hieron. The *Purdiaea* forest has no extraordinary accumulation or deficiency of certain fern groups, although typical terrestrial groups (Thelypteridaceae, Dryopteridaceae, Aspleniaceae) are poorly represented. The same was observed on ridges in the other study sites. In terms of species numbers, the pteridophyte flora at the three study sites did not differ significantly, although slopes were significantly richer than ridges (Table 19.1).

A bryophyte inventory of the RBSF yielded 505 bryophyte species (Gradstein et al. 2007), one of the highest numbers ever recorded from such a rather small tropical area. The species list underscores the richness of the northern Andes as one of the world's main hotspots of biodiversity. Interestingly, species with smaller ranges gain in importance towards higher elevations in the RBSF (Nöske et al. 2003). About 30% of the species from above 2150 m a.s.l., from the *Purdiaea* forest and the páramo have restricted range sizes (endemic, northern Andean, Andean taxa), twice as many as at lower elevation. This trend is noteworthy as two-third of all species occur below 2150 m a.s.l. The increase in species with smaller ranges towards higher elevation has also been observed in vascular plants and other organisms (Balslev 1988; Kessler 2002) and is explained by the reduced and often fragmented habitat surface area available in mountains as compared with lowlands.

With 55 terrestrial bryophyte species the *Purdiaea* forest is similar to other South Ecuadorian ridge forests in terms of bryophyte diversity (Table 19.2). A 1:2 or 1:3 ratio between mosses and liverworts appears to be characteristic for the terrestrial habitat. Towards the forest canopy species richness decreases and the moss/liverwort ratio shifts dramatically in favor of liverworts (mainly Lejeuneaceae and Frullaniaceae). In the outer *Purdiaea* crowns the percentage of mosses is reduced to 4%. Similar trends were observed on slopes and ridges of the other study sites.

Table 19.1 Number of pteridophyte species on ridges and slopes in the three study sites

	RBSF	El Tiro	Tapichalaca
Ridge	49	48	56
Slope	81	91	87
Total	93	94	102

Table 19.2 Bryophytes in the *Purdiaea* forest and nearby slopes: species numbers and ratio of mosses to liverworts growing terrestrially, on stem bases or in the canopy

	Slopes	Ridges – *Purdiaea* forest			
	Terrestrial	Terrestrial	Stem base	Inner canopy	Outer canopy
Species numbers	55	55	51	43	33
Mosses:liverworts	1:3	1:2	1:5	1:11	1:24

Bryophyte communities of RBSF and El Tiro were described in detail by Parolly and Kürschner (e.g. 2004a, b, 2005). A notable floristic feature of the *Purdiaea* forest separating it from the two other study sites is the abundant occurrence of the very rare liverwort *Pleurozia heterophylla* Steph. ex Fulf., worldwide known only from three localities (RBSF, Mt. Roraima in the Guyana Highlands, Honduras; Gradstein et al. 2001). It is a character species of the *Frullanio serratae–Holomitrietum sinuosi* subassociation *pleurozietum heterophyllae* (Parolly and Kürschner 2004a). Another rare species is *Pleurozia paradoxa* Jack, having the same general range as *Purdiaea nutans* (Fig. 19.3). Further noteworthy floristic records from the *Purdiaea* forest are the liverwort *Fuscocephaloziopsis subintegra* Gradst. and Vána (which is new to science) and the rare moss *Macromitrium perreflexum* Steere. The latter is only known from southern Ecuador and occurs in all sites but with greatest abundance in the *Purdiaea* forest and subpáramo vegetation. According to Parolly and Kürschner (2004a), *Macromitrium perreflexum* and *Pleurozia paradoxa* are characteristic species of the high montane epiphytic bryophyte community (*Macromitrio perreflexi–Pleuroziaetum paradoxae*) endemic to the RBSF, the Podocarpus National Park and surrounding areas.

19.3 Conclusions

The *Purdiaea nutans* forest in southern Ecuador should be considered an azonal ridge vegetation type. Its dominance on mountain crests and exposed plateaus in the San Francisco valley of southern Ecuador cannot be explained by a single factor but is apparently caused by a unique combination of abiotic features, related to the topography, geology, pedology and macroclimate. The gap in the cordillera presumably functions as a funnel for trade winds and may have caused the very wet climate of the RBSF. The very broad ridges and plateaus and the poor soils offer a suitable environment for the development of *P. nutans* on the southern slopes of the Río San Francisco valley. From our point of view, historical fire events are the most likely cause for the massive occurrence and dominance of this species.

The pteridophyte and bryophyte compositions of the *Purdiaea* forest underscore the ridge-top characteristics of the forest. However, the species richness and composition of this forest in terms of cryptogams do not differ significantly from those of the two other investigated ridge sites.

Chapter 20
Climate Variability

J. Bendix(✉), R. Rollenbeck, P. Fabian, P. Emck, M. Richter, and E. Beck

20.1 Introduction

In this paper variability refers to the range of values between particular climate maxima and minima over a period of time (Mitchell 1976). Atmospheric variability in space and time can significantly alter the response of the ecosystem. For instance, comprehensive studies emphasise notable relations of reproduction phenology to climate in tropical rain forests (e.g. Hamann 2004). Three types of relevant spatio-temporal heterogeneities are dicussed in this chapter:

1. Long-term and quasi-periodic oscillations of meteorological conditions;
2. Seasonal changes of selected meteorological parameters (clouds and precipitation), as a supplement to information already presented in Chapter 8;
3. The diurnal course of rainfall.

20.2 Results and Discussion

20.2.1 Quasi-Periodic Variability and Extreme Events

Specific year-to-year variability, quasi-periodic oscillations and extremes of the meteorological parameters are of ecological importance for the study area. In this section we briefly discuss:

1. The influence of the quasi-periodic El Niño–Southern Oscillation (ENSO) phenomenon = El Niño (EN) and La Niña (LN) events;
2. Specific circulation patterns which lead to very low temperatures in the RBSF.

The analysis of rainfall and temperature data of several EN/LN events since 1972 in Ecuador has shown that the phenomenon mainly affects the coastal plains with floods (EN) or droughts (LN; e.g. Bendix 2000, 2004). Teleconnections to the highland and the eastern escarpment of the Andes are irregular and less clear. For the eastern slopes, a slight tendency exists towards reduced (EN) or enhanced (LN) rainfall and positive (EN) or negative (LN) temperature anomalies. However, positive rainfall anomalies are still observed in the terminal EN phase (April to May). Unfortunately, no long-term time series of rainfall and temperature are available for the RBSF area. The latest strong EN/LN event (1998/1999) shows positive average temperature anomalies (+197% of 1σ standard deviation) at the ECSF meteorological station in the terminal EN phase (April to May). In contrast to the observed deviation, enhanced average temperatures are registered also during LN 1999 (October 1998 to May 1999; +90% of 1σ standard deviation). It is striking that the central LN phase (December 1998 to January 1999) shows significantly reduced minimum temperatures (−160% of 1σ standard deviation). Rainfall at the ECSF/Cerro meteorological stations during the terminal EN and whole LN is nearly normal (EN −38%/−5%; LN +12/+8% of 1σ standard deviation). Only January 1999 (central LN) reveals considerable positive rainfall anomalies. Monthly totals of rainfall during 1980–2000 are available for the station San Ramón (3 °59′ 02″ S, 79 °05′12″ W, 1820 m a.s.l.) in the Rio San Francisco valley close to ECSF. This period encompasses four EN (1972/1973; 1982/1983; 1986/1987, 1997/1998) and one LN (1998/1999). During all ENs, the San Ramon station reveals generally drier conditions than normal. LN 1998/1999 is characterised by slightly enhanced precipitation, especially in January, April and May 1999.

It is well known that extratropical cold air surges can affect the Tropics of South America (*friajes*; e.g. Breuer 1974) causing low temperatures in the RBSF area. The effects of such cold events, which last for 4–5 days on average, have been observed for the study area during the austral winter (May to October). During 1998 and 2001, 23 events could be registered. The coldest days and lowest temperatures ever observed in the ECSF area since the measurements began were related to cold air surges and featured the absolute minimum temperature of 0.7 °C at the climate divide (main ridge) in 3400 m a.s.l. and 3.9 °C at the Cordillera del Consuelo at 2930 m a.s.l. (11 July, 3 August 1999). The course of average hourly temperature and the synoptic situation during a strong cold air surge in August 1999 are presented in Fig. 20.1.

Cold events are introduced by the formation of a high pressure ridge from the South Pacific anticyclone over the Andes to the Pampa region (14 August) when cold air can intrude the Amazon area from the south (*friaje*). Such circulation patterns initiate the establishment of the south Atlantic anticyclone close to the east coast of southern South America (16–20 August), which is related to an equatorward cold air mass transport at the front of the high pressure system. The resulting cold SE trade winds affect the study area with a temporary complete cessation of rainfall and a significant drop in temperature, especially in the elevated parts.

20.2.2 Seasonal Variability

Cloudiness is a key factor in the climate system because of manifold feedbacks to the other meteorological parameters (Stephens 2005) and hence, of ecological

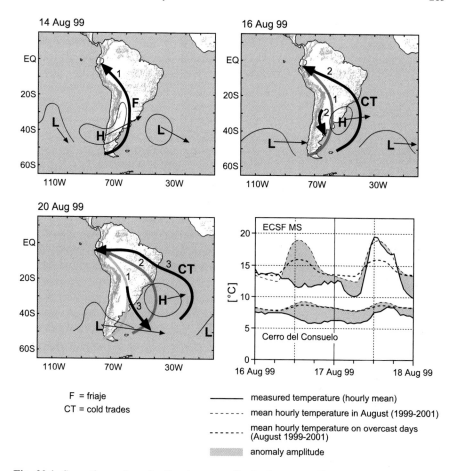

Fig. 20.1 Synoptic weather situation (pressure distribution, streamflow at 850 hPa) during the cold event (14–20 August 1999) and air temperature development in the RBSF in comparison with average temperature

importance. Fog water deposition on the vegetation of tropical montane cloud forests is important for the hydrological cycle (Bruijnzeel 2001; Fleischbein et al. 2005). Modifications of the radiation balance (e.g. heat, UV-B and water stress by high transpiration rates in the absence of clouds) are of ecological importance (e.g. Alados et al. 2000). Some authors demonstrated that enhanced UV-B intensities during clear-sky conditions in some cases decrease, advance or delay (depending on species) the time of flowering (e.g. Caldwell et al. 1998). Flowering and fruit production are synchronised with drought periods and sunny conditions because the absorption of photosynthetically active radiation (PAR) by clouds in the wet season limits the reproduction potential, due to the low accumulation of resources through photosynthesis (e.g. Wright et al 1999; Hamann 2004).

The study area is characterised by a strong gradient of cloudiness in space and time (Bendix et al. 2004b, 2006a). The main chain of the Cordillera Real clearly separates the basin of Loja with reduced frequencies from the area of the Rio San Francisco where the average cloud frequency can exceed 85%. Cloudiness increases along the altitudinal gradient from the valley bottom of the Rio San Francisco to the Cerro del Consuelo. It is especially high when the slopes are exposed to the predominating easterly streamflow due to blocking effects. In contrast, the leeward escarpment is mostly characterised by a reduced cloud amount. The seasonal cloud distribution completely changes from Loja to the Cerro del Consuelo (distance in a straight line <20 km). The area west of the main Cordillera is characterised by a maximum of cloud frequency in austral summer (December–February, DJF) and a secondary peak in the austral spring (September–November, SON). The transition zone between both cloud regimes is marked by the stations of El Tiro and Cajanuma where a third peak in cloud frequency begins to emerge in the austral winter (June–August, JJA). This peak becomes specifically dominant in the Amazon-exposed part of the study area (Cerro del Consuelo). Cloud frequency is high throughout the year at the altitudinal level around 3000 m a.s.l.

The same tendency of seasonal variability in space and time can be observed for rainfall (Fig. 20.2). Generally, the main cordillera separates the moist Amazon-exposed easterly slopes from the drier inner-Andean areas of Loja and Catamayo. However, clear differences can be observed between the austral winter (JJA) and summer (DJF). The highest rainfall amounts in the austral winter (JJA) occur at

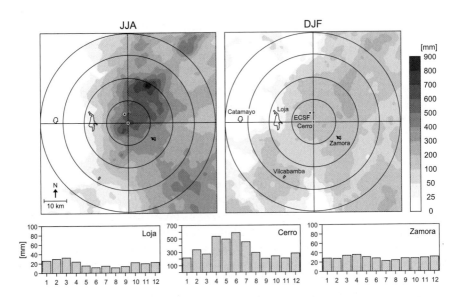

Fig. 20.2 Average distribution of rainfall for the austral winter (*JJA* June+July+August) and summer (*DJF* December+January+February) derived from X-band Radar (*top*, period March 2002 to July 2003); and registered rainfall at Loja (2160 m a.s.l.), Cerro del Consuelo (2930 m a.s.l.) and Zamora (970 m a.s.l.) meteorological stations (*bottom*)

steep slopes well exposed to the prevailing easterlies northeast of the study area. This applies also to the extremely exposed Cerro del Consuelo, the highest peak of the central study area, which shows lower average cloud top heights and standard deviations during this season (see Bendix et al. 2006b). The rainfields in the LAWR maps show a homogenous structure but rainfall slightly decreases towards the Amazon lowlands. It should be stressed that extreme amounts of rainfall recorded by rain gauges in the highest part of the Cordillera about 15 km south of the study area are not reproduced by the radar maps because of the beam obstruction of this sector (Rollenbeck and Bendix 2006). Cloud and rain formation in the austral winter (JJA) is of predominantly orographic character and is linked to the high wind speed in the upper parts of the study area (see Chapter 8 in this volume) which imply high condensation rates.

The analysis of wind direction and rain intensity/rain frequency confirms the complex structure of seasonal variability in precipitation dynamics. Figure 20.3 illustrates that easterly streamflow predominates in the study area throughout the year. While in the austral winter almost no westerly streamflow occurs, the portion of weather situations with winds from the western sector increases up to ~40% in October–November. It is striking that the easterly streamflow from May to September is related to a general high frequency of rainfall per hour. The average rain intensity decreases especially during the austral winter (JJA). In contrast, a lower precipitation frequency is also observed at the end of the year when westerly circulation patterns become significant. The low frequency is related to higher rain rates (especially in November) which points to a more convective situation.

A tracking analysis of rain cells using a series of consecutive radar images illustrates the variation in the rainfall dynamics of different seasons. The rain cells in August (Fig. 20.4, right) mainly originating from lower cap clouds move in an east–west direction but are blocked at the main crest of the Cordillera. On the leeward side, reminders are rapidly dissipated due to the lee effect. Hence, cell trajectories terminate in this area. During such situations, the windward slopes are exposed to high amounts of orographically enhanced rainfall, especially in the upper parts of the study area.

The example of westerly cell propagation (Fig. 20.4, left) is related to convective cells over the whole area west and east of the crestline of the Cordillera Real. However, the reduction in propagation speed and the change in direction points out that the elevated topography also plays an important role for the rainfall dynamics from higher clouds during westerly circulation patterns.

Clear sky conditions are the most effective between October and January by air flows from the northwest, which are termed "Veranillos del Niño" (Fig. 20.3, lower panel). An associated strong subsidence over the cordillera causes extraordinary aridity particularly in the highlands. The large saturation deficit results in an extreme water stress for the vegetation (during 1998–2001 the regional absolute minimum of 11.8% relative humidity recorded at 3400 m a.s.l. is linked with a Veranillos del Niño event). As Veranillos del Niño can last up to three weeks and return relatively regularly, it is quite probable that apart from strategies against the common extreme humid conditions and high wind speeds, vegetation has also been

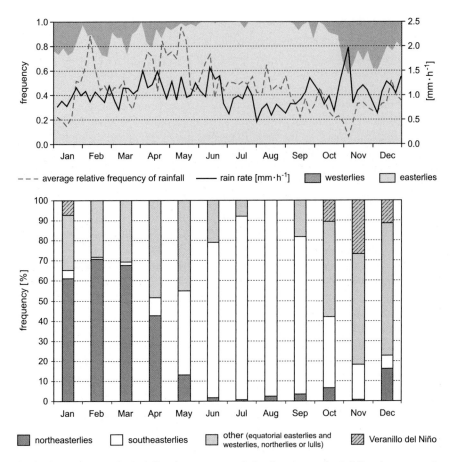

Fig. 20.3 Frequency of wind direction, average relative frequency of rainfall and average rain rate at the Páramo meteorological station (January 1998 to February 2005, *top*). *Bottom* Mean monthly frequency of synoptic winds at 850 hPa over west equatorial South America during 1998–2001, interpreted from daily average wind patterns between 65°S–30°N and 150°W–15°E from NCEP Re-analysis data

forced to adapt strategies against peak water stress situations. After several days of Veranillo del Niño conditions, even the mountain rain forests in the east become desiccated to such a degree that they can catch fire – a circumstance all too often exploited by the local population.

20.2.3 Diurnal Variability

Variability in weather patterns is not only present throughout the year but also throughout the diurnal cycle. This can be exemplarily illustrated by radar images of the spatio-temporal rainfall distribution (Fig. 20.5).

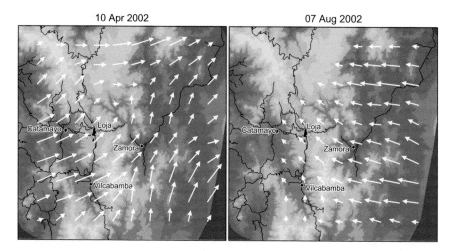

Fig. 20.4 Representative trajectories of rain cells for a weather situation with easterly impact in JJA (*right*, 7 August 2002) and westerly impact in FMA (*left*, 10 April 2002) derived from LAWR storm cell tracking. The length of the arrows is proportional to the displacement of the centre of individual cells. These displacement patterns are the result of the absolute wind driven movement of the cells, including the internal motion of the cloud systems. It should be stressed that the arrows do not represent the wind field itself

In the austral winter (JJA), rainfall affects the entire study area at the windward eastern escarpment, with a maximum in the early morning hours (0600 hours, local time) and in the upper parts of the study area close to the ECSF station. The diurnal cycle in the austral summer (DJF) shows an inverse behaviour. Rainfall is stronger in the late afternoon (1800 hours, local time), spatial patterns differ as well. While early morning precipitation is centred in the upper part of the central study area and is mainly restricted to the area east of the main Cordillera, the afternoon peak shows a clear cellular-convective structure which is partly related to thermal valley-breeze systems. However, the centres of main rainfall activity are the lower parts of the study area east of the Cordillera as well as the basins of Loja and Vilcabamba west of the main Cordillera crest.

Examination of the diurnal cycle in the free atmosphere over the ECSF research station by using a vertically pointing rain radar profiler (MRR) reveals a unique situation of the study area (see Bendix et al. 2006a). Average rainfall is generally low throughout the day. However, two clear maxima can be recognised with a main peak around sunrise (0530–0630 hours, local standard time; LST) and a secondary peak in the early afternoon hours at 1430–1530 hours LST. A third period of above average precipitation is observed at 2300–2330 hours LST. The relatively low rainfall frequency and the high maximum at e.g. 1730–1800 hours or 2030–2100 hours LST points to the occasional occurrence of stronger convective events, triggered by the up-slope breeze system. The precipitation peak around sunrise reveals the greatest variability in rainfall and rain frequency, being characterised by different stratiform and convective mechanisms of rainfall formation.

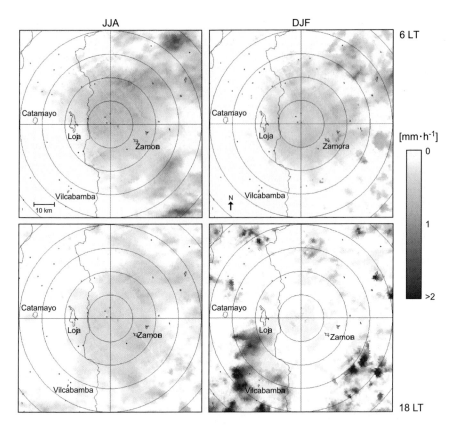

Fig. 20.5 Distribution maps of rainfall for selected periods in austral winter (*JJA* June+July+August 2002) and summer (*DJF* December+January+February 2002/2003) at 0600 hours and 1800 hours, local time (LT), derived from X-band Radar. The images are based on the average rainrate from 0500 hours to 0655 hours LT and from 1700 hours to 1855 hours LT for all days of the given periods

Inspecting the average vertical profiles of rainfall for the two peak times (Bendix et al. 2006a, Fig. 6), the maximum in the vertical profile is observed for the fifth level at 1347 m above the ECSF station (= 3207 m a.s.l.). The comparison of the situation at 0600 hours and 1500 hours LST reveals interesting differences in the vertical stratification. The shape of the vertical profile during the afternoon maximum is quite similar to the daily mean, but shows a slightly higher decrease in rainfall of 39% from the maximum zone at ~3200 m a.s.l. down towards to the ECSF station. This decrease is clearly less pronounced (16%) during the pre-dawn/dawn maximum where a significant reduction of rainfall amount is observed only in the lowest two levels. The shape of the profiles and further investigations of droplet spectra and satellite images (Bendix et al. 2006b) points to different rain generating processes. The afternoon peak is clearly related to convective processes. Rainfall is formed in the upper atmopheric levels (>3200 m a.s.l.) and droplets mainly originating from the stratiform area of the cells partly evaporate on their way to the valley bottom. The weak gradient for the morning peak at the valley bottom (which turns to an afternoon maximum at the crest levels) is most likely a product of complex mesoscale dynamics (Bendix et al. 2006b). As a result, rain clouds can overflow the bordering ridges of the Cordillera del Consuelo providing rains of higher intensity at the ECSF research station. We hypothesise that the seeding of fog and low stratus clouds, which form due to nocturnal radiation processes in the valley, by the upper-level rain clouds, can lead to continuous droplet growth right down to the near surface layer.

20.2.4 Weather Variability and Tree Phenological Response

A high degree of intra- and interspecific synchronisation of phenological traits was noticed for flowering and fruiting of 12 tree species in the study area. Just a short summary of results can be given in this paper. The whole study is presented by Bendix et al. (2006c).

Apart from one species that flowered more or less continuously, two groups of trees could be observed, of which one flowered during the less humid months (SO) while the second group started to initiate flowers towards the end of that phase and flowered during the wettest season (AMJJ). Phenological events of most of the plant species showed a similar periodicity of 8–12 months which followed the annual oscillation of relatively less and more humid periods and thus were in phase or in counter-phase with the oscillations of the meteorological conditions. Periods of unusual cold or dryness, presumably resulting from underlying longer-term trends or oscillations affected the homogeneity of quasi-12-month flowering events, fruit maturation and also the production of germinable seeds. Some species indicate underlying quasi-two-year oscillations which are synchronised with the development of e.g. air temperature, others reveal an underlying decrease or increase of flowering activity over the observation period, influenced e.g. by solar irradiance.

20.3 Conclusions

With regard to meteorological variations in space and time, it is shown that the ENSO phenomenon does not cause extraordinary weather anomalies in the RBSF area. In general, LN situations seem to induce somewhat stronger effects. Extratropical cold air surges can affect the study area, mainly in the austral winter with a significant reduction of air temperature, especially in the crest areas. Both findings underline the strong influence of the Atlantic circulation on the study area and the effect of the Andes which shelter the eastern Andean slopes from Pacific airmasses most of the time.

Rainfall and cloudiness reveal marked season-specific spatial patterns. Unexpectedly, the analysis of cloudiness, rainfall dynamics and wind field reveal that the JJA maximum in precipitation is mainly a result of an orographic lifting of moist air masses on the eastern Andean slopes. It is striking that the diurnal course of rainfall shows a unique behaviour for the RBSF area. A pre-dawn/dawn maximum and a secondary afternoon peak is observed for rainfall at the ECSF station. The morning maximum is probably a result of mesoscale dynamics in combination with local effects. However, the exact mechanisms must be investigated in future studies.

The phenological cycle for the most investigated trees is clearly associated with the seasonal course of the weather (temperature, radiation, precipitation). However, most species are affected by longer-term cycles which are probably related to climatic oscillations.

Chapter 21
Growth Dynamics of Trees in Tropical Mountain Ecosystems

A. Bräuning(✉), J. Homeier, E. Cueva, E. Beck, and S. Günter

21.1 Introduction

Tropical mountain rainforests are commonly regarded as a stable ecosystem in which life processes face almost invariable environmental conditions. However, as discussed for the eastern Cordillera of Ecuador in Chapter 8, even the perhumid tropics experience a more or less pronounced seasonality of precipitation patterns. Intra annual fluctuations of temperature are small and usually do not depart more than 1–3 K from the annual mean. In contrast, diurnal temperature variations of 5.1 K and 11.1 K were recorded at elevations of 2670 m a.s.l. and 1950 m a.s.l., respectively. Short-term climate irregularities, e.g. during periods with prevailing westerly winds, can alter the normal climate conditions considerably and thus are of great significance for plant life. Furthermore, tremendous short-distance variations of the climate are caused by effects of altitude and topography (Richter 2003).

Seasonality in precipitation is reflected by phenological phenomena and by changes in the growth rates of woody plants. The cambial activity of many tropical rainforest tree species is usually not interrupted by a dormancy period like in temperate climate zones. Therefore, many tropical woods lack distinct anatomical boundaries between growth zones of consecutive calendar years. In such cases, changes in growth rates can be detected by point dendrometers or band dendrometers that register variations in the radial extension or in the circumference of a tree stem. To evaluate rates of wood formation, several hundred trees from different species were equipped with dendrometers.

During the last years, an increasing number of studies on the growth of tropical trees showed that the generalized picture of continuous growth is an oversimplification. While the majority of tree species shows uninterrupted, albeit unsteady growth rates throughout the year, a considerable number of species exhibit seasonal growth rhythms. These growth changes are documented by the formation of different types of wood tissues or even by the formation of distinct growth boundaries. Nevertheless, in most cases the challenge remains to associate these visible growth boundaries to climate features or to internal biological rhythms. Besides climate, internal periodicities of flowering, fruiting, leaf shedding, and external disturbances like stand dynamics can significantly alter the growth behavior of tropical trees.

Table 21.1 Average monthly growth rates of ten selected tree species of an evergreen mountain rain forest in South Ecuador determined with dendrometers which were mounted at breast height. The increments of the stems at breast height were recorded in biweekly or monthly intervals over a three-year period (2000–2003)

Species	Number of trees	Range of circumference at breast height (cm)	Average monthly diameter growth rate (mm)
Heliocarpus americanus	25	15–115	1.18
Piptocoma discolor	27	15–162	0.63
Ficus sp.	17	16–120	0.70
Cedrela cf. *montana*	16	30–115	0.41
Clethra revoluta	36	20–113	0.41
Myrica pubescens	12	18–47	0.38
Inga sp.	8	26–86	0.35
Tabebuia chrysantha	23	15–90	0.35
Isertia laevis	30	17–81	0.30
Vismia tomentosa	21	17–80	0.25

In summary, temporal growth heterogeneities in trees are caused by internal physiological processes that are largely influenced by external environmental factors. In this chapter we try to correlate the growth reactions of trees from the tropical mountain forest of the RBSF and the Podocarpus National Park area with specific triggering factors.

To get a general idea on the range of growth rates in the mountain rainforest of South Ecuador, average growth rates of ten tree species growing between 1800 m and 2200 m a.s.l. are presented in Table 21.1. Pioneer species like *Heliocarpus americanus*, *Piptocoma discolor*, and *Ficus* sp. show the highest growth rates. The evergreen *Clethra revoluta*, *Myrica pubescens*, *Inga* sp., and the deciduous species *Cedrela* cf. *montana* and *Tabebuia chrysantha* grow moderately. *Isertia laevis* and *Vismia tomentosa* (both evergreen) exhibit the lowest growth rates.

21.2 Temporal Growth Heterogeneities Caused by Internal Phenological Processes

Most tree species in humid tropical environments are evergreen. This also holds for the mountain forests of southern Ecuador. However, in the study area two important forest constituents are obligatorily deciduous: *Tabebuia chrysantha* (Bignoniaceae) and *Cedrela* cf. *montana* (Meliaceae). Besides, some tree species can be facultatively deciduous or show variable leaf-shedding rates like e.g. *Ficus* spp (Moraceae). Depending on local site and weather conditions, some individuals of *Ficus* regularly drop their leaves, e.g. at the end of the less humid season (October to March; Fig. 21.1). However, defoliation of *Ficus* is never as complete as with *T. chrysantha* and *C. montana*. While the latter two species drop their leaves simultaneously, leaf shedding of *Ficus* occurs rather randomly. Between April 2002 and April 2003, leaf shedding was observed more or less continuously (Fig. 21.1).

How is the temporal growth linked to a certain type of phenological behavior? Figure 21.2 compares growth patterns of different tree species that belong to the various phenological types. For the evergreen pioneer species *Piptocoma discolor* (Asteraceae), the growth rates of all 23 studied individuals are almost linear (Fig. 21.2a). A slight deceleration of growth is only visible at the end of the very rainy period during July and August, which is also the coolest season. The decrease of the growth rate may therefore be associated with low temperatures rather than with a lack of water. Decreases and increases in growth rates are not synchronous between individuals and can not be attributed to regular climatic events. *Piptocoma* frequently suffers considerable damage from herbivorous insects, and transient partial defoliation may be an explanation for the irregularities in the growth curves. Similar growth characteristics were also observed with *Clethra, Isertia, Vismia, Inga,* and *Myrica*.

In contrast, the deciduous *T. chrysantha* (Fig. 21.2b) and *C. montana* exhibit extended periods of suspended growth. Interestingly, these growth interruptions are not restricted to the leafless months (August to mid-October), but persist during the entire season of heavy rains. Defoliation commenced in June (Fig. 21.1), but secondary growth already ceased in April or even some weeks earlier. This might be attributed to a dampening of the photosynthetic productivity due to the low light intensities in the rainy season from April to June (Graham et al. 2003). *T. chrysantha* as well as *C. montana* prefer more open areas such as forest edges and forest gaps and can be considered as light-demanding species. Low light intensities may also be an environmental trigger for leaf senescence. In addition, comparable to the carbon allocation of broad-leaved trees in temperate regions, the accumulation of reserves for bud break and sprouting in the axes and roots prevails over growth.

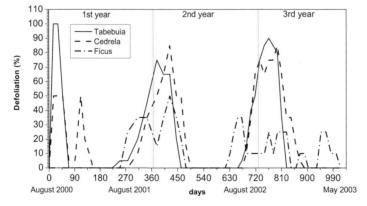

Fig. 21.1 Leaf shedding of a selected number (Table 21.1) of *Cedrela* cf. *montana*, *Ficus* cf. *motana* and *Tabebuia chrysantha* in the RBSF in the course of three years

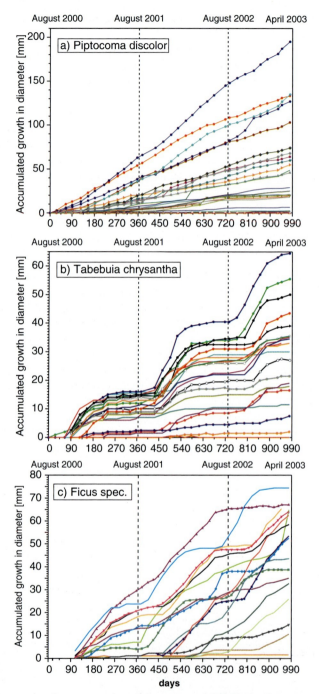

Fig. 21.2 Accumulated stem extension of three types of life-forms of trees over three years. **a** Evergreen pioneer (*Piptocoma discolor*, 27 individuals), **b** obligatory leaf shedding species (*Tabebuia chrysantha*, 23 individuals), **c** facultative deciduous (*Ficus* sp., 17 individuals). Growth was measured with mechanical dendrometers, which were controlled biweekly.

Decelerated growth during the period of lower light intensity was also observed with the fast-growing pioneer species *Heliocarpus americanus* (Tiliaceae), which in spite of its evergreen phenology has a higher rate of leaf shedding during the wetter months.

A third type of growth pattern was observed with the facultatively leaf-shedding *Ficus* (Fig 21.2c). Due to the highly individual leaf shedding behavior during the drier months from December to March, some trees suspend growth during the rainy season while others show a contrary growth rhythm. Continuous growth over two or three years is also possible, since *Ficus* never defoliates completely.

21.3 Growth Heterogeneities Along Altitudinal Gradients

The influence of elevation on tree growth was studied along an altitudinal gradient at 1800–2450 m a.s.l. Twenty to 30 individuals of 12 common tree species with a minimum diameter at breast height (dbh) of 10 cm were equipped with band dendrometers (Homeier 2004) that were controlled in monthly intervals since June 1999. In addition, diameter growth rates of all trees with a minimum dbh of 10 cm were registered within 11 permanent plots. The investigated tree species exhibit distinct differences in growth performance over the measuring period of four years. Pioneer species like *H. americanus* and *P. discolor* achieve average increments in circumference of 64–70 mm (Fig. 21.3). This is five to six times more than for the three slowest growing species *Clusia* cf *ducuoides* (Clusiaceae), *Purdiaea nutans* (Cyrillaceae), and *Podocarpus oleifolius* (Podocarpaceae; Fig. 21.4). The remain-

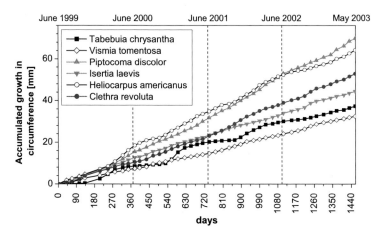

Fig. 21.3 Cumulative circumferential increments of six fast-growing common mountain forest species (*C. revoluta*: 27 individual trees/1 of them died within the study period; *H. americanus*: 22/1; *I. laevis*: 23/1; *P. discolor*: 25/1; *V. tomentosa*: 19/1; *T. chrysantha*: 20/0) over a period of four years

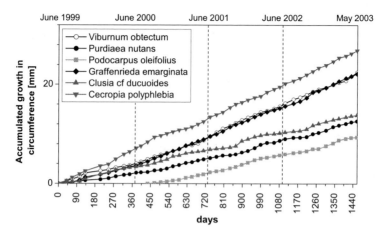

Fig. 21.4 Cumulative circumferential increments of six slowly growing common mountain forest species (*C. polyphlebia*: 17/0; *C. ducuoides*: 21/0; *G. emarginata*: 30/1; *P. oleifolius*: 17/0; *P. nutans*: 22/0; *V. obtectum*: 18/0) over a period of four years

ing species show mean annual increments in circumference between 5–13 mm. Besides the deciduous *Tabebuia*, some of the evergreen species (*Clusia* cf *ducuoides*, *Heliocarpus americanus*) show slight seasonal growth variations too, which could be the result of rainfall seasonality. The diameter growth rates from our study area (minimum: *Purdiaea nutans* 0.6 mm year^{-1}; maximum: *Piptocoma discolor* 5.7 mm year^{-1}) are within the range of growth rates for neotropical trees reported from lowland rainforests (0.5–11.6 mm year^{-1} in Ecuador, Korning and Balslev 1994; 0.2–4.6 mm year^{-1} in Costa Rica, Lieberman and al. 1985) or other tropical premontane forests (1.6–12.6 mm year^{-1} in Costa Rica, Homeier 2004). In the RBSF, the studied tree species prefer different habitats. The fast-growing species are mostly found below 2200 m a.s.l. Slow growing species like *Graffenrieda emarginata* (Melastomataceae), *Podocarpus oleifolius*, and *Clusia* cf. *ducuoides* are found across the whole altitudinal range at 1800–2500 m a.s.l. *Purdiaea nutans* is mainly found at higher elevations.

In general, growth rates decrease with elevation (Fig. 21.5). In the lower study plots, maximum annual increments of 8–11 mm are registered whereas in the two uppermost plots which are dominated by *Purdiaea nutans*, maximum annual circumference increments are 2 mm only. The decrease in tree growth rates with elevation can be partly explained by a change in species composition, since fast-growing species like *Piptocoma discolor*, *Heliocarpus americanus*, *Ficus* spp or *Inga* spp only occur below 2200 m a.s.l. (Homeier 2004). However, analyses of soil and foliar nutrients demonstrated that the most evident factor influencing tree growth in our study area is nutrient supply (see Chapter 17 in this volume). Slower mineralization rates due to lower temperatures and occasional water logging at higher elevation result in nutrient limitation for tree growth.

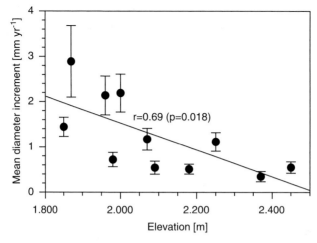

Fig. 21.5 Changes in annual mean diameter growth rates (plot averages of the years 2001 and 2002 ± 1 SE) as a function of elevation. Within the permanent plots of 400 m² the diameter growth of all stems with a dbh ≥ 10cm was monitored. Stem numbers of the plots varied between 12 and 31

Besides climate, micro-site parameters strongly influence growth. Below 2000 m a.s.l., topography gets more complex due to alternations of deeply incised creeks (Spanish: *quebradas*) and ridges. In comparison with soils on the ridges the soils of the quebradas are comparatively moist and nutrient-rich. This landscape pattern is reflected by the increasing growth variability in the study plots (Fig. 21.5; Homeier 2004). A good example is *Piptocoma discolor*, which grows single or in small groups on crests, slopes, and in gorges within a rather narrow altitudinal range in the RBSF area. From Fig. 21.6 it is obvious that this species performs best in the continuously wet gorges whereas the wind-exposed crests allow only a very low growth.

21.4 Growth Heterogeneities, Stand Dynamics and Competition

Temporal variations in growth rates can also be provoked by stand disturbances and associated changes in the social status of trees. To study this phenomenon, monthly increments of 222 trees from nine species were monitored between June 2004 and June 2005. The social status and the competition by overtopping and neighboring trees were registered according to the IUFRO classification.

The impact of crown competition was not consistent with all species, suggesting that other factors could override its effect (Table 21.2). As expected, some species showed higher diameter growth without crown competition (*Ficus subandina*, *Podocarpus oleifolius*, and *Tabebuia chrysantha*), but some species did not respond

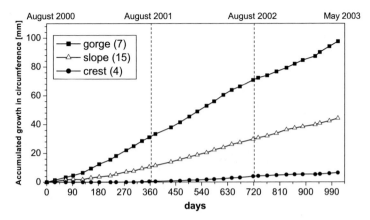

Fig. 21.6 Circumference growth of *Piptocoma discolor* individuals at different sites of the research area between 1800 m and 2200 m a.s.l. The curves represent the mean accumulated growth of four individuals growing on crests, 15 individuals on slopes, and seven individuals in a gorge. The slopes of the growth curves of the three site types are different at the $P = 0.001$ level as tested by a linear mixed model

Table 21.2 Mean annual growth of trees with and without overtopping crown competition. According to the results of a *t*-test for heteroscedastic samples, the differences are not significant ($P > 0.1$)

Species	Without overtopping competitors			With overtopping competitors		
	Mean (mm)	SD	n	Mean (mm)	SD	n
Cedrela sp.	17.8	10.5	8	18.3	10.2	8
Clusia cf. *ducuides*	1.7	2.3	35	1.6	2.3	12
Ficus subandina	22.3	23.8	8	21.8	11.8	6
Hyeronima asperifolia	4.2	5.6	17	9.6	12.0	8
Hyeronima moritziana	1.3	1.4	14	1.0	0.6	6
Inga acreana	5.5	10.5	8	8.0	8.8	10
Nectandra cf. *membranacea*	3.1	5.3	14	5.4	7.1	17
Podocarpus oleifolius	2.9	3.1	9	1.0	1.4	7
Tabebuia chrysantha	5.0	6.0	21	2.7	3.8	14

(*Clusia* cf. *ducuides*, *Hyeronima moritziana*) or even showed reduced increment (*Cedrela* cf. *montana*, *Hyeronima asperifolia*, *Inga acreana*, *Nectandra membranacea*). These results were unexpected, since trees without competitors should in general grow better (Finegan and Camacho 1999; Dolanc et al. 2003).

Cedrela and *Tabebuia* are both deciduous and mostly mid-successional species. Thus, different competition strategies cannot explain the different growth responses that can be seen in Fig. 21.7. However, it must be mentioned that the exact age of the trees was not known and that therefore age effects could have overlaid the effects of crown competition. Mature trees usually suffer less from competition and

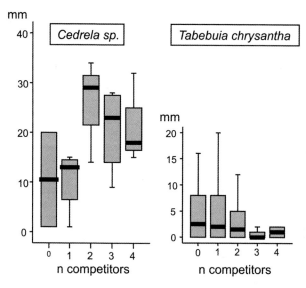

Fig. 21.7 Relationship between number of crown competitors and annual diameter growth (mm) for *Cedrela* cf. *montana* ($n = 20$) and *Tabebuia chrysantha* ($n = 35$). *Black horizontal bars* represent the median. *Box limits* give the 25–75% quartiles, and *short horizontal bars* indicate maximum and minimum values

thus cannot always benefit from release due to limited growth reaction capacity (Messier and Nikinmaa 2000; Sterck and Bongers 2001). Other explanations include stress effects by increased light intensities or allocation phenomena by higher investment in underground biomass (King 1991). In summary, we found that the radial growth of *Tabebuia chrysantha*, *Podocarpus oleifolius*, and other species reacts very sensitive to competition. Since these species have a high timber value, it can be concluded that silvicultural measures like improvement thinning can foster diameter growth by reducing crown competition (see Chapter 26 in this volume).

Another experiment was performed on *Piptocoma discolor*. The potential cover area of each individual was calculated from a circle with the radius of the largest branch around the stem. Then, the actual crown area was determined from its vertical projection on the ground. Finally, the ratio of the actual to the potential crown area was calculated. The 27 studied individuals were grouped into five categories from A to E: <25%, 25–49%, 50–74%, 75–99%, and 100%. Figure 21.8 shows the mean growth rates as a function of the extension of the crown. A decrease in the potential crown area by 25% or even 50% impairs stem growth by only about 10% and 20%, respectively. A further decrease in the crown area, however, results in a dramatic decline in growth rates, and trees with a potential crown cover of less than 25% show barely any stem growth at all.

The growth of the studied individuals might also have been influenced by additional environmental factors that were neglected in our approach. Therefore, our

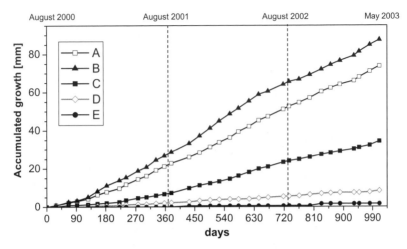

Fig. 21.8 Stem extension growth of *Piptocoma discolor* as a function of the relative potential crown area. For further explanations see text

observations only give a first impression on the influence of the stand dynamics on tree growth in the extremely heterogeneous tropical mountain forest.

21.5 Growth Rate Heterogeneities and Climate

To construct long tree-ring chronologies for the reconstruction of climate during the past centuries, it is crucial to understand the connection between seasonal climate variations and wood formation. In comparison with temperate zones, the seasonal formation of different anatomical types of wood and their relation to environmental triggers is still hardly known in the tropics. To disentangle the connection between climate, tree growth and wood anatomy, we installed point dendrometers on several individuals of *Prumnopitys montana* and *Alnus acuminata*. The study site is situated at 2100 m a.s.l. at the western slope of the Cordillera Real, where the rainfall seasonality is more pronounced than in the RBSF forest reserve (Richter 2003). Radial changes of the stem diameters were registered at 30 min intervals. Small wood samples for wood anatomical studies were collected with the help of an increment puncher (Forster et al. 2000) every two weeks. The extracted wood cores include the cambial zone between the bark and the freshly formed xylem cells. Temperature, rainfall, and relative humidity were recorded by a meteorological station that was installed at a distance of about 1.5 km from the study site at the same altitude. Examples for *A. acuminata* are presented by Bräuning and Burchardt (2006), therefore we focus here on growth changes in the conifer *P. montana*.

Prumnopitys does not form distinct annual rings, but its wood is characterized by the occurrence of density fluctuations that reflect variations in cell wall thickness of the tracheids. Figure 21.9 shows the growth curve of a tree, together with a series of consecutive wood microsections and daily sums of rainfall and mean daily relative air humidity. It is possible to identify characteristic density fluctuations between the different microsections. Small tangential variations between these features are caused by irregularities in the stem as the samples were taken in a horizontal distance of 3 cm. It can be seen that, during a relatively dry period between end of May and middle of September 2004, no wood was formed and the cambium was inactive. Starting with the rainy season at the beginning of October, new cells were produced and the stem diameter substantially increased. By the end of December, cambial activity ceased again despite abundant rainfall, but a pronounced thickening of the cell walls in the outermost part of the stem occurred.

During the dry period in September 2004, the stem diameter of *Prumnopitys* was actually shrinking (Fig. 21.10). The daily diameter minima were recorded at 1200–1400 hours, when water stress was maximal. In the course of the night, the water status of the stem recovered and maximum diameters were reached during the morning hours at 0600–0800 hours. Within these ten days, the daily amplitude of the diameter change, i.e. the difference between the maximum (swelling) and the following minimum (shrinking), increased linearly with the duration of the dry period ($r = 0.80$, $P < 0.01$).

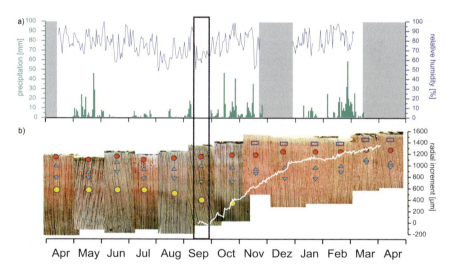

Fig. 21.9 Connection between climatic variables, wood anatomy, and tree growth in *Prumnopitys montana* in the forest of El Bosque de Vilcabamba (2100 m a.s.l.) during the period April 2004 to April 2005. **a** Course of mean daily air humidity (*line*) and daily sums of rainfall (*columns*). Gaps in the recordings are indicated in *gray*. **b** Dendrometer curve of daily minimum diameter values for the period September 2004 to March 2005 (*white curve*) with series of microsections that were taken in regular intervals. Wood tissues that represent isochronous formations are marked with identical symbols for better visibility. The period from the middle to the end of September 2004 highlighted by the *open column* is shown in detail in Fig. 21.10

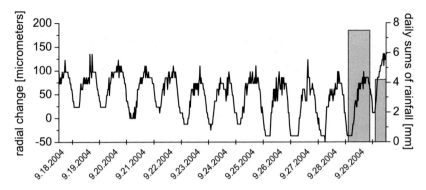

Fig. 21.10 Dendrometer curve from *Prumnopitys montana* during the period 18–30 September, 2004. The dry period ended with a rainfall event on 29 September (gray column)

21.6 Conclusions

In contrast to the growth behavior of *Podocarpus* at the RBSF (see Fig. 21.4), *Prumnopitys* at "El Bosque" shows a clear seasonality in radial growth and in the formation of different types of wood structures in correspondence with the climatic seasonality. In temperate climate zones near the upper tree line, stem shrinkage during drought periods in the growing season also occurs (Herzog et al. 1995). However, stem shrinkage has never been recorded before at tropical mountain rainforest trees. In the investigated tropical mountain ecosystem, a high variety of microsites with different microclimates and the exceptional geomorphological dynamics continuously initiate successional processes. The high diversity of tree species with different phenological rhythmicity and a broad variety of anatomical wood structures result in a multitude of phenomena in response to environmental events. Temporal heterogeneities of tree growth occur on different spatial and temporal scales and can be utilized as indicators of environmental impacts on perhumid tropical mountain rainforests.

Chapter 22
Temporal Heterogeneities – Matter Deposition from Remote Areas

R. Rollenbeck(✉), P. Fabian, and J. Bendix

22.1 Introduction

The tropical montane cloud forests of southern Ecuador have developed on poor soils characterized by low nutrient availability and high acidity (Beck and Müller-Hohenstein 2001; Brummitt and Lughadha 2003; Chapter 13 in this volume). Their nutrient balance might be affected by atmospheric inputs via precipitation, as indicated by our results. Especially the high input of fog and wind-driven rain can cause a significant deposition of ionic loads, because this precipitation type has a longer impact time on vegetation surfaces than falling raindrops.

The atmospheric nutrient input by rain water was determined for a micro-catchment on a long-term basis from 1998 to 2005 (Wilcke et al. 2001a; Chapter 13 in this volume). In the lower parts of the research area a measurement transect covering the whole altitudinal range from 1800 m to 3200 m was added later (2002–2005; Fabian et al. 2005) to determine the variations with increasing elevation. Because of the significant contribution of fog and cloud water to the water balance in the higher elevations, these precipitation types were investigated too (Rollenbeck et al. 2005).

The research site is dominated by trade winds whose directions vary between E/NE (February) and E/SE (July; Richter 2003). Thus local emission sources like road traffic or small industry play a minor role for chemical components precipitated at the site. The only major city, Loja, is located downwind at the other side of the mountain chain of the study area hence its plume exerts marginal impact. Small settlements, the road to Zamora and its 10 000 inhabitants 15 km upwind are unlikely to exert more than small episodic pollution events. Overall it can be expected that the dominating tradewind flow provides air masses which originate in the Amazon lowlands.

22.2 Methodology: Measurements, Analysis and Synthesis of Data

To determine variations along the altitudinal gradient, rainwater was collected by means of UMS-RS 200 polyethylene samplers of 20 cm diameter, installed at five altitudes between 1800 m and 3185 m (see Table 22.1). Fog collectors according to Schemenauer and Cereceda (1994), consisting of 1×1 m polypropylene nets of 2×1 mm mesh width fixed at 1 m above the ground, were operated alongside the rain collectors. Their orientation was perpendicular to the predominant wind direction. Sampling efficiency and calibration was assessed by means of an active fog collector (NES 210) and a scatterometer (VPF 730) operating along with the meshgrid fog collectors at the reference station at an altitude of 1960 m. Regular rain and fog water samples were taken at approximately weekly intervals, from which 120 ml aliquots were shipped to Germany for analysis. The measurements of pH and conductivity were carried out by Methron 730 665/682 and WTW-LF90 respectively. Cation analyses were performed by the inductivity-coupled plasma method (Perkin Elmer Optima 3000) while anions, in particular sulfate and nitrate, were analysed by ion chromatography (Dionex DX-210).

From March 2002 onwards, pH values, electrical conductivity and chemical ion composition (K^+, Na^+, NH_4^+, Ca^{2+}, Mg^{2+}, SO_4^{2-}, NO_3^-, PO_4^{3-}) were regularly measured in weekly rain and fog water samples collected along the altitudinal transect explained in Table 22.1.

22.3 Results and Discussion

22.3.1 Local/Altitudinal Gradients

The ionic loads of rain and fog precipitation in this area are generally low. Concentrations in fog water are slightly higher but rain water contributes more matter input to the nutrient balance, because the total water input by rain is much higher. As expected fog water shows altitudinal variations not observed in rainwater (Fig. 22.1): Whereas the pH values are more or less equal for all altitudes, electric conductivity

Table 22.1 Sampling sites along the altitudinal gradient

Name	Altitude (m a.s.l.)	Rain	Fog/cloudwater
Rio	1800	x	x
ECSF	1960	x	x
Tom	2000	x	–
Plataforma	2270	x	x
TS1	2660	x	x
Antenas	3180	x	x

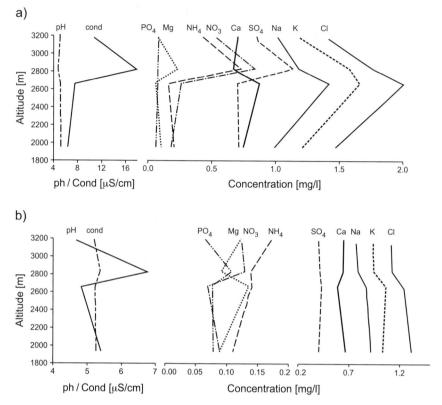

Fig. 22.1 Altitudinal gradients of pH, conductivity and ion concentrations in rain water **a** and fog water **b**. Average data are based on 3800 weekly samples from 2002–2005 taken at six sampling sites

shows a peak value of 11 μS/cm that is 25% higher than the average at the cloud condensation level, which has a mean height of 2800 m a.s.l. The components responsible for this peak are the two nitrogen species of nitrate and ammonia and furthermore sulfate, reaching average concentrations of 1.13 mg/l (Fig. 22.2).

22.3.2 Seasonal Patterns

To identify processes that potentially cause these seasonal patterns, time series of the different components were investigated: in rain water sulfate, sodium and chloride concentrations show significant oscillations. Conductivity in rain water varies between 2.3 μS/l and 15.0 μS/l. In fog water the range is 2.0–40.0 μS/l. The pH value has a range of 4.8–6.1 in rain water and 4.8–5.8 in fog water. In fog water nitrogen species like ammonia and nitrate are also highly variable.

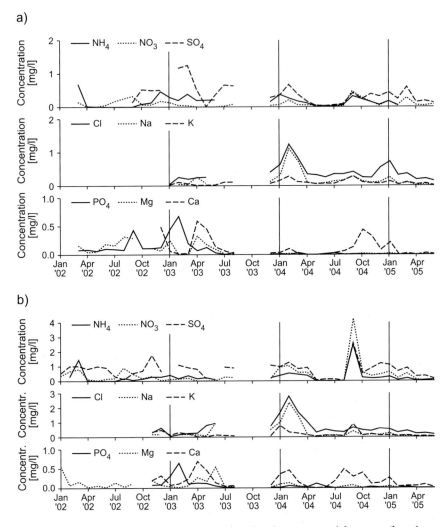

Fig. 22.2 Time series of average ion concentrations in rain water **a** and fog water (**b**; volume-weighted monthly means). No data are available for August to November 2003

The time series for rain water (Fig. 22.2a) reveals elevated concentrations at the beginning of each year, where sodium, chloride and sulfate are more dominant. In September 2004 a significant peak occurred, which was caused by elevated nitrogen and ammonia contents. This pattern is even more pronounced in fog water (Fig. 22.2b). A sulfate peak in November 2002 could be associated with volcanic gases, when the plume of the volcano El Reventador (which erupted 3 November 2002, about 500 km N/NE of the study area) passed over the site (Fabian et al. 2005).

In all three years, the dry period from November to March brought higher inputs of sulfate, sodium and chloride. This may have been caused by maritime air masses

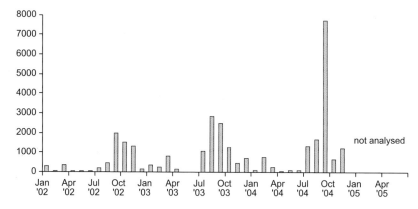

Fig. 22.3 Time series of fire frequency. The accumulated number of fires for each 10-day trajectory reaching the study site is grouped into monthly totals. Fire observations from NOAA, GOES and MODIS sensors are published by the Brazilian weather service (www.dpi.inpe.br)

from the Pacific Ocean, which was transported by the more frequent westerlies in this season.

Another distinct peak was visible in September 2004 which coincided with local observations of a highly polluted air mass, in which smoke was visible and could be smelled. Similar observations were made in 2005, for which analysis data are not yet available. High concentrations of nitrate, ammonia and sulfate point to the impact of smoke plumes stemming from forest fires in the Amazon basin. This hypothesis has already been made by Wilcke et al. (2001a). Besides CO_2, a large array of substances is emitted from burning forests, such as carbon monoxide, hydrocarbons, halocarbons, SO_2, COS, NO_x, HCN and aerosols (Koppmann et al. 1997; Ferek et al. 1998; Andreae and Merlet 2001). In the plume secondary substances such as ozone are produced photochemically (Fujiwara et al. 1999; Marufu et al. 2000). Likewise, the reactive sulfur and nitrogen compounds are converted to sulfate and nitrate respectively, which are soluble in cloud droplets (Chang et al. 1987). Scavenging by rain is considered fast, with atmospheric residence times of SO_2^{2-} and NO_3^- of about 2 days only. However, recent investigations show (Andreae et al. 2004) that heavy smoke from fires reduces cloud-drop size, thus causing the suppression of low-level rainout, allowing transport to higher levels and thus over larger distances (Koren et al. 2004).

It is now well established that substances emitted into the atmosphere from burning forests, along with these photochemically generated reaction products, can be transported over very long distances, even traversing oceans (Roelofs et al. 1997; Forster et al. 2001): Products of burning tropical biomass have been found over the Atlantic and Pacific oceans (Andreae et al. 2001). Forest fire emissions from Siberia and Canada have been identified, respectively, in Japan (Tanimoto et al. 2000) and in Europe (Andreae et al. 2001; Spichtinger et al. 2001). Forest fire smoke from Russia has even circled the world (Damoah et al. 2004).

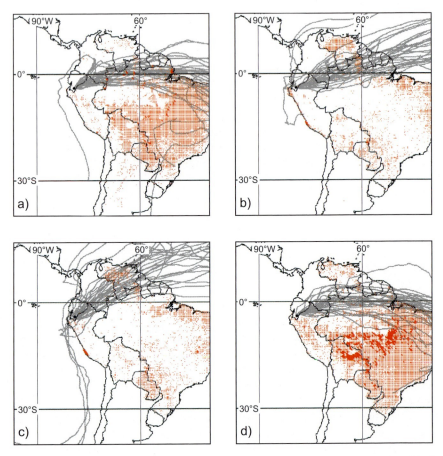

Fig. 22.4 Ten-day daily backward trajectories, calculated for the receptor site at 3000 m altitude for September 2002 (**a**), January 2003 (**b**), January 2004 (**c**) and September 2004 (**d**). *Dots* mark areas of biomass burning during the particular month, as derived from satellite data (see Fig 22.3)

Transport mechanisms of similiar efficiency are not known for the highly turbulent atmosphere over the Amazon basin, but they seem to play a role here. Therefore, satellite products were used to determine the frequency of forest fires in the upwind regions (Brazil, Peru, Venezuela, Colombia, Paraguay, Bolivia) of our study site and, indeed, September 2004 was the month of the most extreme biomass burning activity in the analysed dataset (Fig. 22.4).

To determine the probability of air mass transport from the forest fire sites to our study region, 10-day backward trajectories were computed for the receptor site at 3000 m altitude. We used a trajectory model developed by Stohl et al. (2003) which is based on re-analysed meteorological data (ECMWF). A software tool was developed that detects the coincidence of fire and the synchronous overpass of trajectories reaching the study site. Data from January 2002 to November 2004 was processed (Fig. 22.4).

The highest coincidence between eastern trajectories and biomass burning occurs in September, whereas higher frequencies of western trajectories in January to April are potential sources of sodium and chloride, stemming from marine aerosols. Westwind situations are observed more frequently in October and November, but precipitation rates, especially fog input, are rather low during this time of year. Cloud and rain generating processes are limited to the coastal areas; hence rainout from atmospheric aerosol is more effective there and consequently does not reach the study site.

22.4 Conclusion

The results of this investigation show that matter input from remote areas influences the chemical characteristics of precipitated water. The input concentration is controlled by several interfering processes, which are the frequency of emission events (biomass burning, volcanism, sea salt) and a climatic situation which links these events with the study region and the occurrence of precipitation, especially fog input in the research area. With the increasing intensity and frequency of agricultural activity in the tropical forests of South America, an alteration of the nutrient balance in the remote areas of the mountain forests of the Andes is a possible secondary consequence and may also serve as an indication of the ongoing changes, especially in the Amazon basin.

Chapter 23
Gap Dynamics in a Tropical Lower Montane Forest in South Ecuador

J. Homeier(✉) and S.-W. Breckle

23.1 Introduction: Tropical Forest Gap Dynamics

Canopy gaps are integral parts of every primary mature forest ecosystem, no matter whether it is a temperate or tropical forest. They can be caused by various factors, including landslides, strong winds and injury or death of individual trees. Landslides and uprooting may open-up the soil to the mineral horizon; other causes may not affect soil structure such drastically. Many fast growing, shade-intolerant, pioneer species are adapted to canopy disturbance and they need light gaps to establish and to reach maturity. Late-successional canopy tree species often tolerate shade as juveniles, but most of them also require disturbance of the forest canopy to reach reproductive maturity (Hartshorn 1980; Whitmore 1989).

Gap size is considered as a key factor to determine colonization by trees (e.g. Arriaga 2000; Brokaw 1985; Denslow 1980; Whitmore 1989, 1996).

Smaller gaps often show a species composition that is distinctly different from larger gaps (e.g. Arriaga 2000).

Some of the parameters which are associated with gap size are light availability and quality (Barton et al. 1989; Denslow et al. 1998), herbivory (Pearson et al. 2003) and nutrient availability (Denslow et al. 1998). Canopy gaps could be regarded neither as a homogeneous environment nor as sharply delimited (Brandani et al. 1988; Dalling and Hubbell 2002; Núnez-Farfan and Dirzo 1988; Whitmore 1996). Plant species composition and diversity changes from the gap edge to the center (Hartshorn 1989; Fig. 23.1).

Garwood (1989) proposed that, in small gaps, the growth of seedlings and suppressed saplings is promoted, whereas large gaps are dominated by pioneer species, which arrive mainly by seed rain or, to a limited amount, are in the soil seed bank (Vasquez-Yanes and Orozco-Segovia 1993).

As a result, species composition and, thus, competition effects are related to gap size: Within smaller gaps no own succession series starts in contrast to bigger gaps, which allow for a secondary succession to establish. Two aspects determine the further development of a canopy gap: first, the colonizing species, which are adapted to the special conditions within gaps; and second, the subset of species, which was present and survived gap creation. Competition between both species

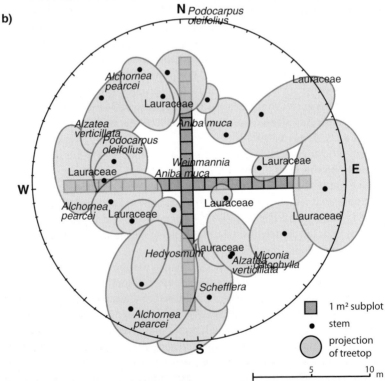

Fig. 23.1 a Photo of one of the studied canopy gaps (gap no. 10) at 2000 m a.s.l. located in ridge forest with a canopy opening of ca. 60 m². **b** Outline of the same gap, showing the canopy opening with the surrounding trees (dbh >10 cm) and the permanent regeneration transect

groups may be rather severe for the limited resources light, nutrients, water and space. Thus, gap fate is partly determined by niche partitioning (e.g. light partitioning; Poorter and Arets 2003) and also by chance (Brokaw and Busing 2000).

23.2 Results and Discussion: Gap Dynamics of the Lower Montane Forests of the RBSF

The forests of the study area are highly dynamic, partly driven by landslides (see Chapter 24 in this volume), which are common on slopes with more than 30° inclination, as consequence of the local geology and steepness. However, up to around 2200 m a.s.l. canopy gaps caused by tree falls or fallen branches play an important role for forest dynamics, too. Frequently landslides occur together with tree falls and amplify their disturbing impact. With increasing elevation stand height decreases and canopy becomes more open, so that site conditions inside canopy gaps above 2200 m a.s.l. are not very different from the conditions in the surrounding stands.

Because of the generally lower stand heights and smaller crown dimensions in tropical mountain forests, the gaps caused by tree falls normally do not reach the size of tropical lowland forest gaps. The typical canopy gap-size found for the investigated forest is between 20 m^2 and 150 m^2 (Fig. 23.1). Canopy height of the surrounding trees is an important parameter determining gap microclimate. Comparing same-sized canopy openings, the forest ground receives a higher amount of radiation in forest stands with lower canopy.

To investigate the role of canopy gaps for tree regeneration in the lower montane forests of the RBSF, we chose six gaps formed in 2002 to install permanent transects (each containing 41 subplots of 1 m^2, Fig. 23.1). All studied gaps were caused by tree falls and had canopy openings between 60 m^2 and 130 m^2.

They were located in two distinct forest types (valley forest versus ridge forest, or forest type I versus forest type II, according to Homeier 2004; see Chapter 10.1 in this volume), all at elevations from 1950 m to 2100 m a.s.l.. The two forest types are distinct in structural parameters (e.g. stand height, tree basal area, tree growth dynamics), tree species composition and diversity (Homeier 2004; Chapter 18 in this volume).

The cross-shaped transects cover different light conditions from completely open canopy at the gap centre over the gap edge to the closed canopy. Microclimatic site parameters like soil temperature (Onset TidBit), air temperature and air humidity (Onset HoboPro) were quantified. All tree regeneration (height ≥5 cm) rooting inside the transects was inventoried with plant height in a first census in 2003.

Plant determination in most cases was only possible to family or genus. A second census was done one year later to get information about site-specific mortality and recruitment.

In the gap centers of both ridge gaps and valley gaps, climatic conditions are more fluctuating, concerning air temperature and organic layer temperature. The variation in temperature, like the variation in vapor pressure deficit (VPD) is high in the gap center, whereas the climatic conditions underneath a closed canopy are more constant (Fig. 23.2). The daily amplitude of air temperature in the gap center is about two times higher than below canopy and may reach more than 20 K on sunny days. The temperature in the organic layers can exceed 40 °C; the amplitude may exceed 30 K in contrast to closed canopy, where it is less than 5 K. The VPD at noontime can increase in gaps up to 1–2 kpa in comparison with the closed canopy where it remains still below 1 kpa on sunny days.

The five most important plant families registered within the tree regeneration on the plots are Rubiaceae, Melastomataceae, Lauraceae, Arecaceae and Mimosaceae (Fig. 23.3). In studies on tree species composition the first three of them were found to be the species-richest tree families of the area (Homeier 2004; Homeier et al. 2002). Some families like Arecaceae and Melastomataceae occur principally in the ridge gaps, whereas others like Piperaceae and Rubiaceae are more common in the valley gaps. The Cyrillaceae (with *Purdiaea nutans* as the only species) are restricted to the ridge (see Chapter 10.1 in this volume) and Mimosaceae occur almost exclusively in the ravines.

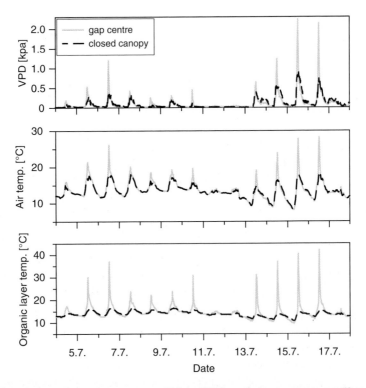

Fig. 23.2 Microclimate [vapour pressure deficit (VPD) and air temperature at 50 cm above ground and temperature in the organic layer] of two sites (gap centre vs closed canopy) within a ridge gap (gap no. 10) at 2000 m a.s.l. for two weeks in July 2003

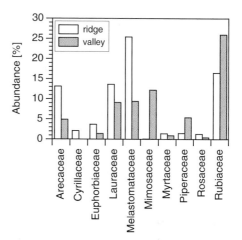

Fig. 23.3 The ten most important families involved in tree regeneration in ridge- and valley-gaps

Typical neotropical pioneer taxa such as *Piptocoma discolor* (Asteraceae), *Cecropia* spp (Cecropiaceae) and *Heliocarpus americanus* (Tiliaceae) occur only in the valley gaps. In contrast, species composition of ridge gaps is quite similar to species composition of the mature ridge forest (forest type II).

We found obvious differences in tree regeneration between the two forest types, the numbers of plant individuals in the valley gaps were lower than in the ridge gaps (Fig. 23.4). Average plant numbers were more than two-fold higher for the ridge forest transects (Table 23.1). One year after the first census, mean plant density had decreased at both sites.

Valley gap plants on average were significantly taller than ridge gap plants in both study years, but the differences between single gaps were not significant.

Relative annual growth rates from 2003 to 2004 were quite similar for both gap types (Table 23.1). Relative increments showed highest variations within the smaller height classes of woody regeneration at both sites. Many of the smaller plants exhibited negative increments (losing height by predation), whereas most of the saplings bigger than 1 m displayed positive increments. Plants at both sites had grown on average around 5 cm in height within one year.

Absolute annual recruitment and absolute annual mortality were both significantly higher on the ridges than in the valley. Relative mortality rates were between 20.2% for the ridge gaps and 21.9% for the valley gaps. Recruitment in all transects was lower than mortality, thus leading to the observed lower plant densities after one year.

23.3 Conclusions

Gaps are supposed to play a major role in maintaining plant species diversity in tropical forests (e.g. Brokaw 1985; Denslow 1980; Schnitzer and Carson 2001; Whitmore 1989). However, recruitment limitations (due to unfavorable soil conditions

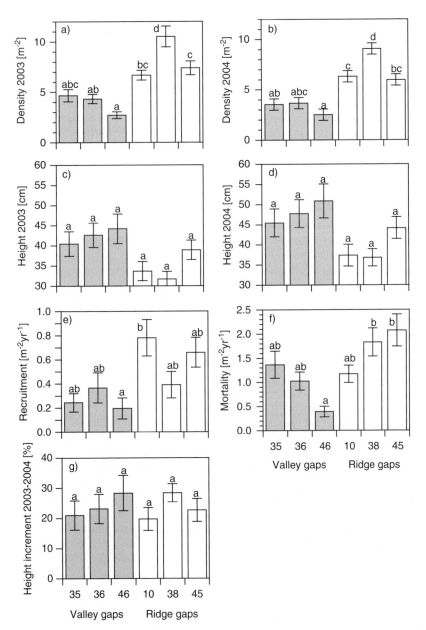

Fig. 23.4 Woody plant regeneration dynamics in three ridge-gaps and three valley-gaps: **a, b** mean plant density in 2003 and 2004; **c, d** mean plant height in 2003 and 2004; **e** one year recruitment from 2003 to 2004; **f** one year mortality from 2003 to 2004; **g** mean annual height increment for all tree regeneration within gap transects. For **a–f** means of 41 subplots of 1 m² are illustrated (±1 SE). Significant differences are indicated by *different letters*. **a, b, e, f**: ANOVA, Scheffé test; **c, d, g**: ANOVA, Tukeys HSD test for unequal *N*

Table 23.1 Differences between valley gaps and ridge gaps in density, height, mortality, recruitment and growth of tree regeneration (Student t-test, $P<0.05$)

	Valley gaps	Ridge gaps	Significance
Plant density 2003 (m^{-2})	3.89	8.19	$P<0.01$
Plant density 2004 (m^{-2})	3.23	7.12	$P<0.01$
Plant height 2003 (cm)	42.2	34.2	$P<0.01$
Plant height 2004 (cm)	47.7	38.9	$P<0.01$
Mortality 2003–2004 (m^{-2})	0.93	1.69	$P<0.01$
Recruitment 2003–2004 (m^{-2})	0.27	0.61	$P<0.01$
Mean annual height increment 2003–2004 (%)	23.7	24.4	$P=0.84$

or dispersal limitation) within the heterogeneous mosaic of environmental conditions (see Chapter 18 in this volume) can overlay gap patterns (Hubbell et al. 1999), especially in the RBSF ridge-forest where light is not as limiting for regeneration as in lowland rain forests with their tall canopy.

Species composition of the gaps from the two compared forest types is quite different: The valley forest (forest type I) shows distinct succession phases with fast growing pioneer species adapted to the gap conditions, whereas species composition of ridge forest gaps is quite similar to mature ridge forest.

We suppose that in the RBSF gap effects on regeneration and composition of other life forms, e.g. shrubs and lianas, may be stronger than on trees, as Schnitzer and Carson (2001) found for a lowland forest, but this has to be proven by further investigation.

The valley soils generally exhibit higher nutrient contents compared with ridges (Homeier 2004). Tree basal area growth (dbh ≥ 10 cm) responds positively to better nutrient supply in the valley forest (Homeier 2004), but surprisingly the average growth of regenerating plants shows no differences between valley forest and ridge forest.

The observed thinning of plants is the result of competition for various resources. Higher plant densities in the ridge gaps could be explained by the better light availability due to the lower forest canopy.

Vegetative regeneration has no importance in the studied montane forest, in contrast to results from a premontane Costa Rican forest (Wattenberg 1996), where a rather high percentage of new growth came from vegetative sprouting which may be one reason for a fast growth. However, annual precipitation of the Costa Rican site is two-fold higher than in our study area, which may foster vegetative sprouting.

Chapter 24
Landslides as Important Disturbance Regimes – Causes and Regeneration

R.W. Bussmann(✉), W. Wilcke, and M. Richter

24.1 Introduction

Landslides are part of rapid mass movement events, and considered processes of denudation. The classification scheme in Fig. 24.1 identifies different processes with regard to the type of movement in relation to the type of material (Varnes 1978). Other classification systems (e.g. Ahnert 1996; Crozier 1986a; Dikau et al. 1996; Hutchinson 1988) consider the medium of transport (gravitation, pore water, snow, soil ice, etc.) or the velocity of the phenomenon (rapid, intermediate, slow).

In the steep mountain forests of Ecuador, many gaps without vegetation or with vegetation in young succession stages, and numerous incomplete soil profiles indicate erosion and mass movements (Frei 1958; Wilcke et al. 2003). In the research area, the visible remains of landslides in undisturbed forest covered 3.7% of the 1117-ha study area (Wilcke et al. 2003).

The effect of landslides on soil properties depends on the type of the landslide, e.g. on the depth of the failure plane and the type of soil movement. Landslides can cause nutrient-poor topsoil to be removed and nutrient-richer subsoil or substrate to be exposed to the surface or vice versa. For humid southern Ecuador, Schrumpf et al. (2001) postulated that landslides improve soil fertility because they bring deeper, less-weathered and therefore nutrient-richer material to the surface. Other authors reported that landslide soils were less fertile than surrounding undisturbed soils (Dalling and Tanner 1995; Guariguata 1990; Lundgren 1978; Zarin and Johnson 1995a, b). A common feature is that the landslide area is structured into mass depletion and accumulation zones that may differ in soil fertility (Walker et al. 1996). Soil properties on landslides can also vary unsystematically on a small scale because a heterogeneous mixture of substrates is produced (Adams and Sidle 1987; Guariguata 1990; Lundgren 1978; Wilcke et al. 2006).

Investigations into the vegetation regeneration on landslides are scarce. Stern (1995) studied the regeneration of a single landslide in northern Ecuador (1995), while Kessler (1999) addressed landslide succession in Bolivia, and Erickson et al. (1989) in the central and southern Andes. Garwood (1981) and Garwood et al. (1979) analyzed species colonization in Panama, and Guariguata (1990) in Puerto Rico. Batarya and Valdivia (1989; Himalayas in India), Restrepo and Vitousek

Type	Rock	Debris	Soil
Fall	rockfall	debris fall	soil fall
Topple	rock topple	debris topple	soil topple
Rotational landslide	rock slump	debris slump	soil slump
Translational landslide	rock slide	debris slide	earth slide, soil slide
Flow	rock flow	debris flow	earth flow mud flow
Complex slope movements	e.g. multi-storied slides, landslides breaking down into mudslides or flows at the toe		

Fig. 24.1 Classification of rapid mass movements. Event types occurring within the RBSF are indicated in dark gray and those occurring in the Cordillera Real and its escarpments in light gray (modified after Varnes 1978)

(2001; Hawaii), and Ohl and Bussmann (2004) included geomorphologic processes in their considerations. Keefer (1984) studied earthquake-triggered landslides. Stern (1995) and Kessler (1999) hypothesized that landslides maintain species diversity. They create secondary forests dominated by colonizing species that are not able to survive in mature stands.

24.2 Types of Mass Movement in the RBSF and its Surroundings

A classification of the landslide events in the research area is not easy, due to different transitory traits. Each of the several types is linked to special landforms, petrographic issues, or areas of human impact. Slow mass movements by gelifluction, soil creeping, or streaming of blocks are absent, but several of the fast events listed in Fig. 24.1 occur.

At RBSF rockslides and earth flows are restricted to the Loja–Zamora road, where it undercuts solid rock masses or profound regolith. In the first case, material (from pebbles up to boulders) falls down as a free movement from steep roadside cliffs (see example in Fig. 24.2a) while, in the second case, soft and wet fine mate-

Fig. 24.2 Pictures of a recent rock slide blocking the Loja–Zamora road, near RBSF (**a** November 2004) and a debris flow (**b** several events in 1999, photo October 2005). **c** A section of the east-facing slope of Quebrada Milagro within the RBSF terrain stressed by a narrow system of recently active, in parts multiple slides, with El Tiro Pass in the background. **d** Same photo as **c**, but contrasted and containing explanatory signs (see text)

R.W. Bussmann et al. Landslides as Important Disturbance Regimes

rial deforms readily by streaming down slowly. These are man-made hazards where initial succession traits are quasi-permanently interrupted and set back.

Debris topples are rare phenomena of block failure restricted to cliffs and columns of lacustrine and limnic sediments of the Mandango series filling the basins of Vilcabamba and Loja outside RBSF. In soil slumps, blocks tilt backward after an initial failure, slide downhill, and fall into various pieces caused by transverse tension cracks. There are rare rotational slides on pasture land in the humid areas of southern Ecuador. Finally, debris flows consist of stony material bedded in a muddy slurry which surges downward a pre-existing drainage way. Sometimes the fine part of the deposits is washed out by following rainfalls, leaving the coarser part behind, e.g. at the mouth of a V-shaped valley near RBSF (Fig. 24.2b).

During 1962–1999, 8.5% of the total RBSF terrain was affected by slide processes, as determined by differently old aerial photos and an intensive ground check. 3.7% of the area showed recent activity in 2000. Landslides occur in size ranges with a length of 10–250 m and a breadth of 3–50 m. Most of the natural slides form narrow slips, only a few meters in width (Fig. 24.2c), while slides caused by road construction are much broader (up to 200 m) and lead to rock falls. Of the almost 450 visible traces of different landslides, about 425 are of natural origin, with 12 of them extending over 200 m.

Among 249 recent landslides between 1989 and 1999, 40% are confined to one single event, while 50% result from multiple events (Stoyan 2000). One-tenth is marked by continuing mass movement during episodes of high precipitation. A (rarely more than 1 m high) scarp at the head of the slide characterizes the basic type. The shallow depth of less than 1 m is determined by a substratum discontinuity between the humus layer and the surface sediments, from which the mineral soil develops, or between the surface sediment and the massive parent rock. Only few and rarely occurring deep-reaching slides include part of the parent rock. Length (L), depth (D), and width (W) classify the type of movement (Crozier 1986a). The D/L ratio of the natural slides in RBSF is mostly <0.08 and the L/W ratio never >5. All slips can be regarded as translational landslides (see Fig. 24.3). Some events show a ratio <0.025, which hints at mudflows. Many of the multiple landslides are

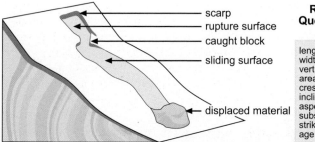

Fig. 24.3 Sketch with description of a translational landslide in early February 1999

mixed forms. Translation slides develop as a rapid event in an initial step. Subsequent smaller mudflows and/or gully erosion modify the previous slip surface. If the latter is deepened by surface run-off, subsequent slides tend to enlarge the landslide retrogressively. The displaced material accumulates at the toe. If the substrate is completely water-saturated, the run-out body deforms to lobes by liquefaction.

The most important trigger for landslides is the enormous steepness of the study area and the resulting geomorphologic instability. Landslides are favored by the surface-parallel orientation of the organic layer, which can reach of up to 700 t ha^{-1} (Wilcke et al. 2002), and of surface sediments covering the parent rock. For the few deeper-reaching slides, fine and surface-parallel stratification of the dominating slightly metamorphosed clayey/sandy sediments might also play a role.

Rainfall and seismic events are occasional triggers. At RBSF highest precipitation rates occur during the wettest season (April–August). If a rainstorm hits already saturated soil, the landslide risk is high. If such an event is accompanied by one of the common earthquakes, slide processes show peak activity.

While landslides at RBSF are frequent phenomena on phyllites, they seem to occur to a lesser degree on granite (e.g. at Bombuscara) and on sandstone (e.g. at Tapichalaca, and in narrow strata within RBSF). Higher infiltration rates of sandy and coarse-porous soils can be a main reason for the lower mass movement impact. Landslides can be frequent even under drier conditions on phyllitic basements (e.g. above Vilcabamba/Yambala). Here, slide events seem to be much more connected to years of high rainfall intensities, like the La Niña year in 1999. The rainfall amount was 140% of the average, with several heavy downpours stronger than the highest intensities at RBSF. While slip phenomena in the semihumid area are restricted to long-term periods, the perhumid area of RBSF shows a "steady state cycle" with quasi-permanent mass movement processes, which are driving forces for ecosystem functioning.

Within the RBSF terrain the distribution patterns of landslides vary considerably. This can be seen in a susceptibility map for landslide hazards computed by Brenning (2005) using a spatial logistic regression model. The model uses geomorphometric parameters derived from a digital elevation model and data on land cover, land use, and infrastructure to predict the probability of future landslide occurrences conditional on the distribution of past landslide events. It was fitted by stepwise backward variable selection to reduce the set of explanatory variables. Fig. 24.4 indicates highest landslide risk for the San Francisco valley. This relatively drier area is predominantly affected by artificial undercutting of slopes releasing mainly rockslides of larger dimension but to a lesser extent by natural slips. The area around Cerro de Consuelo is also hardly impacted, although its surroundings belong to the steepest terrain and are extremely rainy. Here, as well as on the flattened secondary crest lines, denudation of more weathering-resistant quartzite by sheet wash processes is the primary type of mass movement. Its influence on landslide activities finds a response further down by infiltrating into the substrata of the forested lateral slopes. Specific morphometric properties like size, curvature, and slope as well as destabilizing effects of the previous distribution of landslide scarps are local factors conditioning the susceptibility to sliding.

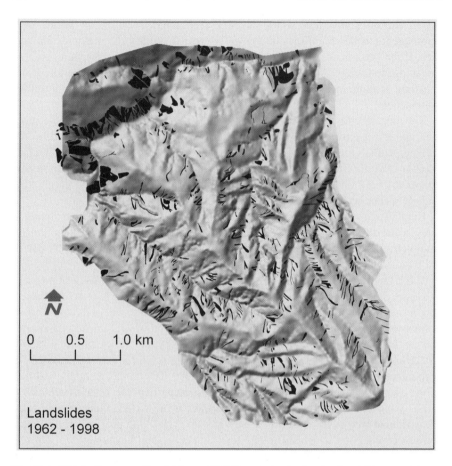

Fig. 24.4 Distribution map of landslides on RBSF terrain (evaluation by Ronald Stoyan, elaboration by Dr. Alexander Brenning, Erlangen). While most of the bigger landslides in the lower part of the terrain are caused by man (road and canal construction) those of the middle and upper part are natural phenomena

Figure 24.2c shows the landslide impact on a section of the eastward facing slope of Quebrada Milagro. A section of the same photo in Fig. 24.2d indicates different locations of slide activity zones between ridge b and c: while none of the slips extends up to the flattened part of the crest they plunge into the bottoms of the catchments contributing to the form of V-shaped valleys (see 1* below ridge a). They also form "tributary slides" running from the adjacent slopes into the main channels (e.g. 3*). One slide stage (1) with many slides originating from a recent year, as well as one with a more complete herbaceous plant cover from a former year (3), demonstrates activity peaks while slides of intermediate ages are much rarer. Older stages leading to shrub or even forested stages are less visible. High landslide activity seems to be caused by periods of extreme rainfall with highest pore water pressure. July 2002 was the wettest month measured (area input: 1065 mm m^{-1}) and many of the stage 1 landslides in Fig. 24.2d might result from this date.

24.3 Some Soil Properties of Landslides

To assess the impact of landslides on soil fertility, the properties of selected shallow translational debris slides – the most common type of landslide in the RBSF as outlined above – were compared with those of adjacent undisturbed soils (Wilcke et al. 2003). A chronosequence of four small landslides, approx. 0.5, approx. 2–3, approx. 8–10, and approx. 20 years old, was selected on a 30–50°, forest-covered, east-facing slope at 1900–2300 m a.s.l. The slides were mainly triggered by heavy rainfall and ranged from 100 m to 300 m in length and from 10 m to 30 m in width. The L/W ratios of 0.04–0.15 indicated an intermediate stage between translational debris slides and earth flows (Crozier 1986b). The inclination of all slides was >35°, with a highest value up to 65° in the scar area. The age of the landslides was estimated using aerial photographs from 1976, 1989, and 1998, interviews with local residents, and the stage of vegetation succession.

We sampled O and A horizons along transects across all landslides. The transects included soils above and below the landslide areas as references and three soils on the landslides themselves. The small scar area in the uppermost part of the landslide had no soil or only a thin cover. The head of the landslide was the upper part, the foot was the lower part, and the track (the path of the landslide mass) represented the connection between head and foot. The different landslide zones were identified visually. The head zone was located in a depression and the foot zone in an elevated position relative to the surrounding undisturbed area. All soil samples were analyzed for physical and chemical properties using standard methods described by Wilcke et al. (2003).

All the soils were Typic Dystropepts (Soil Survey Staff, personal communication in 1998) or Dystric Cambisols (FAO–Unesco 1997) whether inside or outside the landslide areas. The stone content of the A horizons ranged from a field-estimated <10% by volume to >80%. The soils above and on the youngest landslides had the largest stone contents: 50% in the A horizon of the reference soil above the landslide and >80% in all soils on the landslide. At the other landslide sites the stone content was less, and we did not observe systematic differences in stone contents between the soils on the landslide and the undisturbed reference soils. The greater stone content of the A horizon on the youngest landslide suggests that it reached deeper into the soil than the other landslides.

On the youngest landslide, no organic layer was present. Only the oldest landslide had an organic layer at each of the three studied positions. Furthermore, we observed an Oa horizon, i.e. the most altered organic horizon that has lost all plant structure and contains >30% organic matter, at only one sampling site. This indicates that development of a complete organic layer on the studied landslides takes >20 years. This is more than under the closed forest canopy, where turnover times of the organic layers at 1900–2000 m a.s.l. vary from 9 years to 16 years (Wilcke et al. 2002). The reasons for this include increased post-landslide erosion on the incompletely vegetated landslides and smaller litter fall because of the smaller biomass on the landslides. Although the thickness of the organic layers tended to increase with increasing landslide age, the differences in mean organic layer thickness among the studied four landslides were not significant.

The most obvious change in soil morphology caused by the land sliding was the decrease in organic layer thickness compared with the undisturbed reference soils. The reason for the lack of significant differences in other morphological characteristics may be related to the variability of soil properties on landslides (Adams and Sidle 1987; Guariguata 1990; Lundgren 1978). Our data set is probably too small to encompass this variability.

There were few consistent trends in nutrient concentrations of the studied soil horizons with increasing age of the landslides. Exceptions were:

1. The exchangeable Ca concentrations in the A horizons of the soils in the head area increased with age: $1.1\,\text{mmol}_c\,\text{kg}^{-1}$ (youngest landslide) $< 5.8\,\text{mmol}_c\,\text{kg}^{-1}$ (oldest landslide).
2. Total P concentrations in the A horizons of the soils in the head area of the three older landslides, which increased in the order $0.76\,\text{g P kg}^{-1}$ (youngest landslide) $< 1.5\,\text{g P kg}^{-1}$ (oldest landslide).

These results showed that the soils in the head area were greatly depleted in exchangeable Ca and total P as a consequence of the land sliding, and that inputs from the atmosphere and litter from the adjacent forest and successional plants resulted in progressive increases in their concentrations with time. Zarin and Johnson (1995a, b) reported increases in the concentrations of C, N, and exchangeable base metals on landslides with increasing age of the landslide in Puerto Rico.

The organic layers of the soils on the landslides generally had greater concentrations of Ca and lower concentrations of C, Mg, N, and P than those of the undisturbed reference soils, but the differences were significant only for N and P (Table 24.1).

The mean contents of total C and N were significantly greater in the undisturbed soils. Thus, the O horizons on the landslides contained less organic matter and more mineral soil than those of the reference soils (Table 24.2). Furthermore, the mean C/N and C/P ratios of the organic layer were greater at all sampled positions on the landslides than in the reference soils (ratios calculated from Table 24.1).

The soils in the accumulation zone contained consistently more nutrients in their organic layers than those in the depletion zone where an O horizon was present only on the oldest landslide. However, the mean nutrient stocks in the organic layer of the soils in the accumulation zone was less than that of the undisturbed reference soils although the differences were not significant because of the large variation among the replicates.

In summary, the soils on the landslide had fewer nutrients in their organic layers than the reference soils, which were probably less plant available because of slower mineralization, as indicated by greater C/N and C/P ratios. On the landslides, the

Table 24.1 Mean total concentrations of elements in undisturbed soils ($n=7$) and in soils at three positions on landslides ($n=4$)

Horizon		C (g kg^{-1})	Ca	K	Mg	N	P
Oa	Undisturbed	475	0.81	1.4	0.44	17.0	0.48
	Head	475	0.91	1.3	0.37	11.0	0.41
	Track	440	1.1	1.9	0.27	12.0	0.35
	Foot	453	0.95	1.6	0.32	9.5	0.35
A	Undisturbed	32	0.18	20	1.5	2.0	0.79
	Position 2	23	0.20	26	0.87	1.6	1.1
	Position 3	16	0.30	16	1.6	1.1	0.47
	Position 4	28	0.27	19	2.7	1.7	1.2

[a] Not all landslides had organic layers at all positions; means were calculated from the organic layers present.

Table 24.2 Mean total stocks of elements in the organic layers of undisturbed soils ($n=7$) and of the soils at three positions on the landslides ($n=4$, not all landslides had organic layers at all positions, means were calculated from the organic layers present)

Location	Mass (t ha^{-1})	C (kg ha^{-1})	Ca	K	Mg	N	P
Undisturbed	291	134 000	132	459	80	5440	134
Head	46	21 800	41	65	17	522	19
Track	101	44 600	112	192	27	1260	37
Foot	102	45 700	89	182	31	991	36

topsoil of the accumulation zone tended to be more fertile than those in the depletion zone. The greater fertility of the accumulation zone coincides with our qualitative impression that the vegetation succession was more advanced in the lower than in the upper parts of the studied landslides.

24.4 Plant Succession and Regeneration

At RBSF most natural landslides are very similar in shape because of the dominance of the translational slides and are colonized quickly either at the borders of the slide or around vegetation islands that slipped down without being overturned. Other patches of high vegetation cover are created by the clonal, looping runner-shoot building growth of most of the individual pioneers that managed to establish seedlings first (Gleicheniaceae, Lycopodiaceae, Bambusoideae and Ericaceae). On landslides well protected against wind and direct sunlight, seedlings of the surrounding flora established themselves after a few months. In contrast, landslides exposed to wind and direct sunlight were bare of any vegetation for months after the slide event. Differences in vegetation along the altitudinal gradient and between ridge crests and slopes were found. Small scale differences in microclimate and water relations are important factors for the patchy distribution of different vegetation types (see Chapters 12, 18 in this volume). The main floristic change occurred at an elevation of about 2100 m. This altitude corresponds to the change in the vegetation zonation in the surrounding forests (Bussmann 2001). On the landslides at higher elevations some species typical for paramo vegetation (*Paepalanthus meridensis* Klotzsch ex Körn. – Eriocaulaceae; or *Xyris subulata* Ruiz and Pav. – Xyridaceae) occur. The influence of different soil chemistry combined with the influence of changing altitude would offer an explanation. Slightly different pH values and a different percentage of exchangeable Ca^{2+} (Ohl and Bussmann 2004; see Chapters 12, 18 in this volume) are characteristic for the different ridges. The amount of Ca^{2+} correlates negatively with the abundance of Al^{3+} ions, which are toxic to plants and could therefore be responsible for the differences in floristic composition. Succession followed always a distinct pattern: The first (cryptogam) stage (Fig. 24.5a) was rather similar at all altitudes with mosses and lichens covering the ground. The percentage cover of the layer of lichens and mosses highly depended on soil and water conditions at a very small spatial scale and was therefore not useful as an indicator for succession. The duration of this stage is highly variable depending on the erosion of the site. The occasional presence of lignified plants, in particular *Baccharis genistelloides* (Lam.) Pers. (Asteraceae) and Lycopodiaceae (Fig. 24.5b) already on first-stage sites pointed to an advanced age of at least 5–10 years. The second stage developed with the extension of the scattered plant individuals and ramets that established in early succession. *Lycopodiella glaucescens* (C. Presl) B. Øllg. and *Lycopodium clavatum* L. spread more quickly than Gleicheniaceae but built stands of less density. Locally, dense covers of Gleicheniaceae and Lycopodiaceae dominated the vegetation cover in the second and third stage (Fig. 24.5c). The dominant role of *Lycopodiella glaucescens* decreased with increasing total vegetation cover. Seedlings of bushes like *Tibouchina lepidota* (Bonpl.) Baill., *Graffenrieda harlingii* Wurdack (both Melastomataceae) or *Bejaria aestuans* Mutis ex L. (Ericaceae) were frequently found under dense layers of *Sticherus*. Seedlings of *Purdiaea nutans* Planch. (Cyrillaceae), the dominant species in the upper mountain forest, were also found, but apparently these never matured to shrubs or trees in any of the early successional stages. Various woody Asteraceae like *Ageratina dendroides* (Spreng.) R.M. King and H. Rob., *Munnozia senecionidis* Benth. or *Liabum kingii* H. Rob. were present. In the upper strata *Tibouchina lepidota* (Melastomataceae) and other species built bushes or small trees (Fig. 24.5d, e).

Species richness during the first two stages of regeneration was low due to the dominance of a few species of ferns or grasses. During the third stage of succession, species composition still differed from the surrounding forest, but diversity was high (Ohl and Bussmann 2004). The second stage with a dense cover of Gleicheniaceae has not been described from northern Ecuador (Stern 1995). Stern found a dominant species of *Chusquea*, 3.5 years after the slide event at the lower zone of a landslide. The bamboo occurred at sites with a reasonable upper layer of organic debris. In contrast, Gleicheniaceae were found on landslides in Bolivia (Kessler 1999) and Hawaii (Restrepo and Vitousek 2001). There, the role of Gleicheniaceae seems similar, although in Hawaii the regeneration patterns are more and more influenced by invasive exotic species. Gleicheniaceae did not

Fig. 24.5 Successional stages on landslides. **a** Cryptogam stage, **b** Cryptogam stage with a few Asteraceae invading, **c** Gleicheniaceae stage, **d, e** slow take-over by woody species

hinder the establishment of bushes. As a result of landslide succession a mosaic-like forest structure with younger and older forest stages was described by Kessler (1999) from the mountain forest in the Bolivian Andes. He observed irregularly formed and spaced patches of senescent forest with single trees having already collapsed. This could explain the clustered occurrence of landslides, as the risk of slipping in zones of senescent forest is higher than in zones of mature forest. In this case the effect of landslides on the ecosystem would be very important for the natural regeneration of the system.

Anthropogenic landslide succession showed remarkable differences to the natural situation (Hartig 2000). Anthropogenic slides were usually very extensive. On anthropogenic landslides grasses largely replaced the Gleicheniaceae and built a very dense layer often limiting the establishment of bush species. Especially the number of orchid species was tremendously high on man-made slides (Gross 1998). Generally succession on man-made slides was much slower than on natural landslides.

24.5 Conclusions

Since the RBSF terrain is considered an extremely landslide-prone area, results of studies of this type of disturbance regime are of the highest importance to understand regional traits of ecosystem functioning.

The shallow landslides, which are common in the tropical mountain forest area of Ecuador, mainly affect the mass and composition of the O horizons. Landslides remove the organic layer and the uppermost part of the A horizons in the head area and redeposit the material in the foot area. This results in a considerable loss of nutrients in the topsoil on the landslides. Even on the ca. 20-year-old landslides, the organic layer had still not reached pre-landslide thickness. Thus, more than 20 years are needed to restore the organic layer completely in the landslide area. Although soil fertility of the accumulation zone was greater than in the depletion zone, the loss of the organic layer still implied a considerable loss in total nutrient content of the soils compared with the undisturbed reference soils. We conclude that the studied landslide type reduced topsoil fertility in the whole landslide area and provided less favorable conditions for plant growth than undisturbed soils.

The duration of the succession process is highly variable. The initial stage of the succession is a community of non-vascular plants interspersed with scattered individuals of vascular plants. By means of runner-shoots they form vegetation patches that start growing into each other. Gleicheniaceae dominate the second succession stage. In the third stage, bushes and trees colonize, sheltered by the ferns, and a secondary forest develops with pioneer species that are not found in the primary forest vegetation. The common phenomenon of the natural landslides leads to an increase in structural and species diversity on a regional scale.

Acknowledgements The authors thank Ronald Stoyan (Erlangen), Hector Valladarez (Loja), Carlos Valarezo (Loja), Syafrimen Yasin (Bayreuth), Wolfgang Zech (Bayreuth), Constanze Ohl (Halle) and Pablo Lozano (Loja), for contributing to the results presented here.

Chapter 25
Sustainable and Non-Sustainable Use of Natural Resources by Indigenous and Local Communities

P. Pohle(✉) and A. Gerique

25.1 Introduction

By now it is sufficiently well understood that any attempt to preserve primary forest in the tropics is destined to fail if the interests and use claims of the local population are not at the same time, and in the long term, taken into account. Therefore, in addition to strict protection of the forests, an integrated concept of nature conservation and sustainable land use development needs to be sought (e.g. Ellenberg 1993). The DFG research project presented here will figure out the extent to which traditional ecological knowledge and indigenous biodiversity management strategies can be made available for a long-term land use development. The project chose to use a specialized approach, namely the investigation of indigenous/local knowledge systems as part of the ethnoecological methodology (e.g. Münzel 1987; Posey and Balée 1989; Warren et al. 1995; Müller-Böker 1999; Nazarea 1999). In biodiversity-rich places local people usually have a detailed ecological knowledge e.g. of species, ecosystems, ecological relationships and historical or recent changes of them. Numerous case studies have shown how traditional ecological knowledge and traditional practices serve to effectively manage and conserve natural and man-made ecosystems and the biodiversity contained within (e.g. Posey 1985; Toledo et al. 1994; Berkes 1999; Fujisaka et al. 2000; Pohle 2004). In ongoing interdisciplinary and integrative research projects like BIOTA AFRICA (Biodiversity Monitoring Transect Analysis in Africa, German Federal Ministry of Education and Research), STORMA INDONESIA (Stability of Rainforest Margins in Indonesia, German Research Foundation, Collaborative Research Centre 552) or within the interdisciplinary programme of the National Centre of Competence in Research North–South implemented by the Swiss National Science Foundation (SNSF), investigations on traditional ecological knowledge and biodiversity management are an integral part.

25.2 The Tropical Mountain Rainforests of Southern Ecuador – a "Hot Spot" of Biodiversity

The area under study, the Podocarpus National Park and its surroundings, is especially noteworthy for its species diversity and belongs to one of the so-called "hot spots" of biodiversity worldwide (Barthlott et al. 1996; Myers et al. 2000; cf. Chapter 1 in this volume). The tropical mountain rainforests of southern Ecuador are of crucial importance for the preservation of genetic resources and play an important role as an ecosystem and habitat for flora and fauna. At the same time humans have lived here and sustain themselves since centuries. However, in more recent times (the past four or five decades), these mountain forest ecosystems, which have been described as particularly sensitive (Die Erde 2001), have come under enormous pressure from the expansion of agricultural land (especially pastures), the extraction of timber, the mining of minerals, the tapping of water resources and other forms of human intervention. According to Hamilton et al. (1995), 90% of the original forest cover in the Andes can be regarded as either destroyed or altered.

25.3 Indigenous and Local Ethnic Communities

The surroundings of the Podocarpus National Park are settled by indigenous Shuar and Saraguro communities as well as Mestizo-Colonos (cf. Chapter 3 in this volume).

The Shuar area of settlement extends from the lower levels of the tropical mountain rainforest (approx. 1400 m a.s.l.) down to the Amazonian lowland (*Oriente*) in the region bordering Peru. The Shuar, who belong to the Jívaro linguistic group (Amazonian Indians), are typical forest dwellers who practice shifting cultivation, mainly in a subsistence economy. Their staple crop is manioc which they plant together with taro and plantains on small rotating plots in forest gardens. They also fish, hunt and gather forest products. During the past decades some Shuar have also begun to raise cattle and some are engaged in timber extraction as well.

The Saraguros, highland Indians who speak Quichua, live as agro-pasturalists for the most part in the temperate mid-altitudes (1800–2800 m a.s.l.) of the Andes (*Sierra*) in southern Ecuador. As early as the nineteenth century the Saraguros kept cattle to supplement their traditional "system of mixed cultivation", featuring maize, beans, potatoes and other tubers (Gräf 1990). Now, cattle ranching has developed into the main branch of their economy.

In the North and East of Podocarpus National Park Mestizo-Colonos are the most dominant ethnic group in numbers. They are colonists of mestizo ethnicity, who came into the area during the past four or five decades to log timber and to practice cattle farming and agriculture. In Latin America the term mestizo is generally used to indicate people of mixed Spanish and indigenous descent. In South Ecuador, however, the Mestizo-Colonos have little tendency to identify themselves as a distinct ethnic group, instead they refer to themselves as Ecuadorian but not as mestizos.

25.4 Aims and Methods of the Ethnoecological Research Project

During 2004, 2005 and 2006 ethnoecological, especially ethnobotanical, and agro-geographical research was undertaken in sample communities of the Shuar (Shaime, Napints, Chumbias), the Saraguros (El Tibio) and the Mestizo-Colonos (Sabanilla). The goals were:

- To document the indigenous and local knowledge of traditionally utilized wild and cultivated plants (the ethnobotanical inventory was undertaken following the "Code of Ethics");
- To analyze current forms of land use including the cultivation of forest and home gardens;
- To evaluate ethno-specific life-support strategies as well as strategies for natural resource management.

The ethnospecific plant knowledge was documented using various ethnobotanical techniques of unstructured and semi-structured inquiry, like participant observation techniques, artefact interviews, plant interviews, checklist interviews and group interviews (cf. Alexiades 1996). The "ethnobotanical inventory technique" was the main procedure to collect ethnobotanical information. An inventory of wild and cultivated plants was compiled, including botanical names as well as indigenous Shuar, Saraguro and Spanish names. The use of forest products such as food, medicine, fodder, construction material etc. was documented. Empirical fieldwork also encompassed the agro-geographical analysis and cartography of land use with special consideration of traditional forest gardening. Information was gathered by semi-structured and thematically focused interviews with residents and local experts. In addition, a multi-temporal analysis of aerial photographs was undertaken to identify and quantify the change in forest cover and land use.

25.5 The Significance of Plant Use for the Shuar, Saraguros and Mestizo-Colonos

The Shuar of the Nangaritza valley have a comprehensive knowledge of plants and their utilization. All households make extensive use of forest products. According to the ethnobotanical survey[1], the actual inventory of traditionally used wild plants of the Shuar includes 211 species (Fig. 25.1). Most of the wild plants are used to supplement the diet (74). Given the lack of state health care, medicinal plants also assume great significance (63). Many plants, too, are used as construction

---

[1] The ethnobotanical survey in sample communities of the Shuar, Saraguros and Mestizo-Colonos was conducted by Andrés Gerique.

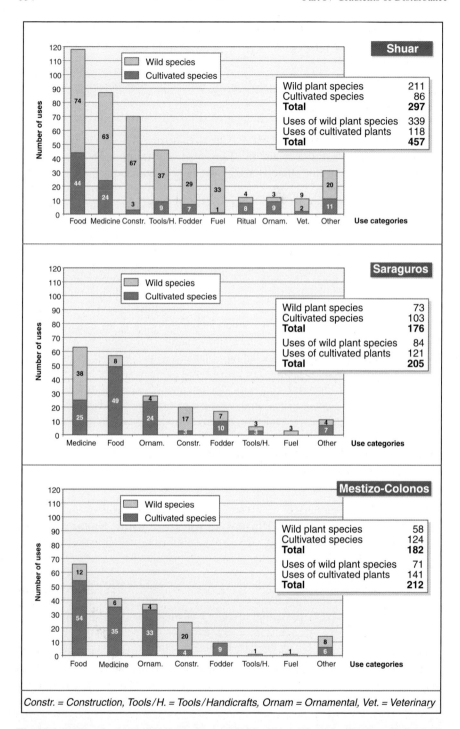

Fig. 25.1 Wild and cultivated plant species used by the Shuar (Chumbias, Napints, Shaime), the Saraguros (El Tibio) and the Mestizo-Colonos (Sabanilla) according to use categories. Note: one species can be found in more than one use category

material (67), as tools and for handicrafts (37), as fuel, fodder or as ritual plants. The Shuar use forest products exclusively for their own needs, there is virtually no commercialization.

The Saraguros from El Tibio have a far less comprehensive knowledge on wild plant species and their utilization. The actual ethnobotanical inventory includes only 73 wild plant species (Fig. 25.1). Most of them are ruderal plants used as medicine (38) or plants used for their wood (17). As agro-pasturalists they have conversed most of the primary forest into pastures, home gardens and fields, leaving forest remains only along mountain ridges or in river ravines. Their actual plant knowledge reflects this traditional way of life. They have a comprehensive knowledge of cultivated plants (103) mainly pasture and crop plant species – even more than the Shuar (86) – but they are less familiar with forest plant species. The latter knowledge is mainly reduced to woody varieties which they extract and sell before clearing the forest.

The Mestizo-Colonos of the surrounding areas of Sabanilla base their economy on cattle ranching. They have converted large areas of forests into pastures. As settlers from the western and most arid area of Loja Province they seem not to be familiarized with the local flora and hence make only little use of it. The actual ethnobotanical inventory – although not completed yet – includes a total of only 58 wild plant species (Fig. 25.1). Timber is the main forest product (20), while some ruderal plants and the fruit of a few tolerated tree species are used as food (12) or as medicine (6). However, the Mestizo-Colonos cultivate 54 different species for food and 35 species for medicinal purposes and have a comprehensive knowledge of pasture species.

25.6 Agrobiodiversity in Shuar and Saraguro Tropical Home Gardens

The tropical home gardens of indigenous and local communities are generally regarded as places of great agrobiodiversity and refuges of genetic resources (Watson and Eyzaguirre 2002). Furthermore, they contribute significantly to securing and diversifying food supplies. This applies wholly to Shuar and Saraguro gardens, which feature a large number of both wild and cultivated species and which play their part in providing subsistence needs. Staple crops, such as maize, tubers and beans, may be cultivated primarily in *chacras* (fields), but home gardens (*huertas*) have an essential role to play in supplementing the diet with fruits and vegetables, furnishing households with medicinal plants and spices, and fodder and timber.

The forest gardens of the Shuar (Fig. 25.2) are characterized by an especially great diversity of species and breeds. In five *huertas* studied (size: approx. 600–1000 m^2), a total of 185 wild and cultivated plant species and breeds were registered. For the most part they serve as nutritional items (58%) or medicines (22%). The main products cultivated are starchy tubers like manioc (*Manihot esculenta*) and taro (*Colocasia esculenta*), along with various breeds of plantains (*Musa* sp.).

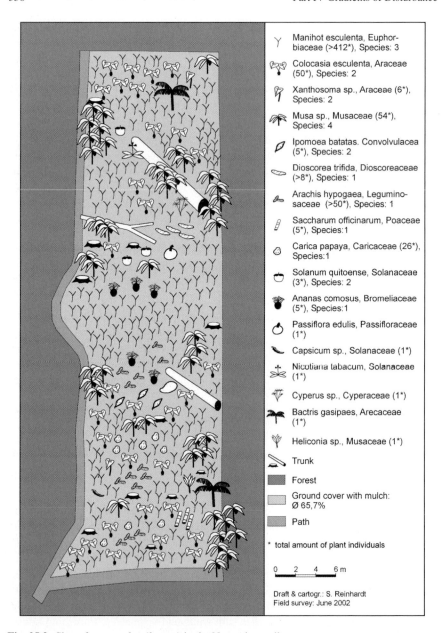

Fig. 25.2 Shuar forest garden (*huerta*) in the Nangaritza valley

Moreover, the planting of a large number of traditional local breeds was documented: e.g. 29 breeds of manioc and 21 breeds of *Musa* sp. – a further indication of the crucial significance that home gardens have for the in situ conservation of botanical genetic resources (Münzel 1989, p. 434).

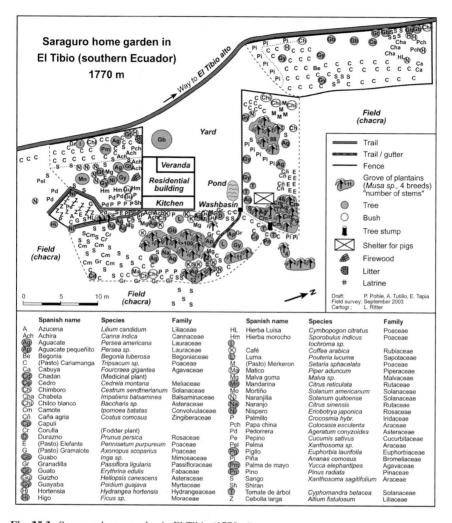

Fig. 25.3 Saraguro home garden in El Tibio (1770 m)

The *huertas* of the Saraguros likewise display a great diversity of useful plants. In one sample home garden studied in El Tibio (Fig. 25.3), 51 species of cultivated plants were identified. In total 95 cultivated plants were registered among the Saraguros of El Tibio. Again, the majority are plants that supply nutritional value (41%), followed by medicinal and ornamental plants (each 20%). The most important cultivated products are plantains, tubers and various types of fruit. Given their relatively dense and tall stands of trees, the multi-tiered arrangement of plants and the great diversity of species, the gardens of the Saraguros can be seen as an optimal form of exploitation in the region of tropical mountain rainforests.

25.7 Indigenous Concepts of Biodiversity Management – Their Contribution to a Sustainable Land Use Development

If the hypothesis is accepted that a multi-facetted economic and cultural interest in the forest on the part of indigenous and local communities offers effective protection against destruction, then a key role must be assigned to the analysis and evaluation of the ethno-specific knowledge about tropical mountain rainforests and their potential uses. Both indigenous groups have developed natural resource management strategies that could be used and expanded, in line with the concept "preservation through use", for future biodiversity management, but this should be done only in an ethno-specific way.

The Shuar traditional way of managing biodiversity is based on a sense of being closely bound culturally, spiritually and economically to the forest. The sustainability of their form of land use has long since been put to the test (Münzel 1977, 1987). As traditional forest-dwellers, sustainable elements of biodiversity management can be found in (Fig. 25.4):

- Their regulated practice of shifting cultivation, which – given the correspondingly long time for regeneration – is thought to conserve the soil and the vegetation. The system of cultivation and fallow on small rotating plots as shown in the sketch map (Fig. 25.5) has much in common with ecological succession in that it uses the successional process to restore the soil and the vegetation after use for farming (Kricher 1997, p. 179). In the Shuar forest gardens the fallow periods last for about 24–30 years while the cultivation periods covers four years.

Fig. 25.4 Napints (1000 m): scattered settlement of the Shuar in the tropical rainforest at the eastern periphery of the Podocarpus National Park. Photo by A. Gerique

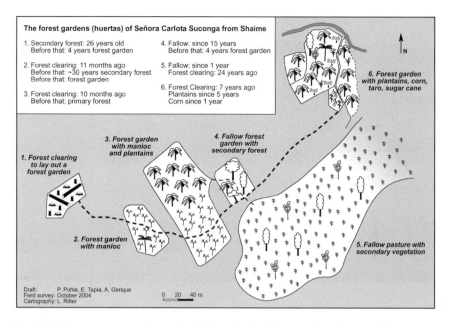

Fig. 25.5 The forest gardens (*huertas*) of Señora Carlota Suconga from Shaime (920 m)

- Their tending of forest gardens according to principles of agroforestry and mixed cropping with a high agrobiodiversity and a particular high breed variety of cultivated plants. As is commonly known, polycultures are more resistant to insect attacks and plant diseases;
- The natural fertilization of soils by mulching and the use of digging sticks and dibbles as a suitable form of cultivating the soil;
- Their sustainable use of a broad spectrum of wild plants in small quantities, satisfying only subsistence needs and avoiding over-harvesting.

If the Shuar's traditionally practiced and clearly sustainable plant biodiversity management is to be preserved, then this is possible only by the following preconditions: the legalization of their territorial claims and, along with that, a comprehensive protection of their territories by demarcating reservations for example. The establishment of a so-called *Reserva Shuar* is indeed currently being planned (Neill 2005). Additionally it is necessary to respect and support the Shuar's cultural identity, not only to avoid the loss of traditional environmental knowledge especially traditional plant lore. To improve livelihood in an economic sense, additional sources of financial income are essential. In this line the promotion of ecotourism, support of traditional handicrafts and the cultivation of useful plants for a regional market could be discussed.

While the Shuar's forest management can be evaluated as preserving plant biodiversity, the sustainability of the Saraguros' use of the environment has yet to be rated. Market-oriented stockbreeding has particularly led in recent decades to the rapid increase of pastures at the expense of forest. In spite of ecological conditions

Fig. 25.6 Richly chequered cultural landscape of the Saraguros on a steep slope of the Río Tibio valley with the scattered settlement of El Tibio (1770 m). Photo by P. Pohle

unfavorable to agricultural pursuits (steep V-shaped valleys, acidic soils, extremely high precipitation) these Andean mountain farmers have at least, by means of their intensive form of pasture management, succeeded in generating a sufficiently stable agrarian and cultural landscape (Fig. 25.6). In contrast to many completely deforested and ecologically devastated areas settled by Mestizo-Colonos (Fig. 25.7), the richly chequered agrarian landscape of the Saraguros presents, not only aesthetically but also ecologically, a fundamentally more positive picture. The comparison of the land use mosaic of El Tibio in the aerial photograph taken in 1969 and the land use map of El Tibio mapped in 2003 shows that changes according to forest cover have taken place, particularly at the slope opposite to the village, whereas the slope of the village site itself shows no dramatic change (Fig. 25.8).

Among the Saraguros, initial attempts have also been elaborated to manage biodiversity in line with the concept "preservation through use". The first thing to be mentioned in this context is the keeping of home gardens with a wide spectrum of wild and cultivated plants, particularly woody species. With regard to the diversity of species, the remnants of forest still largely preserved in ecologically unfavorable locations are significant (Fig. 25.8). In order to stem the further loss of biodiversity, however, it will be necessary to convince the Saraguros that in particular scrub- and wasteland (*matorral*) should be replanted with native tree species. The pressure on the tropical mountain rainforests caused by the pasturing economy will only be reduced, though, when the Saraguros can be shown a profitable alternative to it. As examples of promising endeavors in this context may be regarded:

Fig. 25.7 Deforested and overused agrarian landscape of the Mestizo-Colonos north of Loja. Photo by P. Pohle

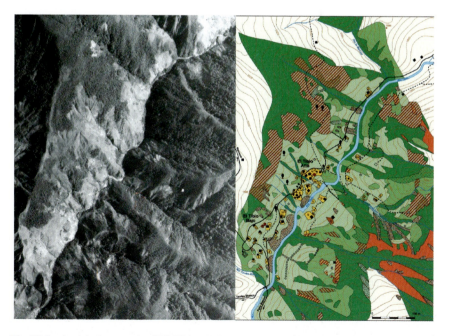

Fig. 25.8 Aerial photograph of El Tibio 1969 and land use map of El Tibio 2003 (Pohle and Tutillo 2003)

- The selective timber production and replanting with native tree species as proposed by foresters (Günter et al. 2004);
- The introduction of silvipastoral or agroforestry systems;
- The market-oriented gardening;
- The cultivation and marketing of useful plants, e.g. medicinal herbs;
- The promotion of "off-farm" employment opportunities;
- The payment for environmental services to protect the watershed area of Loja.

25.8 The Overexploitation of Natural Resources by the Mestizo-Colonos

The Mestizo-Colonos living today in the northern buffer zone of the Podocarpus National Park arrived from the 1960s onwards, encouraged by the national land reform of 1964. Most of the Mestizo-Colonos settling along the road between Loja and Zamora arrived only during the past 25 years. As colonizers they converted large areas of tropical mountain rainforests into almost treeless pastures (Fig. 25.7). To sustain livelihood they were forced to use even very steep and marginal areas for cattle ranching. Fire was the most common (easy and fast) way of clearing the forests to gain pasture land and as a consequence, large areas of forest were burned up, often in an uncontrolled way. The abandoned and scorched areas were immediately taken by the *llashipa* or bracken fern (*Pteridium aquilinium*). During the late 1970s and in the 1980s cattle ranching was displaced as main regional economic activity by timber extraction. This provoked the selective disappearance of many tree species with economic value and the alteration of adjacent forest areas, but as Wunder (1996a) pointed out, the need for new pasture land and not the extraction of timber was the principal reason for deforestation in this area.

At present, the economy of the Mestizo-Colonos is again based on cattle ranching, but the use of fire to open new pastures has declined, even though not disappeared. This has several reasons: First, many areas belong to the *Bosque Protector Corazón de Oro* or are part of the buffer zone of the Podocarpus National Park where it is forbidden to clear large forest areas and peasants face penalties if they do so. The road allows an easy control by the police and by the officers of the Ministry of the Environment in this area. Second, the remaining forests are today under private ownership. Therefore, uncontrolled fire can be a problem for peasants if it affects the neighboring private properties. Since the remaining forest is nowadays further away from their housing and good quality timber is getting scarce in the forests, the Mestizo-Colonos are becoming more receptive to the idea of tolerating a variety of tree species in order to have shadow for cattle, timber for repairing fences, fuel, or material for construction. *Tabebuia chrysantha*, *Cedrela* spp, *Piptocoma discolor*, *Inga* spp and *Psydium guayava* are among the most common tolerated species. Thus, the new pastures with tolerated tree species are easy to distinguish from older ones, which have almost no trees growing on them.

Among the Mestizo-Colonos, the low profitability of the extensive form of cattle ranching has lead to a high in- and out-migration and a correspondingly high fluctuation in land-ownership and possession. To avoid poverty most of them have an alternative income, e.g. from a small shop near the road, or a second occupation as day labourers. As a consequence, in marginal areas the pasture management is obviously neglected, mainly because of the limited availability of labor. Compared with the indigenous ethnic groups, the Mestizo-Colonos are much more heterogeneous. Extreme overexploitation is especially typical for newcomers and colonizers of the first generation mainly living in scattered *fincas* along the road between Loja and Zamora. In contrast, colonizers of the second or third generation living in village communities like in the upper Río Zamora valley have developed a more adapted and sustainable form of land use, in many respects similar to that of the Saraguros.

To avoid a further loss of biodiversity and to reduce the pressure on the tropical mountain rainforests caused by cattle ranching, profitable alternatives have to be offered. Similar measures as those presented above for the Saraguros are necessary: e.g. timber production in the form of reforestation with native tree species, market-oriented gardening, cultivation of medicinal herbs, payment for environmental services. These measures should also take into account an improvement of the carrying capacity of existing pastures by introducing legumes (Leguminosae) and other fodder plants and improving the cattle breeds and the veterinary services. In any case, measures to stop the loss of biodiversity in the area should take into account the difficult economic situation of most Mestizo-Colonos households. The prohibition of using fire and the establishment of protected areas have surely reduced the deforestation rate, but such measures do not face the real causes of deforestation and have increased the animosity against conservation.

25.9 Protecting Biological Diversity – from National Park to Biosphere Reserve

The Podocarpus National Park, covering a total of 146 280 ha, was established in 1982 as southern Ecuador's first conservation area, whose goal is to protect one of the country's last intact mountain rainforest ecosystems, one particularly rich in species and largely untouched by humans. The creation of a national park in the middle of a fairly densely populated mountain region necessarily gave rise to numerous conflicts of interest and use rights, e.g. agrarian colonization, illegal timber extraction, conflicts about landownership and possession, mining activities, tourism.

The experience in international nature conservation during the past decades has shown that resource management, if it is to be sustainable, must serve the goals of both nature conservation and the use claims of the local population. The strategy is one of "protection by use" instead of "protection from use", a concept that has emerged throughout the tropics under the philosophy "use it or lose it" (Janzen 1992, 1994). In the following, strategies are presented that show a way how people can benefit from the national park without degrading the area ecologically: the implementation of extractive reserves, the promotion of ecotourism and as the most promising approach the establishment of a Biosphere Reserve.

Given the high biodiversity of tropical rainforests and the fact that indigenous people in general have a comprehensive knowledge of forest plants and their utilization, it seems possible according to Kricher (1997, p. 357) to view the rainforest as a renewable, sustainable resource from which various useful products can be extracted on a continuous basis. In view of the high number of plant species that are currently collected by extractivists in the surroundings of the Podocarpus National Park (e.g. the Shuar) the preservation of large areas of rainforest would make economic sense as well as serve the interests of conservation and preservation of biodiversity. Thus, the establishment of an extractive reserve could be suggested as an alternative to deforestation.

In line with the concept "protection by use" ecotourism can be structured such that it is compatible with conservation interests and serves the local economy as well. This is also the experience around Podocarpus National Park. The attraction of the park is clearly the tropical rainforest with its specific wildlife, particularly tropical birds, fewer visitors have botanical or eco-geographical interests. However, compared with other national parks in South America (e.g. the Manu National Park of Peru), southern Ecuador and the Podocarpus National Park are not a major tourist destination.

The most promising approach, in which conservational protection and sustainable development are the guiding principles, is the integrated concept of conservation and development exemplified by UNESCO's Biosphere Reserve (UNESCO 1984). The idea behind it is to mark out representative sections of the landscape composed of, on the one hand, natural ecosystems (core area) and, on the other, areas that bear the impress of human activity (buffer zone, development zone; Erdmann 1996). In Ecuador, three biosphere reserves have already been drawn up (Ministerio del Ambiente 2003). The establishment of such a reserve would also be desirable for southern Ecuador, and this was recently submitted for approval. Alongside a strictly protected core area comprising Podocarpus National Park, it might encompass in a buffer zone cultural landscapes that have arisen historically (e.g. Vilcabamba) together with sanctuaries of indigenous communities (e.g. the proposed *Reserva Shuar*) and, in a development zone, areas of recent agrarian colonization.

Biosphere reserves are strongly rooted in cultural contexts and traditional ways of life, land-use practices and local knowledge and know-how. In the buffer and development zones of the Podocarpus National Park measures to be taken could rely on the rich ethno-specific traditions in forest- and land-use practices by indigenous and local communities. In a first step it would be desirable to develop with the participation of the local communities' environmental management plans. On the one hand they could support the ethno-specific cultural tradition and strengthen the social identity of local communities. On the other hand these plans should comprise regulations for hunting, fishing, timber and plant harvesting, for the exclusion of human-created fires but also for house and road construction. In southern Ecuador management plans are still in various phases of discussion and implementation. However, there is still a big gap between vision and reality. An intensive discussion was started about the participation of the population and local NGOs for assigning environmental competence to the regional–local administration. But, the realization of concepts like "cooperative management structure" could not consolidate and find political acknowledgement. According to Gallrapp (2005) it failed to create effective platforms for participation and negotiation, to build a common vision of the participating persons and to create a social awareness in order to implement new structures.

Acknowledgements We wish to thank all the inhabitants of the communities of Shaime, Chumbias, Napints, El Tibio and Sabanilla for their hospitality and generous participation in this study. In particular, we would like to thank Don Vicente and Don Carmelo Medina, María Silvia Chiriap, Angel Tscuenka, Carlota Sukonga and Bartolomé Kukush. In addition, we are grateful to Jhofre Aguirre, Bolivar Merino, Carlos Chimbo, Néstor León, Wilson Quizhpe and Holger Salas from the UNL for their expertise and assistance with data collection and plant identification. Our gratitude goes to Eduardo Tapia for his competence in collecting and processing geographical data and for making our fieldwork so comfortable.

Chapter 26
Natural Forest Management in Neotropical Mountain Rain Forests – An Ecological Experiment

S. Günter(✉), O. Cabrera, M. Weber, B. Stimm, M. Zimmermann,
K. Fiedler, J. Knuth, J. Boy, W. Wilcke, S. Iost, F. Makeschin,
F. Werner, R. Gradstein, and R. Mosandl

26.1 Introduction

26.1.1 The Forest Depletion Process

In tropical forests, the first step in the destruction cycle is usually the over-exploitation of high value timber, leading not only to extinction of the extracted species (Silva Matos and Bovi 2002) but also to the conversion of the forests into pastures in many cases (Wunder 1996b). In many highlands of Ecuador the productivity of the pastures is depleted due to the invasion of bracken fern (see Chapter 28 in this volume), which finally leads again to the conversion of primary forests by local farmers (Paulsch et al. 2001; Hartig and Beck 2003). This process usually is accompanied by loss of biodiversity (Brooks et al. 2001), increased erosion, changes of hydrology (Bruijnzel 2004), and further environmental disturbances. More information on deforestation in Ecuador is given by Mosandl et al. (Chapter 4 in this volume).

26.1.2 Strategies for Conservation, Reforestation, and Sustainable Management

The best way to protect tropical forest biodiversity is to keep populations of *Homo sapiens* away. But people have never stayed away from tropical forests, and they likely never will (Chazdon 1998). Compared with the complete conversion of a forest into agricultural land, sustainable forest management can be considered to have a low-intensity impact on biodiversity (Grau 2002). Several authors consider natural forest management as a suitable tool for biodiversity conservation and the integration of forest conservation with economic development (Bawa and Seidler 1998; Chazdon 1998), especially for buffer zones around protected areas (Hall et al. 2003). Even critical authors, who prioritize conservation strategies leading to an increase in the protected areas in biologically important sites or corridors, believe that sustainable forest management is a promising strategy (Bowles et al.

1998). Alternative land use strategies like reforestation with native species are presented by Weber et al. (Chapter 34 in this volume).

26.1.3 Silvicultural Systems in the Tropics

Although natural forest management techniques were already developed during the past century (Brünig 1996; Dawkins and Philips 1998), it is still very difficult to verify the sustainability of these silvicultural systems, especially how forest management affects biodiversity in primary tropical forest (Bawa and Seidler 1998). Most systems derived from the dipterocarp forests of Asia, which are known for their exceptional richness in high timber value species and abundances are therefore not transferable to the RBSF area and other Andean rain forests. Nevertheless, it should be emphasized that these silvicultural systems were developed for specific target species and for specific regional ecological conditions. Measures like reduced impact logging, the adoption of cutting cycles, or changes in the traditional minimum felling diameter systems in general lead to higher yield, reduced forest damage, and an improvement in biodiversity conservation (Günter 2001; Boltz et al. 2003; Krueger 2004). Although some silvicultural tools and instruments are applicable worldwide, sustainable management of highly diverse tropical forest ecosystems requires case-specific silvicultural approaches meeting the local socioeconomic and ecological requirements of management and conservation (Hutchinson 1993). Improvement felling for example is one of the oldest silvicultural instruments (Dawkins and Philipps 1998), but there is almost no knowledge about its feasibility for sustainable management of tropical mountain rain forests.

From the three main pillars of sustainability, ecological, economic, and sociocultural sustainability, we only refer to ecological aspects in this chapter (social aspects are considered by Pohle and Gerique, Chapter 25 in this volume). We consider a forest management measure as sustainable, if the diversity of plant and animal species is not markedly reduced, the composition of organism assemblies are changed little, and if forest functions such as regulation of water and element cycles are maintained. In the following sections we demonstrate the effect of selective cutting in the RBSF on forest stand parameters (Section 26.3.1), epiphytes (Section 26.3.2) and moths (Section 26.3.3) within the forest as indicator groups for biodiversity, and selected parameters for nutrient cycling (Section 26.3.4). We discuss our results in the context of a general review of the ecological impact of selective logging activities.

26.2 Methods: Silvicultural Treatments as an Ecological Experiment

On the experimental area of 13 ha we registered all individuals with dbh >20 cm out of a list of defined timber species, including *Tabebuia chrysantha*, *Cedrela* sp., and *Podocarpus oleifolius*, being the most valuable ones, followed by

Nectandra membranacea and *Hyeronima asperifolia*, *H. moritziana*, and finally *Inga acreana*, *Clusia cf ducuoides*, and *Ficus subandina* with reduced economic value. Five hectares within the catchment Q2 served as a reference area without any silvicultural treatment. On four ha in Q5 we carried out a "strong" intervention (felling of 32 trees/ha) and in another four ha in Q3 a "slight" intervention (18 trees/ha). Some 215 selected individuals on a list of defined timber species are considered as potential crop trees (PCTs). The most effective crown competitors of these trees were eliminated in June 2004 in order to stimulate growth and natural regeneration of the PCTs.

Before the silvicultural treatments, the stands in Q2, Q3 and Q5 had different basal areas: Q5: $24.3\,m^2$/ha (± 1.8 standard error); Q3: $10.3\,m^2$/ha (± 0.8); and Q2: $17.4\,m^2$ (± 1.7). The tree logging intensity corresponded to the removal of $2.5\,m^2$/ha (± 0.64) or 13.5% of the basal area in Q5 and $0.62\,m^2$/ha (± 0.16) or 10.2% in Q3. The reference area Q2 remained untouched. Note that these values represent all trees and palms with dbh $\geq 20\,cm$. Silvicultural measures usually always go along with damages of the remnant stand. In our case, 4.3% (1.2–12.3%) of the forest surface suffered collision damage from the felled crowns in Q3 and 8.6% (3.8–16.8%) in Q5.

26.3 Results and Discussion

26.3.1 Impact of Silvicultural Treatments on Forest Stand Parameters

The felling of trees has direct consequences on the light conditions in the stand. Figure 26.1 shows the shift of the canopy openness (measured with spherical densiometers) before and after felling in the canopy openness distribution. It is notable that all three stands had very similar canopy openness distributions before the silvicultural interventions, but only the "strong" intervention resulted in a significant shift in the canopy openness.

The number of felled trees per hectare (18 in Q3; 32 in Q5) seemed high, but the basal area removed was relatively small compared with other studies (Uhl et al. 1991; Verissimo et al. 1992). Concerning natural gap fall, Hubbel et al. (1999) did not find a relationship between gap fall dynamics and tree diversity. Cutting activities without skidding therefore probably does not result in a permanent loss of biodiversity when felling intensity and emerging gap sizes are not too high. The combination of cutting and skidding in contrast, leads to more serious damages (Jackson et al. 2002).

The finding that only the relatively strong intervention had a significant effect on the canopy openness can be explained by the small crowns of dominant trees in this forest compared with tropical lowland forests. From a silvicultural point of view this has the advantage that harvesting intensities (in number of trees per hectare) can be higher than in the lowland forests without creating larger disturbances.

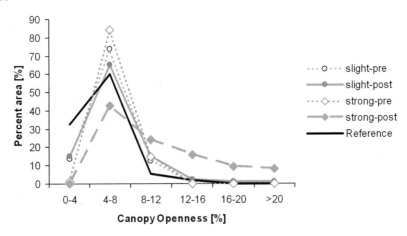

Fig. 26.1 Canopy openness distribution, measured by spherical dendrometers before ("*x*-pre", *open symbols*) and after ("*x*-post", *filled symbols*) the felling of 18 trees/ha ("slight" intervention, *circle symbols*, n=80) and 32 trees/ha ("strong" intervention, *diamond symbols*, n=80). The *Reference* line indicates the canopy openness in the undisturbed area (n=100)

An unresolved problem for the RBSF and many other Andean rain forests remains the skidding of the logs. Trail construction for skidding would have strong impacts on hydrology and erosion in the steep slopes of the neotropical Andes. Horse skidding combined with mobile saw mills or chain saw frames for small-scale logging or cable yarding techniques on a community or regional scale are probably much less problematic.

On the one hand, the removal of competitors, especially in log landings and logging gaps combined with leaving seed trees near these micro-environments generally improves regeneration of woody species (Rheenen et al. 2004). Very large gaps on the other hand usually lead to temporary dominance of short-lived pioneers and rapid growth of forest herbs, which tend to suppress the growth of the more desirable timber species (Thompson et al. 1998). In our study site we generated rather low damages in the forest structure and succeeded in the creation of numerous artificial gaps with slightly increased light incidence, which probably stimulated the natural regeneration. Nevertheless, these effects have to be monitored in a long-term approach.

For all cases where the commercially valuable trees are absent among gap regeneration (Park et al. 2005) alternative silvicultural techniques have to be applied, e.g. enrichment plantings. The most common method of enrichment by line planting (Catinot 1965a, b, c) seems not to be suitable for the extremely heterogeneous geomorphology of Neotropical mountain rain forests. In contrast spot-orientated underplanting of natural or artificial gaps driven by felling impact is a more promising alternative. In addition, even-aged groups of tree cohorts within an uneven-aged matrix provide better options for forest management than the traditional single-tree systems (Pinard et al. 1999).

Table 26.1 Diameter increment of silviculturally enhanced potential crop trees (PCT) and reference trees in the first year after felling. Significant differences at $P = 0.1$ are indicated by *, $P = 0.05$ is represented by **; A Levene test was applied for homogeneity of variances and t-test for heteroscedasticity or homoscedasticity, respectively. Significantly different growth rates between species are indicated by different letters (a–d). Timber values are high (A), medium (B), and low (C)

	Diameter growth PCT (mm/year)	Diameter growth reference (mm/year)	Significance (between species)	Timber value
Cedrela sp.	13.3 ± 1.9	18.05 ± 2.3	bc**	A
Clusia cf *ducuoides*	1.6 ± 0.3	1.7 ± 0.3	a**	C
Ficus subandina	17.8 ± 8.0	20.8 ± 4.5	d**	C
Hyeronima asperifolia	7.2 ± 1.1	5.8 ± 1.5	abc**	B
H. moritziana	2.4 ± 0.5*	1.2 ± 0.3*	a**	B
Inga acreana	14.9 ± 4.3*	6.7 ± 2.1*	c**	C
Nectandra membranacea	5.8 ± 1.8	4.1 ± 1.0	ab**	B
Podocarpus oleifolius	2.9 ± 1.1	2.0 ± 0.6	a**	A
Tabebuia chrysantha	2.5 ± 0.8	3.8 ± 0.7	a**	A

Significant different growth rates were observed between the studied species (Table 26.1), but only five species responded with an enhanced diameter increment (*Hyeronima asperifolia*, *H. moritziana*, *Inga acreana*, *Nectandra membranacea*, *Podocarpus oleifolius*) one year after the activity. In some cases the silvicultural treatments tend to have even negative effects (*Tabebuia chrysantha*, *Cedrela* sp.) at least in the first year after liberation. If the silviculturally enhanced species benefit from reduced crown competition and thereby expand their own crowns in the following years, further stimulation of diameter growth could be possible. As the intensity of the liberation measures, by eliminating only one crown competitor, was relatively moderate, it is probable that only stronger interventions lead to higher increments for some species.

Silvicultural treatments like thinning or refinement stimulate diameter growth in the remaining stand in many cases (Finegan and Camacho 1999; Dolanc et al. 2003), but usually only for a certain period of up to 14 years (de Graaf 1999; Kammesheidt et al. 2003). In order to sustain or even maximize the mean annual growth, repeated improvement fellings are necessary. It should be regarded that intense basal area removal treatments do not always achieve higher growth rates of the remaining trees, but also increase the danger of dominance by lianas or early successional species (Smith and Nichols 2005). Therefore for growth optimization a higher frequency of low impact felling is a more suitable silvicultural tool than lower frequencies with higher felling intensities.

The maximum yield can only be achieved when the cutting cycle length is equal to the culmination of mean annual volume growth, but this point of time is species-specific (Zagt 1997) and site-specific (Günter 2001). In the study area (see Table 26.2) and in general for tropical natural forests (Finegan et al. 1999), the tree species of a single stand usually show extreme variation in diameter growth. Thus, traditional silvicultural systems with only one cutting cycle for all timber species cannot

Table 26.2 Mortality within major taxonomic groups of understory epiphytes during the first 12 months following selective logging

	Mortality (%)	
	Forest interior	Gap edge
Pteridophyta	5.31 ($n=358$)	6.20 ($n=371$)
Monocotyledonae	4.24 ($n=118$)	5.56 ($n=108$)
Dicotyledonae	15.07 ($n=73$)	14.08 ($n=71$)
Total	6.38 ($n=549$)	7.09 ($n=550$)

fit perfectly for fast-growing species like *Cedrela* sp. and at the same time for slow-growers like the podocarps.

26.3.2 Impact of Silvicultural Treatments on the Growth Dynamics of Epiphytes

We studied the response of 1099 individuals of vascular understorey epiphytes on each of 28 host trees along the edges of freshly created gaps and in undisturbed forest, respectively. Initial sampling was conducted in Q5 immediately following felling in Q5. Mortality and leaf number on gap edges were not altered after 12 months, neither regarding any of the major taxonomic groups (ferns, monocotyledons, dicotyledons) nor any single family or genus (Chi-square test, Yates-corrected; Table 26.2). However, the three most abundant genera showed a slightly reduced maximum leaf size after 12 months. In *Asplenium* (Aspleniaceae) the mean leaf size increment on gap edges was 0.01 as compared with 0.17 in undisturbed forest ($n = 158$, $P < 0.005$; Mann–Whitney U-test). In *Elaphoglossum* (Dryopteridaceae) the leaf size increment was −0.04 and 0.03 respectively ($n = 79$, $P < 0.05$). In *Vittaria* (Vittariaceae) the leaf size increment was 0.05 and 0.08 respectively ($n = 170$, $P < 0.05$). Interestingly, the variability in growth parameters (leaf number, leaf size) tended to be higher within experimental groups, indicating that many plant individuals did respond to gap creation, yet inconsistently so as for direction.

Disturbance effects on epiphytes are poorly understood. While some studies find assemblages of vascular epiphytes to be highly sensitive to anthropogenic disturbance (Barthlott et al. 2001; Krömer and Gradstein 2004; Werner et al. 2005), others do not reveal adverse effects (Hietz-Seifert et al. 1996; Larrea 1997; Nkongmeneck et al. 2002). The artificial formation of treefall gaps in the experiment represents a moderate level of disturbance. Surprisingly, no significant changes could be observed at RBSF within 12 months after logging. However, epiphytes are inherently slow-growing (Zotz 2004) and generally require the order of a decade from germination until first reproduction (Zotz 1998; Hietz et al. 2002). Changes consequently occur slowly and can still be detected 25–50 years after selective logging

(Barthlott et al. 2001). Thus, long observation periods are required to show the effects of selective logging on the assemblages of epiphytes.

Forest disturbance may alter populations and assemblage structures of epiphytes rapidly, depending on the magnitude of the disturbance. Changes in α-diversity and species composition will probably last longer than the usual felling cycles of 20–40 years. It remains an open question whether shorter felling cycles, which are usually associated with higher frequencies of small-scale disturbances than longer felling cycles (causing large-scale disturbances with longer intervals), have a less negative impact on vascular epiphyte diversity.

26.3.3 Impact of Silvicultural Treatments on Animal Diversity in the Example of Moths

There was no statistically significant difference between sites in Q3 and Q5 (experimental treatment) compared with near-natural forest sites (Q2 plus six more sites in near-natural ridge forest; Süßenbach 2003; Zimmermann 2005), neither with regard to the number of moth individuals (4024 individuals from 230 species) attracted per night, nor in terms of local species diversity (measured as Fisher's alpha of the logseries distribution as diversity index). However, with regard to species composition, Arctiidae moth ensembles showed a significant response to the experimental selective cutting of trees. Moth ensembles in the quebradas Q3 and Q5 formed a distinct cluster which was significantly separated from near-natural forest sites as well as from 15 succession sites at the forest edge in the same altitudinal belt (1800–2000 m a.s.l.; Hilt and Fiedler (2005); Fig. 26.2). This faunal shift was mainly due to changes in the relative abundances of moth species. Hence, the change of vegetation cover and forest structure associated with the forestry experiment had a detectable influence on the species composition of a diverse herbivorous arthropod group 8–11 months after the disturbance event, whereas local diversity and abundance of moths in the forest understorey was not measurably affected. To control for potential seasonal effects (Hilt et al. 2007) only samples taken during the humid part of the year (February–May) were considered.

The species richness of animals generally declines with increasing disturbance, especially when natural landscapes are converted into agricultural land (Lawton et al. 1998; Dunn 2004; Floren and Linsenmair 2005). However, the impact of logging on animal diversity and population dynamics is discussed controversially (Vitt et al. 1998). A recent review of the effects of selective logging regimes in Borneo revealed quite different responses across various groups of animals (Meijaard et al. 2005). Overall, selective logging appears to be suitable to maintain high levels of tropical biodiversity, but at the cost of the disappearance of certain sensitive forest species. In accordance with our findings for moths, selective logging had no effect on the diversity of bats (Clarke et al. 2005) and ants (Vasconcelos et al. 2000) but caused changes at the levels of population structure or species composition.

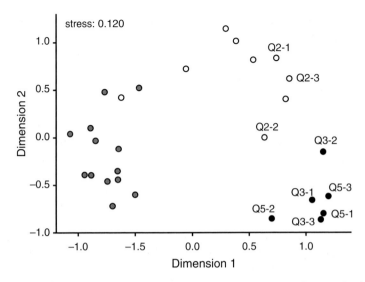

Fig. 26.2 Non-linear two-dimensional scaling plot of 28 samples of Arctiidae moths, based on Gallagher's chord-normalized expected species shared index (= CNESS distance; Trueblood et al. 1994) with the sample size parameter set as $m=5$. The moderate stress value (0.120) indicates a good representation of the original distance matrix in two-dimensional ordination space. *Open circles* Natural forest, *filled circles* sites in forest experiment quebradas, *shaded circles* succession sites at forest edge. An analysis of similarities (ANOSIM) revealed that groups of sites differ highly significantly (global $r = 0.756$, $P < 0.001$). Pairwise comparisons between sites in the experimentally treated quebradas Q3 and Q5 were significant against all other groups of habitats (early succession habitats: $r = 0.907$; late succession habitats: $r = 0.831$; near-natural forest: $r = 0.765$; all $P < 0.01$)

It can be assumed that sustainable forest management, which maintains biodiversity, is possible, but it depends very much on the disturbance intensity and frequency (length of cutting cycles, gap sizes, reduced impact logging, etc.). Studies of disturbance impact on diversity have therefore to be accompanied by quantitative rather then qualitative analysis of the disturbance regimes. Moreover, responses vary considerably between groups of animals (Meijaard et al. 2005) and also strongly depend on the scale of study (Hill and Hamer 2004). Positive transient effects of habitat disturbance on local diversity are rather often seen on a small spatial scale and can be explained by a greater abundance and variety of micro-habitats and niches. This is in line with the general ideas of the "intermediate disturbance hypothesis" (Connell 1978) according to which moderate levels and intensities of disturbance result in highest biodiversity. This is exactly what was observed with two very large moth families in disturbed and succession habitats at the edge of RBSF (Hilt and Fiedler 2005; Hilt et al. 2006) as well as elsewhere in tropical regions (Fiedler et al. 2007). When viewing animal communities at larger scales, however, disturbance effects usually decrease the species pool and thus biodiversity (Hill and Hamer 2004).

26.3.4 Impact of Silvicultural Treatments on Nutrient Cycling and Respiration

Our assessment of the impact of the natural forest management measure on nutrient cycling consisted of three different treatments, each replicated threefold: undisturbed forest (reference), the area surrounding a favored potential crop tree, and a gap in the managed catchment (for the location of the reference microcatchment, MC2, see Fig. 1.2 in Chapter 1). At each of a total of nine study sites we collected litter leachate. Furthermore, incident precipitation was collected at weekly intervals in two gaging stations (consisting of five samplers each) on forest-free areas adjacent to MC2 and MC5, respectively. Lysimeters to collect litter leachate were installed after the management in June 2004 with three collectors at each of the nine study sites. Throughfall was measured with 60 collectors in each catchment. Two months after application of the silvicultural treatments the total soil respiration in 55 gaps was analyzed, divided into reference plots (undisturbed forest, small, medium, large gaps, respectively). For every plot five measurement points were distributed randomly and measured 3–5 times with a portable IRGA, combined with a closed chamber system EGM-4 with SRC-1 (PP Systems).

There were no differences between the natural reference (MC2) and managed forest catchments (MC5) in pH and electrical conductivity of incident precipitation (MC2: 4.7; MC5: 4.7) and throughfall (MC2: 5.8; MC5: 5.9; Fig. 26.3a, b). The major ions in all solutions were base metals (Na, K, Ca, Mg), N-species, organic anions, and Cl^-, most of them being essential plant nutrients (Wilcke et al. 2001a). Thus, both catchments received a similar electrolyte input and showed a similar response of electrolyte concentrations during the passage through the canopy, resulting in an increase in the electrical conductivity because of the wash-off of dry deposition and the leaching of elements from plant tissue. The mean electrical conductivity was lower in the stream water of MC5 (Fig. 26.3d) while the mean pH was the same (pH 6.3). This was already the case before the natural forest management experiment was started in June 2004, indicating a smaller export of ions from MC5 than from MC2. The management measure did not result in a change in the pH and electrical conductivity of the stream water. Although ion inputs are similar among both catchments and ion output is even lower in MC5 than in MC2, the higher electrical conductivity of litter leachates in MC5 at similar pH implies a higher nutrient availability for PCTs (Fig. 26.3c).

Soil respiration was significantly higher in all gap sizes versus the reference in undisturbed forest; surprisingly the highest respiration rates were not found in the largest, but in the medium-sized gaps. However, the differences in the means are relatively low and the variation of all gap sizes and the reference is rather high (Fig. 26.4).

Our results suggest that the natural forest management measure increased the nutrient availability in soil solution compared to the undisturbed reference

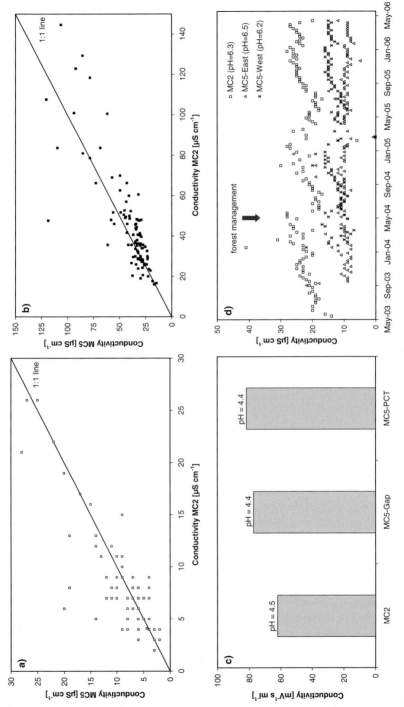

Fig. 26.3 Relationship between electrical conductivity and: **a** incident precipitation in the undisturbed reference microcatchment (MC5) and the managed microcatchment (MC2), **b** throughfall in MC2 and MC5, **c** litter leachate in the three treatments of the natural forest management experiment (*MC2*: n = 123; *MC5-Gap*: n = 76; *MC5-PCT*: n = 77), and **d** stream discharge between April 2003 and November 2005 in MC2 and the two streams of MC5

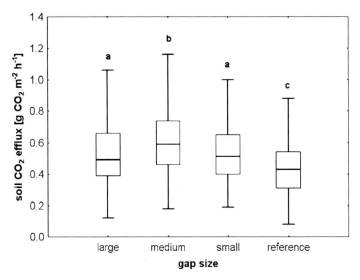

Fig. 26.4 Soil respiration two months after felling versus gap size. Statistical differences from the Kruskal Wallis test at the $P < 0.01$ level ($n = 784$) are indicated by different letters (a, b, c). Small gaps result from felling of one small single tree, medium gaps from one large single tree, large gaps from various large trees

forest. The reason for this finding is probably an increased mineralization rate as a result of increasing soil temperatures after gap creation. Increased mineralization rates resulting in elevated nutrient concentrations and decreased pH of soil solution and surface water have particularly been observed after clear-cutting (Ulrich 1984; Lundström 1993). Furthermore, nutrient uptake might be reduced because of the reduced root competition. Increased nutrient availability in the soils of the managed forest, however, does not result in changing pH or changing electrical conductivity of input and output. Thus, the impact of our management measure seems to be less than in an Australian Eucalyptus stand, where visible changes in the water and element budget occurred when 20% of the trees were removed (Cornish 1993). However, even the removal of 50% of the trees in an Alpine catchment in south Germany only increased the nutrient export with surface runoff in the first year after the management (Bäumler 1995). Thus, the objective of improving growing conditions for PCT seems to be reached without changing the overall forest functioning, as indicated by the element budget. If our preliminary data are confirmed by the on-going more detailed and longer-term analyses of nutrient concentrations in ecosystem fluxes, this would indicate that the forest management measure can be considered as sustainable with respect to the element cycle.

Total soil respiration (TSR) is influenced by a variety of factors like soil temperature, soil moisture, and litter supply (Sulzman et al. 2005). Root respiration as an important part of TSR is mainly influenced by the availability of photosynthetic carbon (Bond-Lamberty et al. 2004). Among the few studies investigating root respiration in the tropics there are no studies at all applying root exclusion methods. Therefore it is not clear how roots and thus soil respiration react on the cutting of roots or trees in tropical mountain forests. In the current study we assume that, due to the silvicultural treatments, the fine roots of damaged trees and the litter from harvested trees enhance the amount of easily available carbon for microorganisms, which increases their activity and thus soil respiration compared with the reference. This enhanced CO_2 efflux may over-compensate the reduced activity of severed roots (Epron et al. 1999). Furthermore we expect microbial activity to be positively influenced by enhanced soil temperature and moisture in the gaps (see also Iost et al., Chapter 14 in this volume). The fact that large and small gaps did not differ significantly in respiration possibly indicates that temperature-related factors like soil moisture in the uppermost litter layer could be one limiting factor for humidity-adapted microorganisms. If CO_2 losses are to be avoided in comparison with undisturbed forests, the generation of large gaps by silvicultural measurements should be avoided.

26.4 Silvicultural Conclusions

The initial silvicultural situation for the RBSF area was typical for many stands in the Neotropical mountain rain forests: as shown in previous chapters the forests represent a hotspot in biodiversity and the abundance of high value timber species of medium diameters is very promising for sustainable management, but the individuals with harvestable dimensions are very scarce. Domestication and an enrichment of the stands with large-sized timber species by avoiding the destruction of the enormous (alpha-) biodiversity and various ecosystem functions is therefore a promising alternative to prevent the ecosystem's conversion into pastures.

For instance, *Cedrela* sp. showed very high growth rates, and since it represents a very high value timber species, it could be a key genus for natural forest management. Also, advantage should be taken of the high abundance of further valuable timber species despite their rather low growth rates, in order to improve the added value and to diversify the products for the timber market. This should lead to more independence from price fluctuations, which finally is one prerequisite for a stable income and economic sustainability.

In the study area only the felling of 32 trees per hectares in contrast to 18 per hectare led to a canopy openness which could facilitate favorable light conditions for the natural regeneration of PCTs. The stronger intervention increased the nutrient availability for the PCTs because of a reduced demand by the remaining vegetation and because of the stimulation of mineralization in the gap created by the felled tree. Our results revealed also that only larger disturbances affect soil respiration and mineralization processes and that less intense disturbances rather lead to changes in beta- than in alpha-diversity of epiphytes and moths. This indicates that ecologically sustainable management of Neotropical mountain rain forests is possible if the disturbance is low and if the disturbed area is imbedded within a matrix of undisturbed forest which allows recruitment of the affected taxa. Silvicultural research should focus on the detection of gap size and distribution which enhances the PCT regeneration on the one hand but still only marginally affects nutrient fluxes and biodiversity on the other hand.

Chapter 27
Permanent Removal of the Forest: Construction of Roads and Power Supply Lines

E. Beck(✉), K. Hartig, K. Roos, M. Preußig and M. Nebel

27.1 Introduction

The construction of roads and power supply lines causes disturbances which differ from forest clearing for farming purposes by several major issues:

1. The disturbance is linear and narrow but extends over a wide distance.
2. Not only is the topsoil affected but usually the entire soil layer is stripped and, especially in a mountainous area, even the slopes are clipped and ablated and broadenings are banked.
3. The forest is usually cleared mechanically and not by the use of fire.
4. Once the traffic route is completed, disturbance continues either by regular cutting of the vegetation along the wayside or by landslides of various dimensions caused by erosion of the steep embankments.

In the area of the RBSF several campaigns of road construction are recorded as well as the establishment of a power supply line from the neighboring hydro-electrical power plant to the capitals of Loja and Zamora. Remnants of an ancient trail traversing the forests of the RBSF at about 2100–2300 m can be detected – presumably the former connection route between both provincial capitals. The area is meanwhile completely covered by forest, but there are still indications (charcoal, pollen from cereals) that homesteads or rest houses had been put up along the trail.

27.2 The Study Area

In 1962 the former country road from Loja (2100 m) via the pass El Tiero (2700 m) was closed and a new route was established for the new road in the upper Rio San Francisco valley. Several landslides had buried the old road but unfortunately landslide activity also severely affects the new country road, especially during the heavy rains. The road surface of the old road is mostly sealed gravel. Therefore break-up of the road surface and regeneration of a vegetation cover can be monitored. Lichens, liverworts and mosses are the first colonizers, and from the vegetation of the roadside also vascular plants, mostly rhizomatous grasses like *Cortaderia* cf. *jubata* or *Melinis minutiflora* invade the planum. Distinctive components of the initial stages are the lichens *Dictyonema glabratum*, *Cladonia* cf. *cervicornis* ssp. *verticillata* and *Stereocaulon* sp., the liverworts *Riccardia amazonica* and *Marchantia chenopoda* and the moss *Pogonatum tortile*. Creeping pteridophytes, e.g. *Lycopodium magellanicum* and *Lycopodiella glaucescens* and the Gleicheniaceae *Sticherus* spp are characteristic of later successional stages. Colonization of the sealed planum takes a relative long time, due to the compressed substrate and recurrent stepping by the local people.

27.3 Results and Discussion

27.3.1 The Vegetation of the Embankments

The vegetation cover of the road and trail embankments is very variable and an interpretation of apparent types as successional stages in the sense of a space-for-time substitution is debatable. Great differences exist between the upper and the lower embankments as the upper resembles rather the slide face of a landslide, while the lower can be compared with a mudflow. On the upper face, the soil material is often completely removed and rocky material or compacted gravel forms the surface. Due to heavy erosion, formation of a plant cover is extremely slow and, even after 40–50 years only patchy vegetation, dominated by cryptogams, is observed. At the same time, a dense bush cover or even a secondary forest may have developed on the soil material heaped up on the lower embankment.

27.3.2 Colonization of Bare Ground

Bryophytes and lichens (e.g. *Dictyonema glabratum*, Fig. 27.1) are the first cryptogams which get a foothold on bare ground and dominate the initial phase of colonization. Liverworts (e.g. *Riccardia amazonica*) are the typical colonizers of moist and shaded places while mosses and lichens prefer more dry and open conditions.

27.3.2.1 Liverworts and the Role of Mycorrhization

Mycorrhizal fungi with their fine and dense hyphal network strongly influence the soil structure, e.g. glomeromycote fungi produce the glycoprotein glomalin that

Fig. 27.1 *Dictyonema glabratum*, growing in a carpet of the moss *Pogonatum tortile* on a road embankment in the San Francisco valley at 1850 m a.s.l

agglutinates soil particles (Frei et al. 2003; Piotrowski et al. 2004). Among the bryophytes, only liverworts have obligate mycorrhiza-like interactions, suggesting that liverworts and their mycorrhizal partners possibly play an essential role in the stabilization of disturbed habitats, especially in the humid tropics. The colonization process was monitored in relevés, the involved bryophytes were identified, and the importance of their mycorrhization was assessed. Concerning geology and altitude, two sites were chosen for the investigations. The plots along the path above the Rio Bombuscaro (1000–1050 m a.s.l.) east of Zamora were on deeply weathered granite soils. The plots on the Camino Canal near ECSF (1800–1900 m) were on soils with a phyllitic base rock.

All plots were 1×1 m, shaded and had an inclination of between 75° and 90°. The development of the vegetation was recorded four times at 4-month intervals. Five plots at the Camino Canal and seven plots in Bombuscaro had already some vegetation and were therefore termed "succession plots". Six plots, the "initial plots", were without any vegetation. In five of those which were adjacent to successional plots, the 5 cm thick topsoil layer was removed. Total plant cover, cover of bryophytes and cover of vascular plants were separately recorded and only bryophytes were identified to species level. Cover abundance was determined exactly if values were lower than 15%. Values exceeding 15% were recorded in 5% steps. The frequency of mycorrhization was assessed using three small samples from each liverwort species.

Expectedly, total plant cover and cover of bryophytes and vascular plants increased on all plots during the observation period (Table 27.1). Total cover reached at least 70% at the end of the investigation period. The succession plots had

Table 27.1 Development of cover abundance (%) from the first to the last relevé. Symbols indicate the increase in cover abundance: + weak increase (1–20%), ++ moderate increase (21–40%), +++ strong increase (>40%). *Index*, cover abundance of liverworts divided by that of mosses

Plot no.	Total		Bryophytes		Vascular plants		Index
Initial plots							
S9b	0 → 70	+++	0 → 45	++	0 → 40	++	5
S8b	0 → 70	+++	0 → 60	+++	0 → 20	+	7
S10b	0 → 80	+++	0 → 70	+++	0 → 15	+	13
S12b	0 → 80	+++	0 → 80	+++	0 → 10	+	6
S1	0 → 85	+++	0 → 85	+++	0 → 15	+	5
S13b	0 → 95	+++	0 → 90	+++	0 → 15	+	24
Succession plots							
S3	10 → 90	+++	10 → 80	+++	1 → 25	++	2
S2	15 → 80	+++	15 → 80	+++	2 → 10	+	6
S4	50 → 85	++	40 → 70	++	10 → 45	++	4
S6	55 → 90	++	40 → 65	+	15 → 70	+++	6
S5	60 → 80	+	50 → 65	+	15 → 40	++	10
S7	60 → 95	++	55 → 85	++	10 → 20	+	10
S8a	60 → 85	++	60 → 80	+	5 → 10	+	5
S10a	65 → 90	++	55 → 75	+	10 → 45	++	8
S9a	70 → 95	++	30 → 60	++	50 → 90	++	6
S13a	70 → 95	+	50 → 75	++	30 → 65	++	14
S12a	85 → 100	+	70 → 100	++	20 → 35	+	13
S11	90 → 100	+	75 → 90	+	20 → 40	+	8

a slightly higher average total cover (90%) than the initial plots (80%). Provided that conditions for growth of bryophytes were favorable (i.e. high air and substrate humidity, sufficient diaspore rain) the vegetation cover of some initial plots in Bombuscaro was nearly re-established within 9 months (Fig. 27.2). The composition of species, however, was far from constant after this period. At the end of the observation period, the bryophytes reached at least 60% cover on the succession plots and 45% on the initial plots. The range of abundance of the vascular plants was considerably wider (10–90%) than for the bryophytes (45–100%) and average values differed substantially between the succession plots (41%) and the initial plots (18%).

The succession on the investigated plots is biphasic. Phase 1 (the actual pioneer phase) is characterized by the dominance of bryophytes. All investigated plots with low initial total cover (0–20%) show a strong increase of bryophyte cover. Phase 2 (stabilization phase) is characterized by upcoming vascular plants. On plots with an initial total cover of more than 50%, the increase in the abundance of vascular plants sometimes exceeds that of the bryophytes. However, at the end of the observation period, bryophytes cover a significantly greater area than vascular plants.

A comparison of the three bryophyte groups shows that liverworts are most important. Cover abundance of the liverworts exceeds that of the mosses 2- to 24-fold. Hornworts are negligible. The common pioneer vegetation is dominated by only a few species, in our case thalloid liverworts: *Riccardia amazonica* is present in 89% of the

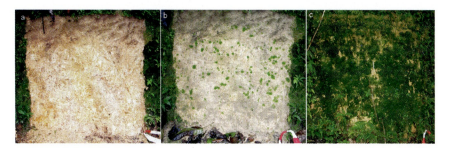

Fig. 27.2 Documentation of the (re-)colonization of plot S12b (see Table 27.1). Photographs were taken on 20 July 2004 (**a**), 05 October 2004 (**b**) and 21 April 2005 (**c**)

plots, frequently reaching the highest cover abundances. *Symphyogyna brongniartii* is present in 72% of the plots but hardly gets dominance. Members of the foliose genera *Calypogeia* and *Lophocolea* are present in many plots but at a low cover abundance. Among the mosses, only *Pogonatum tortile* plays an important role in some plots. Algae could be detected only on a few plots under relatively wet conditions.

Clear trends are visible concerning the mycorrhization of the liverworts. *Riccardia amazonica*, the most important species was never mycorrhized. Producing enormous amounts of generative and vegetative propagules, it was one of the very first colonizers in all initial plots forming dense patches within a few months. In a few plots its climax was already outrun within the observation period. *Paracromastigum bifidum*, another non-mycorrhized species, proved to be a fast and dominant colonizer in two plots in Bombuscaro. Liverwort species of a later successional stage were regularly mycorrhized. *Symphyogyna brongniartii* was the most frequent one, whose cover abundance increased steadily but never reached high values. This species is dependent on generative reproduction and apparently needs more time for colonization. Likewise *Marchantia chenopoda*, in spite of its ability for asexual reproduction, was encountered only on few plots, where it never reached dominance. The same holds for the foliose genera *Calypogeia*, *Cephalozia* and *Lophocolea*. A successional decrease in cover abundance of non-mycorrhized liverworts was balanced by an increase in species with fungal symbiosis. The type of the mycorrhizal partner seems to have little influence on the colonization process. Similar colonization rates were observed with Glomeromycetes (in *Symphyogyna* and *Marchantia*) or Ascomycetes (in *Calypogeia* and *Lophocolea*) as the fungal symbionts. Even young plants were already mycorrhized, indicating the strong dependence of these liverworts on their fungi. An interesting aspect is the quantity of rhizoids of the respective species. The non-mycorrhized species had only a few rhizoids while the mycorrhized species produced long and dense rhizoids. Together with proliferation of the mycelium, rhizoid formation contributes to the stabilization of the topsoil.

First experiments with homogenates of the above species show good results in initiating the re-colonization of steep patches of bare soil. Spraying such homogenates, liverworts could contribute to stop erosion in the area.

27.3.3 Colonization by Vascular Plants

Several species of orchids (*Epidendrum lacustre, E. carpophorum, Elleanthus aurantiacus*), a few bushes (*Escallonia paniculata, Gaultheria erecta, Tibouchina lepidota*), grasses (*Andropogon bicornis, Cortaderia jubata, Melinis minutiflora*) and the fern *Blechnum cordatum* are the first vascular pioneers (Fig. 27.3). *Elleanthus aurantiacus*, e.g. gets a foothold by a dense network of stolon-like roots with a *Velamen radicum*. The bushes and grasses colonize cracks, erosion ravines and small ledges where fine soil material accumulates.

Ferns, mainly of the genus *Sticherus*, cover the nearly vertical flanks of upper embarkments. *Sticherus rubiginosus* heavily propagates vegetatively by runners, forming a curtain-like cover (Ohl and Bussmann 2004).

27.3.4 Regeneration of a Vegetation Cover on Accumulated Soil Material

On less steep upper embankments, where some soil material has been left, recovery of the vegetation is much faster and several successional stages can be found along

Fig. 27.3 Stabilization of an upper embankment by phanerogams, lichens and mosses/liverworts: *Andropogon bicornis* (*A*), *Cortaderia jubata* (*C*), *Melinis minutiflora* (*M*), *Dictyonema glabratum* (*Dg*), *Gaultheria erecta* (*G*)

a trail or road. Such stages differ from the plant cover of abandoned pastures by the absence of species (e.g. bracken) which are typical of areas where the forest has been cleared by fire.

Successional stages which are apparently stable for a longer time are:

1. A dense herbal vegetation with a high abundance of grasses and orchids (Table 27.2);
2. Two bush stages as potential transient stages to
3. Secondary forests.

The plant cover of these successional stages is not homogeneous and therefore several phenotypes may be differentiated by the dominant herbs and bushes.

Table 27.2 Characteristic compositions of the herbal stage. Several vegetation subtypes could be classified based on the dominant species and the species composition (from 1a to 7). Cover abundances according to the Braun–Blanquet scale

Type	Species	Subtype										
		1a	1b	2a	2b	3	4a	4b	4c	5	6	7
Mosses, lichens and herbs	*Baccharis genistelloides*	1	1	1	1	1	1	1	1	1	1	–
	Andropogon bicornis	2	1	2	1	2	3	3	3	–	–	1
	Andropogon leucostachys	3	3	2	3	–	–	–	–	2	2	–
	Cortaderia jubata	2	2	–	–	2	–	–	–	–	2	3
	Elleanthus aurantiacus	–	–	1	–	–	1	1	1	1	1	–
	Pteridium arachnoideum	–	–	1	–	–	–	–	–	–	–	–
	Sobralia fimbriata	–	–	1	–	–	–	–	–	–	–	–
	Sticherus rubiginosus	–	–	1	1	–	1	–	–	–	–	–
	Lycopodium sp.	–	1	–	–	1	1	–	–	–	–	–
	Epidendrum carpophorum	–	1	–	–	–	–	1	–	–	1	–
	Frullania sp.	–	2	–	3	–	–	–	–	–	–	–
	Epidendrum lacustre	–	1	–	1	–	–	–	–	–	1	–
	Sphaerophorus sp.	–	1	–	1	1	–	–	1	–	–	–
	Maxillaria aurea	–	–	–	–	–	–	–	1	–	–	–
	Melinis minutiflora	–	–	–	–	–	–	2	3	–	–	3
Bushes	*Bejaria aestuans*	–	–	2	–	–	–	–	–	–	–	–
	Escallonia paniculata	–	1	–	–	–	–	–	1	–	–	–
	Tibouchina lepidota	–	1	–	1	–	–	–	–	–	–	–
	Gaultheria erecta	–	–	–	1	1	–	–	1	–	–	–

27.3.4.1 The Herbal Stage

The dominant species of the herbal regeneration stage (Table 27.2) show xerophytic characters like small leaves (*Baccharis genistelloides*) or densely hirsute leaf surfaces (*Melinis minutiflora*) and exhibit a high potential of vegetative propagation, especially so orchids and grasses. The height of the plant cover commonly exceeds 50 cm, but rarely reaches 1 m. The few bushes may be interpreted as remnants of a former pioneer stage (e.g. *Escallonia paniculata*) or as forerunners of the bush stage (e.g. *Bejaria aestuans*).

27.3.4.2 The Bush Stage

The bush stage is more difficult to categorize and only early and late stages could be differentiated (Fig. 27.4). Twenty-five relevés were examined for consistency and cover abundance using the percent Londo scale (Londo 1976). In a space-for-time substitution (Pickett and White 1986) structural elements were used as indicators of succession. The dominance of grasses and herbs was considered as an indication of an early stage while the occurrence of pioneer tree species was symptomatic of a late successional stage. This graduation agreed fairly well with an index ("diversity index") calculated from the number of bush species multiplied by the sum of their cover abundances. Fourteen relevés could be attributed to the early stage (diversity index 75–290), while 11 represented the late phase (diversity index 420–1850). Fig. 27.4A shows the general decrease in the cover abundances of herbs and grasses from the early to the late stage, and Fig. 27.4B, C show the concomitant increase of bushes, lianas and trees. Pioneer bushes, e.g. *Monochaetum lineatum* and *Gaultheria erecta*, like herbs and grasses, decreased in cover abundance as the succession progressed. *Myrica pubescens* and *Baccharis latifolia*, which do not exhibit a correlation with successional stages, can be addressed as general indicators of disturbance. Both species are frequent on those lower embankments, where secondary forest cannot establish, due to an ongoing disturbance or an unfavorable soil.

The successional bush stage, in contrast to the climax on abandoned agricultural areas, leads to a secondary forest if the disturbance is not permanent and the area borders the natural forest.

27.3.4.3 Secondary Forest

Various secondary forests could be differentiated, which at least partly can also be interpreted as space-for-time series. As secondary forests are not only found as forest regeneration stages following the bush stage, but also as remnants of fire-attacked primary forest, they are treated separately (see Chapter 32 in this volume).

Maintenance of a road or a power supply line as vital lines requires an ongoing mechanical removal of the overgrowing vegetation and therefore the composition of the plant cover remains the same over the time, varying at most with the seasons.

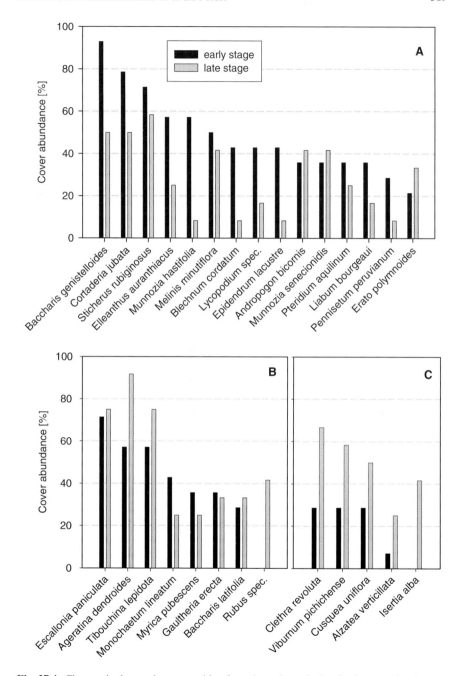

Fig. 27.4 Changes in the species composition from the early to the late bush successional stage. **A** Decrease in the abundance of herbs. **B** Overall increase in bush cover. **C** Increase in the cover abundances of trees and *Chusquea* sp., the typical liana of secondary forests. Only those species are shown with cover abundance >10%, which was not reached by another 34 herbal, 11 bush and 14 tree species (from Hartig and Beck 2000)

This type of the vegetation forms a reservoir for seeds which spread to cleared areas in the vicinity, e.g. to the pastures. Asteraceean species like *Baccharis genistelloides, B. latifolia, Ageratina dendroides* and representatives of the Melastomataceae (*Brachyotum* sp., *Monochaetum lineatum, Tibouchina lepidota*) are typical components of the wayside vegetation which are also major constituents of the mixed bracken and bush vegetation of the abandoned pastures.

27.4 Conclusions

In contrast to the wide areas where the forest has been cleared for farming purposes and which (because of the use of fire) finally develop a kind of climax vegetation composed of bracken fern and bushes, forest can recover alongside linear disturbances, if the disturbance is not maintained. Several successional stages could be differentiated, starting with a cryptogamic stage, characterized by liverworts, lichens and mosses. Although colonization of suitable bare ground especially by liverworts is extremely fast, on bedrock or otherwise sealed surfaces the cryptogamic stage may last for decades until sufficient soil material accumulates to allow growth of a closed plant cover which can re-develop into a secondary forest. Continuation of the disturbance by human maintenance activities results in a wayside vegetation which heavily exchanges propagules with the adjacent open areas. Rigorous road construction in steep terrain, however, has caused sizeable lesions in the vegetation which never recover and result in a permanent landslide activity.

Chapter 28
Forest Clearing by Slash and Burn

E. Beck(✉), K. Hartig, and K. Roos

28.1 Introduction

Burning is still the common method used by settlers (Colonos; see Chapter 3 in this volume) to clear the primary forest for new farming areas. Every fire, irrespective of being lit inside or at the edge of the forest, kills the trees by burning or by the emerging heat. A fringe of dead, but not charred, trees is always found where forest has been cleared by fire. Later, when these heat-killed trees have completely dried up, they can be used to start a new fire.

28.2 Vegetation Succession After Repeated Burning

The remnants of previous burnings – charred trunks and large branches – are frequently left where they have fallen, because of the enormous efforts to remove them from the steep and often remote areas. Many areas have thus to be burned repeatedly until the spaces between the remaining logs are wide enough to plant beans or maize, or the pasture grasses *Setaria sphacelata* (Schumach.) Stapf & C.E.Hubb. ex Chipp. and *Melinis minutiflora* P. Beauv. At just 3–4 weeks after the fire, bracken fern (*Pteridium arachnoideum* (Kaulf.) Maxon was observed, sprouting vigorously on both the burnt and the heat-killed areas, while it was absent in the intact primary forest. This de novo colonization by bracken may result via its readily germinating spores (Conway 1953, 1957; Mitchell 1973) or from already present leaf-producing lateral branches of elongating main rhizomes (Watt 1940; Daniels 1985) which form a dense network (Fig. 28.1) in the soil. Elongation growth and leaf sprouting is significantly stimulated by heat shock up to 70 °C (Roos and Beck, unpublished data).

Bracken and crops develop simultaneously after burning. Sooner or later the crops are replaced by tillering pasture grasses, in particular *Setaria sphacelata*, and the bracken fronds protrude mainly from spaces between the tussocks. When planted manually, *Setaria* grows faster than bracken and forms homogeneous pastures. However, since only the very young leaf blades and the tips of mature leaves

Fig. 28.1 In situ "network" of bracken rhizomes in an area of 1×1 m from a depth of 0 cm to about 80 cm on an abandoned pasture within the RBSF

are eaten by cattle, the carrying capacity of these pastures is low. The same is true for another grass, the stoloniferous, curtain-forming *Melinis minutiflora*, which maintains only one to three green leaves on a shoot. Bracken is not eaten by cattle due to its toxicity (Evans 1986; Hannam 1986; Fenwick 1989). Therefore sooner or later the fern overtops the grasses and, by shading, weakens their growth. On flat slopes, its mainly horizontal fronds produce a closed canopy, preventing the establishment of a shade-intolerant vegetation beneath; but on the steep slopes its canopy is more open. Wind-dispersed seeds and light can protrude to the soil surface and a variety of herbaceous and shrubby plants are found in addition to *Pteridium* in abandoned pasturelands. In particular, Asteraceae (*Baccharis latifolia*, *Ageratina dendroides*) and Melastomataceae (*Monochaetum lineatum*, *Tibouchina laxa*) can successfully compete with the fern (Stuart 1988). For pasture rejuvenation and weed killing, farmers set fire whenever the weather permits. Especially the bushy Asteraceae survive recurrent fires and resprout from their base simultaneously with the emergence of the new fern fronds. A patchy vegetation results in which islands of bracken are separated by the 2–3 m high bushes and in which sporadic tufts of *Setaria* bear witness to the former pasture land. This is a highly stable type of vegetation (Fig. 28.2), due to the high propagation potential of the bushes via seeds and of the fern via rhizomes. It is encountered in many abandoned farming areas and thus may be addressed as a long-lasting serial stage if not a climax. The described successional sequence of stages has been documented phytosociologically by Hartig and Beck (2003).

Many measures have been implemented to control bracken in agricultural areas but, due to the vigor of the rhizome system, none has been sustainable or successful (Lowday 1986; Marrs et al. 2000). Thus bracken is considered one of the world's most powerful weeds (Webster and Steeves 1958) as it destroys arable land that has been managed by fire everywhere from the tropics to the temperate zones.

Tree species characteristic of the former primary forests are very rarely found in the bracken–bush vegetation of abandoned pastures, and therefore a fast regeneration of a forest is very unlikely. The regenerative pressure of bushes, which produce immense amounts of wind-dispersed seeds, by far outstrips that of forest trees.

Fig. 28.2 Pastures (*bright green*) and bracken-dominated former pasture areas (*dark green to brownish*) which have been abandoned. At the crest of the mountain, remnants of the original forest can be detected

Seeds and fruits of the latter are dispersed predominantly by birds and bats (Matt 2001) and a single seed that is dropped by an animal on that kind of bushland has hardly any chance of germination. In addition, a substantial seed input from forest is unlikely as the remnants of the primary forests are usually far away.

28.3 Conclusions

As elsewhere in the tropics, farmers in the Andes of South Ecuador make extensive use of fire to convert primary forest into farming land and to foster their pastures. Repeated burning of the pastureland weakens the competitive strength of the pasture grasses, but increases the competitive strength of the extremely aggressive and fire-tolerant bracken fern. Pastures are finally abandoned when bracken becomes completely dominant. On steep slopes wind-dispersed seeds of several weeds germinate in the shade of the bracken leaves. A long-lasting successional vegetation composed of dense patches of bracken interspersed with individual bushes develops. Since in these areas natural regeneration of the indigenous forest is very unlikely, reforestation may be the only way out of the dilemma caused by the extensive use of fire.

Chapter 29
Gradients and Patterns of Soil Physical Parameters at Local, Field and Catchment Scales

B. Huwe(✉), B. Zimmermann, J. Zeilinger, M. Quizhpe, and H. Elsenbeer

29.1 Introduction

In this chapter, we focus on the influence of natural and anthropogenic disturbance – and recovery cycles – on soil physical properties.

Of particular interest is the interaction of soil state and vegetation development (Fig. 29.1). Soil physical and soil hydrological variables determine edaphological site conditions to a high degree. Landslide generation is controlled by soil mechanical and hydrological factors in combination with stabilizing and destabilizing factors of above- and belowground vegetation (Belloni and Morris 1991; Tibaldi et al. 1995; Schuster et al. 1996; Guimaraes et al. 2003; Vieira and Fernandes 2004). The time–space heterogeneity of soil hydrology plays a crucial role in triggering landslides (Marwan et al. 2003; Fernandes et al. 2004; Vieira and Fernandes 2004).

The hydrology of tropical soils has been studied from several points of view, including agricultural and silvicultural sustainability (Hutjes et al. 1998; Arya et al. 1999), erosion control and the management of water resources (Calder 2001). In the context of expected global climatic changes, hydrological consequences play an increasing role (Douglas et al. 1999; Douville et al. 2002). Due to different climatological, geological and pedological conditions, fundamental problems of soil hydrology have to be studied in detail; they include parameter variability in space and time, hill-slope flow and slope stability, catchment discharge and soil–root–shoot interactions (e.g. Chappell et al. 1998; Marin et al. 2000; Fentie et al. 2002). The impacts of perturbation on vertical conductivity profiles, anisotropy of permeability and spatial parameter fields are of particular hydrological importance (Elsenbeer 2001). Here we present the results from multi-scale soil physical studies of undisturbed, disturbed and regenerating field sites, with a focus on in situ hydraulic conductivity due to its role as an edaphological indicator and key parameter of flow systems.

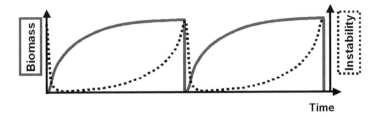

Fig. 29.1 Hypothetical link between biomass and hill-slope stability in tropical mountain forests

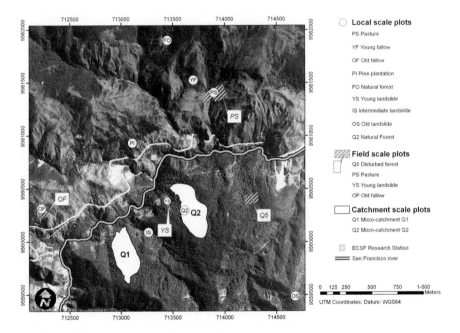

Fig. 29.2 Locations of study sites in the vicinity of the ECSF

29.2 Sites and Methods

The study sites in the vicinity of the research station ECSF comprise three different types of research areas: local scale (~2500 m^2), field scale (~20 000 m^2) and catchment scale (~100 000 m^2; Fig. 29.2). We determined profile geometry, soil texture, bulk density, penetration resistance and saturated hydraulic conductivity. Table 29.1 gives an overview for the field and catchment scale. At the local scale, saturated hydraulic conductivity was measured at three depths with a high spatial resolution, with an experimental design, explained in detail in Section 29.3.2.1.

Soil texture, bulk density and penetration resistance were determined by standard methods. Saturated hydraulic conductivity (Ks, from here on) was measured using constant-head well infiltrometers (McKenzie et al. 2002). Two different equipments were

Table 29.1 Measured soil physical and soil hydraulic parameters at the field and catchment scale. *PG* Profile geometry, *TXT* soil texture, *BD* bulk density, *PR* penetration resistance, *Ks* saturated hydraulic conductivity, *Q1/Q2* micro-catchments

	PG	TXT	BD	PR	Ks
Q1	x	x	x	–	–
Q2	x	x	x	–	x
Pasture	x	x	x	x	x
New landslide	x	x	–	x	x
Old fallow	x	x	–	x	x

in use due to the project history and different research groups involved: the "Guelph" parameter (Elrick and Reynolds 1992) and the "Amoozemeter" (Amoozegar 1989). The methods do not differ in principle but in tube size and algorithms.

Direct comparisons of the two methods showed comparable results. Paired tests (paired *t*-tests, paired Wilcoxon tests) did not indicate systematic errors. However, the correlation between log-transformed conductivities determined with both methods was only modest ($r = 0.507$).

For micro-catchment Q2, the effective hydraulic conductivity was expressed as weighted harmonic mean of hydraulic conductivities in order to hydrologically characterize the entire profile, with the thickness of each horizon being the weight.

GIS Arc-View was used to create maps of the data at the catchment level. Exploratory univariate, bivariate and multivariate analyses yielded insights into data structure and disturbance effects. Calculations were done in R and S (Venables and Ripley 2002).

We further applied the Rosetta software package (Schaap and van Genuchten 2001) to estimate residual water contents, saturated water contents and saturated hydraulic conductivities of soil materials. Rosetta is essentially a trained neuronal network. Input parameters are soil texture and bulk density.

Spatial analysis was performed by calculating empirical variograms. We used the gstat software (Pebesma 2004) implemented in R for all geostatistical calculations. We studied the spatial structure at the local, field and catchment scale. At the local scale, we employed a stratified random sampling design (Pettitt and McBratney 1993), and we used irregular grid schemes at the field and catchment scales.

29.3 Results and Discussion

29.3.1 Soil Structure at the Catchment Level

Height differences and slope gradients are most pronounced at the catchment level. Thus, deterministic and stochastic parameter variability should be greatest at this scale. Hence, we studied parameter interrelations exemplary in micro-catchment Q2.

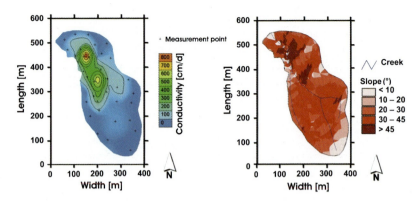

Fig. 29.3 Effective conductivity map of micro-catchment Q2 (*left*) and associated slope map (*right*)

Figure 29.3 shows a map of effective hydraulic conductivities in the undisturbed micro-catchment Q2, which serves as a reference location for comparisons with disturbed sites, together with a map of slope classes. Hydraulic conductivities seem not to be randomly distributed at the catchment level. Two hot spots of conductivity can be detected. Comparison with the slope class map suggests a correlation with topography, which suggests a functional relationship with slope processes like landslides and hill-slope hydrology. On the whole, soils are characterized by high spatial variability of effective conductivities.

The extreme heterogeneity of hydraulic conductivities can be more clearly derived from Fig. 29.4. Conductivities vary over a range of five to six magnitudes (powers of ten). In the organic layer, Ks reached 10 000 cm/d. The decrease in the mineral layers with values around 10–100 cm/d is considerable.

The Rosetta estimates of Ks do not differ significantly from in situ measured Ks values as indicated by the corresponding overlapping notches (Fig. 29.4). The decrease with depth corresponds to an increase in bulk density and stoniness with increasing depth.

However, empirical linear correlation coefficients between measured soil physical parameters and hydraulic conductivity are very low. Multivariate linear regressions (data not shown), cluster analyses and principle component analyses did not reveal additional structures. Also, no spatial trend could be detected by multivariate linear and polynomial regression analysis. Correlations between measured conductivities and the Rosetta estimates were found to be $r = 0.217$ for the A horizon, $r = 0.560$ for the B horizon and $r = 0.495$ for the C horizon of Q2. Clear correlations were observed for estimated residual water content, saturated water content and Ks with soil texture and bulk density, while no correlations were found for the measured Ks values. This is mainly due to these parameters being used by Rosetta for estimating Ks. However, bulk density seems to be the key predictor for Ks on our site. Table 29.2 summarizes these results.

Weak correlations may be due to hidden variables, multivariate interrelationships and nonlinear effects caused by nonlinear processes during soil evolution.

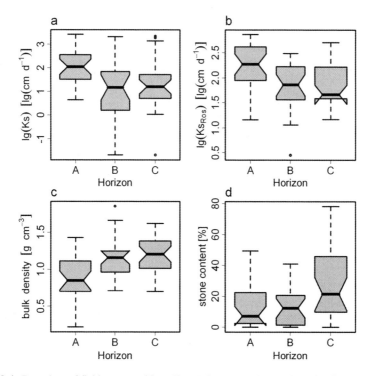

Fig. 29.4 Box plots of field-measured $\log_{10} K_s$ (**a**), Rosetta-estimated $\log_{10} K_s$ (**b**), bulk density (**c**) and stone content (**d**) of A, B and C horizons in the micro-catchment Q2. The *crossbar* within the *box* shows the median, the length of the *box* reflects the interquartile range, the fences are marked by the extremes if there are no outliers, or else by the largest and smallest observations that do not qualify for an outlier. Outliers are defined as data points more than 1.5 times the interquartile range away from the upper or lower quartile. The notches represent the 95% confidence interval for the median, and overlapping notches from two box plots indicate that there is no significant difference between the medians

Land use changes and landslides may have strong impacts on soil mechanics and soil hydrology resulting in considerable shifts of the corresponding parameters. This is discussed in the next section.

29.3.2 Disturbance Effects

Land use changes hydrological and mechanical site conditions in several ways, depending on substrate, topography and kind of land use, among other things. The diversity of and interdependence among these factors renders the soil response to land-use change rather complex and unpredictable. Changes in vegetation and the mechanical impact of cattle, however, result in predictable changes in porosity and hydraulic conductivity in surface horizons compared with undisturbed sites.

Table 29.2 Empirical linear correlation coefficients ($n=27$) of measured (*S* sand, *Si* silt, *C* clay, *BD* bulk density) and estimated soil physical parameters for the A, B, and C horizons of micro-catchment Q2. Ks_est, θr (residual water content), and θs (saturated water content) are estimated with Rosetta

	A horizon				B horizon				C horizon			
	S	Si	C	BD	S	Si	C	BD	S	Si	C	BD
Ks	0.04	0.06	0.13	0.19	0.06	0.09	0.03	0.34	0.28	0.21	0.19	0.32
Ks$_{est}$	0.30	0.11	0.33	0.92	0.20	0.32	0.11	0.83	0.11	0.04	0.20	0.90
θr	0.71	0.30	0.74	0.48	0.84	0.51	0.78	0.71	0.61	0.39	0.65	0.61
θs	0.22	0.07	0.26	0.93	0.68	0.55	0.42	0.94	0.25	0.19	0.18	0.95

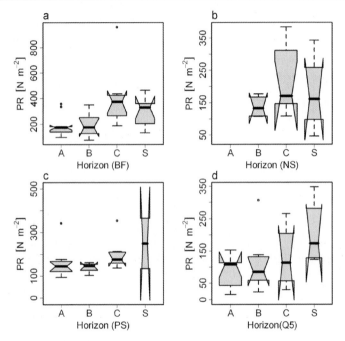

Fig. 29.5 Penetration resistance (*PR*) in A, B, C and S horizons of study sites with different land use (**a** old fallow, **b** new landslide, **c** pasture, **d** micro-catchment Q5)

Reduced water extraction from deeper layers may affect soil aeration as well as soil aggregation, the latter mainly by reduced surface tension of soil water. In the following we present penetration resistance data as a measure for mechanical root penetrability and hydraulic conductivity for characterizing water flow conditions.

Figure 29.5 summarizes the field-scale penetration resistance (for locations, see Fig. 29.2) for the old fallow (OF), new landslide (NS), pasture (PS) and a disturbed forest (Q5). Despite minor thinning in Q5, we consider this site undisturbed with respect to soil hydrology and mechanical behavior. As the A horizon in the NS (new slide) was removed by the slide process, no data is available for the A horizon.

Although the penetration resistance in all horizons appears to be higher on disturbed sites (old fallow, new slide, pasture) than under forest sites, the differences are not significant for the given sample size. The variability of penetration resistance is highest for the forest site (Q5) and lowest for the pasture, presumably due to the homogenizing effect of cattle treading. The corresponding results for hydraulic conductivity (Fig. 29.6) suggest only a weak correlation between these two parameters.

We dedicated a much greater sampling effort to hydraulic conductivity at the local scale, where sample sizes range over 30–150 per depth. The gradient of recovery from anthropogenic disturbance is defined in more detail thanks to two natural secondary succession stages (2 years old, 10 years old), and a 25-year-old pine plantation, and its trajectory is defined by a 20-year-old pasture and natural forest (for locations, see Fig. 29.2) The corresponding gradient of recovery from a natural disturbance encompasses three landslides with ages of 2.0, 7.5 and 27 years, and again natural forest.

Figure 29.7 illustrates the influence of land use on K_s in the surficial soil layers. At the 12.5 cm depth, the old fallow and pasture do not differ significantly, whereas the pine plantation exhibits much higher K_s values, which in turn is undistinguishable from the forest site. K_s of the young fallow is even lower than that of the pasture

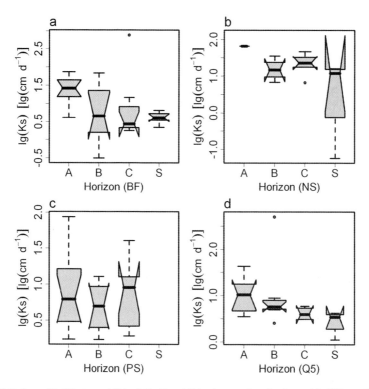

Fig. 29.6 $Log_{10} Ks$ [Ks in cm/d] in A,B, C and S horizons of study sites with different land use (**a** old fallow, **b** new landslide, **c** pasture, **d** micro-catchment Q5)

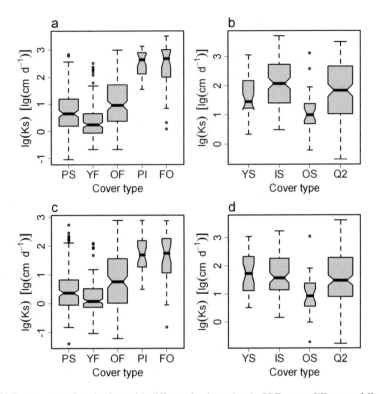

Fig. 29.7 $Log_{10} Ks$ of study sites with different land use (**a, c**). *PS* Pasture, *YF* young fallow, *OF* old fallow, *PI* pine plantation, *FO* natural forest. $Log_{10} Ks$ of study sites of landslides of different age (**b, d**). *YS* Young slide, *IS* intermediate slide, *OS* old slide, *Q2* forested micro-catchment Q2. $Log_{10} Ks$ data collected at two depths (**a, b**: 12.5 cm; **c, d**: 20.0 cm)

due to a variability which cannot be attributed to land use. At the 20 cm depth, forest and the pine plantation again exhibit higher K_s values than the other land covers, but the old fallow already shows some regeneration from its former pasture use.

In contrast to the pronounced drop of K_s caused by human disturbance, the two landslides do not show any significant difference between themselves and in comparison to the natural forest of micro-catchment Q2 (Fig. 29.7); for reasons yet unknown, the oldest slide seems to be an outlier, which we ignore in the following. Landslides do apparently not change the hydrological properties of the mineral soil, a conclusion in line with the interpretation by Wilcke et al. (2003).

29.3.2.1 Spatial Analysis

So far we based our study on explorative uni-, bi- and multivariate statistics providing an impression of physical edaphological properties on different scales. However, as we are interested in landscape processes this may not suffice. In addition, dynamic processes like water and solute transport, relocation of soil by erosion and landslides also depend on spatial distribution, continuity, preferential directions and extreme values of governing parameters and boundary conditions. Thus, spatial autocorrelation figures prominently in this study.

Measuring grids for the different studied scales are shown in Fig. 29.8 for the field and catchment scale, and in Fig. 29.9 for local scale.

29.3.2.2 Variogram Analysis

In most cases variograms revealed pronounced nugget effects on all scales and for all variables, often resulting in "pure nugget" variograms. Clear spatial autocorrelations were found on the pasture site (field scale) for stone and sand content in the A horizon, pH in the A, and B horizons, and $\log_{10}$ Ks-values in the A-horizon. In Q2 we found spatial autocorrelations for bulk densities in the A-, sand content for A- and B-, silt content for the B-, and – surprisingly – Ks estimates by Rosetta in the A-horizon (Fig. 29.10). No autocorrelations were detected for Ks and log(Ks)-values or effective conductivities.

On the local scale, autocorrelations of Ks are weak or do not exist for the land covers pasture, young fallow and intermediate landslide, whereas clear correlation patterns characterize the old fallow and the natural forest. Hence, disturbances not only result in mean differences of Ks between the disturbed and regenerating land covers but also impact on its spatial structure.

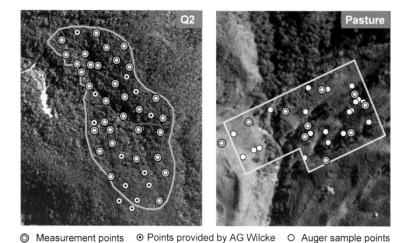

Fig. 29.8 Sample grids of the micro-catchment Q2 and pasture

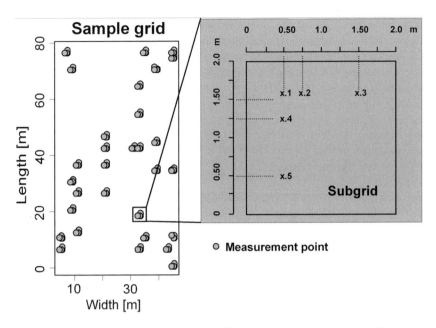

Fig. 29.9 Local scale grid resulting form a stratified random sampling process; within the grid cells a rigid sampling scheme was maintained (*gray box*)

29.3.2.3 Mapping

The overall goal of geostatistics is map generation by best linear unbiased estimators. In addition to mapped target variables, maps of estimation variances can be drawn thus providing a measure of goodness. Here, we focus on effects of interpolation, optimal estimation and simulation on the resulting maps. Owing to shortness of space we stick here to local and catchment scale data in micro-catchment Q2.

Figure 29.10 presents several maps for $\log_{10} Ks$. The semivariogram indicates a clear spatial autocorrelation. Directional analysis did not reveal clear results. Thus, an omnidirectional Gauss model was fitted by a weighted optimization procedure in gstat. Due to the low nugget variance and the horizontal slope near the origin, the gauss model causes self-similarity for small lags. Comparing the inverse distance interpolation with the kriged map one clearly recognizes the resulting smoothing of the Gauss model.

When dealing with transport and turnover in soil we are further interested in spatial connectivity, actual roughness and extreme values of the parameter field. In this respect, geostatistical simulation provides more realistic maps. We used sequential Gauss simulation for simulating conditional (measured values are met in the sample points) and unconditional random fields (Gómez-Hernández and Journel 1993).

In principal, simulated maps reveal similar patterns in a statistical sense compared with the kriged map. However, kriging introduces some smoothing and there

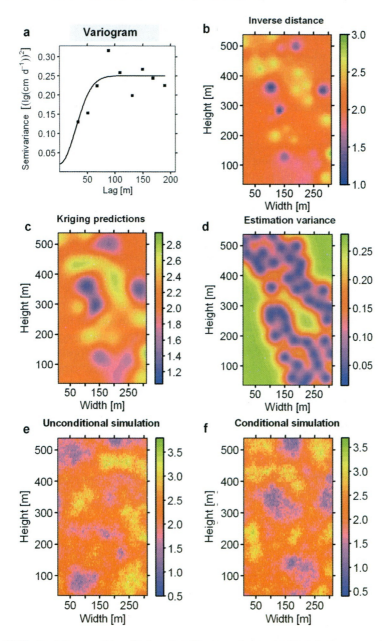

Fig. 29.10 Spatial maps of $\log_{10}$Ks based on Rosetta estimates in the A horizon: **a** variogram with fitted Gauss model, **b** $\log_{10}$Ks map generated with inverse distance weighted interpolation, **c** ordinary kriging estimation, **d** corresponding estimation variance, **e** unconditionally simulated map, **f** conditionally simulated map

is a considerable amount of (white) noise caused by the nugget variance. Further, the span of values is considerably increased by the simulations. This is of crucial importance e.g. when threshold processes are involved.

29.4 Conclusions

In this chapter, we present soil physical and hydraulic data of forest and pasture sites with different stages of ensuing regeneration. Although we detected land-use impacts on several soil physical parameters, our results show rather fuzzy trends. Undoubtedly, this partially derives from the extreme spatial heterogeneity of our target variables, but we cannot rule out small-scale variations in lithology, which may or may not affect our target variables regardless of disturbance or stage of regeneration. It is interesting to note, with a view to soil landscape modeling, that the correlation between in situ measured and pedotransfer-modeled saturated conductivities is only modest. Therefore, the potential for regionalization of Ks values by way of pedotransfer functions remains restricted. Spatial autocorrelations could be detected at all scales, but nugget effects were mostly high. Depending on the variogram model, kriging generates more or less smooth maps. These maps are suitable for estimating e.g. soil volume or soil mass in landscapes. However, when threshold processes and extreme values (e.g. denitrification, emission of trace gases, slope stability) are of interest, conditional or unconditional simulations should be preferred. Furthermore the described correlation structures of Ks at the local scale can be used to simulate flow pathways in the near-surface soil, e.g. for an advanced testing of the hypothesis that fast near-surface flow occurs (see Chapter 12 in this volume), or for an assessment of the likelihood of overland flow under the different land covers.

Acknowledgement We thank Prof. Dr. W. Wilcke from Johannes-Gutenberg-University (Mainz, Germany) for his help with site-specific information and soil data.

Chapter 30
Visualization and Analysis of Flow Patterns and Water Flow Simulations in Disturbed and Undisturbed Tropical Soils

C. Bogner(✉), S. Engelhardt, J. Zeilinger, and B. Huwe

30.1 Introduction

The complexity of flow patterns and occurrence of preferential flow in soils depend upon spatial heterogeneity of the upper boundary condition, heterogeneous distribution of soil hydraulic parameters and soil structure (Flury et al. 1994; Kulli et al. 2003). In a tropical rainforest, the canopy transforms the spatially almost uniform rainfall to heterogeneous patterns of throughfall on soil surface. These patterns continue in the soil amplified by its own heterogeneity. More details on canopy interactions and water relations are given by Wilcke et al. (Chapter 12 in this volume).

One of the most powerful methods to study water flow and solute transport in soils is to perform dye tracer experiments (Ghodrati and Jury 1990; Flury et al. 1994; Flury and Wai 2003). For this purpose the dye Brilliant Blue is frequently used for its low toxicity and good visibility against the background color of most soils (Flury and Flühler 1995; German-Heins and Flury 2000). Usually, pictures of stained patterns serve for a qualitative illustration of preferential flow. Recent works, however, propose a more quantitative approach based on modern image processing techniques (Forrer et al. 2000; Schwartz et al. 1999; Weiler and Flühler 2004).

Most tracer experiments documented in literature were carried out on agricultural soil. To the authors' knowledge, there are only a few studies in stony forest soils (Buchter et al. 1997; Schulin et al. 1987a, b) or in tropical soils (Reichenberger et al. 2002; Renck and Lehmann 2004). The aim of the present research was to study water flow in disturbed and undisturbed tropical soils, with special emphasis on stony forest soils. We focused on undisturbed primary forest and two types of typical disturbances in the study area: natural landslides and pastures as a form of human land use. Thus we investigated a new and an old landslide, two sites in the primary forest and one on pastures. Combining the advantages of computer-based image analysis with extreme value statistics we estimated a risk index for vertical solute propagation in soil, as proposed by Schlather and Huwe (2005). Furthermore, we chose one of the primary forest sites for more detailed studies. We performed two-dimensional small-scale simulations using the model Hydrus-2D (Simunek et al. 1999) and analyzed the influence of soil texture heterogeneity and stone content on water flow.

30.2 Methods

30.2.1 Definition of the Risk Index

Schlather and Huwe (2005) propose a model based on extreme value statistics. It describes the dye coverage function $p(d)$, the number of stained pixels (p) in depth d, as being an estimate of the probability to find stained pixels in at least this depth, up to a multiplicative constant m. The authors fit the tail distribution $1-H$ to the function $p(d)/m$, H being the generalized Pareto distribution:

$$H(d,\xi_r,s) = 1 - \left(1 + \frac{\xi_r d}{s}\right)^{-1/\xi_r} \quad (30.1)$$

where s is the scale parameter and $s > 0$, ξ_r is the form parameter and $\xi_r \in \mathbb{R}$ and d is the depth in the profile (measured in the image in pixels) such that $(1 + \xi_r d/s) > 0$. They propose ξ_r, called the form parameter of the generalized Pareto distribution, as a risk index for the vulnerability of groundwater to pollutants. Given the experimental and pedological conditions, three classes of values for ξ_r can be distinguished:

1. If $\xi_r < 0$, H has an upper end-point and the dye tracer does not exceed a certain depth.
2. If $\xi_r > 0$, H has an infinite upper end-point and the water table is surely reached.
3. If $\xi_r = 0$, H decreases exponentially and the water table is reached, but the transported mass might be negligible.

We suppose that the parameters of the Pareto distribution vary with depth if the flow regime changes (i.e. from preferential to matrix flow). Therefore, ξ_r should only be interpreted as a risk index for groundwater vulnerability for regions where the flow regime stays the same between the soil surface and the water table. This might be the case in sedimentary basins with rather homogeneous geological material and a shallow water table. In our mountainous and heterogeneous study area, the situation is different. Thus we propose to take the form parameter of the Pareto distribution as a risk index for vertical solute propagation in soil and a characteristic of the flow regime for given experimental conditions.

30.2.2 Dye Tracer Experiments and Image Processing

We investigated five different sites: a new landslide (west of Q2), a cambisol on an old landslide (behind the ECSF building), a dystric skeletic cambisol (site we named "primary forest I" in Q2), a dystric leptosol ("primary forest II" on T2 at 2000 m) both in the primary forest and a regosol on the pastures (100 m above road level, opposite the ECSF). See Chapter 1 for more details on the research area and Chapter 9 for further information on soils.

At each study site, a plot of approximately $2\,m^2$ was chosen. We eliminated litter and grass from the soil surface and applied 40 mm of a 10 g/l concentrated Brilliant Blue solution using a spray system similar to the one described by Ghodrati et al. (1990). The irrigation intensity was 55 mm/h. The day after irrigation several vertical soil profiles were excavated and photographed within a metallic frame of $1\,m^2$. After geometrical correction the image size was reduced in a way that 1 cm corresponded approximately to six pixels and some parts like plot surface or a shadow of the frame were cut away. Using Matlab ver. 7.1 software (MathWorks 2005a) and Image Processing Toolbox ver. 5.1 (MathWorks 2005b), we extracted the blue patterns by a color-based segmentation and generated binary images with stained parts in black and non-stained in white. From these images, the dye coverage function $p(d)$ was calculated. Finally, the risk index for vertical solute propagation (for a definition, see Section 30.2.1) was estimated by the software package SoPhy ver. 1.0.25 (Schlather 2005) written in R (R Development Core Team 2005).

30.2.3 Water Flow Simulations

Next to the dye tracer experiment site primary forest I, a 4.0×1.5 m soil profile (Fig. 30.1) was prepared. Soil texture was estimated based on a 10 cm grid. Additionally, bulk density, water retention curves and hydraulic conductivity were measured in the laboratory.

The matric potential-water content function and the hydraulic conductivity function were calculated with Rosetta (Schaap et al. 2001), a computer program for estimating soil hydraulic parameters. Two-dimensional simulations were performed with Hydrus-2D (Simunek et al. 1999), a model for water flow, heat and solute transport in two-dimensional variably saturated media that uses the finite element method to solve the Richards' equation.

Fig. 30.1 Photograph of a 4.0×1.5 m soil profile. The *red squares* show two digitized sections used for water flow simulations. Photograph by J. Zeilinger.

30.2.3.1 Simulation Run 1

In Hydrus-2D, we excluded the upper part of the profile in Fig. 30.1 (grass and litter) and generated a 4.0×1.0 m soil profile with texture distribution determined in the field. For this profile, stones were not taken into account and we focussed on soil texture heterogeneity and its influence on flow patterns. A constant infiltration rate of 2 cm/day (d) was applied on the upper boundary of the profile; the lower boundary condition was free drainage.

30.2.3.2 Simulation Run 2

Two 40×40 cm large sections of the profile were chosen (Fig. 30.1, red squares) and the stones were digitized and simulated as internal zero flux boundaries to investigate their influence on the flow regime. We applied the same constant infiltration rate of 2 cm/d on the upper boundary of each section.

30.3 Results and Discussion

30.3.1 Dye Tracer Experiments

We chose one characteristic profile from each dye tracer experiment to calculate the risk index ξ_r. Fig. 30.2 shows the rectified original photographs and Fig. 30.3 the binary images. Except for the new landslide, we had difficulties to fit one single distribution $1-H$ to the whole dye coverage function $p(d)$, i.e. from the top to the bottom of the soil profile, as different flow regimes occurred in different parts of the profile. So we limited the fit to the lowest part of the profile. The estimated risk indices and the depths from which on we fitted the distribution $1-H$ are given in Table 30.1. In order to plot the estimated distribution $1-H$, we had to optimize the parameter s of the Pareto distribution. This is not yet implemented in the software package SoPhy (Schlather 2005) and was done directly in R (R Development Core Team 2005). Fig. 30.3 shows the function $p(d)$ and the fitted distribution $1-H$ for each profile.

On the *new landslide*, the stained patterns were rather simple. Only a small amount of dye infiltrated below 50 cm and just a few preferential pathways were visible. The estimated risk index is negative indicating a low risk of vertical solute propagation. The pattern complexity increased on the *old landslide*. Here, the dye infiltration was limited to few points and the first 20 cm of the topsoil were bypassed showing highly preferential flow. We found blue-stained plant roots indicating that water flowed along root channels. Deeper in the soil, larger blue stains occurred and the profile was stained down to a depth of 1 m. Accordingly, the risk index is positive which means that there is a high risk of vertical solute propagation.

This pronounced difference between the stained patterns on the new landslide and on the old one is probably due to soil regeneration. Indeed, on the new landslide soil

Fig. 30.2 Vertical soil profiles (1×1 m) of the Brilliant Blue dye tracer experiments. **a** New landslide, **b** old landslide, **c** primary forest I, **d** primary forest II, **e** pastures. Photographs by C. Bogner

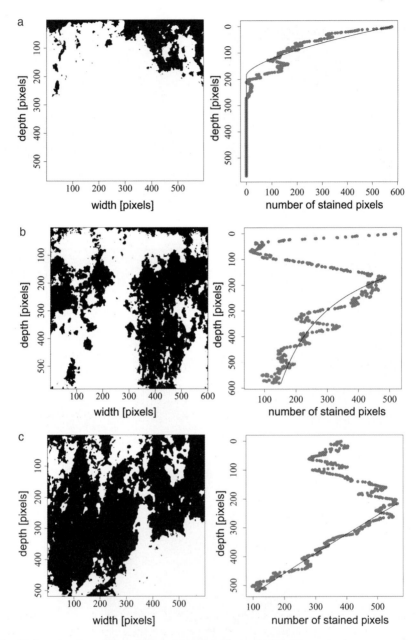

Fig. 30.3 *Left column* Binary images (segmented images where blue-stained parts are colored black and non-stained parts are white) of the stained patterns from dye tracer experiments on the sites **a** New landslide, **b** old landslide, **c** primary forest I, **d** primary forest II, **e** pastures. *Right column* The function $p(d)$, i.e. number of stained pixels with depth (*dotted lines*), and the fitted Pareto distribution (*solid lines*)

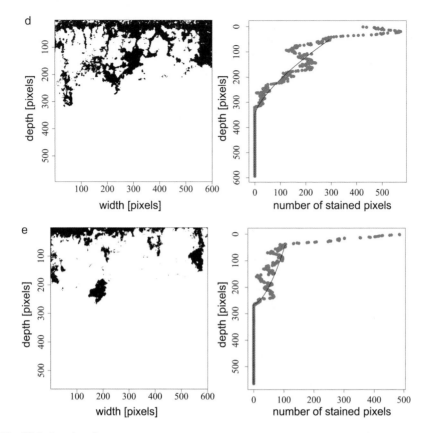

Fig. 30.3 (continued)

Table 30.1 Estimated risk indices for the dye tracer experiments

	New landslide	Old landslide	Primary forest I	Primary forest II	Pastures
Risk index (ξ_r)	−0.62	0.79	−0.91	−0.69	−1.66
Depth (cm)[a]	4	32	50	9	7

[a] Approximate depths from which the distribution $1-H$ was fitted to the dye coverage function $p(d)$

structure was destroyed by mass movement, producing a more or less heterogeneous mixture of soil material and stones. Pedogenetic processes and plant activities recreated soil structure on the old landslide, thus increasing the occurrence of preferential flow especially along bio-macropores such as root channels or earthworm burrows. More details on soil properties of landslides and plant succession can be found in Chapter 24.

At the *primary forest I* site, we found similarly complicated patterns with localized infiltration as on the old landslide. Moreover, the soil at the primary forest I site contained a lot of stones which amplified the development of preferential flow, as will be discussed in Section 30.3.2. The interpretation of the risk index for this site is difficult. It seems contradictory that the risk index is negative although the tracer reached the bottom of the profile (Fig. 30.3c). But one should keep in mind that the risk index only describes whether there is an end point for tracer infiltration or not and tells nothing about the infiltration depth. The function $p(d)$ decreases rapidly in the lower part of the profile and goes towards zero. So, there was an end point for tracer infiltration, but it exceeded the visible depth of the profile. Therefore, the risk of vertical solute propagation is greater than for the new landslide or the primary forest II site despite a more negative risk index. But it is still lower than for the old landslide, as there the dye coverage function did not tend towards zero at the bottom of the profile. For profiles as complicated as the primary forest I site, the second parameter of the Pareto distribution s plays an important role. Schlather and Huwe (2005) mention that, for a given risk index, the parameter s depends monotonically on the maximum depth of the dye tracer front. So for complex patterns we suggest to take s into account, in order to correctly estimate the risk index for vertical solute propagation. This is not yet possible within the software package SoPhy (Schlather 2005) and should be implemented in upcoming versions.

At the *primary forest II* site and on *pastures*, we found a compact top soil of about 30 cm. The dye stained surface on the first mentioned was greater partially because of a higher stone content in the top soil. Indeed, stones constituted preferential flow surfaces and were colored. But we can not exclude that infiltrability on pastures was reduced by compaction of the soil surface due to changes in land use. This could explain the smaller amount of dye penetrated into the soil. Several studies reported a decrease in saturated conductivity or an increase in bulk density after the primary rainforest was slashed and burnt and used as pastures (Elsenbeer et al. 1999; Ziegler et al. 2004; Martinez and Zinck 2004) This form of anthropogenic disturbance is typical for our study area and further work is required to understand its influence on soil hydrologic properties. The calculated risk index indicates a low propagation risk at the primary forest II site and on pastures.

30.3.2 Simulation Study

The small-scale heterogeneities of soil texture led to a non-homogeneous distribution of soil moisture. We observed the development of relatively dry areas in the lower left corner of the profile and areas of stagnation in the right part. This produces different environments for chemical reactions.

Stones have a high influence on the flow regime. This is an important aspect as in our mountainous study area soils have a high stone content. Fig. 30.4 shows flow velocities and pressure heads of simulation run 2 at steady state. We observed complex patterns in their distributions, with pronounced differences between the two sections. A high stone content leads to high velocities near the stones and especially in the gaps between them, creating preferential flow paths. Stones cause a higher grade of

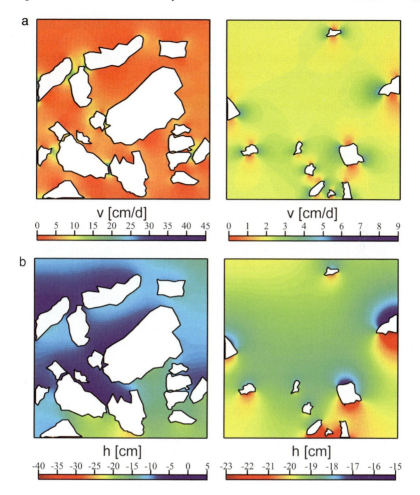

Fig. 30.4 Results of water simulation run 2: distribution of flow velocities (**a**) and matric potential (**b**) in the two digitized sections (Fig. 30.1) at steady state. Boundary conditions: upper = constant infiltration rate of 2 cm/d; lower = free drainage; stones = internal boundaries of zero flux

differentiation in the soil with relatively wet areas on their tops and dryer areas below them. As a consequence, they modify water flow and transport of solutes like nutrients and pollutants and have therefore an influence on chemical processes in soils.

30.4 Conclusions

We studied water flow in disturbed and undisturbed tropical soils with dye tracer experiments. Soil structure, stone content, plant root systems and possibly land use are controlling factors for water flow in soils. Accordingly, the complexity of stained patterns and infiltration depths of the dye varied between the different study sites: rather simple patterns with a few preferential flow paths on the *new landslide* and on *pastures*, complex patterns at the *primary forest II* site and deep infiltration at the *primary forest I* site and on the *old landslide*.

The index proposed by Schlather and Huwe (2005) can serve as a useful characteristic of flow regime and as a risk index for vertical solute propagation in a variety of soils. The estimated indices for the *new landslide*, the *primary forest II* site and the *pastures* were negative. Despite the high irrigation intensity and the important amount of solution applied, the tracer did not exceed a certain depth and there is a low risk of vertical solute propagation to deeper soil regions or contamination of groundwater on these sites. In contrast, the index on the *old landslide* was positive, indicating a high propagation risk. Future studies should investigate the robustness of the risk index concerning the experimental conditions and analyze the role of litter for the development of preferential flow paths.

The detailed simulation study conducted on *primary forest I* site showed that stones were one of the reasons for heterogeneities in soil moisture distribution, creating preferential flow paths and increasing flow velocities.

Preferential water flow results in heterogeneous soil moisture distribution and has therefore several ecological implications. In general, preferential flow means a heterogeneous water supply and the coexistence of zones with high and low oxidation potential. Thus preferential flow influences multiple factors such as root growth, C- and N-mineralization, denitrification or humus accumulation and leads to rapid leaching of nutrients. In a non-homogeneously moist soil, the soil air phase is discontinuous and therefore the oxygen supply of plant roots and microorganisms could be interrupted. Chemical and physico-chemical reactions, for example cation exchange, kinetic sorption processes or nutrient exchange between mobile and immobile water, precipitations and oxidations are also concerned by the occurrence of preferential flow.

Acknowledgements We thank Martin Schlather for his help with the software package SoPhy and R and Benjamin Wolf for having written the scripts in R and Matlab.

Chapter 31
Pasture Management and Natural Soil Regeneration

F. Makeschin(✉), F. Haubrich, M. Abiy, J.I. Burneo, and T. Klinger

31.1 Introduction

There are only a few studies on soils and their bioelement status in primary forest and forest-derived land use for mountain rainforests (Rhoades et al. 2000); most investigations up to now have focused on lowland tropics. Important changes in physical and chemical soil characteristics and biological cycles follow pasture establishment and can affect soil fertility and the interaction of soils with atmosphere and downstream aquatic ecosystems (Reiners et al. 1994; Feigl et al. 1995; Koutika et al. 1997; Deborah et al. 2001; Cerri et al. 2003). Some of these changes, such as increase in pH and exchangeable bases that follow forest cutting and burning for pasture establishment, are practically universal and occur in a wide variety of sites and different soils. Changes to other biogeochemical attributes, such as soil organic matter turnover, are more variable increasing under pasture in some locations and decreasing in others (Neill et al. 1998; Rhoades et al. 2000; Powers and Veldkamp 2005).

A better understanding of the top soil features and the biogeochemical changes that follow forest conversion to pasture is important for managing pastures in a sustainable manner and for interpreting the consequences of deforestation in the mountain rain forests. We examined the geological and mineralogical background, soil reaction and base status, C, N and available P concentrations and natural ^{13}C and ^{15}N abundance in top soils under natural forest, pasture and succession vegetation in the Rio San Francisco Valley of southern Ecuador.

31.2 Study Sites, Geology and Mineralogy

The research was carried out in natural forest in the Estacion Cientifica San Francisco (ECSF) and on pasture and succession sites on the opposite slope of the valley at about 2000 m a.s.l. Each land-use type consisted of five plots each about 30×30 m; there five sub-plots were chosen and sampled (each two sub-samples) according to defined horizons and depth units, respectively. In the pasture we selected foxtail pasture [*Setaria sphacelata* (Schumach.)], and abandoned pastures under successional vegetation with bracken fern {*Pteridium aquilinum* [(L.)

Kuhn]} and shrubs (see Chapter 28 in this volume). The vegetation types studied are dominant landscape types of the mid-elevation region of the Rio San Francisco Valley. The old-growth natural forest sites are located in the direct neighbourhood to the scientific field station San Francisco (see Chapters 1, 10.1). All pastures were established before 1962. The succession sites consist of former pastures that have been abandoned since 1990.

A realistic comparison between soils under different land use assumes that geological and geochemical conditions are similar or within an acceptable range of site factors, respectively (see also Chapter 1). In order to prove this, we conducted an intensive study on the geology and mineralogy of the research area under forest, pasture and succession at about 2000 m a.s.l. in the ECSF and opposite slopes. To classify the different rocks of the Chiguinda unit we analysed the mineralogical and chemical composition of unweathered rock samples. Two dominant groups of rocks which differ in mineralogical and chemical composition were identified: (a) meta-siltstones/sandstones/quartzites and (b) slates/phyllites with partly alternating fine layers. All studied rock samples are mainly composed of quartz, muscovite/illite, chlorite and albite in different ratios. The main differences between the rock groups mentioned above are the variable magnitudes of the easily weatherable minerals muscovite/illite, chlorite and albite. Key elements used to distinguish between this rock groups were Al, K, Mg, Fe, Na and Ca, which are dependent on the mineral content of the rocks. P occurred in minor concentrations depending on the rock types.

The contents of K and Al of the parent rocks and the soil were used to prove the chemical comparability of the soils (Fig. 31.1a) The analyses reveal similar K to Al ratios, but significant differences in concentrations of these elements between the two groups of rocks. Similar ratios of K and Al were detected within the three groups "non-weathered rocks", "weathered rocks" and "soils" (Fig. 31.1b), but with a decrease in slope, resulting from weathering of muscovite, chlorite and albite and subsequent leaching of K, while Al remains in secondary minerals. Figure 31.1c shows the K/Al quantities of all analysed top soils depending on land-use type (forest, pasture, succession). According to the Al and K contents all top soils are located in the range of meta-siltstones/sandstones/quartzites, which means the chemical compositions of the soils of the different land-use types are similar.

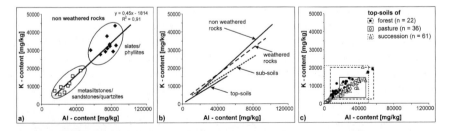

Fig. 31.1 Comparison of the K:Al ratios of non-weathered rocks (**a**), trend of K:Al ratios in the course of weathering to soil formation (**b**) and comparability of the different plots of land use (**c**)

Therefore forest, pasture and succession sites are comparable in a priori soil mineralogy.

31.3 Results and Discussion

31.3.1 Soil Reaction and Base Status

For methodology of soil chemical analysis see Scheuner and Makeschin (2005). The forest soils reveal a strong acidity through the whole top soil profile (Table 31.1). In the organic layers (except Oi) pH values were lower than pH 4 and in the deeper mineral soil pH values increased slightly. Depth function and range of soil reaction in natural forest are comparable with those reported by Wilcke et al. (2003). In the mineral soil, repeated burning of the pastures over decades resulted in significantly higher pH values of about one pH unit. Linked to the reduction in soil acidity, concentrations of exchangeable K, Ca and Mg at 0–10cm of the pasture top soil increased by about four times in comparison with the forest sites; this effect diminished with depth. Dominant base cation in pasture top mineral soil with about 25% was Ca, followed by Mg. In contrast, exchangeable Al, Fe, Mn and H^+ as acidic components were significantly reduced from values >90% in the whole top mineral soil profiles under forest to 39% and 84% at 0–10cm and 10–20cm in pastures, respectively. However, below this uppermost horizons values converged towards those of forest sites. Abandoning the pasture continued to have clear effects on the base status of the soil, also in the organic layer.

31.3.2 Soil Organic Carbon and Nitrogen

Clear cutting and slash burning were used initially to convert forest to pasture, and then burned periodically to eliminate secondary succession vegetation for the maintenance of productive pasture grass. As in all studies conducted in the tropics, there are no quantitative records available on the frequency of burning as well as any other traditional pasture management activities for the study sites. On the soils under forest, the organic layer shows an average thickness of 18cm (8–36cm), which is densely rooted (Fig. 31.2). The forest biomass and the original organic layer of the soils under pasture have been either completely lost or at least significantly reduced; thus in pastures organic layers are almost missing while in succession sites top soil features are re-developing towards conditions in forest.

Both soil organic C (TOC) and total soil N (TN) concentrations showed a clear and systematic decrease with depth in all investigated land-use systems (Table 31.1). Highest TOC and TN values were observed in the organic layers under forest

Table 31.1 pH, exchangeable base cations, organic carbon, nitrogen and extractable phosphorus content in soils. Note: values for Na^+, Al^{3+}, Mn^{2+}, Fe^{3+}, and H^+ are not presented, but included in the calculation of the base saturation (BS). Means (with standard error) for land use followed by the same letter are not significantly different (Tukey HSD test, $P \leq 0.05$)

	pH (H_2O)	K^+ $\mu mol_c\ g^{-1}$	Ca^{2+}	Mg^{2+}	BS %	TOC	TN	C/N ratio	Bray P mg kg^{-1}
Forest									
Oi	5.5 (0.1)[a]	54.5 (3.6)[a]	204.2 (11.2)[a]	184.2 (8.4)[a]	95.9 (0.5)[a]	50.9 (0.2)[a]	1.5 (0.1)[a]	35.1 (1.5)[a]	55.0 (4.3)[a]
Oe	3.9 (0.1)[a]	20.2 (1.4)[a]	47.1 (12.6)[a]	56.4 (5.6)[a]	51.1 (4.4)[a]	47.3 (0.7)[a]	2.3 (0.1)[a]	21.0 (0.6)[a]	56.5 (3.8)[a]
Oa	3.8 (0.1)[a]	13.2 (1.2)[a]	43.9 (21.8)[a]	47.5 (7.8)[a]	41.2 (8.2)[a]	47.3 (1.1)[a]	2.3 (0.1)[a]	20.7 (0.7)[a]	45.3 (5.2)[a]
0–10 cm	3.6 (0)[a]	0.8 (0)[a]	1.5 (0.3)[a]	1.3 (0.2)[a]	31 (0.8)[a]	4.8 (0.4)[a]	0.3 (0)[a]	17.8 (0.5)[a]	4.7 (0.7)[a]
10–20 cm	3.9 (0)[a]	0.4 (0)[a]	0.8 (0.2)[a]	0.4 (0)[a]	3.1 (0.4)[a]	2.4 (0.3)[a]	0.1 (0)[a]	16.9 (0.5)[a]	0.7 (0.1)[a]
20–30 cm	4.1 (0)[a]	0.3 (0)[a]	1.1 (0.3)[a]	0.2 (0)[a]	3.6 (0.8)[a]	2.1 (0.2)[a]	0.1 (0)[ab]	16.9 (0.7)[a]	0.8 (0.2)[a]
Pasture									
0–10 cm	4.6 (0.1)[b]	3.3 (0.4)[b]	25.8 (4.3)[b]	10.7 (2.1)[b]	40.2 (4.7)[b]	8.6 (0.9)[b]	0.5 (0.1)[b]	16.5 (0.5)[b]	3.7 (0.3)[a]
10–20 cm	4.6 (0.1)[b]	1.7 (0.4)[b]	6.7 (2.1)[b]	4.1 (1.3)[b]	15.2 (3.6)[b]	4.5 (0.4)[b]	0.3 (0)[b]	18.3 (0.6)[a]	1.3 (0.2)[a]
20–30 cm	4.6 (0)[b]	1.4 (0.4)[b]	3.1 (1.0)[b]	1.9 (0.8)[b]	9.6 (3.0)[b]	3.2 (0.4)[a]	0.2 (0)[b]	18.6 (0.7)[a]	0.3 (0.1)[a]
Fallow									
Oi	5.2 (0.1)[b]	35.9 (3.9)[b]	233.2 (9.7)[b]	131.6 (9.7)[b]	92.7 (0.5)[a]	50.0 (0.4)[b]	1.1 (0)[b]	45.5 (1.5)[b]	58.3 (5.2)[a]
Oe	4.4 (0.1)[b]	23.8 (2.2)[a]	93.2 (10.6)[b]	47.8 (4.2)[b]	65.2 (4.3)[a]	42.9 (1.1)[b]	1.7 (0.1)[b]	231 (1.1)[a]	48.1 (5.6)[b]
Oa	4.1 (0.1)[b]	14.1 (1.3)[a]	62.8 (19.1)[b]	31 (7.9)[a]	38.6 (7.6)[a]	38.1 (2.0)[b]	1.8 (0.1)[b]	21.3 (0.5)[a]	27.7 (4.4)[b]
0–10 cm	4.2 (0)[c]	3.2 (0.3)[b]	7.3 (2.1)[a]	5.3 (1)[a]	14.7 (3.3)[a]	8.9 (0.7)[b]	0.4 (0)[a]	23.6 (0.4)[c]	4.0 (0.6)[a]
10–20 cm	4.4 (0)[c]	1.1 (0.1)[ab]	1.1 (0.3)[a]	1 (0.1)[a]	3.8 (0.7)[a]	4.4 (0.4)[b]	0.2 (0)[a]	23.7 (0.5)[b]	0.9 (0.2)[a]
20–30 cm	4.7 (0)[b]	0.7 (0)[ab]	0.4 (0)[a]	0.2 (0)[a]	2.1 (0)[a]	2.6 (0.4)[a]	0.1 (0)[a]	22.0 (0.8)[b]	0.2 (0)[a]

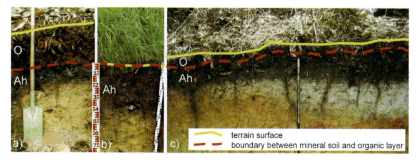

Fig. 31.2 Typical soil profiles under forest (**a**), pasture (**b**) and succession (**c**)

and succession, where succession ranged significantly lower as compared to the forest. All parameters measured show a comparatively high variation within horizons and treatments; this was also reported by Wilcke et al. (2002) for forest soils in the research area. In the mineral soil, TOC showed clearly higher concentrations in pasture and succession within the top 20 cm with up to almost 9% as compared with forest, while means under pasture and succession were similar. TN revealed a similar picture as TOC, but values in the top mineral soil of succession were smaller than in pasture, possibly resulting from fast disappearing effects of low, but measurable pasture fertilization (see also ^{15}N signatures). TOC/TN values in top soils ranged within 15–26 with the exception of the young Oi horizon (Table 31.1). Values in forest and pasture were similar, but succession showed significantly wider ratios in most horizons. Comparable results for forest organic layers for this altitude are reported by Iost et al. (Chapter 14) and Wilcke et al. (Chapter 9) in this volume.

A comparison of organic carbon stocks revealed a clear predominance of mineral soils as compared with the organic layers (data not shown). While average stocks of carbon in the latter amounted to 36.3 t ha^{-1} (forest) and 17.2 t ha^{-1} (succession), C values in the top mineral soil (0–30 cm) added up to 87.9, 161.8 and 103.1 t ha^{-1} in forest, pasture and succession. For total C stocks in top soils (organic layer, mineral soil) values amounted to 124.3 t ha^{-1} (forest), 161.8 t ha^{-1} (pasture) and 120.4 t ha^{-1} (succession). According to the comparison of the land-use units, the initial C losses especially in the organic layers could be re-accumulated above and below ground by similar amounts by vigorous growing pasture grass.

In natural ecosystems or in long-lasting land management systems equilibrium exists between input of organic matter from primary production and losses by autotrophic and heterotrophic respiration (see Chapter 14 in this volume). This equilibrium can be significantly disturbed due to land-use changes like clear cut, slash and burn and/or fire for pasture management. However, stabilization or even an increase in organic matter stocks can be achieved by land-use systems with higher biomass production and higher root litter input into the mineral soil. Research quantifying changes in the soil contents of C following forest to pasture conversion in tropical regions had conflicting results, ranging from net gains in C contents (Feigl et al. 1995; Cerri et al. 2003; Osher et al. 2003) to net losses (Reiners et al. 1994; Veldkamp 1994; Rhoades et al. 2000; Osher et al. 2003). Some studies found no change (Hughes et al. 2000) or initial C losses followed by subsequent gains of organic matter (Moraes et al. 1996, Koutika et al. 1997).

Gains or losses of soil C in planted pastures depend on both the rate of decay of organic C derived from the former forest vegetation and the contribution of C from grass roots and litter. Additionally, the state of soil carbon and nitrogen contents may change depending upon pasture age, as indicated by Veldkamp (1994) and Neill et al. (1997). Soil carbon pools typically change most rapidly soon after conversion and pasture loss rates are likely to slow down with time (Veldkamp 1994; Rhoades, 2000; Cerri et al. 2003). On Hawaiian volcanic tephra (at 1700 m a.s.l.) Towensend et al. (1995) observed a 48% loss of forest-derived C in the top 20 cm of a 40–50 year old pasture. Veldkamp (1994) reports comparable soil C losses in Costa Rican pastures on Andisols. Ecuadorian Andisols lost 28% of C3 carbon in the top 30 cm of a 15 year old *Setaria* pasture and 40% in mixed-species pasture of the same age (Rhoades et al. 2000). Brazilian Ultisols with naturally low soil C content lost 8–35% of their original forest C in 20 years (Neill et al. 1997). Brazilian Oxisols decreased about 40% of the original forest C following 50 years of cropping with sugar cane (Vitorello et al. 1989).

The accumulation of C4-C in the top 30 m beneath *Setaria* and mixed pasture reported by Rhoades et al. (2000) averaged 1.2 t ha^{-1} year^{-1}. Veldkamp (1994) observed a decreasing accumulation rate of C_4-C from 2.4 t ha^{-1} year^{-1} on a 3 year old pasture to 1.1 t ha^{-1} year^{-1} on 18 year old pasture in Costa Rica. According to the results of Rhoades et al. (2000), *Setaria* fine roots amounted to 7.5 t ha^{-1} in the top 30 cm of the pasture soil. This was four times more C than found in root systems of an adjacent old-growth forest. According to these results, on pasture soils the below ground biomass production can be more important for C sequestration than the above ground biomass production.

Compared with other tropical ecosystems the concentrations and the pools of organic carbon on our study sites were rather high. There are only a few investigations about organic matter dynamics in soils of mountain rain forests. Rhoades et al. (2000) observed in the top 30 cm of North Ecuadorian Andisols C stocks of 88.4 t ha^{-1} and 102.7 t ha^{-1} on forest sites and 76.3–97.8 t ha^{-1} on *Setaria* and mixed-species pasture. Osher et al. (2003) measured in the top 30 cm of two Hawaiian Andisols under pasture 144.7 t ha^{-1} and 150.0 t ha^{-1} carbon which are similar to our results. In the top 50 cm of soils in the Brazilian Amazon Basin (Rondônia), Neill et al. (1997) reported C stocks of 52–61 t ha^{-1} for 20–81 year old pastures and 36.9 t ha^{-1} and 43.3 t ha^{-1} for two forest sites.

Neill et al. (1997) showed in the Brazilian Amazon Basin N concentrations in the top 10 cm of forests and pastures between 0.66 mg N g^{-1} and 2.63 mg N g^{-1}. On our investigation sites the N concentrations in the top 10 cm of the mineral soil ranged between 2.4 mg N g^{-1} and 4.1 mg N g^{-1}. The highest concentrations on our sites were measured in the top 10 cm mineral soils under pasture and succession. Cattle dung input could be one of the sources, another one could be in slight mineral fertilizer input. Many investigators who have quantified changes of soil N as a result of conversion of native forests to pasture reported conflicting results ranging from no difference between primary forest and established pastures (Neill et al. 1995; Buschbacher et al. 1988; Piccolo et al. 1996; Rhoades and Coleman 1999) to increase or loss in total N on conversion to pasture (Cerri et al. 2003; Neill 1997). However, in most cases the increase in total N after conversion was less marked, which led to C/N ratios in the pasture soils being higher than in the forest soil. Our results show that N concentrations as well as pools in the mineral soils under pasture were higher than in the forest soils and C/N ratios were similar.

31.3.3 Available Soil P (Bray-P)

Oe and Oa horizons under succession showed a trend to have slightly lower concentrations of available P than forest (Table 31.1); the values ranged between 28 mg kg^{-1} and 57 mg kg^{-1}. Mineral soil P concentrations were ten-fold lower than in the organic layer. Pastures revealed extremely low PO_4-P values as compared to forest and succession. However, due to high variability in the field the latter did not differ significantly. Presumably these outliers are based on spots of cattle dung deposited and leading to nutrient heterogeneity in pastures (Glatzle 1990).

It is reported that phosphorus (like nitrogen) is lost due to fires after clear cutting and in pastures; as a consequence soil fertility is assumed to be negatively affected in the middle and long run (Glatzle 1990; Mackensen 1996). Thus, nutrient deficiencies frequently occur leading both to reduced pasture productivity and quality as well as an invasion of fern and succession vegetation (Friesen et al. 1997; Loker 1997; Jin et al. 2000). Jin et al. (2000) investigated soil chemical changes caused by forest-to-pasture conversion and subsequent livestock grazing; the levels found for available P were both in forest and pastures below instrumental detection limits. They report a trend of increasing pH and P retention capacity with change from forest to pasture cover, but no significant differences between topographic positions or between forest and pasture were found for any measured soil properties. The authors conclude that soil chemical properties may not hinder recruitment and survival of forest species in abandoned pastures. This aspect is also investigated in our research area (see Chapter 34 in this volume); however time after tree establishment seems to be too short for showing nutrient deficiencies. Friesen et al. (1997) investigated P acquisition and cycling in low-P Oxisols in a crop rotations and ley pasture systems experiment on the Colombian eastern plains. Comparison of rooting patterns stresses that, despite low available P at depth, there are important differences in root size and distribution among native savanna, introduced forage and crop species which affect their ability to acquire P from these soils. Differences in crop/forage residue decomposition and P release rates suggest that managing the interaction of residue with soil may help slow P fixation reactions. Despite these differences, soil P fractionation measurements indicate that applied P moves preferentially into labile inorganic P pools, and then only slowly via biomass production and microbes into organic P pools under both pastures and crop rotations.

31.3.4 Abundance of ^{13}C and ^{15}N in Soils

Natural variations of stable isotopes in the environment can often trace the origins of organic matter, the pathways and dynamics of these transformations. Thus, $\delta^{13}C$ and $\delta^{15}N$ signatures of vegetation and soils are a powerful tool to study interactions and processes in environmental research (Nadelhoffer and Fry 1988).

Vegetation shifts from forests to pastures leads to a dominance of plants using C3 to C4 photosynthetic pathway. Depending on the metabolic pathways of terrestrial plants the discrimination against ^{13}C is more or less pronounced (Boutton 1991). According to this, plants can be grouped into three categories by their stable C isotope composition; the mean $\delta^{13}C$ values of C3 and C4 plants range from −25‰ to −33‰ and from −11‰ to −15‰, respectively.

In the study area we determined the ^{13}C and ^{15}N values of fresh C3 and C4 plants, organic layers and top soils of forest, pasture and succession. Leaves and roots of *Graffenrieda emarginata*, as representative for a C3 forest species, leaves and roots of the C4 pasture grass *Setaria sphacelata* as well as leaves and roots of the fern *Pteridium aquilinum* in the succession were analysed. Soil organic carbon (TOC) and total nitrogen (TN) contents were determined with an elemental analyser (Vario El III, Elementar), ^{13}C and ^{15}N with an elemental analyser (NC2500 Carlo Erba continous-flow isotope-ratio mass spectrometer MS Delta plus, Finnigan). The reproducibility of measurements for both elements was better than 0.15‰. The natural ^{13}C:^{12}C ratios are expressed as $\delta^{13}C$, compared with the Peedee Formation (PDB) as a reference. The ^{15}N:^{14}N ratios are expressed as $\delta^{15}N$; the reference is atmospheric nitrogen (N_2).

The $\delta^{13}C$ and $\delta^{15}N$ values of vegetation, organic layers and top soil under forest, pasture and succession are given in Table 31.2 and Fig. 31.3.

The ^{15}N content increased with increasing soil depth in all land-use types, which was confirmed by other authors (Osher 2003). In forest we found an overall enrichment of ^{13}C and ^{15}N with depth. Typically for $\delta^{13}C$ there was a 2–3‰ enrichment of soil $\delta^{13}C$ from vegetation to mineral soil with soil depth (Nadelhoffer and Fry 1988); we found values between 1‰ and 2.5‰ for ^{13}C and up to 9‰ for $\delta^{15}N$ between vegetation and mineral soil. Similar results were also detected by Selles et al. (1986) and Wilcke and Lilienfein (2004).

Soil organic $\delta^{13}C$ values differ significantly depending upon the plant species composition (Fig. 31.3a). In C3-dominated vegetation types like forest the isotopic signature of the plant litter input ranges between −25‰ and −30‰. The pasture mineral soil is enriched in ^{13}C, reflecting the influence of C4 vegetation with a median $\delta^{13}C$ value of −21.3‰ compared with forest at −25.8‰ in mineral soil (Fig. 31.3.b). The $\delta^{13}C$ in mineral soil of pasture and succession reflect a mixture between TOC

Table 31.2 Variations in $\delta^{13}C$ and $\delta^{15}N$ values of forest, pasture and succession fallow in vegetation, organic layers and top soils. Upper values are minimum/maximum, lower values are median (n)

	Forest		Pasture		Succession fallow	
	$\delta^{13}C$ (‰)	$\delta^{15}N$ (‰)	$\delta^{13}C$ (‰)	$\delta^{15}N$ (‰)	$\delta^{13}C$ (‰)	$\delta^{15}N$ (‰)
Vegetation	−30.5/−25.5	−3.9/0.4	−16.3/−11.9	−1.1/3.1	−27.5/−25.6	−2.9/−0.4
	−27.7 (32)	−1.3 (17)	−13.3 (17)	0.9 (17)	−26.4 (14)	−1.3 (14)
Organic layer	−29.6/−25.4	−4.8/4.7			−28.5/−24.8	−3.3/6.1
	−26.9 (59)	−0.3 (59)			−26.4 (34)	0.8 (33)
Top soil	−26.9/−24.1	2.5/6.6	−24.7/−19.1	1.6/6.1	−24.8/−21.0	2.5/6.2
	−25.8 (34)	3.8 (36)	−21.3 (19)	3.2 (19)	−22.7 (25)	5.1 (25)

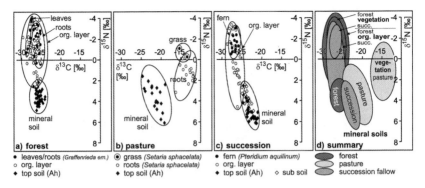

Fig. 31.3 Distribution of ^{13}C and ^{15}N in vegetation, organic layers and mineral soils of forest, pasture and succession

derived from forest plants (C3) and pasture grass (C4 – *Setaria sphacelata*). Compared with forest the mineral soil of succession is enriched in ^{13}C which is a result of the former land use under pasture and the invading forest vegetation after abandonment. The $\delta^{13}C$ signature of the succession vegetation shows a stronger similarity to forest vegetation (Fig. 31.3c).

The $\delta^{15}N$ of *Setaria sphacelata* roots was 0–3‰ and differed consistently from the roots of forest plants like *Graffenrieda emarginata* with −4.0‰ to 0.5‰ (Fig. 31.3 a, b). This may be a result of the application of urea fertilizer or cattle urine and/or droppings. The $\delta^{15}N$ in mineral soil of pasture (1.6–6.1‰) and succession (2.5–6.2‰) shows a wider range than forest soil (2–5‰). On pasture and succession $\delta^{15}N$ values increase with depth. According to Nadelhoffer et al. (1996), Koba et al. (1998) and Liao (2004) $\delta^{15}N$ in mineral soil is related to the degree of humification and increases with higher degrees of decomposition. Liao (2004) found higher $\delta^{15}N$ values in mineral soils of grassland than in forest. According to this author higher $\delta^{15}N$ values in depth could be evidence for older and more humified organic matter.

According to Vitorello et al. (1989) we calculated the part of carbon derived from forest material C_f and that one derived from grass residues C_g in any layer of the soils with: $C_g (\%) = [(\delta - \delta_o)/(\delta_g - \delta_o)] \times 100$; $C_f (\%) = 100 - C_g (\%)$. Where $\delta = \delta^{13}C$ value of sample from pasture soil, $\delta_o = \delta^{13}C$ value of corresponding sample from forest soil (median −26‰) and $\delta_g = \delta^{13}C$ value of grass residues (−13‰).

In at least 44 years old pasture soils, 10–53 % of the TOC is C4-C and thus derived from *Setaria* grass. Rhoades et al. (1998) found similar values for *Setaria sphacelata* pastures in the northern part of Ecuador. Mineral soil under succession has C4-C contents between 10% and 25%. The influence of C3 plants dominating the vegetation at the succession is reflected by the decrease of C4-derived carbon in the top 5 cm of the mineral soil. Below this depth (5–25 cm) the content of C4-derived carbon is not yet strongly influenced by the change of vegetation, as indicated by percentages of C4-derived carbon between 10% and 35% (Fig. 31.4).

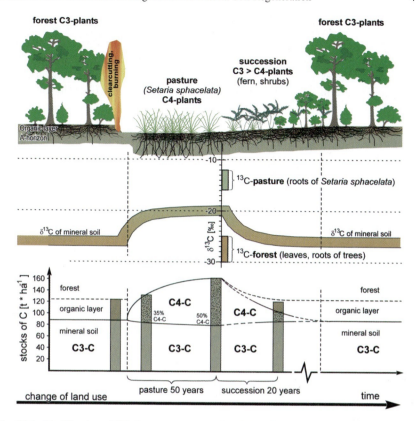

Fig. 31.4 Modification of ^{13}C signatures of mineral soils by change of land use

The forest soil reflects the typical ^{13}C-signature of C3-plants (−26‰). In the case of change of land use to pasture, the input of root-carbon of the grass *Setaria sphacelata* (−13‰) lead to an increase of ^{13}C to 20±2‰ as a result of the entry of C4-C. C4-C abundance in pasture soils depends on time and root production. Over time the δ^{13}C values of pasture soils theoretically can increase to the δ^{13}C values of C4-plants. After abandoning the pasture, woody species invade and begin to dominate (phase of succession) and initiate the re-conversion of the pasture to forest. Entry of C3-C from woody species leads to a decrease in δ^{13}C values to −26‰. Thus, we can use the δ^{13}C signatures of mineral soils as a tool to prove the history of land use.

Summing up, over the investigated gradient δ^{13}C and δ^{15}N values were grouped for forest, pasture and succession (Fig. 31.4). Some 50 years after the conversion of forest to pasture we found between 10% and 50% of C4-derived carbon in the top soils. About 20 years after abandonment of the pasture and development of succession vegetation, the content of C4-derived carbon had decreased again and amounted only to between 10% and 36%. These assumptions are well supported by other authors: After 10–30 years of forest conversion to pasture Cerri et al. (2004) found more than 50% C4-derived carbon in pasture mineral soils. After 80 years approximately 90% of TOC in the top 20 cm originated from pasture grass input. Consequently, we assume $\delta^{13}C$ values to be a reliable tool for differentiation of land uses and $\delta^{15}N$ useful for evaluation of organic matter decomposition in soils.

31.4 Conclusions

The geochemical and mineralogical survey clearly revealed a similarity and thus a priori comparability of the site conditions of the areas investigated under different land use. Results from the investigation show significant effects of land use on soils. These findings are supported by other authors indicating strong initial losses of soil carbon after conversion of natural forest into pastures, presumably also accompanied by additional losses of nitrogen and phosphorus. As forest and pasture fires with medium intensity as a rule may not oxidize the whole organic matter in organic layers, a part of unburned carbon remains as black carbon and accumulates over the decades. In contrast, alkaline ashes produced by fires are either blown over or remain on the soil surface, supplying available K, Ca and Mg and also P to the planted pasture grasses. A major part of the ashes repeatedly re-alkalifies the top soil and leads to higher pH values and an increase in base saturation. Even if the total P content of the pastures was higher than the forest and succession sites, the available P of the pastures was lower. Probably the available P could be almost fully utilized by vigorously growing pasture grasses, so that the concentration in the top soil was very low. According to the state of knowledge, alkalization of the soils enhances organic matter turnover, but these processes and their effects on the microbial community structure are not yet well understood and investigated. The vigorous growth of the planted pasture grasses produces high amounts of above- and below-ground litter, forcing their decomposition and also the decomposition of the remaining organic matter from former forests. In contrast, accumulated black carbon may inhibit matter turnover and microbial activities and can influence microbial and biochemical properties (Klose et al. 2003, 2004). Using $\delta^{13}C$ and $\delta^{15}N$ signatures in soils help to indicate and understand past processes and future development of soils and their functionality for the ecosystems involved.

Chapter 32
Succession Stages of Vegetation Regeneration: Secondary Tropical Mountain Forests

A. Martinez, M.D. Mahecha(✉), G. Lischeid, and E. Beck

32.1 Introduction

Successional stages of a vegetation development leading from bare soil to a plant cover dominated by bushes and a few pioneer trees have been described by Beck et al. (see Chapter 27 in this volume). Under optimal conditions this succession continues to a secondary mountain forest. Analyzing this is an enormous challenge, not only from a scientific but also from a logistic viewpoint. Tropical secondary mountain rain forests appear to be the most difficult vegetation type for a systematic investigation.

In the RBSF at least three types of secondary forest can be differentiated: Forest coming up on clear-felled areas, forests regenerating on mud-flows, and forest recovering from a fire. As these forests are almost impenetrable thickets, a complete inventory of their plant diversity is hardly achievable and therefore additional structural analysis is indispensable.

Seven patches of secondary forests, covering areas of between $1500\,m^2$ and $2200\,m^2$ in a narrow altitudinal range between $1950\,m$ and $2100\,m$ were investigated. Four of these (C1–C4) represent regeneration stages from small-sized clearings during the 1950s and 1960s for the construction/reconstruction of a water pipeline to the nearby power plant. They were found alongside the Camino Canal, where they border the primary forest. One plot (L) developed on the accumulated material of a mudflow, and two (F1, F2) are forests in a Quebrada which are recovering from a fire, as witnessed by several charred trunks. F1 and F2 are flanked by pastures or bracken fields.

As the climatologically conditions of these seven forests are similar, the differences shown below must result from their different ages and from the different modes of disturbance of the original forest.

32.2 Methods

The seven plots were subdivided into 5×5 m subplots which were then individually analyzed for species composition. Species were identified using mainly the *Flora of Ecuador* (Harling and Andersson 1973–2003) and the *Herbarium Reinaldo Espinosa* (Universidad Nacional de Loja). Although rough estimates on the cover abundances of the individual plant species were made, Jaccard distance indices were calculated based on presence/absence data. Further, the nonlinear ordination technique "Isometric Feature Mapping" (Isomap, Tenenbaum et al. 2000) was used for a low-dimensional projection of the data set. The methodological details are described in detail by Mahecha et al. (2007), including a broad methodological comparison of different ordination techniques.

32.3 Results and Discussion

In total about 770 vascular plant species were recorded, half of which could be reliably identified to genus, but only 20% to species; and 8% could not be identified. New species and species new for southern Ecuador were recorded. No minimum area could be established for any one of these forests, which is typical of tropical

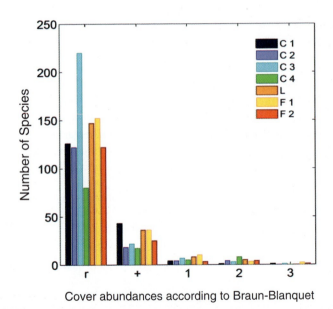

Fig. 32.1 α-Diversity (frequency) and cover abundances of the species recorded in the seven plots of secondary forest

secondary forests. The inventory of the vascular plants shows the extraordinary diversity and the concomitant low cover abundances of the individual species (Fig. 32.1). Another typical feature of secondary tropical forests is the heterogeneity of species distribution, reflected by a high degree of local species turnover. Assessing spatial heterogeneity in terms of β-diversity is crucial for understanding the succession development of secondary tropical forests. Figure 32.2 gives an impression

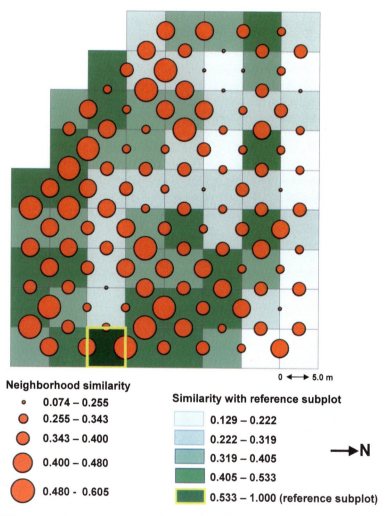

Fig. 32.2 Neighborhood similarity of subplots of Plot C4 based on the Soerensen indices. The colors of the quadrangular subplots indicate the similarity of each subplot with the reference subplot (indicated by the yellow frame), while the circles show the similarity of adjacent plots. Plot C4 is the oldest successional stage of the investigated series

of the heterogeneity of species occurrences at C4. Both the dissimilarity compared with a reference grid cell and the similarity of neighboring spatial plots show the heterogeneity of that patch of secondary forest.

Investigating the relationships between the respective plots requires methods of dimensionality reduction, well known as ordinations (Legendre and Legendre 1998). It was soon recognized that conventional linear ordination methods are not able to capture the inherent patterns of high dimensional species inventories, when dealing not only with a high α-diversity, but also with high degrees of β-diversity (Williamson 1978; Bradfield and Kenkel 1987; Dea'th 1999). Recently, Mahecha and Schmidtlein (2007) showed that such problems can be overcome by nonlinear ordinations (such as Isomap) by emphasizing the local species turnovers only. Indeed, Isomap extracted a well interpretable ordination space comprising the individual plots in one data set (Fig. 32.3a), which explained 78% of the data variance. This ordination further allowed a common visualization of α- and β-diversity (Fig. 32.3b; Mahecha et al. 2007). In the recovered ordination space the series C1–C4 shows, in a space-for-time sequence, the succession on areas where fire played no or only a minor role in forest clearing. Forests C2 and C3 border on the Camino Canal in a horizontal distance of 700 m and their age is about 45 years (C2) and 50 years (C3). Although both forests expectedly share a higher degree of similarity, they are clearly separated by the Isomap ordination. The youngest secondary forest in that series (C1; about 30 years old, which has transitorily been used as an agricultural field) and the oldest (C4; more than 60 years old) represent the corner marks of that series. The forests on plots F1 and F2 have been recovering from a fire for about 15 years and, because their areas are in close vicinity, they show a high degree of similarity. But from their floristic composition, as well as from their structure, they are clearly separated from the series C1–C4 and even more from forest "L" which has developed on a mudflow in a gully. Using the Isomap ordination technique, the peculiarities based on the different origins and ages of these

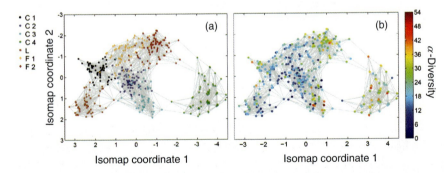

Fig. 32.3 Two-dimensional Isomap ordination based on the subplot floristic composition. Colors indicate different plots (**a**) and α-diversity (**b**), respectively. Reprinted from Mahecha et al. (2007), with permission from Elsevier

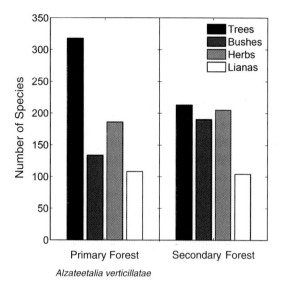

Fig. 32.4 Proportions of the various plant life forms of secondary and primary forest of the RBSF. The plant community "*Alzateetalia verticillatae*" was described by Bussmann (2002) as a formation of the primary lower montane forest of the area

heterogeneous secondary forests can be clearly demonstrated. For example it can be seen that the α-diversity appears as a function of the successional stage of the vegetation as reflected by an increasing species richness with time. However, similar high numbers of species were found in earlier stages of succession, depending on a different mode of disturbance, i.e. on plots affected by fire (Fig. 32.3b).

Our findings further revealed that the investigated secondary forests differ structurally from the corresponding primary forest (Bussmann 2002) by a smaller share of trees (Fig. 32.4) concomitantly with a higher proportion of bushes and tree ferns, whereas the contributions of the plant life forms herbs and lianas are rather similar. According to that composition of plant life forms the secondary forests have a less homogeneous and a more open appearance, as shown by the profiles in Fig. 32.5.

32.4 Conclusions

Floristic as well as structural analysis of a tropical secondary mountain rain forest can be performed using the classic standard Braun–Blanquet relevés and true-to-scale recordings of the transects. However, the mixture of cover and abundance scales in this kind of relevés do not allow any further statistical analysis. Additional biases are caused by highly subjective estimates of cover values in multistrata tropical mountain rain forests. Moreover, since the prerequisite of homogeneous vegetation

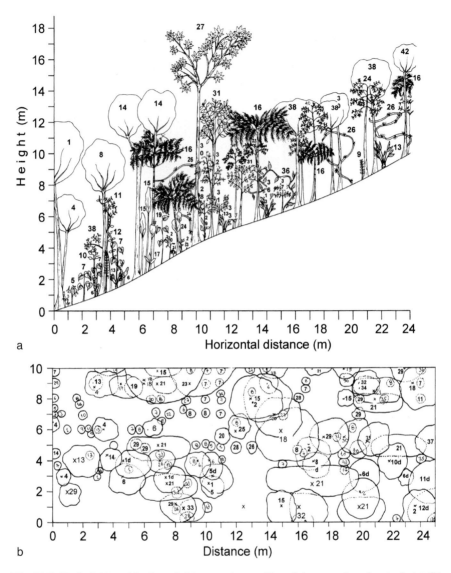

Fig. 32.5 Vertical (**a**) and horizontal (**b**) vegetation profiles of the secondary forest of plot C4 (undisturbed at least since 1962, according to the oldest aerial photograph). Explanation of the plant *numbers* given in the profiles: *1 Heliocarpus americanus, 2 Trichilia* sp., *3 Commelina* sp., *4 Viburnum obtectum, 5 Palicourea* sp., *6 Heliconia escarlatina, 7 Anthurium giganteum, 8 Persea* sp., *9 Philodendron* sp., *10 Piper parareolatum, 11 Miconia* sp. 1, *12 Mikania* sp., *13 Asplundia* sp., *14 Inga acreana, 15 Nectrandra* sp. 1, *16 Cyathea caracasana, 17 Asplenium serra, 18 Hedyosmum* sp., *19 Anthurium dombeyanum, 20 Alchornea glandulosa, 21 Paullinia* sp., *22 Nectandra* sp., *23 Peperomia* sp., *24 Miconia* sp. 2, *25 Hyeronima moritziana, 26 Chusquea* sp., *27 Cecropia gabrielis, 28 Guatteria* sp., *29 Stenospermation* sp., *30 Micropholis guyanensis, 31 Psychotria tinctoria, 32 Monnina* sp., *33* Lauraceae 1, *34 Tapirira guianensis, 35* Lauraceae 2, *36 Cavendishia* sp., *37 Miconia theaezans, 38 Hedyosmum anisodorum, 39 Clusia* sp., *40 Matayba inelegans, 41 Miconia* sp. 3, *42 Dendropanax* sp., *43 Myrsine coriacea, 1d Piptocoma discolor, 2d Miconia* sp. 4, *3d Naucleopsis glabra, 4d Ficus* sp., *5d Parinaria* sp., *6d Trichilia* sp., *7d Wettinia equatorialis, 8d Persea* sp. 2, *9d Guarea* sp., *10d Anthurium trisectum*

responses to environmental gradients and succession trajectories is not fulfilled, data needed for analysis are based only on the presence/absence information for species. Using the Isomap procedure a satisfactory description of the secondary forests was possible in terms of explained variance in connection with a meaningful low dimensional visualizations of the underlying patterns of the traced succession. The simultaneous visualization of α- and β-diversity uncovered systematic patterns of vegetation development as a function of time and of the type of disturbance.

Chapter 33
Reforestation of Abandoned Pastures: Seed Ecology of Native Species and Production of Indigenous Plant Material

B. Stimm(✉), E. Beck, S. Günter, N. Aguirre, E. Cueva, R. Mosandl, and M. Weber

33.1 Introduction

Tropical forests are characterized by a high diversity of tree species together with a low abundance of individual species. This has far-reaching implications on strategies for a sustainable management and conservation of their genetic resources.

Reforestation with native species is considered a preferable option for sustainable development, overcoming some of the ecological drawbacks of the earlier deforestation and concurrently contributing to the conservation of the region's biodiversity. Until the year 2000, 167 000 ha of plantations were successfully established in Ecuador (FAO 2003). Most of the plantations, however, consist of introduced species, i.e. in the coastal region mainly *Eucalyptus* sp. and *Tectona grandis*, and in the Sierra mainly *Eucalyptus* sp., *Cupressus* sp. and *Pinus* sp. Because of ecological problems with these species, more emphasis is nowadays put on plantations with native species (Brandbyge and Holm-Nielsen 1986; Borja and Lasso 1990; Aguirre et al. 2002a, b; Leischner et al. 2004; Predesur 2004).

However, lack of knowledge of the biology of the trees providing seed resources, e.g. their population densities and reproductive phenology, as well as of their germination physiology and the establishment of seedlings poses a severe challenge for any reforestation project.

The permanent availability and supply of high-quality seed and plant material for any kind of planting activity (enrichment planting, reforestation, plantation establishment; see Chapter 34 in this volume) requires the establishment of production standards. To achieve such standards, the monitoring and approval of seed sources of priority species is essential, which is accomplished by seed certification and control of seed procurement.

33.2 Conceptual Aspects for Reforestation Programs

Conservation and sustainable use of forest genetic resources is a major issue in national and international policies (Young et al. 2000). The objective is to secure the adaptedness of forest tree species to the respective environment with its dynamics and potential changes of the ecosystem. Moreover, appropriate concepts must build a basis for improving the production of timber and non-timber commodities but also of other services provided by a forest ecosystem (Graudal et al. 1997). Thus, responsible forest management – aside from maximizing the profit from timber and non-wood forest products (NWFPs) – aims at sustaining the respective forest ecosystem as well. The two basic strategies for conservation of the gene pool complement each other, namely in situ conservation (FAO et al. 2001) and ex situ conservation (FAO, IUFRO 2002).

Therefore forest restoration, e.g. by plantations, can also be an important complementary contribution to a future-oriented "dynamic conservation".

Much emphasis is put on native tree species plantations, however very little attention is still attributed to the provenance of the material from gene-ecological zones and the importance of using autochthonous planting material as well as to the question of how to produce autochthonous plant material at a larger scale. Hansen and Kjaer (1999) stressed the fact that appropriate genetic material may enhance not only the production and quality but also the health and stability as well as the environmental services of plantations.

Variation in forest genetic resources may become apparent between species, populations, individuals and chromosomes. Many tree species are characterized by high levels of intraspecific genetic variation which ensures the plasticity of the gene-pool concomitantly with its ongoing evolution. Ecological conditions in the Andes differ markedly between the eastern and western Cordillera and vary considerably with elevation (see Part III.1 in this volume). Species like *Cedrela montana* or *Tabebuia chrysantha* are widely distributed (Møller Jørgensen and León-Yánez 1999; MBG 2006) and apparently occur in differing populations, depending on the ecology of the habitat. Separated by altitudinal borderlines, gene flow via pollinators and seed-dispersers between these populations is limited.

Hufford and Mazer (2003) pointed out that the geographical distribution of many plant species used in restoration may span a wide range of environmental conditions. Habitat heterogeneity, combined with natural selection, often results in multiple, genetically distinct ecotypes of a single species. Consequently, data are needed to delineate "seed transfer zones", or regions in which plants can be transferred with little or no impact on the population fitness.

Regional programs must be developed which care for a sustainable provision of forest commodities to meet the region's requirements for timber, poles, fuel, fodder, food, medicine and shelter. This must go hand in hand with the rehabilitation of degraded areas and the conservation of genetic resources of trees.

This goal needs time and several short-term steps must be envisaged. One of these is the provision of genetically suitable seed and other plant reproductive material from selected indigenous seed sources. "Suitable seed" means the location, use and maintenance of clearly defined and well documented seed sources. To the best of our knowledge, no species-specific conservation or management plans have been implemented in Ecuador with the objective to protect the great variety of forest tree genetic resources for future use.

33.3 Plant Production as a Prerequisite for Reforestation

Most of the recent Ecuadorian reforestation projects deal with exotic species of the genera *Eucalyptus*, *Pinus* or *Cupressus*, while native species due to the reasons discussed below have not been used in the Andean region so far (see also Chapter 34 in this volume). Knowledge of the reproductive biology of indigenous species is very limited, but indispensable for the supply of high quality tree seeds and the production of adequate numbers of high quality seedlings in tree nurseries. An overview of the key requirements in the process of tree seedling production is shown in Fig. 33.1.

33.3.1 Selection and Survey of Priority Species

Besides the endangeredness of species the main criteria for including species in genetic resource conservation programs are their present use and their potential use (Graudal et al. 1997).

In our research area, based inter alia on a timber market survey by Leischner and Bussmann (2002), a first selection of priority species was made. From the over 200 regional tree species potentially suitable species were selected by the following criteria: high local acceptance, economic value (both timber and non-timber products), endangered or species with a high ecological significance. Some 15 native species with promising potential for reforestation were chosen (see also Table 33.1).

33.3.2 Phenology as a Means for the Prediction of Quality, Quantity and Harvest Time of Seed Crops

Whilst comprehensive studies of reproductive phenology have been performed in tropical lowland rain forests (e.g. Frankie et al. 1974; Croat 1975; Opler et al. 1980; Hamann and Curio 1999; Newstrom et al. 1994; Schöngart el al. 2002), only a few studies exist addressing phenological aspects of tropical mountain rain forests. For example, Koptur et al. (1988) investigated the phenology of bushes and treelets in Costa Rica in an altitudinal range between 1300 m and 1650 m a.s.l. More comprehensive long-term studies of the phenology of flowering and fruiting have been performed in Southeast Asia, showing quite different patterns in one and the same forest (Sakai et al. 1999; Hamann 2004).

Bendix et al. (2006c) discussed various hypotheses addressing individual elements of the climate as ultimate factors that trigger seasonality in the equatorial tropics. They examined the hypothesis that in the area of the RBSF cloudiness is the master trigger of periodic phenological events in plant life which are,

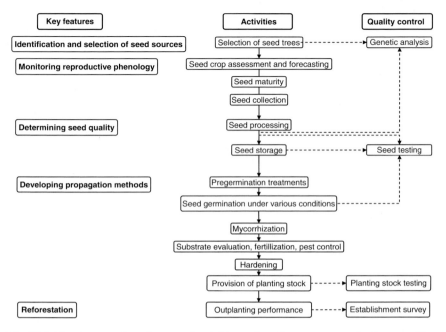

Fig. 33.1 Key features, flow diagram of activities and quality management in forest seedling production

however, modulated by climatic irregularities (see also Chapter 20 in this volume).

In the RBSF phenological studies have been carried out by several working groups (Wolff et al. 2003; Cabrera and Ordoñez 2004; Diaz and Lojan 2004; Bendix et al. 2006c; Cueva et al. 2006). Cueva et al. (2006) examined 400 individuals of more than ten tree species over a period of three to four years with respect to flowering and fruiting (see also Table 33.2). In spite of the perhumid climate the authors observed a high degree of synchronization at the species level, also groups of non-related species showed synchronized flowering and fructification.

33.3.2.1 Seasonality of Flowering

The records of flowering showed two principal patterns. One group of trees, consisting of *Piptocoma discolor*, *Tabebuia chrysantha*, *Myrica pubescens*, *Cedrela montana*, *Purdiaea nutans* and *Inga* sp. flowered during the less humid period of the year, i.e. between August and February, while *Clethra revoluta*, *Heliocarpus americanus*, *Isertia laevis*, *Viburnum* sp. and *Vismia tomentosa* flowered during the wettest season (March to July). Exceptions were male individuals of

Table 33.1 Seed ecology data of selected native South Ecuadorian mountain rainforest tree species. Species nomenclature follows W³TROPICOS Missouri Botanical Garden's VAST nomenclatural database (ver. 1.5). *Prov* Provenance, *Disp* seed dispersal mode, *M.C.* moisture content of seeds. Data compiled from: Beck (personal communication), Briceño (2005), Cabrera and Ordoñez (2004), Cueva (personal communication), Diaz and Lojan (2004), Homeier (personal communication), Jara and Romero (2005), Leischner (2005), Merkl (2000)

Species	Prov[a]	Disp[b]	Flower[c] (months)	Fruiting (months)	Ave. purity (%)	M.C. (%)	Weight[d] (g)	No. seeds/kg[e] (×1000)	No. seeds/fruit	No. fruits/tree (×1000)	No. seeds/tree (×1000)	Ave. germination[f] (%)
Alnus acuminata Kunth	EB	w	VIII–III	IX–V	91	–	0.29–0.31	3400	180–210	–	–	35 (39)
Alnus acuminata Kunth	A	w	–	–	–	–	0.19	5263	–	–	–	58 (38)
Cedrela lilloi C.DC.	SF	w	–	–	–	–	11.9–15.0	83.9–66.0	21–24	0.01–0.5	0.23–11.5	62
Cedrela montana Moritz ex Turcz.	SF	w	X–III	II–VIII	90	12	12.4	81.6	–	–	–	86 (80)
Cedrela montana Moritz ex Turcz.	EB	w	VIII–III	XI–V	90	–	42.5	23.5	27	–	–	86 (69)
Cinchona officinalis L.	EB	w	ay	ay	77	–	0.67–0.97	1310	39–42	–	–	80 (90)
Clethra revoluta Ruiz & Pav.	SF	w	III–VII	IV–X	63	6.5	0.05	19 417	–	–	–	69 (22)
Clethra revoluta Ruiz & Pav.	EB	w	X–III	XI–IV	93	–	0.073–0.078	13 260	40–46	–	–	4 (8)
Cupania sp.	EB	a/g	I–VIII	VI–XI	93	–	445–580	–	3	–	–	(40)
Heliocarpus americanus L.	SF	w	II–VIII	V–X	93	10.4	1.39	732.3	1–3	21.9–125.0	42.5–250.0	27 (20)
Inga acreana Harms	SF	a/g	IX	III–V	98	3.6	550.1	1.8	–	0.3	2.2	84
Isertia laevis (Triana) B. M. Boom	SF	a	I–IX	VI–III	62	14.0	0.15–0.23	5530	200–300	1.0–2.4	200–720	–89
Juglans neotropica Diels	L	a/g	VII–II	IX–VI	–	–	18 200–21 400	0.05	1	0.5–1.0	–	(72)

(continued)

Table 33.1 (continued)

Species	Prov[a]	Disp[b]	Flower[c] (months)	Fruiting (months)	Ave. purity (%)	M.C. (%)	Weight[d] (g)	No. seeds/ kg[e] (×1000)	No. seeds/ fruit	No. fruits/ tree (×1000)	No. seeds/ tree (×1000)	Ave. germination[f] (%)
Myrica pubescens Humb. & Bonpl. ex Willd.	SF	a	V–VI ♂ ay ♀	II–III ♂ ay ♀	98	11.9	12.3	85.7	1	30.8	30.8	76 (*11*)
Myrica pubescens Humb. & Bonpl. ex Willd.	EB	a	ay ♀	ay ♀	96	–	3.45	290	1	–	–	(*15*)
Piptocoma discolor (Kunth) Pruski	SF	w/(a)	VII–X	VIII–II	93	8.6	0.29	3434.4	–	254.7	509	23 (*13*)
Prumnopitys montana (Humb. & Bonpl. ex Willd.) de Laub.	SF	a/g	XI–II	II–V	–	–	–	–	–	–	–	–
Prumnopitys montana (Humb. & Bonpl. ex Willd.) de Laub.	EB	a/g	IX–IV	XII–VI	–	–	74.0	13.5	1	–	–	–
Tabebuia chrysantha (Jacq.) G. Nicholson	SF	w	VIII–XI	X–I	–	–	9.0	111	–	–	–	86 (*59*)
Vismia tomentosa Ruiz & Pav.	SF	a	IX–V	I–VIII	85	11.0	0.59	1724	60	9	540	15

[a] Provenance: A = Alisal, EB = El Bosque, L = Loja, SF = San Francisco.
[b] Seed dispersal mode: a = animal, g = gravity, w = wind.
[c] Flowering period (in calendar months): ay = all year.
[d] Weight of 1000 seeds.
[e] Number of seeds per kilogram of seeds.
[f] Average germination on wet filter paper or wet sand under laboratory conditions; in *italics*: max. germination (%) in greenhouse tests.

Table 33.2 Fruiting periods and percentage of fruiting trees of ten tree species indigenous to the mountain rain forest of South Ecuador

Species	Fruiting period (calendar months)	Trees fruiting per year (%)			
		2000	2001	2002	2003
Piptocoma discolor	XI–I	100	9	90	–
Tabebuia chrysantha	XI–I	53	28	35	–
Myrica pubescens (female)	II–III	–	40	27	36
Inga cf. *acreana*	III–V	–	50	75	0
Cedrela montana	VI–VII	–	50	83	0
Heliocarpus americanus	VI–IX	–	40	80	66
Clethra revoluta	VI–IX	20	10	50	–
Vismia tomentosa	VI–IX	71	0	75	–
Isertia laevis	V–XI	92	78	56	–
Ficus sp. 1	II–VI	–	33	–	–
Ficus sp. 2	I–III	–	–	11	–
Ficus sp. 3	XI–III	–	–	–	11

Myrica pubescens, of which flowers were seen for ten months per year (but also with a maximum during the less humid period, while female individuals usually flowered for less than one month), and *Graffenrieda emarginata* of which pronounced periodicity of flowering could not be observed. However, this species extends over a wide altitudinal range and therefore the collected data may not be sufficient to reveal intraspecific synchronization. Interestingly flowering of the majority of the wet season species (*Clethra revoluta*, etc.) commenced in December or even earlier, when the monthly rainfall was still low. This observation suggests a climatic factor other than precipitation may trigger flowering of that group of plants.

The proportion of flowering individuals of the examined species differed considerably: Of *Heliocarpus* nearly 90% of the individuals flowered at the same time, and of *Piptocoma* the peak of simultaneously flowering trees even exceeded 90% (except in 2001). *Vismia*, *Myrica*, *Clethra*, *Isertia* and *Inga* were further species with reasonably high rates of flowering. In contrast, relatively low percentages of flowering individuals were recorded for *Cedrela* and *Tabebuia*, two of the deciduous species in the forest.

33.3.2.2 Seasonality of Fructification

Due to the differing species-specific time-spans required for seed-set and fruit ripening, the annual phases of fruiting were wider and in most cases not as clearly delimited as for flowering. Nevertheless, again two groups of trees can be differentiated with respect to fructification. *Piptocoma*, *Tabebuia* and *Myrica* were found fruiting still during the less humid season, i.e. in November, December and January. The second group of trees consists of *Cedrela, Clethra, Heliocarpus, Inga* and *Vismia* which exhibited maximum fruiting predominantly during the wet season. Fruiting trees of *Isertia* were present all year round, however, with a pronounced minimum in February/March. While *Vismia* showed a long fruiting period of eight months (in 2002), other species like *Heliocarpus, Tabebuia* or *Piptocoma* had comparatively narrow time-windows of three to four months for fructification.

On average, fructification of the ten species examined was higher in 2002 than in 2001. This holds in particular for *Clethra revoluta* which flowered as usual in 2001 but then barely produced fruits. In contrast, *Piptocoma discolor* and *Vismia tomentosa* produced almost no flowers in that year and consequently fruits were not developed. No mature fruits could be found with *Inga* sp. and *Cedrela montana* in 2003. The latter species had already aborted unripe fruits. Table 33.2 shows the records of fruiting for the selected species over the years 2000–2003.

Concomitantly with the phenological recordings, climate was monitored in detail at both the regional and the local scale. Some correlation of flowering and fruiting with precipitation, cloudiness and irradiance could be found (Bendix et al. 2006c).

33.3.2.3 Comparing the Phenological Behavior of the Same Species at Different Sites

Comparative phenological observations were initiated in 2001 at two locations, at the RBSF and at the Reserva Protector 'El Bosque', approximately 10 km southwest from San Pedro de Vilcabamba (Province of Loja). The two sites are about 30 km apart, at an altitude of 2100 m a.s.l. in the evergreen montane rainforest zone. They differ mainly in the precipitation pattern (see Chapter 1 in this volume) and related ecological parameters which may also have a strong impact on the activity of the biota (e.g. pollinators or seed dispersers).

Phenological data of flowering and fruiting were registered at two-week intervals. Details of methodology and first results have been published in Günter et al. (2004). At both study sites the phenology of four species was compared: *Cedrela montana, Clethra revoluta, Myrica pubescens* and *Prumnopitys montana*.

Flowering of *Cedrela montana* starts more than two months earlier at 'El Bosque' than at RBSF but finishes only one month earlier, i.e. the period of flowering is longer at 'El Bosque'. Also the fruiting season is about two months postponed at RBSF as compared with 'El Bosque', but its time-span is equal at both sites. For *Clethra revoluta* timing of flowering and fruiting differs by about six months in the two study areas (Fig. 33.2). In general the intraspecific variability of the species is lower at 'El Bosque' than at RBSF, but the interspecific synchronization is higher.

Interestingly, not only the rhythms but also the intensities of flowering and fruiting differed at both sites. While the percentage of flowering and fruiting trees decreased in the second year at RBSF, an increase was registered at 'El Bosque'.

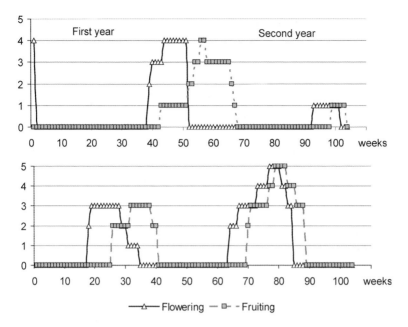

Fig. 33.2 Differences in flowering and fruiting of *Clethra revoluta* between the study areas RBSF (*above*) and 'El Bosque' (*below*). Adapted from Günter et al. (2004)

The seasonality of flowering and fruiting of *Myrica pubescens* was less pronounced at RBSF, as compared with 'El Bosque', reflecting the more pronounced dry and wet seasons in the latter area.

33.3.3 Collection and Germination Potential of Seeds

For practical purposes of seed collection Cabrera and Ordoñez (2004) and Jara and Romero (2005) produced a calendar for selected tree species at RBSF, and Diaz and Lojan (2004) for 'El Bosque'. An example is shown in Fig. 33.3.

Whenever possible, ripe fruits were harvested from the trees. Seeds were detached from the fruits and after drying on paper in the shade were kept in plastic boxes in the dark until sowing. In some cases seeds were also dried while still enclosed in the fruits (*Cedrela*, *Heliocarpus*, *Piptocoma*) which were then scarified for the germination assay.

The germination rates of the seeds varied greatly from species to species (see Table 33.1), but also for the same species from year to year. *Cedrela montana*, *Tabebuia chrysantha* and *Inga acreana* were fast-germinating species while *Isertia laevis*, *Piptocoma discolor* and *Clethra revoluta* required almost a quarter of a year for the seedlings to emerge. *Piptocoma*, *Myrica* and *Isertia laevis* germinated satisfactorily only after pretreatment of the seeds. A still unresolved problem is the propagation of Podocarpaeceae

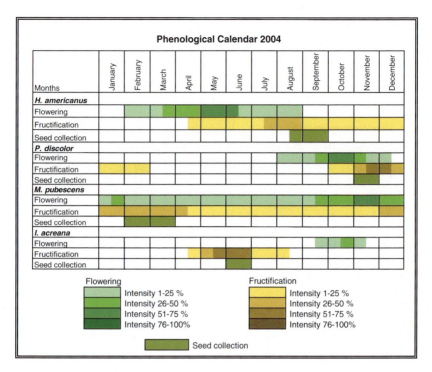

Fig. 33.3 Phenological calendar of four native tree species (after Jara and Romero 2005)

from seeds from provenances of Southern Ecuador. Germination protocols of various species have been elaborated and are presented in detail elsewhere (e.g. Leischner 2005).

33.3.4 Early Development of Seedlings

After germination, seedlings of several species, such as *Isertia*, *Clethra* and *Vismia* did not readily establish but remained tiny and susceptible to getting mouldy, even when subjected to adjusted light intensities in the nursery. Various substrates (with and without fertilizer) were examined in a comprehensive study (Leischner 2005) but the addition of soil from the natural stand was most effective in promoting early growth. This observation suggested that inoculation of a mycorrhiza could play an essential role in the production of young plants for reforestation. To evaluate the possibility of mycorrhizal inoculation Urgiles (2003) set up an exploratory experiment. She inoculated native plant seedlings of *Alnus acuminata* and *Erythrina edulis* growing under controlled environmental conditions with several VA mycorrhiza fungi. An evaluation of the tree seedlings six months after sowing showed that the rate of mycorrhization and biomass production was best after addition of forest soil as inoculum source.

33.3.5 The Problem of Isolated Subpopulations in the Study Area

The rugged morphology of the eastern Cordillera in South Ecuador creates a high patchiness in the vegetation, where similar habitats, e.g. deeply incised gorges are separated by strongly differing ecotones, such as wind-exposed ridges. Figure 33.4 shows the distribution of *Cedrela montana* in four catchment areas of the RBSF valley where clusters of that species are separated from each other by pronounced ridges. Under these marked topographical structures it is questionable whether gene flow between the individual subpopulations could efficiently take place. The situation could be complicated by the phenomenon of functional dioecy in *Cedrela* (Smith et al. 2004).

We also examined the distribution of *Prumnopitys montana* in the four catchments and identified three clusters of this species. *P. montana* is a highly exploited dioecious species and an unfavorable female:male ratio in subpopulations could have a severe impact on its fitness.

It is noteworthy that the remaining subpopulations of *Cedrela* or *Prumnopitys* are at the upper boundary of their natural distribution. We argue that these subpopulations located in the catchments may become endangered by isolation because pollen and seeds may hardly be transported over the surrounding natural barriers (Fig. 33.4).

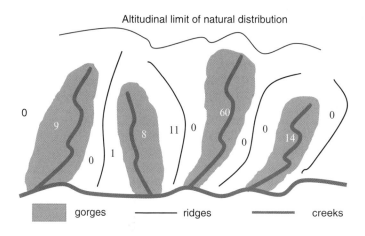

*estimated area of watersheds from left to right: 13, 10, 20 and 8 hectares

Fig. 33.4 Distribution of subpopulations of *Cedrela montana*. Numbers illustrate the number of trees with a dbh >10cm in the four investigated catchments of the RBSF forest (from Günter et al. 2004)

33.3.6 Aspects of the Provenance of Seed Material: Delineation of Gene-Ecological Zones

Based on these observations and on the findings about the conspicuous differences in reproductive phenology between sites the delineation of so-called "gene-ecological" zones is suggested, where the variation of environmental conditions is investigated and uniform zones are identified. To delineate gene-ecological zones (i.e. provenance regions) which can act as seed sources for sustainable forest management programs as well as for the conservation of forest tree genetic variation, the maps of the environmental conditions and the vegetation cover in the Province of Loja were overlaid. After the identification of these zones and their size it can be decided whether the areas will be sufficient for an in situ conservation and sustainable management or whether additional measures have to be undertaken (e.g. establishment of seed orchards, clone archives, amplification of conservation areas; see Günter et al. 2004).

Three environmental parameters (temperature, humidity, soil types) were used for the delineation of gene-ecological zones. A very broad range of environmental conditions can be observed in the Province of Loja (see also Chapter 1 in this volume). The mean annual temperatures vary from 8 °C to 26 °C, the humidity varies between 0 and 12 humid months and 11 major soil types have been stated in the provincial watershed management plan. On the basis of these data four classes of temperature ranges and six classes of humidity were defined. Superimposing the three ecological maps resulted in a map of potential gene-ecological zones (Günter et al. 2004). For the Province of Loja the authors calculated a total of 134 potential gene-ecological zones, 46 of them being smaller than $10\,km^2$ and only 17 being protected in National Parks.

For a proper management urgent aims are the verification of the gene-ecological seed zones, mapping of target individuals and an assessment of potential barriers for gene flow.

33.4 Conclusions and Implications for Reforestation

Tropical forest plants may respond to environmental change through phenotypic plasticity, adaptive evolution, migration to more suitable sites, or extinction. However, the potential to respond is limited by the rapidness of change and the lack of alternative habitats due to past and present trends of deforestation (Bawa et al. 1998).

There are extremely marked differences in flowering and fruiting behavior between the species at the same site and also for the same species at different sites. Despite our attempt to classify species to different groups with similar phenological characteristics, the variability in timing and intensity is still very high within and between groups. According to Bendix et al. (2006c), irregular events may be responsible for the variation in phenology.

Mass flowering, sometimes termed general flowering, is an irregularly occurring phenomenon of perennial plants that is common in the tropics but is also known from temperate forests (Herrera et al. 1998). During more than four years of monitoring phenological events in the Ecuadorian mountain rain forest, mass flowering was not recorded.

Which climatological factors affected flowering and seed development in some species is still an open question. Cold rather than drought presumably affects flowering and seed production as some of the respective species, e.g. *Piptocoma discolor* and *Vismia tomentosa*, are at the upper boundary of their altitudinal range and are typical representatives of the lower, i.e. warmer, montane forest.

Because of high interspecific and interzonal variability seed-harvesting calendars in the study region must be elaborated on a species level and separately for different seed zones. A first approach for the delineation of seed zones in Southern Ecuador has been presented by Günter et al. (2004).

To receive sufficient genetical amplitude for large-scale reforestation it is necessary to harvest a minimum of 50 (better 500) seed trees of one provenance and species (Graudal et al. 1997). Because some species had relatively low flowering and fruiting probabilities (representatives of Podocarpaceae, *Myrica pubescens*, *Vismia tomentosa* among others) it is necessary to monitor a higher number of individuals of a certain seed zone for seed harvesting. This, however, is very difficult in the tropics where many species appear with very low abundance.

Germination experiments with our species showed the need for developing species-specific appropriate propagation techniques and protocols; otherwise planting material for reforestation purposes might not meet the qualitative standards and the required numbers (Cabrera and Ordoñez 2004; Diaz and Lojan 2004; Leischner 2005).

The establishment of competent national and regional tree seed centers (seed banks) in Ecuador is of utmost importance. Inter alia those institutions must be in charge of the founding and appropriate management of a network of seed production stands, which is a prerequisite of sustainable seed and seedling production.

Chapter 34
Reforestation of Abandoned Pastures: Silvicultural Means to Accelerate Forest Recovery and Biodiversity

M. Weber(✉), S. Günter, N. Aguirre, B. Stimm, and R. Mosandl

34.1 Introduction

As a result of high deforestation and non-sustainable land use there exists a substantial and increasing amount of unproductive land in Ecuador, especially concerning over-used, degraded or abandoned pastures (see Chapter 4 in this volume). The loss of productive land is particularly disastrous because forestal and agricultural land use is one of the main generators of income in rural areas. It is interesting to see that 90% of the area annually afforested in Ecuador (3500 ha) are realized in the Andean region, where the socio-economic and ecological functions of the forests are of prominent importance (FAO 2006). However, despite the exorbitant number of 2736 native tree species in Ecuador (Jørgensen and León-Yánez 1999), forestation activities are almost exclusively based on exotic species, mainly pines and eucalypts.

One of the main reasons for this obvious neglect of the native species in reforestation activities is the lack of adequate knowledge on their ecology and silvicultural treatment. Another reason is the non-availability of forest reproductive material in nurseries, a fact that can be led back to the lack of knowledge about the biological basics (phenology, seed germination and storage, etc.; see Chapter 33). This situation is not only true for Ecuador but is also described for many tropical regions (e.g. Butterfield 1996; Butterfield and Fisher 1994; Davidson et al. 1998; Feyera et al. 2002; Holl et al. 2000). For tropical lowland and montane forests a number of studies has been conducted during the past decade to provide a basis for the selection of native species suitable for reforestation of degraded land and to improve the knowledge about their ecological and silvicultural characteristics (Haggar et al. 1998; Knowles and Parotta 1995; Leopold et al. 2001; McDonald et al. 2003; Montagnini 2001; Montagnini et al. 1995, 2003). Most of these studies revealed that species selection and silvicultural means for rehabilitation of degraded land must be very well adapted to the environmental conditions of the sites, which are closely linked to the intensity and duration of former land use. Consequently, Montagnini (2001) postulated the identification of the principal constraints of an area and the definition of specific objectives as the initial steps for successful reforestation.

The two principle options for the reestablishment of forests on degraded land are: reliance on natural regeneration and planting. The adequate method highly depends on the respective site conditions but also on the requirements of the respective land owner. Our investigations aim at providing basic ecological and silvicultural knowledge for the assessment of adequate reforestation options.

34.2 Natural Regeneration

Historically the most common pathway to reforest degraded land has been abandonment and reliance on natural succession. However, very often natural regeneration is prevented by several adverse factors. As a consequence natural regeneration on degraded land does not always operate on a time-scale compatible with human needs (Parotta et al. 1997).

Several authors discuss the lack of seed dispersal due to the remoteness of an area from existing forest edges as a main hindrance for fast regeneration (Cubiña and Aide 2001; Greene and Johnson 1995; Günter et al. 2006; Holl 1999; Lamb 1998; Pokorny 1997). Guariguata and Ostertag (2001) concluded from their results that the regenerative power of neotropical forest vegetation is high if propagule sources are close and land use intensity has not been severe before abandonment.

34.2.1 Methods

To evaluate the effects of distance to forest edges on natural regeneration we studied ten transects within a patch of an abandoned cattle pasture at the RBSF measuring 80×240 m, presently covered by secondary forest (Günter et al. 2006). Each transect consisted of four plots with 25 m^2 each, along a gradient of increasing distance from the forest (−20 m, 0 m, 20 m and 40 m), with an additional reference plot of 400 m^2 inside the surrounding primary forest. In the 25 m^2 plots all woody plants with dbh >2 cm were measured, while in the 400 m^2 plot only those with dbh >10 cm. Parameters analyzed were dbh, species, crown coverage, and tree height (only on the 25 m^2 plots). Aerial photographs proved that the pasture at 2100 m a.s.l. was abandoned approximately 38 years ago and was always surrounded by primary mountain rain forest.

34.2.2 Results and Discussion

Although the patch was completely encircled by forest the speed of natural regeneration as well as the species' composition were not satisfying from a forest user's point of view. The total abundance was high but the abundance of valuable species was insufficient, especially in the plots with a larger distance to the forest edge (Fig. 34.1). Plant families with more valuable species in this region, like Lauraceae (*Nectandra*, *Ocotea*), or Podocarpaceae (*Podocarpus*, *Prumnopitys*), were not represented in the plots far from the forest edge. Considering the time since abandonment height and diameter development of the existing trees was also not sufficient (Table 34.1). In the plots 40 m from the forest edge maximum height was achieved by *Graffenrieda emarginata* and *Clethra revoluta*, but was only 7.0 m! Cubina and Aide (2001) found in their study that, even when abandoned pastures are surrounded by forests, only a subset of species disperses into the pastures and wind-dispersed species are usually overrepresented. They concluded that seed arrival is the major factor limiting recover in abandoned pastures.

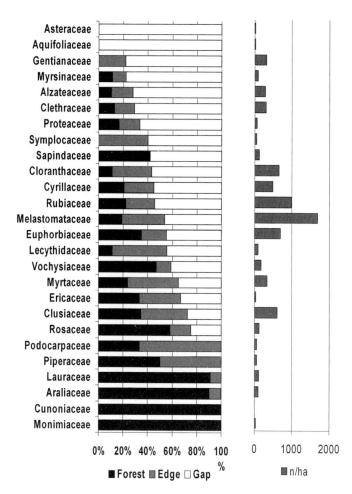

Fig. 34.1 Mean relative (*left*) and absolute (*right*) abundances of the families at all regeneration plots related to distance from forest edge: forest (−20 m), edge (0 m), and gap (40 m; Günter et al. 2006)

Table 34.1 Impact of the distance from forest edge on structural parameters of trees and shrubs with dbh >2 cm (Günter et al. 2006). Different letters indicate statistical differences at these levels: * $P<0.05$, ** $P<0.01$, *** $P<0.001$

Parameter	dbh	Distance from forest edge			
		−20 m	0 m	20 m	40 m
Abundance (n/ha)	>2 cm	7320	8200	8280	6560
	>5 cm	2800	2200	2040	2240
	>10 cm	800	120	0	80
Basal area (m²/ha)	>2 cm	24.7 a*	12.2 b*	11.9 b*	12.4 b*
	>5 cm	21.2 a*	8.1 b*	7.0 b*	7.9 b*
	>10 cm	13.3	1.04	–	0.8
Mean height (m)	>2 cm	5.3 a*	4.3	3.8 b*	4.4
	>5 cm	6.5 a**	5.3	4.7 b**	5.1 b*
	>10 cm	10.1	7.3	–	6.3
Species (n/plot)[a]	>2 cm	12.2 a**	9.2 b**	9.8	8.6 b**
	>5 cm	5.1	3.3	3.2	4.0

[a] No species number per plot was calculated for dbh >10 cm due to the low abundance per plot

Table 34.2 Sörensen indices of similarity between vegetation communities along a gradient of increasing distance from the primary forest edge (Günter et al. 2006)

Distance	dbh	Distance from forest edge			
		−20 m	0 m	20 m	40 m
		Trees with dbh >2 cm			
−20 m	>10 cm	85.1	69.7	61.7	56.4
−20 m	>2 cm	–	67.4	51.8	53.8
0 m	>2 cm	–	–	71.2	65.7
20 m	>2 cm	–	–	–	73.8

Aide et al. (2000) showed in their study that species richness increased very fast after abandonment but species composition was quite distinct from the original forest. Saldarriaga et al. (1988) found in slash-and-burn sites in Venezuela and Colombia that at least 40 years were required for species richness of stems ≥10 cm dbh to attain values similar to that of mature forests. In our study the vegetation communities differed also quite a lot from that of the adjacent natural forest with increasing distance from the forest edge (Table 34.2). It is still questionable whether this is due to a lack of seed dispersal or to the inability of late successional species to germinate on the abandoned pasture.

Guariguata and Ostertag (2001) stated that many forest functions are similar to old-growth conditions long before species composition is. This may be true for many ecological services. However, secondary forests like that analyzed in our study will hardly be able to satisfy the economic demands of a forest owner, even in medium term. In summary we conclude that reliance on natural regeneration is not an acceptable option where fast rehabilitation of the economic productivity is required.

34.3 Plantations

On intensively degraded sites, where seed sources are distant, where soils require fast stabilization, or where the productivity of the land has to be restored within a short time, planting might be the best strategy to rehabilitate forest cover. Lamb (1998) considered timber plantations as one of the few means to reforest large areas of cleared or degraded landscapes. However, he argued that they usually restore the productive capacity of the land but do little to recover biological diversity. In contrast, Guariguata and Ostertag (2001) pointed out that plantations provide valuable contributions to biodiversity as they can accelerate succession of forest understorey communities through the elimination of weed competition while attracting seed dispersers.

34.3.1 Exotic Species Plantations Versus Natives

As already mentioned currently 90% of the plantations established in Ecuador are based on exotic species, predominantly of the genera *Eucalytus*, *Pinus*, and *Cupressus*. Their most prominent advantage is the well founded knowledge about their biology and silviculture and proven experiences. Exotics usually have good growth rates and well known wood properties (Lamb 1998). Likewise site requirements, nursery, and silvicultural methods are well established and forest reproductive material is easily accessible. In some cases exotics can also act as nurse trees (Feyera et al. 2002) for the regeneration of native species or contribute to the restoration of degraded sites (see also Section 34.4).These advantages have to be balanced against the many disadvantages. The most obvious one is that, in regions with high biological diversity like Ecuador, exotic species plantations are highly unnatural and there is a high risk that they may invade natural forests thus affecting not only the vegetation of the planted site but also adjacent natural habitats, as described by several authors (D'Antonio and Meyerson 2002; Lorence and Sussman 1986). Several studies revealed that they can also alter the hydrological properties of soils. Farley et al. (2004) found under plantations with *Pinus radiata* in the Ecuadorian Andes that water retention declined drastically with stand age.

In contrast, native species have biological advantages that make them superior to exotics at many sites: As they are well adapted to the environment they are considered to be less susceptible to stresses. From the ecological point of view they are valuable because they allow also to maintain the native flora and fauna and to conserve the genetic and biological diversity in situ. Furthermore, many native species have a higher market value than exotics or are socially valued by the local communities (Leischner and Bussmann 2002). Therefore, native species are especially interesting for plantation where demand for social or ecological services and benefits is higher than for high volume production. However, there is a huge and urgent need for information on the biological and silvicultural characteristics of native species that are suitable for the reforestation of degraded land. Very little knowledge exists also about the growth performance of natives in plantations.

34.3.2 Methods

In an experimental trial at the RBSF we tested the suitability of five native species (*Alnus acuminata*, *Cedrela* cf *montana*, *Heliocarpus americanus*, *Juglans neotropica*, *Tabebuia chrysantha*) for the afforestation of three successional stages of abandoned pastures: recently abandoned (pasture), abandoned pasture dominated by bracken fern (fern), and abandoned pasture already covered with shrubs (shrub). Because very frequently native species are not tested against exotics we also included the two most commonly used exotic species in that region (*Pinus patula*, *Eucalyptus saligna*). Then 1200 seedlings per species were planted in a generalized randomized block design with 336 plots and eight replications for two treatments: On 50% of the plots ground vegetation has been removed manually only before planting, on the remaining plots additionally every four months after planting.

34.3.3 Results and Discussion

Figures 34.2 and 34.3 show the survival and height of the different species one year after planting. Survival was good on all three sites; only *Alnus* and *Juglans* had survival rates below 80%. Nevertheless, *Alnus* achieved the best height which was even above that of the corresponding exotics. The height of all other species was significantly lower than that of *Alnus* and the two exotics. These results support those of Carpenter et al. (2004b) that nitrogen-fixing trees are an attractive option on poor soils. Fehse et al. (2002) considered *A. acuminata* also as a suitable catalyst species for forest regeneration in high altitude tropical forests.

The manual clearing of ground vegetation did not significantly affect the establishment of seedlings in the first year. In a subsequent experiment established adjacent to the pasture site presented above we also tested the effect of a glyphosate treatment of ground vegetation (Eckert 2006). The chemical treatment dramatically improved the development of the forest plants compared with the other treatments (Table 34.3). Although the biomass of the ground vegetation in the plots with chemical treatment was not much lower than in the manually treated, the share of the tree biomass of the total plot biomass increased from 0.9% (T0) to 26.5% (T2) for *Tabebuia* and from 1.0% to 16.4% for *Cedrela*.

Figure 34.4 shows that the treatment not only affected the total biomass of the plants but also its allocation to above- and belowground components. With the chemical treatment the allocation was quite balanced by both species. However, on both other treatments the species showed distinct behavior. There seemed to be

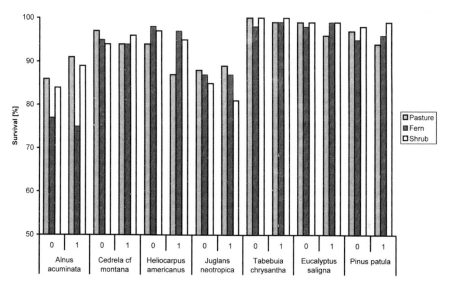

Fig. 34.2 Survival of five native and two exotic species on three sites of different successional states with two different treatments (*0* manual clearing before planting, *1* manual clearing before and every 4 months after planting) one year after planting ($n = 8400$)

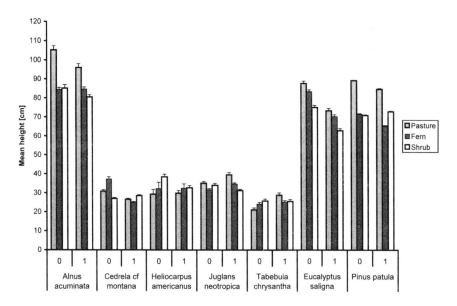

Fig. 34.3 Height of five native and two exotic species one three sites of different successional states and two different treatments (*0* manual clearing just before planting, *1* manual clearing before and every 4 months after planting) one year after planting ($n=8400$). Shown are means and standard error

Table 34.3 Mean height and root-collar diameter of *Tabebuia chrysantha* and *Cedrela montana* as well as total biomass of the ground vegetation and the forest plants under different treatments at 1 year after planting: *T0* manual clearing only before planting, *T1* manual clearing before and every 4 months after planting, *T2* glyphosate application before planting ($n=64$; after Eckert 2006)

Treatment	Compartment	*Tabebuia*	*Cedrela*
T0	Mean height (cm)	26.4	33.9
	Root collar diameter (cm)	1.0	1.2
	Forest plant biomass (g/m^2)	15	17
	Ground vegetation biomass (g/m^2)	1609	1723
T1	Mean height (cm)	28.4	29.3
	Root collar diameter (cm)	1.4	1.2
	Forest plant biomass (g/m^2)	36	13
	Ground vegetation biomass (g/m^2)	995	1075
T2	Mean height (cm)	56.2	49.4
	Root collar diameter (cm)	2.5	2.9
	Forest plant biomass (g/m^2)	297	200
	Ground vegetation biomass (g/m^2)	823	1022

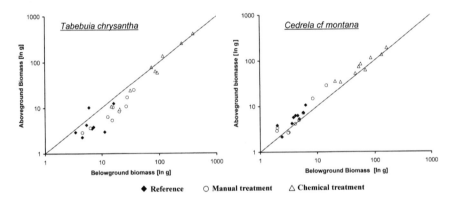

Fig. 34.4 Mean above- and belowground biomass of *Tabebuia chrysantha* (*left*) and *Cedrela cf montana* (*right*) under different treatments

a species-specific reaction to cope with competing ground vegetation: while *Tabebuia* invested more in the roots, *Cedrela* allocated more biomass aboveground. The results led us to the conclusion that root competition is the major factor influencing the early development of the planted trees. Griscom et al. (2005) in contrast concluded from similar experiments on pastures in the dry tropical zone of Panama that the positive effect of the herbicide application on survival and growth of tree seedlings was caused by increased light levels compared with non-herbicided plots. These obvious differences underline the request of Parotta et al. (1997) for a systematic evaluation of the effects of site preparation. Further research is also needed concerning the silvicultural characteristics and site requirements of the native species.

34.4 Enrichment Planting to Improve Productivity and Naturalness of Stands

As described in Section 34.2 secondary forests are often lacking valuable species. Consequently, enrichment planting could also be an interesting means to speed-up the successional processes on abandoned pastures or to enhance the "quality" of a secondary forest by promoting economically or ecologically more valuable timber species. In buffer zones of National Parks or conservation areas the enrichment of stands with rare or highly exploited species could also contribute to conservation. For instance in the region of the RBSF enrichment of secondary forests with *Podocarpus oleifolius* or *Prumnopitys montana* would be a valuable contribution to more natural and productive stands in the buffer zone of the Podocarpus National Park.

On severely degraded land exotics are often the only species that are able to tolerate the extreme site conditions and to restore a forest cover within acceptable time (Sabogal 2005). As several authors have already described (Carpenter et al. 2004a; Feyera et al 2002; Lugo 1997; Parrotta 1992) exotic plantations can also have a nurse effect for native species, thus facilitating the reestablishment of more natural stands and species richness.

34.4.1 Methods

In a *Pinus patula* plantation, adjacent to the reforestation experiment described above, we established an enrichment planting trial (Aguirre et al. 2006). In total, 648 individuals from nine native species were observed under two treatments: planting under closed canopy and in gaps of the pine plantation. Species tested were: *Alnus acuminata*, *Cedrela* cf *montana*, *Cinchona officinalis*, *Cupania* sp., *Heliocarpus americanus*, *Isertia laevis*, *Myrica pubescens*, *Piptocoma discolor*, and *Tabebuia chrysantha*.

34.4.2 Results and Discussion

After 12 months, survival under closed canopy was more than 90% for all species except *P. discolor* (80%) and *C. officinalis* (60%). These two species survived better in the gaps (100% and 70%, respectively). The development of the seedlings in terms of height growth was best in the gaps, with exception of *Cupania* sp. that grew slowly in any of the treatments (Fig. 34.5). Likewise in the main reforestation plots (Figs. 34.2, 34.3) *A. acuminata* showed the best growth performance, together with *M. pubescens*, *P. discolor* and *T. chrysantha*. Plants of all species that survived were generally healthier in gaps than under closed canopy.

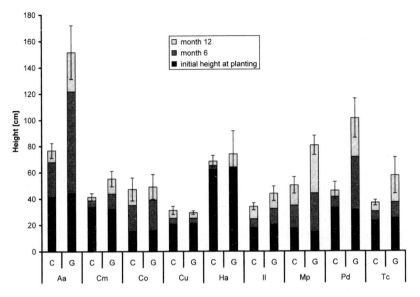

Fig. 34.5 Height increment of nine native species under the closed canopy (*C*) and in gaps (*G*) of a *Pinus patula* stand 12 months after planting. Shown are means and standard error. Species codes: Aa *Alnus acuminata*, Cm *Cedrela cf montana*, Co *Cinchona officinalis*, Cu *Cupania* sp., Ha *Heliocarpus americanus*, Il *Isertia laevis*, Mp *Myrica pubescens*, Pd *Piptocoma discolor*, Tc *Tabebuia chrysantha* (Aguirre et al. 2006)

A comparison of the height of the four native species planted on all experimental sites at RBSF one year after planting (all plots received the same treatment) revealed that the gaps in the pine stand offered the best conditions for all species (Fig. 34.6). We argue that the good results in the gaps in the pine stand were due to improved microclimatic conditions and reduced competition from the ground vegetation, while the demand of the seedlings for light as compared with the fern and shrub sites was not limited. These findings support Carpenter et al. (2004a) who pointed out that pre-preparation of badly degraded sites by planting pines and subsequent intermixing of shade-loving native trees may be more successful than planting such species directly onto open sites.

34.5 Conclusions

Silvicultural interventions can be a valuable tool to accelerate recovery of forest productivity and biodiversity on abandoned land and thus to improve the socio-economic situation of farmers in the tropical mountain rainforest region of Ecuador. A prerequisite for success is that the intended measures do closely consider the actual status and the environmental conditions of the target area. However, they

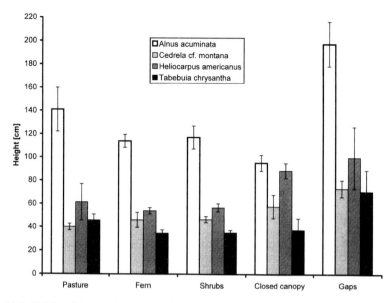

Fig. 34.6 Height of four native tree species 12 months after planting on different sites at the RBSF. Shown are means and standard error

must also consider the demands and financial conditions of the land owners and the temporal scale. From a user's point of view, natural regeneration can be an acceptable option only on areas where a permanent input of seeds is guaranteed due to low distances to forest edges or scattered trees inside the area that can serve as seed sources. It may also be adequate where rehabilitation of biodiversity is the prevailing objective and time is not a limiting factor. If fast recovery of the economic productivity is required it is necessary to invest in planting. To avoid financial losses it is essential to have fundamental knowledge on: (a) the silvicultural characteristics and ecological requirements of the tree species to be used as well as (b) the environmental conditions of the area.

As our results show seedlings of native species can also tolerate harsh conditions on abandoned pastures and partially even compete with exotic species in terms of survival and height growth. However, there is still a great need for further investigation. For instance on abandoned pastures efficient and ecological acceptable treatment of competitive ground vegetation (especially pasture grass or bracken fern) is a crucial factor. Deeper knowledge about the competitive characteristics of native species may help to improve seedling establishment and avoid high costs or losses. On areas where enrichment planting can be applied, knowledge on the light requirements of seedlings of different species and adequate spacing of the upper-storey is essential. At sites with extreme conditions exotics can also be a useful option to create a first forest cover and to act as "nurse trees" for subsequent enrichment planting of native species.

Chapter 35
Successional Stages of Faunal Regeneration – A Case Study on Megadiverse Moths

N. Hilt(✉) and K. Fiedler

35.1 Introduction

So far, little is known about the impact of habitat disturbance and subsequent regeneration on mega-diverse insect assemblages native to Andean mountain rainforests. The speciose insect order Lepidoptera provides an ideal target for such analysis. Butterflies and moths are probably the taxonomically best known insects. Moreover, lepidopterans are functionally connected in two-fold manner with the vegetation they inhabit: through the host-plant affiliations of their phytophagous larvae, as well as through the widespread need for plant-derived resources (e.g. floral nectar, rotting fruits, secondary plant compounds) during the adult phase of the life cycle. However, apart from a few recent studies on moths (for a summary, see Fiedler et al. 2007a, b), most available analyses referred only to the butterfly fauna (or some butterfly guilds such as frugivores; e.g. DeVries et al. 1997; DeVries and Walla 2001), even though moths comprise a far larger fraction of lepidopteran diversity than butterflies. Therefore, we chose two large moth families (Arctiidae, Geometridae) as models to study species diversity and turnover in a successional gradient at the edge of a natural montane rainforest in the Ecuadorian Andes. Species of both families tend to differ in their life histories, habitat fidelity, and resource requirements (e.g. Holloway 1984; Kitching and Rawlins 1999; Minet and Scoble 1999; Holloway et al. 2001). Thus we expected that arctiid and geometrid moth ensembles should respond with differential sensitivity to environmental change in the course of successional processes.

35.2 Methods: Data Collection and Processing

Moth ensembles from natural forest in the RBSF (for information on climate, topography and vegetation, see Chapters 1, 8, 10.1, 10.3 in this volume) have been extensively characterized (see Chapter 11.3 and references therein). We subsequently assembled data on moth faunas of successional stages using the same methods at 15 sites at the margin of the forest reserve. For most of these stages two replicate sites could be found in sufficient proximity with, however, a distance between the sites of at least 100 m. These sites represented seven different stages of vegetation succession, including two abandoned pastures (for details on sites and their topographic positions, see Hilt and Fiedler 2005). Six sites situated in the understorey of natural forest (two replicate sites at every 100 m elevation) represented late stages of vegetation development. Collectively, these 21 sites (situated at 1800–2005 m a.s.l.) depict a successional gradient from almost the bare ground of a landslide to a mature montane forest, including abandoned pastures as a common landscape element in the region. At all sites, vegetation appeared to be homogeneous around the position of the light trap in a radius of at least 50 m. Sampling sites were not identical to those studied by Beck et al. (Chapter 27) with regard to secondary vegetation for the following reasons:

1. We aimed to have two replicate sites of each successional stage.
2. Sites had to be large enough in terms of relatively homogenous vegetation, but also to be situated at some distance to each other, due to constraints of light-trap samples with regard to spatial resolution.
3. Light-trapping could not be performed in very dense thickets.
4. Sites had to be safely accessible also during night time.

For statistical analyses, sampling sites were grouped into three categories [early (E) without substantial woody vegetation, $N=10$ sites; late (L) with shrubs or young secondary forest, $N=5$; understorey of closed-canopy forest (F), $N=6$].

Moths were manually sampled after attraction to weak light sources (Brehm 2002; Brehm et al. 2005; Hilt 2005; see Chapter 11.3 in this volume). Between four to nine nightly catches from each site were pooled and analyzed. Sampling was restricted to periods without strong moonlight (McGeachie 1989). Collected specimens were later sorted and taxonomically identified (see Chapter 5 in this volume).

As a measure of local species diversity we used Hurlbert's rarefaction method for a standardized number of individuals (Arctiidae: 150 moths; Geometridae: 350 moths). Rarefaction is particularly useful if assemblages are sampled with different intensity or success (Gotelli and Colwell 2001; Schulze and Fiedler 2003). To analyze species turnover along the successional gradient we ordinated moth samples by non-metric two-dimensional scaling (Clarke 1993). To alleviate sampling effects (which are especially prevalent if sampling is not complete, if samples contain many rare species, if samples are of different size and diversity; Brehm and Fiedler 2004b) we used the 'chord-normalized expected species shared' index (Gallagher's CNESS index; Trueblood et al. 1994) as a measure of dissimilarity between samples. This measure also allows to explore the influence of commonness, or rarity, of component species of ensembles on similarity values, by adjusting the sample size parameter m from $m=1$ (emphasizing the role of dominant species) to $m=\max$ (emphasizing the information content of rare species; the maximum possible value is the smallest number of individuals sampled at any of the habitats to be compared). Ordinations were calculated separately for both families, but subsequently plotted in the same graph. The significance of faunal differences between predefined groups of sites was assessed by ANOSIM (Clarke and Gorley

2001) with 1000 random permutations. Primary data and more detailed analyses of the data on which the present chapter are based can be found in articles by Hilt (2005), Hilt and Fiedler (2005, 2006), and Hilt et al. (2006).

35.3 Results and Discussion

35.3.1 Species Diversity in a Successional Gradient

We sampled a total of 9211 arctiid individuals representing 287 species, and 23 720 geometrid individuals representing 868 species, in the narrow altitudinal belt at 1800–2005 m a.s.l. In both families more than two-thirds of the species could be identified at species level. Rarefied species richness was high at all sites (Hilt and Fiedler 2005; Hilt et al. 2006) and was always much larger in the Geometridae than the Arctiidae. Highest arctiid diversity was observed in the successional stages, with very little difference between early and late successional stages, while it was distinctly lower in the forest understorey (Fig. 35.1). Geometrid diversity increased distinctly from early to late successional habitats and decreased only marginally in mature forest understorey.

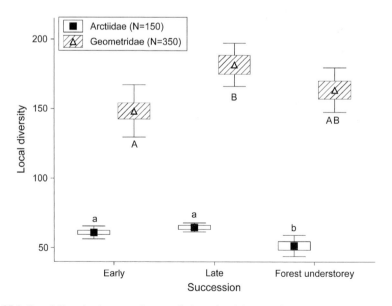

Fig. 35.1 Local diversity (measured as rarefied species richness) of arctiid and geometrid moths in the three different habitat types. *Boxes* labeled with different *letters* differ significantly at $P<0.05$ (one-way ANOVA, followed by Scheffé test; Arctiidae: $F_{2,18} = 9.14$, $P<0.005$; Geometridae: $F_{2,18} = 6.24$, $P<0.01$). *Squares, triangles* Means, *boxes* ±1 SE, *whiskers* = ±1 SD

35.3.2 Moth Diversity in Successional Habitats – A Sensitive, but Idiosyncratic Indicator of Environmental Change

Studies from lowland habitats in other tropical regions suggest that Arctiidae moths may often benefit from anthropogenic habitat disturbance (e.g. Kitching et al. 2000; Schulze 2000). We observed similar patterns for the Arctiidae in the montane forest zone of southern Ecuador (see also Hilt 2005; Hilt and Fiedler 2005). Diversity was high in early successional habitats, even slightly higher in later successional stages, and apparently decreased with forest recovery. In the Geometridae, highest diversity was likewise seen in late successional stages, yet the early successional habitats were clearly impoverished.

At first sight, the intermediate disturbance hypothesis (Connell 1978) might provide an explanatory framework for this pattern. This hypothesis implies that species diversity is highest in communities experiencing intermediate levels of disturbance. Thus, early as well as very late stages of forest recovery should harbor lower species diversity. However, Didham (1997) pointed out that an increase in insect abundance and diversity at forest edges often results from the invasion of generalist species from disturbed habitats outside the forest (Kotze and Samways 2001), also termed the mass effect (Shmida and Wilson 1985; for a review of edge effects, see Murcia 1995). We have no indication that the high moth diversity of late successional habitats were due to such invasions by dominant species that originate from disturbed or devastated habitats. Rather the opposite was true: many forest moth species occasionally showed up in samples at successional sites, as a consequence of their generally high dispersal capacity at the scale of our study (Ricketts et al. 2001). Note that even the most strongly disturbed habitats were situated at a distance of less than 500 m to the natural forest.

Differential dispersal capacity is certainly one of the keys to understand the differences we observed between the two moth families. Many arctiid species are rather large, robust insects and therefore have a good flight capacity (e.g. Kitching and Rawlins 1999; Solis and Pogue 1999; Hilt 2005). In contrast, geometrid moths have more slender bodies, tend to be weaker fliers, and exhibit relatively high habitat fidelity (Holloway 1984). Our results from the montane forest zone in Ecuador thus perfectly match results from other tropical realms (SE Asia, Africa, Australia) according to which Geometridae respond more sensitively to changes in habitat quality (e.g. Chey et al. 1997; Intachat et al. 1997, 1999; Willott 1999; Kitching et al. 2000; Beck et al. 2002; Axmacher et al. 2004a, b) than do many other moths (see also Fiedler et al. 2007a, b).

Other studies indicate that the effects of forest disturbance on species diversity are heavily scale-dependent (Hamer and Hill 2000; Hill and Hamer 2004). At small spatial scales, as in the present study, habitat disturbance indeed frequently increases diversity, whereas at larger scales there is higher diversity in undisturbed forests (Hamer and Hill 2000). For Andean tropical montane forests this still needs to be explored at larger spatial scales.

Two further issues to be considered are resource requirements and stratification. There is increasing evidence that a distinct canopy fauna exists for the Arctiidae

(SE Asia: Schulze et al. 2001; Costa Rica: Brehm 2007), whereas no clear stratification is known from the Geometridae (Beck et al. 2002; Brehm 2007). Thus the apparent decrease of arctiid diversity in the forest understorey may also be influenced, by an unknown degree, by the restriction of our sampling (enforced by logistic constraints) to the understorey layer. Moreover, the larvae of many geometrid species are bound to feed on woody plants, whereas in the Arctiidae herb feeders are more prevalent.

So, diversity patterns in the two moth families considered here do probably not reflect the intermediate disturbance hypothesis in a 'simple' way. Rather, they are the synergistic effect of dispersal capacity, resource requirements, and microhabitat preferences. Since all these are highly variable across insect taxa, it is not surprising that patterns are taxon-specific (even at lower systematic levels; Hilt and Fiedler 2006) and generalizations are critical. Irrespective of these idiosyncrasies, late successional habitats close to natural forest turned out to be valuable sites for both moth families, and such habitats therefore deserve more attention in conservation and management (e.g. as buffer zones around conservation areas, or when aiming at sustainable methods for land use).

35.3.3 Species Composition in a Successional Gradient

Ordination of moth ensembles based on the CNESS index revealed a concordant picture of species turnover for both families (Fig. 35.2). Faunas of successional sites were significantly separated from those of mature forest understorey (ANOSIM; Arctiidae: E vs F: $R=0.664$, $P<0.005$, L vs F: $R=0.395$, $P<0.05$; Geometridae: E vs F: $R=0.814$, $P<0.005$, L vs F: $R=0.492$, $P<0.005$). These effects were stronger in Geometridae than Arctiidae, as shown by the higher values of the test statistics R. Arctiid and geometrid moth samples from the natural forest understorey formed a distinct group which also reflected the altitudinal gradient from the lower sites (3a, 3b at 1800 m) to the higher ones (5a, 5b at 2000 m). A significant separation between the moth ensembles of earlier and later successional sites was only found when the common or moderately abundant species were emphasized, for example by setting the sample size parameter of the CNESS index at $m=1$ (Arctiidae: $R=0.551$, $P<0.005$; Geometridae: $R=0.403$, $P<0.005$; see also Hilt and Fiedler 2005; Hilt et al. 2006). At high levels of the parameter m, i.e. giving larger weight to rare species, this differentiation tended to be obscured, probably due to the very large fraction of rare species that uniquely occurred only in single samples. Moth ensembles at the two abandoned pastures were the most impoverished, and had the most deviant species composition.

Post-hoc analyses revealed that the first ordination axis correlated with environmental factors that mirror the degree of habitat disturbance and regeneration (e.g. visible sky on hemispherical photographs as a measure of canopy closure, or distance of the sites from the natural forest). Samples were ordered along the second dimension according to the elevation of the sites (Hilt and Fiedler 2005; Hilt et al. 2006).

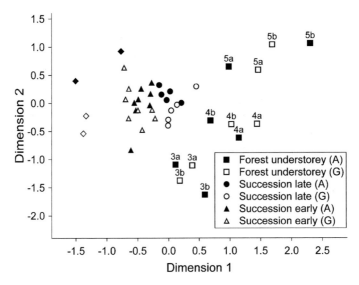

Fig. 35.2 Non-linear two-dimensional scaling plot of the CNESS dissimilarity matrix for arctiid (*A*) and geometrid (*G*) moths with sample-size parameter (*m*) set at an intermediate level (Arctiidae: $m=76$; Geometridae: $m=170$). Ordinations were calculated separately for both families, but subsequently plotted in the same graph. Stress values of ordinations (as a measure of goodness-of-fit between original data matrix and representation in reduced ordination space) were 0.13 for arctiids and 0.12 for geometrid moths. *Numbers* at forest samples refer to elevational levels: *3* 1800 m, *4* 1900 m, *5* 2000 m. *Filled symbols* Arctiidae, *open symbols* Geometridae

35.3.4 Species Turnover During Succession – Striking Parallels Despite Ecological Diversities

Patterns of beta diversity along the successional gradient were unexpectedly similar, despite the many morphological and ecological differences between these moth families. Faunal differentiation was highly concordant at all tested values of the sample size parameter *m* of the CNESS distance coefficient. At high *m* values, i.e. when emphasizing the contribution of rare species, these relationships tended to be even more pronounced (Hilt 2005).

In both families, there was a distinct separation of ensembles at successional sites from those in mature forest understorey. When emphasizing the more common species, the differences between early and late successional habitats were also mirrored with good resolution. Yet, patterns of faunal change were not completely identical between all moth taxa and showed some subtle variation at systematic levels below the family (Arctiidae: Hilt and Fiedler 2006). A clear discrimination among lepidopteran assemblages along habitat gradients was also found in an increasing number of studies in all tropical realms (Kitching et al. 2000; Beck et al. 2002; Schulze 2002; Axmacher et al. 2004b; Fiedler and Schulze 2004; Brehm and

Fiedler 2005). Hence, species turnover in moth communities along such gradients can be a very useful and sensitive measure to assess effects of habitat change on mega-diverse assemblages of plant-feeding animals.

Patterns of beta diversity were best predicted by the degree of habitat openness, and to a lesser extent by elevation. Geometrid moths responded more sensitively than arctiid moths regarding to habitat openness, and statistical differentiation between samples (measured as the R statistics of ANOSIM) was always higher for the Geometridae (Fiedler et al. 2007a, b). Yet, in neither family were faunal differences related to geographical distances between the sampling sites. Thus, even on the small spatial scale of this study, the community differences observed can be related to local ecological effects and are not just a side-effect of dispersal from source areas (such as the neighboring natural forest).

35.4 Conclusion

Tropical arctiid and geometrid ensembles were found to sensitively reflect ecological differences between habitats along the successional gradient. The picture was clearer for geometrid moths (see also Intachat et al. 1997, 1999), and patterns of local diversity were less similar across taxa than those in species turnover. Similar observations were made in the same Ecuadorian study area along an altitudinal gradient through natural forest, where three moth families (Arctiidae, Geometridae, Pyraloidea) and all their larger subordinated taxa revealed idiosyncratic patterns of alpha diversity, but highly concordant patterns of beta diversity (see Chapter 11.3 in this volume). These results indicate that studies of species turnover might generally be more informative in the assessment of environmental effects on 'biodiversity' at the community level, whereas local species diversity, or richness, is more heavily confounded by the specific ecological requirements of the organisms under study. This idea deserves further study in a range of habitats and targeting a broader range of organisms.

Acknowledgements We thank Dirk Süßenbach and Gunnar Brehm, who provided data from their samples from natural forest. Doreen Fetting, Jörg Hager, Claudia Knake, Georg Petschenka, Claudia Ramenda, Caroline Schulze, Annick Servant, Anna Spengler, Katja Temnow, and Anne Walter (all at the University of Bayreuth) helped with sampling and processing moths. Giovanni Onore (Pontifícia Universidad Católica del Ecuador, Quito) and Christoph L. Häuser (Staatliches Museum für Naturkunde, Stuttgart) supported the project logistically. Christoph L. Häuser, Martin R. Honey, Linda M. Pitkin (Natural History Museum, London), and Axel Hausmann (Zoologische Staatssammlung, Munich) provided access to reference collections under their care. Suzanne Rab-Green (American Museum of Natural History, New York) and Benoît Vincent (Paris) kindly identified some species. The Ministerio del Medio Ambiente in Ecuador issued a research permit (No. 002-PNP-DBAP-RLZCH/MA) and the foundation Nature and Culture International (Loja, Ecuador) allowed access to the study sites and their facilities.

Chapter 36
Gradients in a Tropical Mountain Ecosystem – a Synthesis

E. Beck(✉), I. Kottke, J. Bendix, F. Makeschin, and R. Mosandl

36.1 Climate is a Main Driver of Natural Gradients

36.1.1 Global and Regional Atmospheric Forcing

The climate of the research area is mainly influenced by the easterlies of the trade wind regime (see Chapter 8 in this volume) and thus by the Atlantic circulation patterns. The Pacific circulation patterns play a minor role but associated westerlies modify the regional weather in the study area, especially in the austral summer. Due to the dominance of the Atlantic circulation, the Pacific ENSO phenomenon does not cause extraordinary anomalies, whereas La Niña situations induce somewhat stronger effects (see Chapter 20). Weather extremes in the study area are initiated by aperiodic but regularly occurring circulation patterns. Cold air surges (*friajes*) from the south can cause severe plunges in air temperature, which in turn can affect plant life. One unexpected finding is the marked impact of nocturnal mesoscale circulation processes in the near Amazon on cloudiness and rainfall in the Rio San Francisco valley. Also remarkable is the strong spatial gradient of humidity between Loja and the RBSF, over a beeline distance of less than 20 km where the crest of the eastern Cordillera is a clear weather divide. While rainfall at RBSF peaks in austral winter, the basins west of the main Cordillera are characterized by two rainy seasons peaking in March and October, respectively. The rainfields that generate the perhumid climate of the RBSF cannot reach the basin of Loja because they are released from relatively shallow orographic cap clouds in the upper part of the RBSF area, which are not able to pass the main ridge of the eastern Cordillera. The summer rains in the austral summer, especially in the basins of Loja and Vilcabamba, are related to a westerly stream flow initiated by the Pacific circulation.

Ionic concentrations of rain and fog precipitation water in the RBSF area are generally low. But global and to a lesser extent regional circulation processes are also responsible for substantial atmospheric matter intake, mostly via wet deposition along the atmospheric pathway (see Chapter 22). Biomass burning in the Amazon is the major source of atmospheric matter intake via the easterlies. This remote source especially provides significantly enhanced concentrations of nitrate, ammonia and sulfur whereas westerly stream flows in austral summer deposit higher amounts of sulfur, sodium and chloride, most probably from the Pacific. Irregular peaks of sulfur deposition can be associated with remote volcanic activity. Due to the altitudinal distribution of precipitation, matter input increases, however not linearly towards the upper parts of the ecosystem.

36.1.2 Local Climatic Gradients in the Rio San Francisco Valley

Generally, the meteorological elements in the RBSF area reveal the expected changes with increasing elevation (see Chapter 8): a decrease in air temperature and an increase in humidity, wind speed, cloudiness and precipitation. However, these gradients are not simply linear but are clearly modified by the topographic situation of the Rio San Francisco Valley. Because the Cordillera del Consuelo shields the valley from the easterlies, a diurnal valley-breeze system is well established at the lower levels. The shielding effect decreases with altitude and a sudden dramatic transition to high wind speeds and condensation rates towards the crest level takes place. Advective orographic cap clouds and cloud combing by the vegetation is a major water source for the upper parts of the ecosystem, causing a clear surplus of annual precipitation. A remarkable interaction between mesoscale atmospheric dynamics and the altitudinal gradient of precipitation was observed: The valley bottom obtains an extraordinary peak of rainfall around sunrise, not well known for the Tropics, whereas a second daily peak in the late afternoon originates from thermally driven convection. The afternoon rainfall increases with the elevation due to evaporation of rain droplets at the valley bottom, while the morning rainfall is nearly constant over the entire altitudinal range.

36.2 The Altitudinal Gradient of Biodiversity

36.2.1 Plant Life

An evergreen mountain rain forest extends from the valley bottom up to the tree line, which in the RBSF may reach as high as 2850 m a.s.l. This forest was investigated along the entire altitudinal gradient, using a combination of floristic and structural characters. Classification of the forest types follows the general zonation scheme for tropical mountain forests, but provides further differentiation for subunits according to altitude, disturbance and relief situation. Seven major types have been recognized around and in the RBSF area, six of which are evergreen and one semideciduous (see Chapter 10.1). The lowest is a premontane evergreen rainforest (ca. 800–1300 m a.s.l.), followed by lower and upper montane forests (two formations) and an elfin-type woodland which merges into the shrub and dwarf bamboo páramo above the timberline. In total, up to now 1177 species of vascular plants and 525 species of bryophytes have been recorded in the Reserva in an area of 11.2 km². A more or less linear decrease in plant species diversity with altitude applies to the terrestrial plants while the diversity of the epiphytes shows a maximum at 2500 m. The elsewhere rare species *Purdiaea nutans* is the dominant tree between 2150 m and 2600 m due to a unique combination of topographic, geological, pedological, climatic and historical factors. The low and open canopy of this forest promotes the establishment of a rich cryptogamic flora (ferns, bryophytes, lichens) with numerous rare taxa (see Chapter 19).

The timberline ecotone extends between 2800 m and – on rocky outcrops –3300 m a.s.l. (see Chapter 10.4) and thus is much lower than usually in Ecuador. Stormy and nearly permanent easterly winds hinder the growth of trees on the ridges and give rise to a dwarf forest. About 20 tree species constitute this tree line ecotone – an astonishing number taking into account that commonly only two or three tree species (e.g. *Polylepis, Gynoxis*) form the timberline in the neotropics. However, in the Amotape–Huancabamba depression (Weigend 2002, 2004) the mountains do not reach into the altitudinal range of these species and therefore the existence of a real tree line of the elfin forest is an effect of the permanent strong winds.

Arbuscular mycorrhizas are the far-dominant root symbioses of the woody species along the altitudinal gradient (see Chapter 10.5). Ectomycorrhizas occur only with a few Nyctaginaceen species which grow azonally in ravines and along the San Francisco river banks. The altitudinal distribution of mycorrhizal types is thus unlike those in temperate climates where ectomycorrhizas dominate the forests at higher elevations.

36.2.1.1 Effects of Climate on the Forest

Tree ring chronology is a potent method for reconstructing the historic climate. In the humid tropics, little is known about the seasonal formation of wood tissue and their relation to environmental triggers. This holds in particular for southern Ecuador and the RBSF. The first reliable results of dendrometry in the RBSF underline the very complex patterns of different ecological conditions (see Chapter 21). Marked terrain-related differences in radial growth were observed for individual tree species (e.g. *Piptocoma discolor*) with high growth rates in protected gorges and minute increments at wind-exposed crests (see also Chapter 18). On a standing crop basis, mean annual increments expectedly decrease with elevation. This phenomenon can partly be attributed to a change in species composition, as the fast-growing tree species reach their upper limits at around 2200 m. But growth rates also diminish with elevation due to the decreasing temperatures and an increasing shortage of nutrients (see Chapter 17). Some tree species display distinct growth seasonality. Interestingly, growth is almost completely suspended during the months receiving extreme rainfall, which may be an outcome of low biomass production due to cloudiness and a very low light intensity (Wright and van Schaik

1994). This contrasts with the situation in the dry area west of the main Cordillera where wood formation (e.g. of the long-living conifer *Prumnopitys montana*) takes place mainly during the rainy periods.

36.2.2 Insects and Birds

The decrease in biodiversity with increasing elevation holds also for animal life. However, the rate of decrease is neither continuous nor identical in all groups of animals. Nearly one-third of the bird species were recorded in a narrow altitudinal range between 1900 m and 2000 m a.s.l., with strongly decreasing numbers at higher elevations (see Chapter 11.1). When the large number of identified moth species (2129 spp from three families) were lumped together, a faunal attenuation at higher elevations was obvious. However, different families showed strikingly divergent altitudinal diversity patterns. The poor concordance in alpha diversity patterns undermines the concept of biodiversity indication, i.e. extrapolation of species richness estimates from one group of organisms to others (Brehm and Fiedler 2003; see Chapter 11.3). A very specious moth family is that of the Geometridae. Brehm et al. (2005) recorded 1266 morphospecies between 1000 m and 2700 m a.s.l. – a higher number than observed anywhere else in the world. Interestingly, the family did not show a pronounced elevational peak of diversity but a very broad plateau with a high and regular species turnover along the gradient (Brehm et al. 2003a, b). In contrast, pyraloid moth richness peaked around 1000 m a.s.l. (Bombuscaro) and declined with increasing elevation (see Chapter 11.3). A high concordance in altitudinal species turnover suggests that the climatic and biotic gradients associated with altitude affect moth communities in much the same way, despite the multitude of life-histories of their individual component species.

Birds are the best known group of animals in Ecuador. The Eastern escarpment of the Andes is renowned for its outstanding bird diversity (Rahbek et al. 1995; Ridgely and Greenfield 2001) and Paulsch and Müller-Hohenstein (see Chapter 11.1) observed a total of 227 bird species between the years 1999 and 2002 in the RBSF. The number increases considerably when lower and higher elevations are included. Rasmussen et al. (1994) recorded a total of 362 bird species along the (old) Loja–Zamora road (1000–2800 m) that passes the RBSF, 292 species from the Rio Bombuscaro area (950–1300 m) and 210 species from the Cajanuma area (2500–3700 m a.s.l.).

36.2.3 Soil Organisms and Biotic Activities

From soil organisms, oribatid mites and soil microorganisms were investigated within the gradient study. While density of oribatid mites at 1000 m were comparable with other tropical forests, values decreased surprisingly with altitude to extremely low numbers at 3050 m (see Chapter 11.4). This contrasts with the substantially increasing thickness of organic layers with altitude (see Chapter 15). Soil microbial biomass showed a similar negative trend only in the top litter layer, whereas in deeper organic layers an elevational gradient was not observed (see Chapter 14). PLFA analysis did not reveal substantial structural changes in soil microbial consortia (see also Chapter 11.4). As organic matter accumulated in higher altitudes, soil CO_2 efflux decreased (see Chapter 14). Soil water content and soil temperature were identified as the most important abiotic factors for soil life, but nutrient limitations may be likewise crucial (see Chapter 11.4). This assumption is supported by Wilcke et al. (Chapter 9) who noticed systematic negative changes in soil properties and thus in soil fertility along the elevational gradient.

36.2.4 Lacking Knowledge of Biotic Components of the Ecosystem

An excellent knowledge about diversity has been obtained in the RBSF for some groups, e.g. bats, birds, arctiid and geometrid moths, cryptogamic plants and trees (see Chapters, 10.1, 10.3, 10.4, 11, 11.1, 11.2, 11.3,). However, large gaps still remain. Important and species-rich groups such as beetles, ants, wasps, bees, dipterans and many other arthropods, vertebrates and molluscs as well as most of the soil organisms including the fungi could not be investigated in detail hitherto. A great portion of these species are apparently new to science. Megadiverse ecosystems such as tropical mountain forests in Ecuador pose major methodological challenges: Species-richness, difficulties in the identification and incomplete samples are among the major issues (see Chapter 5).

36.3 Ecosystem Aspects of the Biodiversity Hotspot

The Andean tropical rainforests are recognized as global diversity hotspots for vascular plants (Barthlott et al. 2005). However, this notion does not hold in the same way for all kinds of organisms which have been studied so far. Mosses, birds, bats and some groups of insects (e.g. certain families of moths) are likewise extraordinarily specious (Brehm and Fiedler 2003), but others (e.g. soil arthropods) are not outstanding diverse (see Chapter 11.4). Those may peak at elevations lower than the range of the RBSF. However, other groups of organisms have not yet been studied and may be likewise specious.

36.3.1 Latitude and Altitude

Many factors which are effective at various scales must accrue for the achievement of a biodiversity hotspot. On the zonobiome scale Rapoport's rule (Stevens 1989, 1992; Fernandez and Vrba 2005) may explain the increase in alpha diversity in the humid tropics, and adding the vertical dimension (Stevens 1992) of the Andes will further enhance this effect. There is a high species turnover along the altitudinal gradient, peaking usually at medium elevations (Krömer et al. 2005; Rahbek 2005; see Chapter 2). Narrow habitat areas facilitate more species to coexist while the ranges become wider and the species diversity smaller with increasing latitude or altitude. Temperature-dependent upper limits of species-specific areas have been detected for several trees in the RBSF, e.g. *Piptocoma discolor* and *Vismia tomentosa* (Bendix et al. 2006c).

36.3.2 Terrain, History and Spatio-Temporal Dynamics

To become a hot hotspot of biodiversity other factors must modify the general traits and thus increase diversification. On the zonobiome scale one of these factors is known as the Amotape–Huancabamba Zone (see Fig. 10.4.1 in this volume). This region has been reported as a zone of particular diversity of vascular plants with a special set of endemics (Weigend 2002, 2004). Its biodiversity is higher than that of the northern and southern Andes. The Amotape–Huancabamba Zone can be considered as an ample species-rich corridor where plant groups from the northern and the central Andes coexist. Moreover, the depression of the Andean massif facilitates also a migration of species from the drier western range to the humid eastern escarpment and vice versa. On the landscape scale, the extremely rugged terrain of the eastern Cordillera in South Ecuador provides a great variety of habitats, which in turn results in a fragmentation and isolation of populations and a boosted genetic drift, as in the example of the recent radiation of orchids of the genera *Teagueia* and *Scelochilus* (Gentry and Dodson 1987; Jost 2004). Habitat changes were driven mainly by a climate change since the late Pleistocene, when a grass-Páramo covered the higher regions of the eastern Cordillera. Pollen analysis showed a marked increase in tree and shrub pollen during the transition of the late Pleistocene to the Holocene period (see Chapter 10.2), when atmospheric warming allowed the development of an upper mountain forest. The current Subparamo vegetation, e.g. at the El Tiro Pass became established during the mid- to late Holocene. On a shorter time-scale the dynamics of a mosaic climax ("dynamic equilibrium model"; Huston 1994) stabilizes the coexistence of successional stages with their functional plant types and thus enhances biodiversity. Confirming the altitudinal dimension of Rapoports law, gap dynamics in the lower and the upper montane forest differ considerably. Whereas the classic tropical succession is found at lower elevations, typical pioneers are missing in the upper forest (see Chapter 23). Succession also takes place at a larger scale in the research area when the heavy rains on the extremely steep slopes cause an extent of landslides which is unique in the tropics (see Chapter 24). Another effect of the high amount of precipitation is permanent water soaking of the substrate, which favors plants with shallow root systems.

36.3.3 Resource Limitations

A quite different factor that stabilizes the outstanding biodiversity can be attributed to the limitation of resources, in particular macronutrients ("soil nutrient hypothesis"; Kapos et al. 1990; Huston 1994; Woodward 1996) and light. Wilcke et al. (Chapter 9) verified this hypothesis, showing an increase in soil organic matter, carbon stocks and C/N ratios with altitude, while mineralization rates and mean concentrations of all macronutrients were negatively correlated with elevation. The scarcity of abiotic resources prevents the dominance of particular species by excessive biomass production and concomitant suppression of others. Even the organic layer appears rather poor in nutrients (see Chapters 13, 14) wherefore the establishment of mycorrhizae is of particular importance especially for perennial plants. High colonization rates were not only found in arbuscular mycorrhizas of nearly all trees at all altitudes but also in hemiepiphytic ericads, epiphytic orchids and ferns and even in terrestrial liverworts (see Chapters 10.5, 27). The density of litter-decomposing animals in the studied forest ecosystem is generally low and declines with altitude (see Chapter 11.4). Especially the scarcity of primary decomposers is striking, taking into consideration that the structure of the soil food web otherwise resembles that of temperate forests. Presumably, low litter quality and high energy demand restrict decomposer communities in tropical mountain rain forests.

Shortness of light favors the plant life forms lianas and epiphytes which considerably contribute to the biodiversity of the area. Favored by the perhumid climate the wealth of epiphytic species is staggering: a maximum of 98 epiphytic species have been counted on a single tree (Werner et al. 2005)! As one of the very important results of the research reported here, particular plant and animal key species could not be identified in the natural forest. Rather, a variety of up to 160 tree species share one hectare with only very few individuals of any one species present (Martinez 2006). Scattered populations of discrete tree species in a dense tropical forest pose the question of the specificity of their biotic counterparts, e.g. pollinators, seed dispersers and mycorrhiza fungi. Vice versa the outstanding diversity of insects, e.g. moths, should also be reflected by a similar diversity of fodder plants. Although several specific interactions have already been detected, e.g. bats or humming birds with particular flowers (Wolff 2006), biotic interaction is still a field for further investigation (see Chapter 2). Seasonally emerging and cropping insect populations may find specific counterparts in seasonally flowering, fruiting or leaf-flushing plant species. However, vertebrates must adjust to the vegetation's seasonality (Cueva et al. 2006), i.e. must develop a broader spectrum of less specific interactions. It should be borne in mind that it is the multitude of rather weak biotic interactions which stabilizes the outstanding biodiversity of the ecosystem.

36.3.4 Biodiversity Feedback on Ecosystem Processes and Stability

Ecosystems circulate interior matter and energy mainly via food webs and exchange both with their surroundings. In an anthropocentric perspective, special features of an ecosystem, the so-called ecosystem services, are of particular interest, e.g. productivity and carbon sequestration, mineralization, soil formation, water filtration and alleviation of erosion. Humans as components of the biosphere and players in an ecosystem contribute to and profit from these services, and they are scared by a potential change of the services, which apparently are exacerbated by the processes known as "global change". Therefore the question of the strength or resilience of an ecosystem in response to human impact is gaining more and more importance.

As most ecosystem processes are accomplished via food webs, the question is whether biodiversity itself (i.e. the complexity of the foodwebs and their interactions) plays a significant role in the functioning of an ecosystem; in other words: whether there is a feedback of the degree of biodiversity, for example on element cycling and matter balance and thus on the services of a given ecosystem (Ulrich 1987; Ernst et al. 2000). There are theoretically at least two categories of ecosystem components: the essential ones and apparently redundant ones (see Gitay et al. 1996). The latter, however, may either facilitate the ecosystem processes (a greater variety of litter decomposers may accelerate this process more than a high population of only one type) or may be considered as backup components or stabilizers which become essential in various extreme situations. Schulze compared the components and functioning of an ecosystem with the parts of a car and their intended function upon normal use and in emergency situations (Schulze 1989; Schulze et al. 1999). Of course ecosystems must be more dynamic in composition and internal functions than a car, in order to follow the variation in their abiotic elements, e.g. with the seasons. However, adaptedness to a recurrent, even though fundamental change in the environmental conditions must be differentiated from the resilience of an ecosystem to unpredictable changes in its abiotic setting and biotic components.

Especially in the case of a biodiversity hotspot the question arises which species are necessary (and sufficient) for the "normal" functioning of the ecosystem and which may function in alternative processes or bypasses ("exchangeability" of species), which act as emergency components and which may be estimated as actually redundantand dispensable. Before trying to assess this question, we should appreciatively admit that all these species are just there, in smaller or larger populations, irrespective of whether we are presently able to attribute to them a value for the ecosystem. And we must admit as well that their presence has reasons rooted in history, evolution, adaptation, competitive strength and so on. Some of these reasons may be connected with more obvious functions, while others may not.

To typify an ecosystem the flux of matter (including water) and energy is commonly used, and the balance of these fluxes is considered an indicator of its stability, in particular with respect to the above-mentioned ecosystem services. However, characterization of the ecosystem is by its biotic components and their interactions, i.e. by the interactions of individuals in populations. These interactions may be strong in an ecosystem which is poor in species, but may appear weaker and more variable in a highly species-rich ecosystem. A nectar-licking bat (three of at least 21 bat species in the RBSF belong to this guild; see Chapter 11.2) must change its diet during the course of the year, as most of the perennial plants (e.g. its fodder plants *Macrocarpia* and several species of Marcgraviaceae) show a pronounced seasonality of flowering. Weak interactions, however, may often be asymmetric. If the fitness of the plant species relies on pollination by a specific bat (strong interaction[1]), disappearance of the bat will be detrimental for that plant species, whereas extermination of the plant will not necessarily drive the bat out of the ecosystem, as the high diversity of alternative plant species may be sufficient for its alimentation (weak interaction). A high phytodiversity therefore favors weak biotic plant–animal interactions, i.e. counteracts a too-tight specialization, because of the high risk of missing the right partner in a low-density population. In line with this notion, recent work on host-plant relationships revealed that at least in lowland forests tropical herbivores are not generally more specialized than temperate zone ones, supporting the preponderance of weak and asymmetric interactions (Novotny et al. 2006). Asymmetry of interactions, however, weakens the concept of redundant species, as from the viewpoint of "species 1" a high biodiversity may stabilize its fitness via weak interactions, whereas for "species 2" a strong interaction, i.e. a large population concomitant with a low diversity, may be crucial.

As discussed above, keystone species with an above-average function could not be identified in the RBSF, but a pronounced change in the species composition along the altitudinal gradient was recorded. For example, important tree species like *Vismia* and *Piptocoma* reach their upper limits within the forest zone of the RBSF (Bendix et al. 2006c). The same applies for many animals that have been studied in the elevational gradient around the RBSF (for birds, see Chapter 11.1;for insects see Chapter 11.3). What does the lack of these species in the upper mountain rain forest mean for the ecosystem? Are their roles taken over by other species that are better adapted to lower temperatures (exchangeability of species), or is their presence or absence indicative of the attenuation of the biological diversity with the elevation resulting in an altitudinal change in the ecosystem?

The studies of Wilcke et al. (Chapter 13), Iost et al. (Chapter 14) and Moser et al. (Chapter 15) showed an accumulation of organic soil material at higher elevations which cannot only be attributed to a temperature-dependent decrease in the mineralization rate, but is also caused by poorer litter "quality", i.e. a lower nutrient content as indicated by an increasing C/N ratio. This concurs with a decrease in the population density and species composition of litter decomposers and points to a change in the ecosystem rather than to a higher resilience (buffer capacity) of the lower mountain forest ecosystem. As the lower mountain forest is richer in species (see Chapter 23) the higher biodiversity of the lower mountain forest leads to a faster flux of the resources through the ecosystem which in turn favors the diversity of plant species.

---

[1] The flowers of these plants open only for one or two nights. While bats are the only pollinators during the night, hummingbirds may substitute for bats, if a flower is open for more than one night. Nevertheless the interaction between the plant and its (main) pollinator is rather specific (Wolff 2006).

At least in this case, a feedback of the biodiversity as such on the ecosystem processes can be inferred. An assessment of whether this effect stabilizes the ecosystem "tropical lower mountain rain forest" still lacks any empirical data. However, the fast recovery of a highly diverse secondary forest in that area (see Chapter 32) inhabited by an impressive diversity of herbivorous insects (see Chapter 35) suggests a substantial resilience of that ecosystem at least against small-scale disturbances, i.e. as long as recolonization by organisms lost through disturbance is easy.

As yet, we do not have experimental data and therefore all conclusions are drawn from correlations. Experiments on the feedback of biodiversity on ecosystem processes have been performed with herbaceous species (compiled by Schläpfer and Schmid 1999; Schläpfer et al. 1999) but are completely lacking for woody vegetation types. In addition, such experiments can be done only with a relative small number of species and on small plots. For a mega-diverse ecosystem like the tropical Andean mountain forest with its different elevational types we are still far from understanding how its tremendous species richness (and the concomitant, largely unknown multitude of functional interactions between the organisms as carriers of and actors in ecosystem processes) feeds back on emergent properties at the ecosystem level.

36.4 Gradients of Disturbance, Succession and Regeneration

36.4.1 Disturbance as a Driving Factor

The natural ecosystem contrasts in many vital traits with its anthropogenic substitutes, the pastures and the abandoned farming areas. From the viewpoint of ecosystem services like microclimate, soil structure and soil organic matter turnover, the active pastures are in the focus of the interest, whereas with respect to biodiversity this man-made vegetation (Hartig and Beck 2003) is of minor significance. However, the abandoned pastures and other areas where the forest has been destroyed by escaping fires represent territories where the course of biodiversity regeneration can be traced in all its facets. Irrespective of the state at which regeneration commenced, a major difference to the natural ecosystem is the dominance of a few key species which determine the course of the successional process. On small-scale disturbances, e.g. along trails where fire has not been used for forest clearing, a succession can be observed which is somehow similar to that on natural landslides (Ohl and Bussmann 2004; see Chapters 24, 27, 32). Conforming to the key species concept and in contrast to the primary forest, all types of secondary forest are characterized by representatives of the genus *Chusquea* which may even be the dominant species. *Chusquea*, which is a taxonomically extremely difficult genus, occurs in several plant life forms, mainly as tall lianas, but also as bushes or trailing herbs (Martinez 2006). In agreement with the moderate disturbance theory such secondary forests are extremely rich in plant species, but also in specific groups of animals: The results for the species-rich moth families of Arctiidae and Geometridae show that patches of regenerating forests can have high importance for maintaining high biodiversity (see Chapter 35) and that ecotones at tropical forest margins, due to their high niche heterogeneity, are often very biodiverse habitats (Ricketts et al. 2001). This also holds particularly true for birds: At lower altitudes, forest types under human influence and succession stages are extraordinary rich in bird species (see Chapter 11.1). Functional relationships between the distribution of bird species and the vegetation structures were found in different forest types along the gradient of land-use. As a consequence of an ongoing habitat conversion, succession stages, secondary forests and forest remnants are becoming increasingly important for the conservation of tropical biodiversity (e.g. Ricketts et al. 2001).

Whereas fire was apparently rare during the late Pleistocene, charcoal data point to frequent fires during the mid-Holocene period and even an increase during the late Holocene in the San Francisco area, suggesting an increase in human activity, however probably at lower elevations. Charcoal data at lower elevations in the ECSF forest witness human activity in the forest area which however declined in the latest Holocene (845 ± 45 years BP). Pollen analysis indicates a local regeneration of the mountain forest probably due to decreasing human activities after the Inca invasion.

36.4.2 The Bracken Fern Problem

Whereas the described succession leads to a mainly non-linear increase in biodiversity and to a concomitant loss of significance of earlier key species, the development of the abandoned pastures takes quite another route. In these areas, where fire is recurrently used as a measure of pasture maintenance, another key species takes over and dominates the vegetation: the very aggressive bracken fern (*Pteridium arachnoideum*). The competitive strength of this fern is strongly enhanced by each burning, due to its practically invulnerable subterranean network of rhizomes from which new leaves are readily produced after a fire. Its big leaves rapidly form closed canopies and thus widely suppress upcoming seedlings of other plant species. It is only because of the steepness of the terrain that light can reach the ground, allowing the ubiquitous seeds of a few weedy bush species to germinate, grow and slowly penetrate the canopy. These are bushes with an enormous seed production, mainly from the Asteraceae, Ericaceae and Melastomataceae. Together with bracken these bushes form a stable plant community, invasion of which by seeds of trees appears very unlikely. Once bracken has taken over, recultivation of the area appears extremely difficult. Continued cutting of the fronds combined with the application of herbicides may be promising, however this kind of treatment is expensive and needs to be continued for years (Marrs et al. 1998) until the rhizomes die. An experiment of reforestation as a potential method for recultivation has been started with some initial success (see Chapter 34). In contrast to the regeneration of the forest on small-scale clearings the wide abandoned areas are poor in animal species, as neither bracken nor the bushes are attractive fodder plants for herbivores.

36.4.3 Sustainable and Non-Sustainable Forms of Land Use

The Saraguro and Shuar have successfully settled in historical time within the vicinity of the research area and have developed different modes of sustainable coexistence with the forest (see Chapter 3). The Saraguro are small-scale pastoralists, inhabiting the mid-altitudes. Their well structured agricultural landscape consists of pastures and small-scale crop fields interspersed with remnants of the natural forest. The Shuar settle at lower altitudes, running a small-scale agroforestry system based on a sustainable mode of shifting cultivation. They conserve a high level of biodiversity in silviculture, in home gardens and in their daily life (see Chapter 25).

In contrast to both groups the newcomers, the so-called Mestizo-Colonos clear the forest by slash and burn and maintain their pastures by regular burning. This, however, leads to the bracken dilemma described above, to the abandonment of the pastures after several years and to the need of further farmland mainly taken from the untouched rainforest. Systematic investigations of the land-use gradient by Makeschin et al. (Chapter 31) revealed strong changes in soil chemical properties after forest clearing by slash and burn. Clearing the forest in that way results in severe losses of bioelements, especially carbon. The fires, however, induce alkalinization of the top soil, leading to an increased availability of potassium, calcium and magnesium which can be utilized by the planted fodder grasses. Soil microbial activity and mineralization are also enhanced, fostering growth of the pastures and partial restoration of the carbon pool in the soil. Nevertheless, while recording these positive effects of burning on soil chemistry and biology, one must not forget the general deterioration from removing the entire organic layer and with it the habitats of soil organisms. The concomitant effects on the structural and functional diversity are not yet very well known. What is known is the subsequent invasion and spreading of bracken fern which severely reduces pasture productivity (see Chapter 28).

36.4.4 Management Options

As described in the previous sections, land use by the Colonos is not sustainable and the amount of degraded land (especially abandoned pastures) is increasing at an unrelieved rate. To overcome this problem several strategies have to be included in an integrative land use concept. An issue of outstanding importance is to desist from the use of fire as an agricultural tool. In addition to that, enhancement of pasture productivity is indispensable. Carefully intensifying the pasture regime and input–output related fertilization with nitrogen and phosphorus could be a first step in approaching a sustainable pasture management. Besides the need for improvements in *pasture* management and strategies for restoring pastures, various measures in the field of forestry may be promising. Introducing especially N-fixing and multipurpose tree species into pastures would lead to *agroforestry systems*, which could efficiently reduce the bracken fern problem, increase the long-term productivity of cattle farming and create new utilization potentials. Concomitantly, pressure on the remaining primary forests would be reduced.

Conservation of the primary forests in protected areas is doubtless the best option to maintain biodiversity. However, absolute protection requires effective control, low human pressure and an adapted buffer zone management. In this context UNESCO's Biosphere Reserve concept seems to be an interesting option for southern Ecuador.

Another promising option is *natural forest management* to increase the economic value of natural forests. As shown by Günter et al. (Chapter 26) improvement felling in natural tropical mountain rain forests can foster the growth of commercially valuable tree species without reducing the biodiversity or deteriorating the element cycling. Additionally, *enrichment planting* in forest gaps using valuable native species could be a very efficient method to increase the economic value of natural forests. However, estimates, including interest calculations and the validation of risks, are necessary in order to assess whether natural forest management can economically compete with cattle farming and the conventional land use practices. If this final balance sheet favors the agricultural practices, the actual deforestation processes will continue.

Natural succession of abandoned pastures could be a potential measure for the restoration of biodiversity, however only on a very long time-scale and without a preference of socioeconomic interesting tree species.

Reforestation of abandoned pastures can help to restore degraded areas, to meet the timber demand of the local people and to alleviate pressure on the natural forests. In many situations plantations of exotic tree species can create a first tree cover on degraded areas, as they may perform better than many native species. As shown by Stimm et al. (Chapter 33), some native species, e.g. *Alnus acuminata* can compete with exotic trees. As described by these authors, enrichment planting with native species under the shelter of exotic tree species is a possibly successful way to reintroduce biodiversity into the forests. Especially seedlings of late successional species like *Cedrela* spp. or *Tabebuia chrysantha* seem to perform better under a canopy than on open land. Reforestation with native tree species seems to be a very promising management option, but requires better knowledge about the biology of these species and their performance under different site conditions. Autochthonous material of native species should be used to establish ecologically diverse and economically appreciated ecosystems. Seeds should be harvested from many individuals (at least 50) to ensure sufficient genetic amplitude for large-scale reforestation.

References

Aboal JR, Morales D, Hernández M, Jiménez MS (1999) The measurement and modelling of the variation of stemflow in a laurel forest in Tenerife, Canary Islands. J Hydrol 221:161–175

Acosta-Solis M (1984) Los páramos Andinos del Ecuador. Publiciones Cientificas MAS, Quito

Adams PW, Sidle RC (1987) Soil conditions in three recent landslides in southeast Alaska. For Ecol Manage 18:93–102

Aguirre N, Ordoñoz L, Hofstede R (2002a) Identificación de fuentes semilleras de especies forestales Alto-Andinas. (PROFAFOR, Reporte forestal 6) Ministerio del Ambiente y ECOPAR, Quito

Aguirre N, Ordoñoz L, Hofstede R (2002b) Comportamiento inicial de 18 especies forestales plantadas en el páramo. (PROFAFOR, Reporte forestal 7) Ministerio del Ambiente y ECOPAR, Quito

Aguirre N, Günter S, Weber M, Stimm B (2006) Enrichment of *Pinus patula* plantations with native species. Lyonia 10:33–45

Aguirre Z, Cabrera O, Sanchez A, Merino B, Maza B (2003) Composición floristica, endemismo y etnobotanica de la vegetación del Sector Oriental, parte baja del Parque Nacionál Podocarpus. Lyonia 3:5–14

Ahnert F (1996) Introduction to geomorphology. Arnold, London

Aiba S, Kitayama K (1999) Structure, composition and species diversity in an altitude–substrate matrix of rain forest tree communities on Mount Kinabalu, Borneo. Plant Ecol 140:139–157

Aide TM, Zimmerman JK, Pascarella JB, Rivera L, Marcano-Vega H (2000) Forest regeneration in a chronosequence of tropical abandoned pastures: implications for restoration ecology. Restor Ecol 8:328–338

Alados I, Olmo FJ, Foyo-Moreno I, Albados-Arboledas L (2000) Estimation of photosynthetically active radiation under cloudy conditions. Agric For Meteorol 102:39–50

Alexander IJ, Högberg P (1986) Ectomycorrhizas of tropical angiospermous trees. New Phytol 102:541–549

Alexander IJ, Lee SS (2005) Mycorrhizas and ecosystem processes in tropical rain forest: implications for diversity. In: Burslem DF, Pinard MA, Hartley SE (eds) Biotic interactions in the tropics: their role in the maintenance of species diversity. Cambridge University Press, Cambridge, pp 165–203

Alexiades MN (1996) Collecting ethnobotanical data: an introduction to basic concepts and techniques. In: Alexiades MN (ed) Selected guidelines for ethnobotanical research: a field manual. New York Botanical Garden, New York, pp 53–94

Allan NJR (1986) Accessibility and altitudinal zonation models of mountains. Mt Res Dev 6:185–194

Alvarado PJ (2002) Historia de Loja y su provincia. Industria Grafica Senefelder, Loja

Amoozegar A (1989) Compact constant head permeameter for measuring saturated hydraulic conductivity of the vadose zone. Soil Sci Soc Am J 53:1356–1361

Anderberg AA, Zhang X (2002) Phylogenetic relationships of Cyrillaceae and Clethraceae (Ericales) with special emphasis on the genus *Purdiaea* Planch. Organisms, Divers Evol 2:127–137

Anderson JM (1973) The breakdown and decomposition of sweet chestnut (*Castanea sativa* Mill) and beech (*Fagus sylvatica* L.) leaf litter in two deciduous woodland soils. I. Breakdown, leaching and decomposition. Oecologia 12:251–274

Anderson JM (1975) The enigma of soil animal species diversity. In: Vanek J (ed) Progress in soil zoology. (Proc 5th Int Coll Soil Zool, Prague) Junk, The Hague, pp 51–58

Andrade P (1996) Aves de Cajanuma, Parque Nacional Podocarpus. Arcoiris, Loja

Andreae MO, Merlet P (2001) Emission of trace gases and aerosols from biomass burning. Global Biogeochem Cyc 15:955–960

Andreae MO, Artaxo P, Fischer H, Freitag SR, Grégoire JM, Hansel A, Hoor P, Kormann R, Krejci R, Lange L, Lelieveld J, Lindinger W, Longo K, Peters W, deReus M, Scheeren B, Silva Dias MAF, Ström J, Velthoven PFJ van, Williams J (2001) Transport of biomass burning smoke to the upper troposphere by deep convection in the equatorial region. Geophys Res Lett 28:951–954

Andreae MO, Rosenfeld D, Artaxo P, Costa AA, Frank GP, Longo KM, Silva Dias MAF (2004) Smoking rain clouds over the Amazon. Science 303:1337–1342

Anhuf D, Landwehr C, Hetzel F (1997) Untersuchungen zum Transpirationsverhalten halbimmergrüner Feuchtwälder in Westafrika. Würzburger Geogr Arb 92:27–50

Arguropoulou MD, Asikidis MD, Iatrou GD, Stamou GP (1993) Colonization patterns of decomposing litter in a maquis ecosystem. Eur J Soil Biol 29:183–191

Arich M, Billington C, Edwards M, Laidlaw R (1997) Tropical montane cloud forests: an urgent priority for conservation. WCMC Biodivers Bull 2

Armenteras D, Gast F, Villareal H (2003) Andean forest fragmentation and the representativeness of protected natural areas in the eastern Andes, Colombia. Biol Conserv 113:245–256

Arriaga L (2000) Gap building phase regeneration in a tropical montane cloud forest of northeastern Mexico. J Trop Ecol 16:535–562

Arya LM, Dieroll TS, Sofyan A, Widjaja-Adhi IPG, Van Genuchten MT (1999) Significance of macroporosity and hydrology for soil management and sustainability of agricultural production in a humid-tropical environment. Soil Sci 164:586–601

Ashton KG, Tracy MC, Queiroz A de (2000) Is Bergmann's rule valid for mammals? Am Nat 156:390–415

Ataroff M (1998) Importance of cloud-water in Venezuelan Andean cloud forest water dynamics. Proc Int Conf Fog Fog Collect 1:25–28

Atkinson D (1994) Temperature and organism size – a biological law for ectotherms. Adv Ecol Res 25:1–58

Axmacher JC (1998) Vergleich verschiedener Ansätze zur physiognomischen Klassifikation eines tropischen Bergregenwaldes. PhD thesis, University of Bayreuth, Bayreuth

Axmacher JC, Fiedler K (2004) Manual versus automatic moth sampling at equal light sources – a comparison of catches from Mt Kilimanjaro. J Lepid Soc 58:196–202

Axmacher JC, Holtmann G, Scheuermann L, Brehm G, Müller-Hohenstein K, Fiedler K (2004a) Diversity of geometrid moths (Lepidoptera: Geometridae) along an Afrotropical elevational rainforest transect. Divers Distrib 10:293–302

Axmacher JC, Tünte H, Schrumpf M, Müller-Hohenstein K, Lyaruu HVM, Fiedler K (2004b) Diverging diversity patterns of vascular plants and geometrid moths during forest regeneration on Mt Kilimanjaro, Tanzania. J Biogeogr 31:895–904

Bååth E, Anderson TH (2003) Comparison of soil/fungal ratios in a pH gradient using physiological and PLFA-based techniques. Soil Biol Biochem 35:955–963

Backhaus K, Erichson B, Plinke W, Weiber R (1996) Multivariate Analysemethoden – Eine anwendungsorientierte Einführung. Springer, Berlin Heidelberg New York

Bailey VL, Peacock AD, Smith JL, Bolton H Jr (2002) Relationships between soil microbial biomass determined by chloroform fumigation extraction, substrate-induced respiration and phospholipid fatty acid analysis. Soil Biol Biochem 34:1385–1389

Baldock JW (1982) Geology of Ecuador – explanatory bulletin of the national geological map of the Republic of Ecuador. Ministerio de Recursos Naturales y Energéticos, Dirección General de Geología y Minas, Quito

Balslev H (1988) Distribution patterns of Ecuadorian plant species. Taxon 37:567–577

Balslev H, Øllgaard B (2002) Mapa de vegetación del sur de Ecuador. In: Aguirre ZM, Madsen JE, Cotton E, Balslev H (eds) Botánica Austroecuatoriana – estudios sobre los recursos vegetales en las provincias de el Oro, Loja y Zamora–Chinchipe. Abya-Yala, Quito, pp 51–64

Balslev H, Valencia R, Paz y Mino G, Christensen H, Nielsen I (1998) Species count of vascular plants in one hectare of humid lowland forest in Amazonian Ecuador. In: Dallmeier F, Comiskey JA (eds) Forest biodiversity in North, Central and South America, and the Caribbean. (Man and the biosphere series, vol 21) UNESCO, Paris, pp 585–594

Barkman JJ (1979) The investigation of vegetation texture and structure. In: Werger MJA (ed) The study of vegetation. Junk, The Hague, pp 125–160

Barry RG (1992) Mountain weather and climate. Routledge, London

Barthlott W, Lauer W, Placke A (1996) Global distribution of species diversity in vascular plants: towards a world map of phytodiversity. Erdkunde 50:317–327

Barthlott W, Schmitt-Neuerburg V, Nieder J, Engwald S (2001) Diversity and abundance of vascular epiphytes: a comparison of secondary vegetation and primary montane rain forest in the Venezuelan Andes. Plant Ecol 152:145–156

Barthlott W, Mutke J, Rafiqpoor MD, Kier G, Kreft H (2005) Global centres of vascular plant diversity. Nova Acta Leopoldina NF 92:61–83

Barton AM, Fetcher N, Redhead S (1989) The relationship between treefall gap size and light flux in a Neotropical rain forest in Costa Rica. J Trop Ecol 5:437–439

Basnet K (1992) Effect of topography on the pattern of trees in Tabonuco (*Dacryodes excelsa*) dominated rain forest of Puerto Rico. Biotropica 24:31–42

Basset Y, Novotny V, Miller SE, Pyle R (2000) Quantifying biodiversity: experience with parataxonomists and digital photography in Papua New Guinea and Guyana. BioScience 50:899–908

Batarya SK, Valdiya KS (1989) Landslides and erosion in the catchment of the Gaula river, Kumaun Lesser Himalaya, India. Mt Res Dev 9:405–419

Baumann F (1988) Geographische Verbreitung und Ökologie südamerikanischer Hochgebirgspflanzen. Beitrag zur Rekonstruktion der quartären Vegetationsgeschichte der Anden. Phys Geogr 28

Bäumler R (1995) Dynamik gelöster Stoffe in verschiedenen Kompartimenten kleiner Wassereinzugsgebiete in der Flyschzone der Bayerischen Alpen – Auswirkungen eines geregelten forstlichen Eingriffs. Bayreuther Bodenk Ber 40

Bawa KS, Dayanandan S (1998) Global climate change and tropical forest genetic resources. Clim Change 39:473–485

Bawa KS, Seidler R (1998) Natural forest management and conservation of biodiversity in tropical forests. Conserv Biol 12:46–55

Beaman JH (1962) The timberlines of Iztaccihuatl and Popocatepetl, Mexico. Ecology 43:377–385

Beard JS (1955) The classification of tropical American vegetation types. Ecology 36:89–100

Beard JS (1973) The physiognomic approach. In: Whittaker RH (ed) Handbook of vegetation science. Junk, The Hague, pp 357–386

Beccaloni GW, Gaston KJ (1995) Predicting the species richness of Neotropical forest butterflies: Ithomiinae (Lepidoptera: Nymphalidae) as indicators. Biol Conserv 71:77–86

Beck A, Kottke I, Oberwinkler F (2005) Two members of the Glomeromycota form distinct ectendomycorhizas with *Alzatea verticillata*, a prominent tree in the mountain rain forest of southern Ecuador. Mycol Progr 4:11–22

Beck A, Haug I, Oberwinkler F, Kottke I (2007) Structural characterisation and molecular identification of arbuscular mycorrhiza morphotypes of *Alzatea verticillata* (Alzateaceae) a prominent tree in the tropical mountain rain forest of South Ecuador, Mycorrhiza DOI 10.1007/s00572-007-0139-0

Beck E, Müller-Hohenstein K (2001) Analysis of undisturbed and disturbed tropical mountain forest ecosystems in southern Ecuador. Erde 132:1–8

Beck J, Schulze CH, Linsenmair KE, Fiedler K (2002) From forest to farmland: diversity of geometrid moths along two habitat gradients on Borneo. J Trop Ecol 17:33–51

Behling H (1993) Untersuchungen zur spätpleistozänen und holozänen Vegetations- und Klimageschichte der tropischen Küstenwälder und der Araukarienwälder in Santa Catarina (Südbrasilien). Diss Bot 206

Belloni L, Morris D (1991) Earthquake-induced shallow slides in volcanic debris soils. Geotechnique 41:539–551

Belote J (1998) Los Sarguros del sur del Ecuador. Abya-Yala, Quito

Belote J, Belote LS (1997/1999) The Saraguros, 1962–1997: a very brief overview. Available at: www.saraguro.org/overview.htm

Belote J, Belote LS (1999) The Shuar and the Saraguros. Available at: www.saraguro.org/shuar.htm

Belote J, Belote LS (2000) Corn and cattle: the development of a dual strategy. Available at: www.sraguro.org/dual.htm

Bendix A, Bendix J (2006) Heavy rainfall episodes in Ecuador during El Niño events and associated regional atmospheric circulation and SST patterns. Adv Geosci 6:43–49

Bendix J (2000) A comparative analysis of the major El Niño events in Ecuador and Peru over the last two decades. Zbl Geol Palaeontol I1999:1119–1131

Bendix J (2004) Extremereignisse und Klimavariabilität in den Anden von Ecuador und Peru. Geogr Rdsch 56:10–16

Bendix J, Lauer W (1992) Die Niederschlagsjahreszeiten in Ecuador und ihre klimadynamische Interpretation. Erdkunde 46:118–134

Bendix J, Rafiqpoor MD (2001) Studies on the thermal conditions of soils at the upper tree line in the páramo of Papallacta (eastern cordillera of Ecuador). Erdkunde 55:257–276

Bendix J, Rollenbeck R (2004) Gradients of fog and rain in a tropical cloud forest of southern Ecuador and its chemical composition. In: Proceedings of the third international conference on fog, fog collection and dew, 11–15 October, Cape Town. Available on CD-ROM, H7

Bendix J, Rollenbeck R, Palacios E (2004) Cloud classification in the tropics – a suitable tool for climate–ecological studies in the high mountains of Ecuador. Int J Remote Sensing 25:4521–4540

Bendix J, Rollenbeck R, Göttlicher D, Cermak J (2006a) Cloud occurrence and cloud properties in Ecuador. Clim Res 30:133–147

Bendix J, Rollenbeck R, Reudenbach C (2006b) Diurnal patterns of rainfall in a tropical Andean valley of southern Ecuador as seen by a vertically pointing K-band Doppler radar. Int J Climatol 26:829–847

Bendix J, Homeier J, Cueva Ortiz E, Emck P, Breckle SW, Richter M, Beck E (2006c) Seasonality of weather and tree phenology in a tropical evergreen mountain rain forest. Int J Biometeorol 50:370–384

Bengel C (2003) Kartierung des Blattflächenindex in einem kleinen Wassereinzugsgebiet unter tropischem Bergregenwald in Ecuador. PhD thesis, University of Bayreuth, Bayreuth, 84 pp

Benjamini Y, Hochberg Y (1995) Controlling the false discovery rate: a practical and powerful approach to multiple testing. J R Statist Soc B 57:289–300

Bergmann W (1993) Ernährungsstörungen bei Kulturpflanzen. Fischer, Stuttgart

Berkes F (1999) Sacred ecology: traditional ecological knowledge and resource management. Philadelphia Press, Philadelphia

Bernhard-Reversat F (1975) Nutrients in throughfall and their quantitative importance in rain forest mineral cycle. In: Golley FB, Medina E (eds): Tropical ecological systems. Springer, Berlin Heidelberg New York, pp 153–159

Beven KJ, Lamb R, Quinn PF, Romanowicz R, Freer J (1995) TOPMODEL. In: Singh VP (ed) Computer models of watershed hydrology. Water Resources, London, pp 627–668

Bibby CJ, Burgess ND, Hill DA (1995) Methoden der Feldornithologie: Bestandserfassung in der Praxis. Neumann/Radebeul, Stuttgart

BirdLife International, Conservation International (2005) Áreas importantes para la conservación de las aves en los Andes tropicales: sitios prioritarios para la conservación de la biodiversidad. (Serie de conservación de birdlife 14) BirdLife International, Quito

Bitterlich W (1948) Ein neues Messverfahren zur Aufnahme stehender Holzmassen. Oesterr Forst Holzwirtsch 3:89–90

Bloom AJ, Chapin FS III, Mooney HA (1985) Resource limitation in plants – an economic analogy. Annu Rev Ecol Syst 16:363–392

Bogh A (1992) Composition and distribution of the vascular epiphyte flora of an Ecuadorian montane rain forest. Selbyana 13:25–34

Bojsen BH, Barriga R (2002) Effects of deforestation on fish community structure in Ecuadorian amazon streams. Freshwater Biol 47:2246–2260

Boltz F, Holmes TP, Carter DR (2003) Economic and environmental impacts of conventional and reduced-impact logging in tropical South America. A comparative review. For Policy Econ 5:69–81

Bonaccorso FJ (1979) Foraging and reproductive ecology in a Panamanian bat community. Bull Fla State Mus Biol Sci 24:359–408

Bond-Lamberty B, Wang C, Gower ST (2004) A global relationship between the heterotrophic and autotrophic components of soil respiration. Global Change Biol 10:1756–1766

Bonell M, Barnes CJ, Grant CR, Howard A, Burns J (1998) Oxygen and hydrogen isotopes in rainfall-runoff studies. In: Kendall C, McDonnell JJ (eds) Isotope tracers in catchment hydrology. Elsevier, Amsterdam, pp 347–390

Borja Cristina, Lasso S (1990) Plantas nativas para reforestation en el Ecuador. Fundación Natura (EDUNAT III)/AID, Quito

Boscolo M, Vincents JR (1998) Promoting better logging practices in tropical forests. (Policy research working paper 1971) World Bank Development Research Group, New York, 35 pp

Bosmann AF, Hooghiemstra H, Cleef AM (1994) Holocene mire development and climatic change from a high Andean *Plantago rigida* cushion mire. Holocene 43:233–243

Boutton TW (1991). Stable carbon isotope ratios of natural materials. II. Atmospheric, terrestrial, marine and freshwater environments. In: Coleman DC, Fry B (eds) Carbon isotope techniques in biological sciences. Academic, San Diego, pp 173–185

Bowles IA, Rice RE, Mittermeier RA, Fonseca GAB da (1998) Logging and tropical forest conservation. Science 280:1899–1900

Box JD (1983) Investigation of the Folin–Ciocalteau phenol reagent for the determination of polyphenolic substances in natural waters. Water Resour 17:511–525

Bradfield DE, Kenkel NC (1987) Nonlinear ordination using flexible shortest path adjustments of ecological distances. Ecology 68:750–753

Brandani A, Hartshorn GS, Orians GH (1988) Internal heterogeneity of gaps and species richness in Costa Rican tropical wet forest. J Trop Ecol 4:99–119

Brandbyge J, Holm-Nielsen LB (1986) Reforestation of the high Andes with local species. (Reports from the Botanical Institute, 13) University of Aarhus, Aarhus

Braun H (2002) Die Laubheuschrecken (Orthoptera, Tettigoniidae) eines Bergregenwaldes in Süd-Ecuador – faunistische, bioakustische und ökologische Untersuchungen. PhD thesis, University of Erlangen-Nürnberg, Erlangen

Braun H (2007) Tettigoniidae. In: Liede-Schumann S, Breckle SW (eds) Provisional checklists of fauna and flora of the San Francisco valley and its surroundings (Reserva San Francisco/Prov. Zamore–Chinchipe, southern Ecuador). Ecotrop Monogr 4 (in press)

Braun-Blanquet J (1928) Pflanzensoziologie – Grundzüge der Vegetationskunde. Springer, Berlin Heidelberg New York

Bräuning A, Burchardt I (2006) Detection of growth dynamics in tree species of a tropical mountain rain forest in southern Ecuador. In: Dendrosymposium (ed) TRACE – tree rings in archaeology, climatology and ecology, vol 4. (Proceedings of the Dendrosymposium 2005) Dendrosymposium, Fribourg, pp 127–131

Breckle SW (2002) Significance of systematics and taxonomy for biology. In: Breckle SW (ed) Walter's vegetation of the earth – the ecological systems of the geobiosphere. Springer, Berlin Heidelberg New York, pp 4–5

Breckle SW, Breckle U, Homeier J, Scheffer A (2005) Mineral deficiencies in a pine plantation in southern Ecuador. Ecotropica 11:79–85

Brehm G (2002) Diversity of geometrid moths in a montane rainforest in Ecuador. PhD thesis, University of Bayreuth, Bayreuth. Available at: opus.ub.uni-bayreuth.de/volltexte/2003/20

Brehm G (2004) A new species of *Oenoptila* Warren, 1895 from the Andes (Lepidoptera: Geometridae, Ennominae). Ent Z 114:278–280

Brehm G (2005) A revision of the *Acrotomodes clota* Druce, 1900 species-group (Lepidoptera: Geometridae, Ennominae). Ent Z 115:75–80

Brehm G (2007) Contrasting patterns of vertical stratification in two moth families in a Costa Rican lowland rain forest. Basic Appl Ecol 8:44–54

Brehm G, Axmacher JC (2006) A comparison of manual and automatic moth sampling methods (Lepidoptera: Arctiidae, Geometridae) in a rain forest in Costa Rica. Environ Ent 35:757–764

Brehm G, Colwell RK, Kluge J (2007) The role of environment and mid-domain effect on moth species richness along a tropical elevational gradient. Global Ecol Biogeogr 16:205–219

Brehm G, Fiedler K (2003) Faunal composition of geometrid moths changes with altitude in an Andean montane rain forest. J Biogeogr 30:431–440

Brehm G, Fiedler K (2004a) Bergmann's rule does not apply to geometrid moths along an elevational gradient in an Andean montane rain forest. Global Ecol Biogeogr 13:7–14

Brehm G, Fiedler K (2004b) Ordinating tropical moth samples from an elevational gradient: a comparison of common methods. J Trop Ecol 20:165–172

Brehm G, Fiedler K (2005) Diversity and community structure of geometrid moths of disturbed habitat in a montane area in the Ecuadorian Andes. J Res Lepid 38:1–14

Brehm G, Homeier J, Fiedler K (2003a) Beta diversity of geometrid moths (Lepidoptera: Geometridae) in an Andean montane rainforest. Divers Distrib 9:351–366

Brehm G, Süssenbach D, Fiedler K (2003b) Unique elevational diversity patterns of geometrid moths in an Andean montane rainforest. Ecography 26:456–466

Brehm G, Pitkin LM, Hilt N, Fiedler K (2005) Montane Andean rain forests are a global diversity hotspot of geometrid moths. J Biogeogr 32:1621–1627

Brehm G, Colwell RK, Kluge J (2007) The role of environment and mid-domain effect on moth species richness along a tropical elevational gradient. Global Ecol Biogeogr 16:205–219

Brenning A (2005) Spatial prediction models for landslide hazards: review, comparison and evaluation. Nat Hazards Earth Syst Sci 5:853–862

Breuer T (1974) Die Bewölkungsverhältnisse des südhemisphärischen Südamerika und ihre klimageographische Aussagemöglichkeiten. PhD thesis, University of Bonn, Bonn

Briceño MA (2005) Evaluación;n de fuentes semilleras de bosque tropical de montaña mediante ensayos de germinación y sobrevivencia en vivero. PhD thesis, Universidad Técnica del Norte, Loja

Brokaw N (1985) Gap-phase regeneration in a tropical forest. Ecology 66:682–687

Brokaw N, Busing RT (2000) Niche versus chance and tree diversity in forest gaps. Trends Ecol Evol 15:183–188

Brooks TM, Mittermeier RA, Mittermeier CG, Da Fonseca GAB, Rylands AB, Konstant WR, Flick P, Pilgrim J, Oldfield S, Magin G, Hilton-Taylor C (2002) Habitat loss and extinction in the hotspots of biodiversity. Conserv Biol 16:909–923

Brose U, Martinez ND (2004) Estimating the richness of species with variable mobility. Oikos 105:292–300

Brosset A, Charles-Dominique P (1990) The bats from French Guiana: a txonomic, faunistic and ecological approach. Mammalia 54:509–560

Brosset A, Charles-Dominique P, Cockle A, Cosson JF, Masson D (1996) Bat communities and deforestation in French Guiana. Can J Zool 74:1974–1982

Brown S (1997) Estimating biomass and biomass change of tropical forests: a primer. (FAO forestry paper 134) FAO, Rome

Brown S, Iverson LR (1992) Biomass estimates for tropical forests. World Resour Rev 4:366–384

Bruehl C, Mohamed M, Linsenmair KE (1999) Altitudinal distribution of leaf litter ants along a transect in primary forests on Mount Kinabalu, Sabah, Malaysia. J Trop Ecol 15:265–277

Bruijnzeel LA (1990) Hydrology of moist tropical forests and effects of conversion: a state of knowledge review. Unesco (Division of Water Sciences, International Hydrological Programme), Paris

Bruijnzeel LA (2001) Hydrology of tropical montane cloud forests: a Reassessment. Land Use Water Res 1:1.1–1.18 (available at: www.luwrr.com)

Bruijnzel LA (2004) Hydrological functions of tropical forests: not seeing the soil for the trees? Agric Ecosys Environ 104:185–228

Bruijnzeel LA, Hamilton LS (2000) Decision time for cloud forests. (IHP humid tropics programme series, 13) IHP-UNESCO, Paris

Bruijnzeel LA, Proctor J (1995) Hydrology and biogeochemistry of tropical montane cloud forests: what do we really know? In: Juvik JO, Scatena FN (eds) Tropical montane cloud forests. (Ecological studies 110) Springer, Berlin Heidelberg New York, pp 38–78

Bruijnzeel LA, Veneklaas EJ (1998) Climatic conditions and tropical montane forest productivity: the fog has not lifted yet. Ecology 79:3–9

Bruijnzeel LA, Waterloo MJ, Proctor J, Kuiters AT, Kotterink B (1993) Hydrological observations in montane rain forests on Gunung Silam, Sabah, Malaysia, with special reference to the Massenerhebung effect. J Ecol 81:145–167

Brummitt N, Lughada EN (2003) Biodiversity: where's hot and where's not. Conserv Biol 17:1442–1448

Brun, R (1976) Methodik und Ergebnisse zur Biomassenbestimmung eines Nebelwald-Ökosystems in den Venezolanischen Anden. Proc IUFRO World Congr Div I 16:490–499

Brundrett M (2004) Diversity and classification of mycorrhizal associations. Biol Rev 79:473–495

Brünig EF (1996) Conservation and management of tropical rainforests: an integrated approach to sustainability. CAB International, Wallingford

Buchter B, Hinz C, Leuenberger J (1997) Tracer transport in a stony hillslope soil under forest. J Hydrol 192(1–4):314–320

Burghouts TBA, Van Straalen NM, Bruijnzeel LA (1998) Spatial heterogeneity of element and litter turnover in a Bornean rain forest. J Trop Ecol 14:477–505

Burslem D, Pinard M, Hartley S (2005) Biotic interactions in the tropics. Their role in the maintenance of species diversity. Cambridge University Press, Cambridge

Buschbacher RJ, Uhl C, Serrao EA (1988) Abandoned pastures in Eastern Amazonia – nutrient stocks in soil and vegetation. J Ecol 76:682–699

Bush MB, Colinvaux PA, Wiemann MC, Piperno DR, Liu KB (1990) Late Pleistocene temperature depression and vegetation change in Ecuadorian Amazonia. Quat Res 34:330–345

Bussmann RW (2001a) Epiphyte diversity in a tropical Andean forest – Reserva Biológica San Francisco, Zamora Chinchipe, Ecuador. Ecotropica 7:43–59

Bussmann RW (2001b) The montane forests of Reserva Biológica San Francisco. Erde 132:9–25

Bussmann RW (2002) Estudio fitosociólogico de la vegetación en la reserva Biológica San Francisco (ECSF) Zamora–Chinchipe. Herb Loja 8:1–106

Butler L, Kondo V, Barrows EM, Townsend EC (1999) Effects of weather conditions and trap types on sampling for richness and abundance in forest macrolepidoptera. Environ Ent 28:795–811

Butterfield RP (1996) Early species selection for tropical reforestation: a consideration of stability. For Ecol Manage 81:161–168

Butterfield RP, Fisher RF (1994) Untapped potential: native species for afforestation. J For 92:37–40

Buttle JM, McDonald DJ (2000) Soil macroporosity and infiltration characteristics of a forest podzol. Hydrol Process 14:831–848

Cabarle BJ, Crespi, M, Dodson CH, Luzuriaga C, Rose D, Shores JN (1989) An assessment of biological diversity and tropical forests for Ecuador. USAID, Quito

Cabrera M, Ordoñez HE (2004) Fenología, almacenamiento de semillas y propagación a nivel de vivero de diez especies forestales nativas del sur del Ecuador. Graduation thesis, Universidad Nacional de Loja, Loja

Calder IR (2001) Canopy processes: implications for transpiration, interception and splash induced erosion, ultimately for forest management and water resources. Plant Ecol 153:203–214

Caldwell MM, Björn LO, Bornman JF, Flint SD, Kulandaivelu G, Teramura AH, Tevini M (1998) Effects of increased solar ultraviolet radiation on terrestrial ecosystem. J Photochem Photobiol B46:40–52

Carpenter FL, Nichols D, Sandi E (2004a) Early growth of native and exotic trees planted on degraded tropical pasture. For Ecol Manage 196:367–378

Carpenter FL, Nicols JD, Pratt TR, Young KC (2004b) Methods of facilitating reforestation of tropical degraded land with the native timber tree *Terminalia amazonia*. For Ecol Manage 202:281–291

Carvalho LMT de, Fontes MAL, Oliveira-Filho AT de (2000) Tree species distribution in canopy gaps and mature forest in an area of cloud forest of the Ibitipoca Range, south-eastern Brazil. Plant Ecol 149:9–22

Catinot R (1965a) Sylviculture en forêt dense africaine, 1ère partie. Bois For Trop 100:5–18

Catinot R (1965b) Sylviculture tropicale en forêt dense africaine, 2ème partie. Bois For Trop 101:3–16

Catinot R (1965c) Sylviculture tropicale en forêt dense africaine, 3ème partie. Bois For Trop 102:3–16

Cavalier J (1992) Fine-root biomass and soil properties in a semideciduous and a lower montane rain forest in Panama. Plant Soil 142:187–201

Cavelier J (1996) Environmental factors and ecophysiological processes along altitudinal gradients in wet tropical mountains. In: Mulkey SS, Chazdon RL, Smith AP (eds) Tropical forest plant ecophysiology. Chapman and Hall, New York, pp 399–439

Cavelier J, Jaramillo M, Solis D, León, D (1997) Water balance and nutrient inputs in bulk precipitation in tropical montane cloud forests in Panama. J Hydrol 193:83–96

Čermák J, Kučera J (1990) Scaling up transpiration data between trees, stands and water-sheds. Silva Carel 15:101–120

Cerri CE, Coleman K, Jenkinson DS, Bernoux M, Victoria R, Cerri CC (2003) Modelling soil carbon from forests and pasture ecosystems of Amazon, Brazil. Soil Sci Soc Am J 67:1879–1887

Cerri CE, Paustian K, Bernoux M, Victoria RL, Melillo JM, Cerri CC (2004) Modelling changes in soil organic matter in Amazon forest to pasture conversion with the Century model. Global Change Biol 10:815–832

Chang JS, Brost RA, Isaksen ISA, Madronich S, Middleton P, Stockwell WR, Walcek CJ (1987) A three-dimensional Eulerian acid deposition model: Physical concepts and formulation. J Geophys Res 92:14681–14700

Chapin FS III (1980) The mineral nutrition of wild plants. Annu Rev Ecol Syst 11:233–260

Chappell NA, Franks SW, Larenus J (1998) Multi-scale permeability estimation for a tropical catchment. Hydrol Process 12:1507–1523

Charles-Dominique, P (1986) Inter-relations between frugivorous vertebrates and pioneer trees: *Cecropia*, birds and bats in French Guiana. In: Estrada A, Fleming TH (eds) Frugivores and seed dispersal. Junk, Dordrecht, pp 119–134

Chave J, Andalo C, Brown S, Cairns MA, Chambers JQ, Eamus D, Folster H, Fromard F, Higuchi N, Kira T, Lescure JP, Nelson BW, Ogawa H, Puig H, Riera B, Yamakura T (2005) Tree allometry and improved estimation of carbon stocks and balance in tropical forests. Oecologia 145:87–99

Chaves JA (2001) Comparación de avifaunas en dos bosques nublados del Chocó ecuatoriano. In: Nieder J, Barthlott W (eds) Epiphytes and canopy fauna of the otonga rain forest (Ecuador). Results of the Bonn–Quito epiphyte project, funded by the Volkswagen Foundation, vol 2. University of Bonn, Bonn, pp 311–326

Chazdon RL (1998) Tropical forests – log 'em or leave 'em? Science 281:1295–1296

Chen JM, Black TA (1991) Measuring leaf area index of plant canopies with branch architecture. Agric For Meteorol 57:1–12

Chey VK, Holloway JD, Speight MR (1997) Diversity of moths in forest plantations and natural forests in Sabah. Bull Entomol Res 87:371–385

Chiarucci A, Enright NJ, Perry GLW, Miller BP, Lamont BB (2003) Performance of nonparametric species richness estimators in a high diversity plant community. Divers Distrib 9:283–295

Claassen N, Steingrobe B (1999) Mechanistic simulation models for a better understanding of nutrient uptake from soil. In: Rengel Z (ed) Mineral nutrition for crops. Fundamental mechanisms and implications. Food Products, Binghampton, pp 327–367

Clark DB, Clark DA, Brown S, Oberbauer SF, Veldkamp E (2002) Stocks and flows of coarse woody debris across a tropical rain forest nutrient and topography gradient. For Ecol Manage 164:237–248

Clarke FM, Rostant LV, Racey PA (2005) Life after logging: post logging recovery of a neotropical bat community. J Appl Ecol 42:409–420

Clarke KR (1993) Non-parametric multivariate analyses of changes in community structure. Aust J Ecol 18:117–143

Clarke KR, Gorley RN (2001) PRIMER v5. Primer-E Ltd, Plymouth

Cleef AM, Rangel O, Salamanca CS (2003) The Andean rain forests of the Parque Los Nevados transect, Cordillera Central, Colombia. In: Hammen T van der, Santos AG dos (eds) Studies on tropical Andean ecosystems 5. Cramer, Berlin, pp 79–141

Colinvaux PA (1988) Late-glacial and Holocene pollen diagrams from two endorheic lakes of the Inter-Andean plateau of Ecuador. Rev Palaeobot Palynol 55:83–89

Colinvaux PA, Bush MB, Steinitz-Kannan M, Miller MC (1997) Glacial and postglacial pollen records from the Ecuadorian Andes and Amazon. Quat Res 48:69–78

Colwell RK (2006) EstimateS 8.00. Available at: http://viceroy.eeb.uconn.edu/estimates

Colwell RK, Hurtt GC (1994) Non-biological gradients in species richness and a spurious Rapoport effect. Am Nat 144:570–595

Colwell RK, Lees DC (2000) The mid-domain effect: geometric constraints on the geography of species richness. Trends Ecol Evol 15:70–76

Colwell RK, Rahbek C, Gotelli NJ (2004) The mid-domain effect and species richness patterns: what have we learned so far? Am Nat 163:E1–E23

Condit R (1998) Tropical forest census plots: methods and results from Barro Colorado Island, Panama and a comparison with other plots. Springer, Berlin Heidelberg New York

Condit R, Pitman N, Leigh E Jr, Chave J, Terborgh J, Foster BB, Núñez VP, Aguilar S, Valencia R, Villa G, Muller-Landau HC, Losos E, Hubbell SP (2002) Beta-diversity in tropical forest trees. Science 294:666–669

Connell JH (1978) Diversity in tropical rain forests and coral reefs. Science 199:1302–1310

Conway E (1953) Spore and sporeling survival in bracken *Pteridium aquilinum* (L.) Kuhn. J Ecol 41:289–294

Conway E (1957) Spore production in bracken. J Ecol 45:273–284

Cornish PM (1993) The effects of logging and forest regeneration on water yields in a moist eucalypt forest in New South Wales, Australia. J Hydrol 150:301–322

Cosson JF, Pons JM, Masson D (1999) Effects of forest fragmentation on frugivorous and nectarivorous bats in French Guiana. J Trop Ecol 15:515–534

Cotton E, Matezki S (2003) Ecuadorian novelties in *Blakea* and *Topobea* (Melastomataceae). Brittonia 55:73–81

Coûteaux MM, Sarmiento L, Bottner P, Acevedo D, Thiéry JM (2002) Decomposition of standard plant material along an altitudinal transect (65–3968m) in the tropical Andes. Soil Biol Biochem 34:69–78

Cozzarelli IM, Herman JS, Parnell RA Jr (1987) The mobilization of aluminum in a natural soil system: effects of hydrological pathways. Water Resour Res 23:895–874

Croat T (1975) Phenological behavior of habitat and habitat classes on Barro Colorado Island (Panama Canal Zone). Biotropica 7:270–277

Crockford RH, Richardson DP (1990) Partitioning of rainfall in a eucalypt forest and pine plantation in southeastern Australia. Hydrol Process 4:131–167

Crockford RH, Richardson DP (2000). Partitioning of rainfall into throughfall, stemflow and interception: effect of forest type, ground cover and climate. Hydrol Process 14:2903–2920

Crozier MJ (1986a) Landslides: causes, consequences, and environment. Crown Helm, Dover, pp 252

Crozier MJ (1986b) Classification of slope movements. In: Crozier MJ (ed) Landslides: causes, consequences, and environment. Croom Helm, London, pp 3–31

Cubina A, Aide TM (2001) The effect of distance from forest edge on seed rain and soil seed bank in a tropical pasture. Biotropica 33:260–267

Cueva E, Medina E (1986) Nutrient dynamics within Amazonian forest ecosystems. I. Nutrient flux in fine litter fall and efficiency of nutrient utilization. Oecologia 68:466–472

Cueva E, Homeier J, Breckle SW, Bendix J, Emck P, Richter M, Beck E (2006) Seasonality in an evergreen tropical mountain rainforest in southern Ecuador. Ecotropica 12:69–85

D'Antonio C, Meyerson LA (2002) Exotic plant species as problems and solutions in ecological restoration: a synthesis. Restor Ecol 10:703–713

Dalitz H, Homeier J (2004) Visual plants – image based tool for plant diversity research. Lyonia 6:49–62

Dalitz H, Homeier J, Salazar HR, Wolters A (2004) Spatial heterogeneity generating plant diversity? In: Breckle S-W, Schweizer B, Fangmeier A (eds) Proceedings of the second symposium of the AFW Schimper-Foundation. AFW Schimper Foundation, Stuttgart, pp 199–213

Dalling JW, Hubbell SP (2002) Seed size, growth rate and gap microsite conditions as determinants of recruitment success for pioneer species. J Ecol 90:557–568

Dalling JW, Tanner EVJ (1995) An experimental study of regeneration on landslides in montane rain forest in Jamaica. J Ecol 83:55–64

Damoah R, Spichtinger N, Forster C, James P, Mattis I, Wandinger U, Berle S, Wagner T, Stohl A (2004) Around the world in 17 days – hemispheric scale transport of forest fire smoke from Russia in May 2003. Atmos Chem Phys 4:1311–1321

Daniels RE (1985) Studies in the growth of *Pteridium aquilinum* (L.) Kuhn. (bracken): regeneration of rhizome segments. Weed Res 25:381–388

Davidson RF, Gagnon D, Mauffette Y, Hernandez H (1998) Early survival, growth and foliar nutrients in native Ecuadorian trees planted on degraded volcanic soil. For Ecol Manage 105:1–19

Davis MA, Wrage KJ, Reich PB, Tjoelker MG, Schaeffer T, Muermann C (1999) Survival, growth, and photosynthesis of tree seedlings competing with herbaceous vegetation along a water–light–nitrogen gradient. Plant Ecol 145:341–350

Dawkins HC, Philip MS (1998) Tropical moist forest silviculture and management. CAB International, Wallingford

De'ath G (1999) Extended dissimilarity: a method of robust estimation of ecological distances from high beta diversity data. Plant Ecol 144:191–199

Deborah AM, Smith CK, Gholz HL, Oliveira FAO (2001) Effects of landuse change on soil nutrient dynamics in Amazonia. Ecosystems 4:625–645

Dekker M, Graaf NR de (2003) Pioneer and climax tree regeneration following selective logging with silviculture in Suriname. For Ecol Manage 172:183–190

Denslow JS (1980) Gap partitioning among tropical rainforest trees. Biotropica 12:47–55

Denslow JS, Ellison AM, Sanford RE (1998) Treefall gap size effects on above- and below-ground processes in a tropical wet forest. J Ecol 86:597–609

DeVries PJ, Walla TR (2001) Species diversity and community structure in neotropical fruitfeeding butterflies. Biol J Linn Soc 74:1–15

DeVries PJ, Murray D, Lande R (1997) Species diversity in vertical, horizontal, and temporal dimensions of a fruit-feeding butterfly community in an Ecuadorian rainforest. Biol J Linn Soc 62:343–364

Diaz ML, Lojan M (2004) Fenologia y propagacion en vivero de especies forestales nativas del bosque protector "El Bosque". Graduation thesis, Universidad Nacional de Loja, Loja

Didham RK (1997) An overview of invertebrate responses to forest fragmentation. In: Watt AD, Stork NE, Hunter MD (eds) Forests and insects. Chapman and Hall, London, pp 303–320

Die Erde (2001) Themenheft 'Tropische Wald-Ökosysteme'. Die Erde 132

Diels L (1937) Beiträge zur Kenntnis der Vegetation und Flora von Ecuador. (Bibliotheca botanica, vol 116) E. Schweizerbart'sche Verlagsbuchhandlung, Stuttgart

Dierschke H (1994) Pflanzensoziologie: Grundlagen und Methoden. Ulmer, Stuttgart

Dijk AIJM van, Bruijnzeel LA (2001) Modelling rainfall interception by vegetation of variable density using an adapted analysis model. 2. Model validation for a tropical upland. J Hydrol 247:239–262

Dikau R, Brunsden D, Schrott L, Ibsen ML (1996) Landslide recognition identification, movement and causes. Wiley, Chichester

Doff Sotta E, Meir P, Malhi Y, Donato Nobre A, Hodnett M, Grace J (2004) Soil CO_2 efflux in a tropical forest in the central Amazon. Global Change Biol 10:601–617

Dolanc CR, Gorchov DL, Cornejo F (2003) The effects of silvicultural thinning on trees regenerating in strip clear-cuts in the Peruvian Amazon. For Ecol Manage 182:103–116

Dorr LJ, Stergios B, Smith AR, Cuello NL (2000) Catalogue of the vascular plants of Guaramacal National Park, Portuguesa and Trujillo states, Venezuela. (Contributions from the United States National Herbarium, vol 40) Smithsonian Institution, Washington

Douglas I, Bidin K, Balamurugan G, Chappell NA, Walsh RPD, Greer T, Sinun W (1999) The role of extreme events in the impacts of selective tropical forestry on erosion during harvesting and recovery phases at Danum Valley, Sabah. Phil Trans R Soc Lond Ser B 354:1749–1761

Doumenge C, Gilmour D, Pérez MR, Blockhus J (1995) Tropical montane cloud forests: conservation status and management issues. In: Hamilton LS, Juvik JO, Scatena FN (eds) Tropical montane cloud forests. (Ecological studies, vol 110) Springer, Berlin Heidelberg New York, pp 24–37

Douville H, Chauvin F, Planton S, Royer JF, Salas-Melia D, Tyteca S (2002) Sensitivity of the hydrological cycle to increasing amounts of greenhouse gases and aerosols. Clim Dynam 20:45–68

Drechsel P, Zech W (1991) Foliar nutrient levels of broad-leaved tropical trees: a tabular review. Plant Soil 131:29–46

Dünisch O, Morais RR (2002) Regulation of xylem sap flow in an evergreen, a semi-deciduous, and a deciduous Meliaceae species from the Amazon. Trees 16:404–416

Dunn RR (2004) Managing the tropical landscape: a comparison of the effects of logging and forest conversion to agriculture on ants, birds, and Lepidoptera. For Ecol Manage 191:215–224

Dyer LA, Gentry GL, Greeney H, Walla T (2007) Caterpillars and parasitoids of an Ecuadorian cloud forest. Available at: http://www.caterpillars.org

Dziedzioch C (2001) Artenzusammensetzung und Ressourcenangebot kolibribesuchter Pflanzen im Bergregenwald Südecuadors. PhD thesis, University of Ulm, Ulm

Eckert T (2006) Biomasseuntersuchung an jungen Aufforstungspflanzen von *Tabebuia chrysantha* und *Cedrela* cf *montana* in Südecuador. MSc thesis, Technical University of Munich, Munich

Edwards PJ (1982) Studies of mineral cycling in a montane rain forest in New Guinea. V. Rates of cycling in throughfall and litter fall. J Ecol 70:807–827

Edwards PJ, Grubb PJ (1977) Studies of mineral cycling in a montane rain forest in New Guinea. I. The distribution of organic matter in the vegetation and soil. J Ecol 65:943–969

Edwards PJ, Grubb PJ (1982) Studies of mineral cycling in a montane rain forest in New Guinea. IV. Soil characteristics and the division of mineral elements between vegetation and soil. J Ecol 70:649–666

Ehrlich PR (1994) Energy use and biodiversity loss. Phil Trans R Soc Lond 344:99–104

Eisenhauer N, Partsch S, Parkinson D, Scheu S (2007) Invasion of a deciduous forest by earthworms: changes in soil chemistry, microflora, microarthropods and vegetation. Soil Biol Biochem 39:1099–1110

Ellenberg H (1975) Vegetationsstufen in perhumiden bis perariden Bereichen der tropischen Anden. Phytocoenologia 2:368–387

Ellenberg H, Muller-Dombois D (1967) Tentative physiognomic-ecological classification of plant formations of the earth. Ber Geobot Inst Eidgenoess Tech Hochsch Stift Rübel 37:21–73

Ellenberg L (1993) Naturschutz und Technische Zusammenarbeit. Geogr Rund 45:290–300

Elrick DE, Reynolds WD (1992) Methods for analyzing constant head well permeater data. Soil Sci Soc Am J 56:320–323

Elsenbeer H (2001) Hydrologic flowpaths in tropical rainforest soilscapes – a review. Hydrol Proc 15:1751–1759

Elsenbeer H, Newton BE, Dunne T, Moraes JMd (1999) Soil hydraulic conductivities of latosols under pasture, forest and teak in Rondonia, Brazil. Hydrol Process 13(9):1417–1422

Emck P (2006) The climate of the Cordillera Real, southern Ecuador. PhD thesis, FAU Erlangen, Erlangen

Encyclopaedia Britannica (2006) Ecuador – the people: ethnic and linguistic composition. Available at: www.britannica.com/eb/article-25844

Engen S, Lande R (1996) Population dynamic models generating species abundance distributions of the gamma type. J Theor Biol 178:325–331

Engwald S (1999) Diversität und Ökologie der vaskulären Epiphyten eines Berg- und eines Tieflandregenwaldes in Venezuela. PhD thesis, University of Bonn, Bonn

Enriques S, Duarte CM, Sand-Jensen K (1993) Pattern in decomposition rates among phytosynthetic organisms: the importance of detritus C:N:P content. Oecologia 94:457–471

Epron D, Farque L, Lucot E, Badot PM (1999) Soil CO_2 efflux in a beech forest: the contribution of root respiration. Ann For Sci 56:289–295

Erdmann KH (1996) Der Beitrag der Biosphärenreservate zu Schutz, Pflege und Entwicklung von Natur- und Kulturlandschaften in Deutschland. In: Kastenholz HG, Erdmann KH, Wolff M (eds) Nachhaltige Entwicklung. Zukunftschancen für Mensch und Umwelt, Berlin

Erickson GE, Ramirez CF, Concha JF, Tisnado MG, Urquidi BF (1989) Landslide hazards in the central and southern Andes. In: EE Brabb and Harrold BL (eds) Landslides: extent and economic significance. Balkema, Rotterdam, pp 111–117

Ern H (1974) Zur Ökologie und Verbreitung der Koniferen im östlichen Zentralmexiko. Mitt Dtsch Dendrol Ges 67:164–198

Ernst D, Felinks B, Henle K, Klotz S, Sandermann H, Wiencke C (2000) Von der numerischen zur funktionellen Biodiversität. Gaia 9:140–145

Evans WC (1986) The acute diseases caused by bracken in animals. In: Smith RT, Taylor JA (eds) Bracken: ecology, land use and control technology, 1985. Parthenon, Leeds, pp 121–132

Everitt BS (1995) Cluster analysis. Arnold, London

Ewel JJ, Bigelow SW (1996) Plant life-forms and tropical ecosystem functioning. In: Orians GH, Dirzo R, Cushman JH (eds) Biodiversity and ecosystem processes in tropical forests. Springer, Berlin Heidelberg New York, pp 101–126

Eynden V van den (2004) Use and management of edible non-crop plants in southern Ecuador. PhD thesis, Gent University, Gent

Fabian P, Kohlpaintner M, Rollenbeck R (2005) Biomass burning in the Amazon – fertilizer for the mountaineous rain forest in Ecuador. Environ Sci Pollut Res 12:290–296

Faegri K, Iversen J (1989) Textbook of pollen analysis. Wiley, New York

FAO (1993) Forest resources assessment 1990: tropical countries. Food and Agricultural Organization of the United Nations, Rome

FAO (1994) Forest resources assessment 1990: country briefs. Food and Agricultural Organization of the United Nations, Rome

FAO (1999) State of the world's forests 1999. Food and Agricultural Organization of the United Nations, Rome

FAO (2001) State of the world's forests 2001. Food and Agricultural Organization of the United Nations, Rome

FAO (2003) State of the world's forests 2003. Food and Agricultural Organization of the United Nations, Rome

FAO (2006) Global forest resources assessment 2005. Progress towards sustainable forest management. Food and Agricultural Organization of the United Nations, Rome

FAO, IUFRO (2002) Multilingual glossary forest genetic resources. Available at: iufro.ffp.csiro.au/iufro/silvavoc/glossary/index.html

FAO, UNESCO (1997) Soil map of the world. Revised legend with corrections and update. ISRIC, Wageningen

FAO, DFSC, IPGRI (2001) Forest genetic resources conservation and management, vol 2. In managed natural forests and protected areas (in situ). International Plant Genetic Resources Institute, Rome

FAOSTAT (2006) http://faostat.fao.org/site/418/DesktopDefault.aspx?PageID=418

Farley KA, Kelly EF, Hofstede RGM (2004) Soil organic carbon and water retention after conversion of grassland to pine plantations in the Ecuadorian Andes. Ecosystems 7:729–739

Fauth JE, Bernardo J, Camara M, Resetarits WJ, Van Buskirk J, McCollum SA (1996) Simplifying the jargon of community ecology: a conceptual approach. Am Nat 147:282–286

Fehse J, Hofstede R, Aguirre N, Paladines Ch, Kooijman A, Sevink J (2002): High altitude tropical secondary forests: a competitive carbon sink? For Ecol Manage 163:9–25

Feigl BJ, Melillo J, Cerri CC (1995) Changes in the origin and quality of soil organic matter after pasture introduction in Rondônia, Brazil. Plant and Soil 175:21–29

Fentie B, Yu B, Silburn MD, Ciesiolka CAA (2002) Evaluation of eight different methods to predict hillslope runoff rates for a grazing catchment in Australia. J Hydrol 261:102–114

Fenton MB, Archarya L, Audet D, Hickey MBC, Merriman C, Obrist MK, Syme DM, Adkins B (1992) Phyllostomid bats (Chiroptera: Phyllostomidae) as indicators of habitat disturbance in the Neotropics. Biotropica 24:440–446

Fenwick GR (1989) Bracken (*Pteridium aquilinum*) – toxic effects and toxic constituents. J Sci Food Agric 46:147–173

Ferek RJ, Reid JS, Hobbs PV, Blake DR, Liousse C (1998) Emission factors of hydrocarbons, halocarbons, trace gases and particles from biomass burning in Brazil. J Geophys Res 103:32107–32118

Fernandes NF, Guimaraes RF, Gomes RAT, Vieira BC, Montgomery DR, Greenberg H (2004) Topographic controls of landslides in Rio de Janeiro: field evidence and modeling. Catena 55:163–181

Fernandez MH, Vrba ES (2005) Rapoport effect and biomic specialization in African mammals: revisiting the climatic variability hypothesis. J Biogeogr 32:903–918

Feyera S, Beck E, Lüttge U (2002) Exotic trees as nurse-trees for the regeneration of natural tropical forests. Trees 16:245–249

Fichtner K, Schulze E-D (1990) Xylem water flow in tropical vines as measured by a steady state heating method. Oecologia 82:355–361

Fiedler K, Schulze CH (2004) Forest modification affects diversity, but not dynamics of speciose tropical pyraloid moth communities. Biotropica 36:615–627

Fiedler K, Brehm G, Hilt N, Süssenbach D, Onore G, Bartsch D, Häuser CH (2007a) Moths (Lepidoptera: Arctiidae, Geometridae, Hedylidae, Pyraloidea, Sphingidae, and Uraniidae). In: Liede-Schumann S, Breckle SW (eds) Provisional checklists of fauna and flora of the San Francisco valley and its surroundings (Reserva San Francisco/Prov. Zamore–Chinchipe, southern Ecuador). Ecotrop Monogr 4 (in press)

Fiedler K, Hilt N, Brehm G, Schulze CH (2007b) Moths at tropical forest margins – how megadiverse insect assemblages respond to forest disturbance and recovery. In: Tscharntke T, Leuschner C, Zeller M Guhardja E, Bidin A (eds) The stability of tropical rainforest margins: linking ecological, economic and social constraints of land use and conservation. Springer, Berlin Heidelberg New York, pp 39–60

Fierer N, Schimel, Joshua P, Holden PA (2003) Variations in microbial community composition through two soil depth profiles. Soil Biol Biochem 35:167–176

Finckh M, Paulsch A (1995) *Araucaria araucana* – Die ökologische Strategie einer Reliktkonifere. Flora 190:365–382

Finegan B, Camacho M (1999) Stand dynamics in a logged and silviculturally treated Costa Rican rain forest, 1988–1996. For Ecol Manage 121:177–189

Finegan B, Camacho M, Zamora N (1999) Diameter increment patterns among 106 tree species in a logged and silviculturally treated Costa Rican rain forest. For Ecol Manage 121:159–176

Fisher RA, Corbet AS, Williams CB (1943) The relation between the number of species and the number of individuals in a random sample of an animal population. J Anim Ecol 12:42–58

Fleischbein K (2004) Wasserhaushalt eine Bergwaldes in Ecuador: experimenteller und modellhafter Ansatz auf Einzugsgebietsebene. (Gießener Geologische Schriften 71) Lenz-Verlag, Gießener

Fleischbein K, Wilcke W, Boy J, Valarezo C, Zech W, Knoblich K (2005) Rainfall interception in a lower montane forest in Ecuador: effects of canopy properties. Hydrol Process 19:1355—1371

Fleischbein K, Wilcke W, Valarezo C, Zech W, Knoblich K (2006) Water budgets of three small catchments under montane forest in Ecuador: experimental and modeling approach. Hydrol Process 20:2491–2507

Fleming TH (1988) The short tailed fruit bat. A study of plant animal interactions. University of Chicago Press, Chicago

Fleming TH, Breitwisch RL, Whitesides GW (1987) Patterns of tropical vertebrate frugivore diversity. Annu Rev Ecol Syst 18:91–109

Flenley JR (1995) Cloud forest, the Massenerhebung effect, and ultraviolet insolation. In: Hamilton LS, Juvik JO, Scatena FN (eds) Tropical montane cloud forests. (Ecological studies, vol 110) Springer, Berlin Heidelberg New York, pp 150–155

Floren A, Linsenmair KE (2005) The importance of primary tropical rain forest species diversity: an investigation using arboreal ants as an example. Ecotropica 8:559–567

Flury M, Flühler H (1995) Tracer characteristics of brilliant blue FCF. Soil Sci Soc Am J 59(1):22–27

Flury M, Wai NN (2003) Dyes as tracers for vadose zone hydrology. Rev Geophys 41(1):2-1-2-37

Flury M, Flühler H, Jury WA, Leuenberger J (1994) Susceptibility of soils to preferential flow of water: a field study. Water Resour Res 30(7):1945–1954

Forman RTT (1997) Land mosaics – the ecology of landscapes and regions. Cambridge University Press, Cambridge

Forrer I, Papritz A, Kasteel R, Flühler H, Luca D (2000) Quantifying dye tracers in soil profiles by image processing. Eur J Soil Sci 51(2):313–322

Förstel H (1996) Hydrogen and oxygen isotopes in soil water: use of 18O and D in soil water to study the soil-plant-water system. In: Boutton TW (ed) Mass spectrometry of soils. Dekker, New York, pp 285–310

Forster C, Wandinger U, Wotawa G, James P, Mattis I, Althausen D, Simmonds P, O'Doherty S, Jennings SG, Kleefeld C, Schneider J, Trickl T, Kreipl S, Jäger W, Stohl A (2001) Transport of boreal forest fire emissions from Canada to Europe. J Geophys Res 106:22887–22906

Forster T, Schweingruber FH, Denneler B (2000) Increment puncher: a tool for extracting small cores of wood and bark from living trees. IAWA J 21:169–180

Forti MC, Neal C (1992) Hydrochemical cycles in tropical rainforests: an overview with emphasis on central Amazonia. J Hydrol 134:103–115

Francis CM (1990) Trophic structure of bat communities in the understory of lowland dipterocarp rainforest in Malaysia. J Trop Ecol 6:421–431

Francou B, Vuille M, Favier V, Caceres B (2004) New evidence for an ENSO impact on low-latitude glaciers: Antizana 15, Andes of Ecuador, 0° 28 S. J Geophys Res 109:D18106. DOI 10.1029/2003JD004484

Frankie GW, Baker HG, Opler PA (1974) Comparative phenological studies of trees in tropical wet and dry forests in the lowlands of Costa Rica. J Ecol 62:881–913

Franklin E, Hayek T, Fagundes EP, Silva LL (2004) Oribatid mite (Acari: Oribatida) contribution to decomposition dynamic of leaf litter in primary forest, second growth, and polyculture in the central Amazon. Braz J Biol 64:59–72

Frei E (1958) Eine Studie über den Zusammenhang zwischen Bodentyp, Klima und Vegetation in Ecuador. Plant Soil 9:215–236

Frei M, Böll A, Graf F, Heinimann HR, Springman S (2003) Quantification of the influence of vegetation on soil stability. In: Lee CF, Tham LG (eds) Proceedings of the international conference on slope engineering, 8–10 December 2003, Hong Kong, China. University of Hong Kong, Hong Kong, pp 872–877

Friesen DK, Rao IM, Thomas RJ, Oberson A, Sanz JI (1997) Phosphorous acquisition and cycling in crop and pasture systems in low fertility tropical soils. Plant Soil 196:289–294

Fujisaka S, Escobar G, Veneklaas EJ (2000) Weedy fields and forests: interactions between land use and the composition of plant communities in the Peruvian Amazon. Agric Ecosyst Environ 78:175–186

Fujiwara M, Kita K, Kawakami S, Ogawa T, Komela N, Sarasprinya S, Surigsto A (1999) Tropospheric ozone enhancements during the Indonesian fire events in 1994 and 1997 as revealed by ground-based observations. Geophys Res Lett 26:2417–24120

Gallrapp D (2005) Naturschutz im Wandel? Vision und Realität integrativer Naturschutzkonzepte, am Beispiel des Podocarpus-Nationalparks in Südecuador. Tübinger Geographische Studien 142:417–441

Garwood NC (1981) Earthquake-caused landslides in Panama: recovery of the vegetation. Res Rep Nat Geogr Soc 21:181–184

Garwood NC (1989) Tropical soil seed banks: a review. In: Leck MA, Parker VT, Simpson RL (eds) Ecology of soil seed banks. Academic, San Diego, pp 149–209

Garwood NC, Janos DP, Brokaw N (1979) Earthquake-caused landslides: a major disturbance to tropical forests. Science 205:997–999

Gaston KJ, Reavey D, Valladares GR (1992) Intimacy and fidelity: internal and external feeding by the British macrolepidoptera. Ecol Entomol 17:86–88

Gaston KJ, Scoble MJ, Crook A (1995) Patterns in species description: a case study using the Geometridae (Lepidoptera). Biol J Linn Soc 55:255–266

Gaston KJ, Blackburn TM, Spicer JI (1998) Rapoport's rule: time for an epitaph? Trends Ecol Evol 13:70–74

Gauld ID, Gaston KJ, Janzen DH (1992) Plant allelochemicals, tritrophic interactions and the anomalous diversity of tropical parasitoids – the nasty host hypothesis. Oikos 65:353–357

Gentry AH (1977) Endangered plant species and habitats of Ecuador and Amazonian Peru. In: Prange GT, Elias TS (eds) Extinction is forever. New York Botanical Garden, New York, pp 136–149

Gentry AH (1988) Changes in plant community diversity and floristic composition on environmental and geographical gradients. Ann Miss Bot Gard 75:1–34

Gentry AH (1995) Patterns of diversity and floristic composition in neotropical montane forests. In: Churchill SP, Balslev H, Forero E, Luteyn JL (eds) Biodiversity and conservation of neotropical montane forests. New York Botanical Garden, New York, pp 103–126

Gentry AH, Dodson CH (1987) Diversity and biogeography of neotropical vascular epiphytes. Ann Missouri Bot Gard 74:205–233

German-Heins J, Flury M (2000) Sorption of brilliant blue FCF in soils as affected by pH and ionic strength. Geoderma 97(1–2):87–101

Ghodrati M, Jury WA (1990) A field-study using dyes to characterize preferential flow of water. Soil Sci Soc Am J 54(6):1558–1563

Ghodrati M, Ernst FF, Jury WA (1990) Automated spray system for application of solutes to small field plots. Soil Sci Soc Am J 54(1):287–290

Gitay H, Wilson JB, Lee WG (1996) Species redundancy: a redundant concept? J Ecol 84:121–124

Glatzle A (1990) Weidewirtschaft in den Tropen und Subtropen. Ulmer, Stuttgart

Glesener RR, Tilman D (1978) Sexuality and components of environmental uncertainty – clues from geographic parthenogenesis in terrestrial animals. Am Nat 112:659–673

Godbold DL, Fritz HW, Jentschke G, Meesenburg H, Rademacher P (2003) Root turnover and root necromass accumulation of Norway spruce (*Picea abies*) were affected by soil acidity. Tree Physiol 23:915–921

Godfray GHC (2005) Taxonomy as information science. Proc Cal Acad Sci 56[Suppl 1]:170–181

Godoy R, Groff S, O'Neill K (1998) The role of education in neotropical deforestation: household evidence from amerindians in Honduras. Hum Ecol 26:649–671

Goldstein G, Andrade JL, Meinzer FC, Holbrook NM, Cavelier J, Jackson P, Celis A (1998) Stem water storage and diurnal patterns of water use in tropical forest canopy trees. Plant Cell Environ 21:397–406

Goller R (2004) Biogeochemical consequences of hydrologic conditions in a tropical montane rain forest in Ecuador. PhD thesis, University of Bayreuth, Bayreuth

Goller R, Wilcke W, Leng M, Tobschall HJ, Wagner K, Valarezo C, Zech W (2005) Tracing water paths through small catchments under a tropical montane rain forest in south Ecuador by an oxygen isotope approach. J Hydrol 308:67–80

Goller R, Wilcke W, Fleischbein K, Valarezo C, Zech W (2006) Dissolved nitrogen, phosphorus, and sulfur forms in the ecosystem fluxes of a montane forest in Ecuador. Biogeochemistry 77:57–89

Gómez-Hernández JJ, Journel AG (1993) Joint sequential simulation of multigaussian fields. In: Soares A (ed) Geostatistics Troia 1992. Kluwer, Dordrecht, pp 85–94

Gosler A (2004) Birds in the hand. In: Sutherland WJ, Newton I, Green RE (eds) Bird ecology and conservation. A handbook of techniques. Oxford University Press, Oxford, pp 85–118

Gotelli NJ, Colwell RK (2001) Quantifying biodiversity: procedures and pitfalls in the measurement and comparison of species richness. Ecol Lett 4:379–391

Gotellis NJ (2004) A taxonomic wish-list for community ecology. Phil Trans R Soc Lond Ser B 359:585–597

Graaf NR de, Poels RLH, Rompaey RSAR van (1999) Effects of silvicultural treatment on growth and mortality of rain forest in Surinam over long periods. For Ecol Manage 124:123–135

Graaf NR de, Filius AM, Huesca Santos AR (2003) Financial analysis of sustained forest management for timber perspectives for application of the CELOS management system in brazilian Amazon. For Ecol Manage 177:287–299

Gradstein SR, Frahm JP (1987) Die floristische Höhengliederung der Moose entlang des Bryotorop-Transektes in NO-Peru. Beih Nova Hedwigia 88:105–113

Gradstein SR, Churchill SP, Salazar Allen N (2001) A guide to the bryophytes of tropical America. Mem NY Bot Gard 86:1–577

Gradstein SR, Bock C, Mandl N, Nöske NM (2008) Bryophyta (Hepaticae) (Liverworts and Hornworts). In: Liede-Schumann S, Breckle SW (eds) Provisional checklists of fauna and flora of the San Francisco valley and its surroundings (Reserva San Francisco/Prov. Zamore–Chinchipe, southern Ecuador). Ecotrop Monogr 4 (in press)

Gradstein SR, Kessler M, Lehnert M, Abiy M, Homeier J, Mandl N, Makeschin F, Richter M (2007) Vegetation, climate, and soil of the unique Purdiaea forest of southern Ecuador. Basic Appl Ecol (submitted)

Graf K (1992) Pollendiagramme aus den Anden: Eine Synthese zur Klimageschichte und Vegetationsentwicklung seit der letzten Eiszeit. PhD thesis, University of Zurich, Zurich

Gräf M (1990) Endogener und gelenkter Kulturwandel in ausgewählten indianischen Gemeinden des Hochlandes von Ecuador. (Reihe Kulturwissenschaften, 13) Tuduv-Studien, Munich

Graham EA, Mulkey SS, Kitajima K, Phillips NG, Wright SJ (2003) Cloud cover limits net CO2 uptake and growth of a rainforest tree during tropical rainy seasons. Proc Natl Acad Sci USA 100:572–576

Graham GL (1983) Changes in bat species diversity along an elevational gradient up the Peruvian Andes. J Mamm 64:559–571

Graham GL (1990) Bats versus birds: comaprison among Peruvian volant vertebrate faunas along an elevational gradient. J Biogeogr 17:657–668

Granier A, Huc R, Colin F (1992) Transpiration and stomatal conductance of two rain forest species growing in plantations (*Simarouba amara* and *Goupia glabra*) in French Guyana. Ann Sci For 49:17–24

Granier A, Huc R, Barigah ST (1996) Transpiration of natural rainforest and its dependence on climatic factors. Agric Forest Meteor 78:19–29

Grau HR (2002) Scale dependent relationships between treefalls and species richness in a neotropical montane forest. Ecology 83:2591–2601

Graudal L, Kjaer ED, Thomsen A, Larsen AB (1997) Planning national programmes for conservation of forest genetic resources. (Technical note no. 48) Danida Forest Seed Centre, Humlebaek

Greene DF, Johnson EA (1995) Long-distance wind dispersal of tree seeds. Can J Bot 73:1036–1045

Gregory RD, Gibbons DW, Donald PF (2004) Bird census and survey techniques. In: Sutherland WJ, Newton I, Green RE (eds) Bird ecology and conservation. A handbook of techniques. Oxford University Press, Oxford, pp 17–56

Grieve IC, Proctor J, Cousins SA (1990) Soil variation with altitude on Volcán Barva, Costa Rica. Catena 17:525–534

Grimm EC (1987) CONISS: a Fortran 77 program for stratigraphically contrained cluster analysis by the method of the incremental sum of squares. Comput Geosci 13:13–35

Griscom HP, Ashton PMS, Berlyn GP (2005) Seedling survival and growth of native tree species in pastures: implications for dry tropical forest rehabilitation in central Panama. For Ecol Manage 218:306–318

Groisman PY, Legates DR (1994) The accuracy of United States precipitation data. Bull Am Meteorol Soc 75:215–227

Gross A (1998) Terrestrische Orchideen einer Hangrutschung im Bergwald Süd-Ecuadors: Verteilung, Phytomasse, Phänologie und Blütenmerkmale. PhD thesis, University of Ulm, Ulm

Grubb PJ (1977) Control of forest growth and distribution on wet tropical mountains, with special reference to mineral nutrition. Annu Rev Ecol Syst 8:83–107

Grubb PJ (1995) Mineral nutrition and soil fertility in tropical rain forests. In: Lugo AE, Lowe C (eds) Tropical forests: management and ecology. (Ecological studies vol 112) Springer, Berlin Heidelberg New York, pp 308–330

Grubb PJ, Withmore TC (1966) A comparison of montane and lowland rain forest in Ecuador. II. The climate and its effects on the distribution and physiognomy of the forests. J Ecol 54:303–333

Grubb PJ, Lloyd JR, Pennington TD, Whitmore TC (1963) A comparison of lowland and montane rain forest in Ecuador: I. The forest structure, physiognomy, and floristics. J Ecol 51:567–601

Guariguata MR (1990) Landslide disturbance and forest regeneration in the upper Luquillo mountains of Puerto Rico. J Ecol 78:814–832

Guariguata MR, Ostertag R (2001) Neotropical secondary forest succession: changes in structural and functional characteristics. For Ecol Manage 148:185–206

Guffroy J (2004) Catamayo precolombino, investigaciones arqueologicas en la provincia de Loja (Ecuador). IRD, Paris

Guimaraes RF, Montgomery DR, Greenberg HM, Fernandes NF, Gomes RAT, Carvalho OA de (2003) Parameterization of soil properties for a model of topographic controls on shallow landsliding: application to Rio de Janeiro. Eng Geol 69:99–108

Günter S (2001) Ökologie und Verjüngung von Mahagoni (*Swietenia macrophylla* King) in Naturwäldern Boliviens. PhD thesis, University of Goettingen, Goettingen

Günter S, Mosandl R (2003) Nachhaltige Naturwaldbewirtschaftung in Bergregenwäldern Südecuadors. Eine Option zur Erhaltung von Biodiversität. In: Mosandl R, El Kateb H, Stimm B (eds) Waldbau – weltweit. Beiträge zur internationalen Waldbauforschung. Forst Forsch Muench 192:10–23

Günter S, Stimm B, Weber M (2004) Silvicultural contributions towards sustainable management and conservation of forest genetic resources in Southern Ecuador. Lyonia 6:75–91

Günter S, Weber M, Erreis R, Aguirre N (2006): Influence of distance to forest edges on natural regeneration of abandoned pastures – a case study in the tropical mountain rain forest of Southern Ecuador. Eur J For Res. Available at: www.springerlink.com/content/j586261x03135613/fulltext.html

Hafkenscheid R (2000) Hydrology and biogeochemistry of tropical montane rain forests of contrasting stature in the Blue Mountains, Jamaica. PhD thesis, Free University of Amsterdam, Amsterdam

Hagedorn A (2001) Extent and significance of soil erosion in southern Ecuador. Erde 132:75–92

Hagedorn A (2002) Erosionsprozesse in Südecuador unter besonderer Berücksichtigung des Oberflächenabtrags. PhD thesis, University of Erlangen, Erlangen

Haggar JP, Briscoe, CB, Butterfield RP (1998) Native species: a resource for the diversification of forestry production in the lowland humid tropics. For Ecol Manage 106:195–203

Hairston NG, Smith FE, Slobodkin LB (1960) Community structure, population control, and competition. Am Nat 94:421–425

Hajibabaei M, Janzen DH, Burns JM, Hallwachs W, Hebert PDN (2006) DNA barcodes distinguish species of tropical Lepidoptera. Proc Natl Acad Sci USA 103:968–971

Hall JS, Harris DJ, Medjibe V, Ashton PMS (2003) The effects of selective logging on forest structure and tree species composition in a Central African forest: implications for management of conservation areas. For Ecol Manage 183:249–264

Hall M (1977) El volcanismo en el Ecuador. Biblioteca Ecuador, Quito

Halloy S (1990) A morphological classification of plants, with special reference to the New Zealand alpine flora. J Veg Sci 1:291–304

Hamann A (2004) Flowering and fruiting phenology of a Philippine submontane rain forest: climatic factors as proximate and ultimate causes. J Ecol 92:24–31

Hamann A, Curio E (1999) Interactions among frugivores and fleshy fruit trees in a Philippine submontane rain forest. Conserv Biol 13:766–773

Hamer KC, Hill JK (2000) Scale-dependent effects of habitat disturbance on species richness in tropical forests. Conserv Biol 14:1435–1440

Hamilton L, Juvik J, Scatena F (1995) The Puerto Rico tropical cloud forest symposium: introduction and workshop synthesis. Ecol Stud 110:1–19

Hamilton WD (1980) Sex versus non-sex versus parasite. Oikos 35:282–290

Hamilton WD (2001) Evolution of sex. (Narrow roads of gene land, vol 2) Oxford University Press, Oxford

Hammen T van der, Correal Urrego G (1978) Prehistoric man on the Sabana de Bogota: data for an ecological prehistory. Palaeogeogr Palaeoclimatol Palaeoecol 25:179–190

Hammen T van der, Ward PS (2005) Ants from the Ecoandes expedition: diversity and distribution. Stud Trop Andean Ecosyst 6:239–248

Hannam DAR (1986) Bracken poisoning in farm animals with special reference to the North York Moors. In: Smith RT, Taylor JA (eds) Bracken: ecology, land use and control technology, 1985. Parthenon, Leeds, pp 133–138

Hansen BCS, Rodbell DT, Seltzer GO, Leon B, Young KR, Abbott M (2003) Late-glacial and Holocene vegetational history from two sides in the western Cordillera of southwestern Ecuador. Palaeogeogr Palaeoclimat Palaeoecol 194:79–108

Hansen CP, Kjaer ED (1999) Appropriate planting material in tree plantings: opportunities and critical factors. In: CONAF (ed) International expert meeting on the role of planted forests for sustainable forest development. CONAF, Santiago

Hansen RA (2000) Effects of habitat complexity and composition on a diverse litter microarthropod assemblage. Ecology 81:1120–1132

Hanson PJ, Edwards NT, Garten CT, Andrews JA (2000) Separating root and soil microbial contributions to soil respiration: A review of methods and observations. Biogeochemistry 48:115–146

Harling G, Andersson L (1973–2003). Flora of Ecuador, vols 1–68. University of Goeteborg, Goeteborg

Harner M (1984) The Jívaro: people of the sacred waterfalls. University of California Press, Los Angeles

Hartig K (2000) Pflanzensoziologische Untersuchungen von anthropogen gestörten Flächen im tropischen Bergwald Südecuadors. PhD thesis, University of Bayreuth, Bayreuth

Hartig K, Beck E (2003) The bracken fern (*Pteridium arachnoideum* (Kaulf.) Macon) dilemma in the Andes of Southern Ecuador. Ecotropica 9:3–13

Hartshorn GS (1980) Neotropical forest dynamics. Biotropica 12:23–30

Hartshorn GS (1989) Gap-phase dynamics and tree species richness. In: Holm-Nielsen LB, Nielsen IC, Balslev H (eds) Tropical forests: botanical dynamics, speciation and diversity. Academic, London, pp 65–73

Hatton TJ, Moore SJ, Reece PH (1995) Estimating stand transpiration in a *Eucalyptus populnea* woodland with the heat pulse method: measurement errors and sampling strategies. Tree Physiol 12:219–227

Haug I, Lempe J, Homeier J, Weiß M, Setaro S, Oberwinkler F, Kottke I (2004) *Graffenrieda emarginata* (Melastomataceae) forms mycorrhizas with *Glomeromycota* and with a member of *Hymenoscyphus ericae* aggr. in the organic soil of a neotropical mountain rain forest. Can J Bot 82:340–356

Haug I, Weiß M, Homeier J, Oberwinkler F, Kottke I (2005) Russulaceae and Thelephoraceae form ectomycorrhizas with members of the Nyctaginaceae (Caryophyllales) in the tropical mountain rain forest of southern Ecuador. New Phytol 165:923–936

Hauhs M (1985) Wasser- und Stoffhaushalt im Einzugsgebiet der Langen Brahmke (Harz). Ber Forschungszentr Waldoekosyst Waldsterb 17

Häuser CL, Boppré M (1997) Pyrrolizidine alkaloid-related pharmacophagy in Neotropical moths (Lepidoptera). In: Ulrich H (ed) Tropical biodiversity and systematics. Museum Alexander Koenig, Bonn, pp 291–296

Häuser CL, Holstein J, Steiner A (2004) Das globale Artregister Tagfalter – GART: Ein Webbasiertes Informationssystem. Mitt Dtsch Ges Allg Angew Entomol 14:145–148

Häuser CL, Steiner A, Holstein J, Scoble MJ (2005) Digital imaging of biological type specimens – a manual of best practice. European Network for Biodiversity Information, Stuttgart

Häuser CL, Fiedler, K, Bartsch D, Brehm G, Kling M, Süßenbach D, Onore G (2008) Lepidoptera: Papilionoidea (Butterflies). In: Liede-Schumann S, Breckle SW (eds) Provisional checklists of fauna and flora of the San Francisco valley and its surroundings (Reserva San Francisco/Prov. Zamore–Chinchipe, southern Ecuador). Ecotrop Monogr 4 (in press)

Hausmann A (2001) The geometrid moths of Europe, vol 1. Apollo, Stenstrup

Hawkins BA, DeVries PJ (1996) Altitudinal gradients in the body sizes of Costa Rican butterflies. Acta Oecol 17:185–194

Hayek L-A, Buzas MA (1997) Surveying natural populations. Columbia University Press, New York

Heaney A, Proctor J (1989) Chemical elements in litter in forests on Volcán Barva, Costa Rica. In: Proctor J (ed) Mineral nutrients in tropical forest and savanna ecosystems. (Special publication no 9, British Ecological Society) Blackwell, Oxford, pp 255–271

Heaney A, Proctor J (1990) Preliminary studies on forest structure and floristics on Volcán Barva, Costa Rica. J Trop Ecol 6:307–320

Heinrich B, Mommsen TP (1985) Flight of winter moths near 0 °C. Science 228:177–179

Henderson A, Churchill SP, Luteyn JL (1991) Neotropical plant diversity. Nature 351:21–22

Heneghan L, Coleman DC, Zou X, Crossley DA, Haines BL (1999) Soil microarthropod contributions to decomposition dynamics: tropical-temperate comparisons of a single substrate. Ecology 80:1873–1882

Heppner JB (1991) Faunal regions and the diversity of Lepidoptera. Trop Lepid 2[Suppl 1]:1–85

Herrera CM, Jordano P, Guitián J, Traveset A (1998) Annual variability in seed production by woody plants and the masting concept: reassessment of principles and relationship to pollination and seed dispersal. Am Nat 152:576–594

Hertel D, Leuschner C (2007) Fine root mass and fine root production in tropical moist forests as dependent on soil, climate and elevation. In: Bruijnzeel LE, Juvik JO (eds) Mountains in the mist: science for conserving and managing tropical montane cloud forests, Hawaii University Press, Honolulu (in press)

Hertel D, Leuschner C, Holscher D (2003) Size and structure of fine root systems in old-growth and secondary tropical montane forests (Costa Rica). Biotropica 35:143–153

Herzog KM, Häsler R, Thum R (1995) Diurnal changes in the radius of a subalpine Norway spruce stem: their relation to the sap flow and their use to estimate transpiration. Trees 10:94–101

Herzog SK, Kessler M, Bach K (2005) The eleavtional gradient in Andean bird spescies richness at the local scale: a foothill peak and a high-elevation plateau. Ecography 28:209–222

Hietz P, Ausserer J, Schindler G (2002) Growth, maturation and survival of epiphytic bromeliads in a Mexican humid montane forest. J Trop Ecol 18:177–191

Hietz-Seifert U, Hietz P, Guevara S (1996) Epiphyte vegetation and diversity on remnant trees after forest clearance in southern Veracruz, Mexico. Biol Conserv 75:103–111

Hill JK, Hamer KC (2004) Determining impacts of habitat modification on diversity of tropical forest fauna: the importance of spatial scale. J Appl Ecol 41:744–754
Hilpmann J (2003) Waldkundliche Untersuchungen an zwei Höhentransekten im tropischen Bergregenwald Südecuadors. BSc thesis, TU Dresden, Dresden
Hilt N (2005) Diversity and species composition of two different moth families (Lepidoptera: Arctiidae vs. Geometridae) along a successional gradient in the Ecuadorian Andes. PhD thesis, University of Bayreuth, Bayreuth. Available at: http://opus.ub.uni-bayreuth.de/volltexte/2006/201/
Hilt N, Fiedler K (2005) Diversity and composition of Arctiidae moth ensembles along a succession gradient in the Ecuadorian Andes. Divers Distrib 11:387–398
Hilt N, Fiedler K (2006) Arctiid moth ensembles along a successional gradient in the Ecuadorian montane rain forest zone: how different are subfamilies and tribes? J Biogeogr 33:108–120
Hilt N, Brehm G, Fiedler K (2006) Diversity and ensemble composition of geometrid moths along a successional gradient in the Ecuadorian Andes. J Trop Ecol 22:155–166
Hilt N, Brehm G, Fiedler K (2007) Temporal dynamics of rich moth ensembles in the montane forest zone in southern Ecuador. Biotropica 39:94–104
Hobie SE, Jensen DB, Chapin FS III (1994) Resource supply and disturbance as control over present and future plant diversity. In: Schulze E-D, Mooney HA (eds) Biodiversity and ecosystem function. Ecol Stud 99:385–408
Hoch G, Körner C (2003) The carbon charging of pines at the climatic treeline: a global comparison. Oecologia 135:10–21
Hoch G, Körner C (2005) Growth, demography and carbon relations of *Polylepis* trees at the world's highest treeline. Funct Ecol 19:941–951
Hofstede R, Ambrose K, Baéz S, Cueva K (2007) Biodiversity-based livelihoods in Ceja Andina forests: multi-stakeholder learning processes in Northern Ecuador for the sustainable use of cloud forests. In: Bruijnzeel LA, Juvik J, Scatena FN, Hamilton LS, Bubb P (eds) Second international symposium mountains in the mist: science for conserving and managing tropical montane cloud forests. University of Hawaii Press, Honolulu (in press)
Holdridge LR, Grenke WC, Hathway WH, Liang T, Tosi JA (1971) Forest environments in tropical lifezones – a pilot study. Pergamon, Oxford
Holl KD (1999) Factors limiting tropical rain forest regeneration in abandoned pasture: seed rain, seed germination, microclimate, and soil. Biotropica 31:229–242
Holl KD, Loik ME, Lin EHV, Samuels IA (2000) Tropical montane forest restoration in Costa Rica: Overcoming barriers to dispersal and establishment. Restor Ecol 8:339–349
Holloway JD (1984) The larger moths of Gunung Mulu National Park; a preliminary assessment of their distribution, ecology and potential as environmental indicators. Sarawak Mus J 30:149–190
Holloway JD, Nielsen ES (1999) Biogeography of the Lepidoptera. In: Kristensen NP (ed) Handbook of zoology, Lepidoptera: moths and butterflies, vol 1. de Gruyter, Berlin, pp 423–462
Holloway JD, Kibby G, Peggie D (2001) The families of Malesian moths and butterflies. Brill, Leiden
Hölscher D, Abreu Sá LD de, Bastos TX, Denich M, Fölster H (1997) Evaporation from young secondary vegetation in eastern Amazonia. J Hydrol 193:293–305
Homeier J (2004) Baumdiversität, Waldstruktur und Wachstumsdynamik zweier tropischer Bergregenwälder in Ecuador und Costa Rica. (Dissertationes botanicae 391) PhD thesis, University of Bielefeld, Bielefeld
Homeier J (2005a) *Graffenrieda emarginata* (Ruiz & Pav) Triana. In: Schütt P, Schuck HJ, Lang U, Roloff A (eds) Enzyklopädie der Holzgewächse, vol 42. Ecomed, Munich
Homeier J (2005b) *Purdiaea nutans* Planch. In: Schütt P, Schuck HJ, Lang U, Roloff A (eds) Enzyklopädie der Holzgewächse, vol 42. Ecomed, Munich
Homeier J, Werner FA (2008) Spermatophyta. In: Liede-Schumann S, Breckle SW (eds) Provisional checklists of fauna and flora of the San Francisco valley and its surroundings (Reserva San Francisco/Prov. Zamore–Chinchipe, southern Ecuador). Ecotrop Monogr 4 (in press)
Homeier J, Dalitz H, Breckle SW (2002) Waldstruktur und Baumartendiversität im momentanen Regenwald der Estación Científica San Francisco in Südecuador. Ber Reinhold-Tüxen Ges 14:109–118

Hooghiemstra H (1984) Vegetational and climatic history of the HIGH PLAIN of Bogota, Colombia: a continuous record of the last 3.5 million years. Diss Bot 1984:1–368

Hörmann G, Branding A, Clemens T, Herbst M, Hinrichs A, Thamm F (1996) Calculation and simulation of wind-controlled canopy interception of a beech forest in northern Germany. Agric For Meteorol 79:131–148

Howe, HF, Smallwood J (1982) Ecology of seed dispersal. Annu Rev Ecol Syst 13:201–228

Hubbell SP (2001) The unified neutral theory of biodiversity and biogeography. Princeton Univ Press, Princeton

Hubbell SP, Foster RB, O'Brien ST, Harms KE, Condit R, Wechsler B, Wright SJ, Loo de Lao S (1999) Light-gap disturbances, recruitment limitation, and tree diversity in a Neotropical forest. Science 283:554–557

Hufford KM, Mazer SJ (2003) Plant ecotypes: genetic differentiation in the age of ecological restoration. Trends Ecol Evol 18:147–155

Hughes RF, Kauffman JB, Jaramillo VJ (2000) Ecosystem-scale impacts of deforestation and land use in a humid tropical region of Mexico. Ecol Appl 10:515–527

Humboldt A von (1806) Ideen zu einer Physiognomik der Gewächse. Cotta, Tübingen

Hunter MD, Price PW (1992) Playing chutes and ladders: heterogeneity and the relative roles of bottom-up and top-down forces in natural communities. Ecology 73:724–732

Hurlbert SH (1971) The nonconcept of species diversity: a critique and alternative parameters. Ecology 52:577–586

Huston MA (1994) Biological diversity – the coexistence of species in changing landscapes. Cambridge University Press, Cambridge

Hutchinson ID (1993) Puntos de partida y muestreo silvicultural de bosques naturales del tróico húmedo. (Colección silvicultura y manejo de bosques naturales no 7. Seria técnica: informe técnico no 204) CATIE, Turrialba

Hutchinson JN (1988) Morphological and geotechnical parameters of landslides in relation to geology and hydrology, general report. In: Bonnard C (ed) Landslides. (Proceedings of the fifth international symposium on landslides) 1:3–35

Hutjes RWA, Kabat P, Running SW, Shuttleworth WJ, Field C, Bass B, Dias M, Avissar R, Becker A, Claussen M, Dolman AJ, Feddes RA, Fosberg M, Fukushima Y, Gash JHC, Guenni L, Hoff H, Jarvis PG, Kayane I, Krenke AN, Liu C, Meybeck M, Nobre CA, Oyebande L, Pitman A, Pielke RA, Raupach M, Saugier B, Sschulze ED, Sellers PJ, Tenhunen JD, Valentini R, Victoria RL, Vorosmarty CJ (1998) Biospheric aspects of the hydrological cycle – preface. J Hydrol 213:1–21

Hutley LB, Doley D, Yates DJ, Boonsaner A (1997) Water balance of an Australian subtropical rainforest at altitude: the ecological and physiological significance of intercepted cloud and fog. Aust J Bot 45:311–329

Ibisch P (1996) Neotropische Epiphytendiversität – das Beispiel Bolivien. PhD thesis, Martina Galunder, Wiehl

Ibisch P, Boegner A, Nieder J, Barthlott W (1996) How diverse are neotropical epiphytes? An analysis based on the 'Catalogue of the flowering plants and gymnosperms of Peru'. Ecotropica 2:13–28

Illig J, Langel R, Norton RA, Scheu S, Maraun M (2005) Where are the decomposers? Uncovering the soil food web of a tropical montane rain forest in southern Ecuador using stable isotopes (15N). J Trop Ecol 21:589–593

Illig J, Sandmann D, Schatz H, Scheu S, Maraun M (2008) Orbatidae (Mites). In: Liede-Schumann S, Breckle S-W (eds) Provisional checklists of fauna and flora of the San Francisco valley and its surroundings (Reserva San Francisco/Prov. Zamore–Chinchipe, southern Ecuador). Ecotrop Monogr 4 (in press)

Illig J, Schatz H, Scheu S, Maraun M (2007b) Decomposition rates and microarthropod colonization of litterbags with different litter types (*Graffenrieda emarginata*, *Purdiaea nutans*) in a tropical montane rain forest in southern Ecuador. J Trop Ecol (in press)

INEC (1962) II censo nacional de población y I de vivienda 1962. Instituto Nacional de Estadística y Censos, Quito

INEC (1974) III censo nacional de población y II de vivienda 1974. Instituto Nacional de Estadística y Censos, Quito
INEC (1982) IV censo de población y III de vivienda 1982. Instituto Nacional de Estadística y Censos, Quito
INEC (1990) V censo de población y IV de vivienda 1990. Instituto Nacional de Estadística y Censos, Quito
INEC (2001) VI censo de población y V de vivienda 2001. Instituto Nacional de Estadística y Censos, Quito
INEC (2004) Plan migración, comunicación y desarrollo, Ecuador–España (2004) El proceso emigratorio en la provincia de Loja.
Intachat J, Holloway JD, Speight MR (1997) The effect of different forest management practices on geometrid moth populations and their diversity in Peninsular Malaysia. J Trop For Sci 9:411–430
Intachat J, Holloway JD, Speight MR (1999) The impact of logging on geometroid moth populations and their diversity in lowland forests of peninsular Malaysia. J Trop For Sci 11:61–78
Jackson SM, Fredericksen, TS, Malcolm, JR (2002) Area disturbed and residual stand damage following logging in a Bolivian tropical forest. For Ecol Manage 166:271–283
Jacobson NL, Weller SJ (2002) A cladistic study of the Arctiidae (Lepidoptera) by using characters of the immatures and adults. (Thomas Say Public Entomol) Entomol Soc Am, Lanham
James SA, Clearwater MJ, Meinzer FC, Goldstein G (2002) Heat dissipation sensors of variable length for the measurement of sap flow in trees with deep sapwood. Tree Physiol 22:277–283
Janos DP (1980) Vesicular-arbuscular mycorrhizae affect lowland tropical rain forest plant growth. Ecology 6:151–162
Janzen DH (1982) Seed removal from fallen guanacaste fruits (*Enterolobium cyclocarpum*) by spiny pocket mice (*Liomys salvini*). Brenesia, 19/20:425–429
Janzen DH (1992) A south–north perspective on science in the management, use and economic development of biodiversity. In: Sandlund OT, Hindar K, Brown AHD (eds) Conservation of biodiversity for sustainable development. UN, Oslo, pp 27–52
Janzen DH (1994) Wildland biodiversity management in the tropics: where are we now and where are we going? Vida Silv Neotrop 3:3–15
Janzen DH, Hallwachs W (2007) Philosophy, navigation and use of a dynamic database (ACG Caterpillars SRNP) for an inventory of the macrocaterpillar fauna, and its foodplants and parasitoids, of the Area de Conservación Guanacaste (ACG), north-western Costa Rica. Available at: http://janzen.sas.upenn.edu
Jara AK, Romero JM (2005) Aspectos fenológicos y calidad de semillas de cuatro especies forestales nativas de bosque tropical de montaña, para restauración de hábitats. PhD thesis, Universidad Técnica particular de Loja, Loja
Jarvis PG (1995) Scaling processes and problems. Plant Cell Environ 18:1079–1089
Jenkinson DS, Brookes PC, Powlson PS (2004) Measuring soil microbial biomass. Soil Biol Biochem 36:5–7
Jiménez A, López F (1999) Guia de las aves del bosque nublado de San Francisco, Parque Nacional Podocarpus. Arcoiris, Loja
Jiménez MS, Morales D, Kučera J, Čermak J (1999) The annual course of transpiration in a laurel forest of Tenerife. Estimation with *Myrica faya*. Phyton 39:85–90
Jin VL, West LT, Haines BL, Peterson CJ (2000) P retention in tropical pre-montane soils across forest-pasture interfaces. Soil Science 165:881–889
Joergensen RG, Meyer B, Mueller T (1994) Timecourse of the soil microbial biomass under wheat: a one year field study. Soil Biol Biochem 26:987–994
Jordan CF, Heuveldop J (1981) The water budget of an Amazonian rain forest. Acta Amazonica 11:87–92
Jordan CF, Kline JR (1977) Transpiration of trees in tropical rain forest. J Appl Ecol 14:853–860
Jordan E (1983) Die Verbreitung von *Polylepis*-Beständen in der Westkordillere Boliviens. Tuexenia 3:101–112

Jørgensen PM, Leon-Yanez S (1999) Catalogue of the vascular plants of Ecuador. Monogr Syst Bot Mis Bot Gard 75
Jørgensen PM, Ulloa Ulloa C (1994) Seed plants of the high Andes of Ecuador – a checklist. Aarhus Univ Rep 34
Jost L (2004) Explosive local radiation of the genus *Teagueia* (Orchidaceae) in the upper Pastaza watershed of Ecuador. Lyonia 7:42–47
Kalko EKV, Handley CO, Handley D (1996) Organisation, diversity and long term dynamics of a neotropical bat community. In: Cody M, Smallwood J (eds) Longterm studies of vertebrate communities. Academic, Los Angeles, pp 503–553
Kammesheidt L, Dagang AA, Schwarzmüller W, Weidelt HJ (2003) Growth patterns in treated and untreated plots. For Ecol Manage 174:437–445
Kaneko N, Salamanca E (1999) Mixed leaf litter effects on decomposition rates and soil microarthropod communities in an oak-pine stand in Japan. Ecol Res 14:131–138
Kapos V, Pallant E, Bien A, Freskos S (1990) Gap frequencies in lowland rainforest sites on contrasting soils in Amazonian Ecuador. Biotropica 22:218–225
Kappelle M, Uffelen JG, Cleef AM van (1995) Altitudinal zonation of montane *Quercus* forests along two transects in the Chirripó National Park, Costa Rica. Vegetatio 119:119–153
Karr JR (1981) Surveying birds in the tropics. In: Ralph CJ, Scott JM (eds) Estimating the number of terrestrial birds. (Studies in avian biology 6) Cooper Ornithological Society, Lawrence, pp 548–553
Kattan GH, Alvarez-López H, Giraldo M (1994) Forest fragmentation and bird extinctions: San Antonio eighty years later. Conserv Biol 8:138–146
Kauffman S, Sombroek W, Mantel S (1998) Soils of rainforests. In: Schulte A, Ruhiyat D (eds) Soils of tropical forest ecosystems: characteristics, ecology, and management. Springer, Berlin Heidelberg New York, pp 9–20
Keefer DK (1984) Landslides caused by earthquakes. Geol Soc Am Bull 95:406–421
Kelliher FM, Köstner BMM, Hollinger DY, Byers JN, Hunt JE, Mcseveny TM, Meserth R, Weir PL, Schulze E-D (1992) Evaporation, xylem sap flow and tree transpiration in a New Zealand broad-leaved forest. Agric For Meteorol 62:53–73
Kelting DL, Burger JA, Edwards GS (1998) Estimating root respiration, microbial respiration in the rhizosphere and root-free soil respiration in forest soils. Soil Biol Biochem 30:961–968
Kennerly JB (1973) Geology of the Loja Province. Rep Inst Geol Sci 23
Kessler M (1992) The vegetation of south-west Ecuador. In: Best BJ (ed) The threatened forests of south-west Ecuador. Biosphere, Leeds, pp 79–100
Kessler M (1995) Polylepis-Wälder Boliviens: Taxa, Ökologie, Verbreitung und Geschichte. Diss Bot 246
Kessler M (1999) Plant species richness and endemism during natural landslide succession in a per humid montane forest in the Bolivian Andes. Ecotropica 5:123–136
Kessler M (2001) Pteridophyte species richness in Andean forests in Bolivia. Biodivers Conserv 10:1473–1495
Kessler M (2002a) The elevational gradient of Andean plant endemism: varying influences of taxon-specific traits and topography at different taxonomic levels. J Biogeogr 29:1159–1165
Kessler M (2002b) Environmental patterns and ecological correlates of range size among bromeliad communities of Andean forests in Bolivia. Bot Rev 68:100–127
Kessler M, Kessler PJA, Gradstein SR, Bach K, Schmull M, Pitopang R (2005) Tree diversity in primary forest and different land use systems in Central Sulawesi, Indonesia. Biodivers Conserv 14:547–560
King DA (1991) Correlations between biomass allocation, relative growth rate and light environment in tropical forest saplings. Funct Ecol 5:485–492
Kinner DA, Stallard RF (2004) Identifying storm flow pathways in a rainforest catchment using hydrological and geochemical modelling. Hydrol Process 18:2851–2875
Kitayama K, Aiba S-I (2002) Ecosystem structure and productivity of tropical rain forests along altitudinal gradients with contrasting soil phosphorus pools on Mount Kinabalu, Borneo. J Ecol 90:37–51

Kitching IJ, Rawlins JE (1999) The Noctuoidea. In: Kristensen NP (ed) Lepidoptera: moths and butterflies. (Handbook of zoology, vol 1) Gruyter, Berlin, pp 355–401

Kitching RL, Orr AG, Thalib L, Mitchell H, Hopkins MS, Graham AW (2000) Moth assemblages as indicators of environmental quality in remnants of upland Australian rain forest. J Appl Ecol 37:284–297

Klose S, Wernecke K, Makeschin F (2003) Microbial biomass and enzyme activities in coniferous forest soils as affected by lignite-derived deposition. Biol Fert Soils 38:32–44

Klose S, Wernecke K, Makeschin F (2004) Microbial activities in forest soils exposed to chronic depositions from a lignite power plant. Soil Biol Biochem 36:1913–1923

Knowles OH, Parrotta JA (1995) Amazonian forest restoration: an innovative system for native species selection based on phenological data and field performance indices. Commonw For Rev 74:230–243

Koba K, Tokuchi N, Hobble EA, Iwatsubo G (1998) Natural abundance of nitrogene-15 in a forest soil. Soil Sci Soc Am J 62:778–781

Kögel I (1986) Estimation and decomposition pattern of the lignin component in forest humus layers. Soil Biol Biochem 18:589–594

Köhler L (2002) Die Bedeutung der Epiphyten im ökosystemaren Wasser- und Nährstoffumsatz verschiedener Altersstadien eines Bergregenwaldes in Costa Rica. Ber Forschungszentr Waldökosysteme A 181:1–134

König N, Baccini P, Ulrich B (1986) Der Einfluß der natürlichen organischen Substanzen auf die Metallverteilung zwischen Boden und Bodenlösung in einem sauren Waldboden. Z Pflanzenernähr Bodenk 149:68–83

Koopowitz H, Thornhill AD, Andersen M (1994) A general stochastic model for the prediction of biodiversity losses based on habitat conversion. Conserv Biol 8:425–438

Koppmann R, Khedim A, Rudolph J, Poppe D, Andreae MO, Helas G, Welling M, Zenker T (1997) Emissions of organic trace gases from savanna fires in southern Africa during the 1992 Southern African Fire Atmosphere Research Initiative and their impact on the formation of tropospheric ozone. J Geophys Res 102:18879–18888

Koptur S, Haber WA, Frankie GW, Baker HG (1988) Phenological studies of shrubs and treelet species in tropical cloud forests of Costa Rica. J Trop Ecol 4:323–346

Koren I, Kaufman YJ, Remer LA, Marius JV (2004) Measurement of the effect of Amazon smoke on inhibition of cloud formation. Science 303:342–345

Körner C (1999) Alpine plant life. Functional plant ecology of high mountain ecosystems. Springer, Berlin Heidelberg New York

Körner C (2000) Why are there global gradients in species richness? Mountains might hold the answer. Trends Ecol Evol 15:513–514

Korning J, Balslev H (1994) Growth rates and mortality patterns of tropical lowland tree species. J Trop Ecol 10:151–166

Köstner BMM, Falge EM, Alsheimer M, Geyer R, Tenhunen JD (1998a) Estimating tree canopy water use via xylem sapflow in an old Norway spruce forest and a comparison with simulation-based canopy transpiration estimates. Ann Sci For 55:125–139

Köstner BMM, Granier A, Čermák J (1998b) Sapflow measurements in forest stands: methods and uncertainties. Ann Sci For 55:13–27

Kottke I (2002) Mycorrhizae – rhizosphere determinants of plant communities. In: Waisel Y, Eshel A, Kafkafi U (eds) Plant roots: the hidden half, 3rd edn. Dekker, New York, pp 919–932

Kottke I, Haug I (2004) The significance of mycorrhizal diversity of trees in the tropical mountain forest of southern Ecuador. Lyonia 7:50–56

Kottke I, Nebel M (2005) The evolution of mycorrhiza-like associations in liverworts: an update. New Phytol 167:330–334

Kottke I, Beck A, Oberwinkler F, Homeier J, Neill D (2004) Arbuscular endomycorrhizas are dominant in the organic soil of a neotropical montane cloud forest. J Trop Ecol 20:125–129

Kottke I, Haug I, Setaro S, Suárez JP, Weiß M, Preußing M, Nebel M, Oberwinkler F (2007) Guilds of mycorrhizal fungi and their relation to trees, ericads, orchids and liverworts in a neotropical mountain rain forest. Basic Appl Ecol. DOI: 10.1016/j.baae.2007.03.007

Kotze DJ, Samways MJ (2001) No general edge effects for invertebrates at Afromontane forest/grassland ecotones. Biodivers Conserv 10:443–466

Koutika LS, Bartoli F, Andreux F, Cerri CC, Burtin G, Choné T, Philippy R (1997) Organic matter dynamics and aggregation in soils under rain forest and pastures of increasing age in the eastern Amazon Basin. Geoderma 76:87–112

Krabbe N (2000) The birds of Ecuador. Voice Recordings, Quito

Krashevskaya V (2007) Testacea. In: Liede-Schumann S, Breckle SW (eds) Provisional checklists of fauna and flora of the San Francisco valley and its surroundings (Reserva San Francisco/Prov. Zamore–Chinchipe, southern Ecuador). Ecotrop Monogr 4 (in press)

Krell FT (2004) Parataxonomy vs taxonomy in biodiversity studies – pitfalls and applicability of 'morphospecies' sorting. Biodivers Conserv 13:785–812

Kricher J (1997) A neotropical companion. An introduction to the animals, plants and ecosystems of the new world tropics. Princeton University Press, Princeton

Krömer T, Gradstein SR (2004) Species richness of vascular epiphytes in two primary forests and fallows in the Bolivian Andes. Selbyana 24:190–195

Krömer, T, Kessler M, Gradstein SR, Acebey A (2005) Diversity patterns of vascular epiphytes along an elevational gradient in the Andes. J Biogeogr 32:1799–1809

Krömer T, Kessler M, Herzog SK (2006) Distribution and flowering ecology of bromeliads along two climatically contrasting elevational transects in the Bolivian Andes. Biotropica 38:183–195

Kron KA, Powell EA, Luteyn JL (2002) Phylogenetic relationships within the blueberry tribe (Vaccinieae, Ericaceae) based on sequence data from matK and nuclear ribosomal ITS regions, with comments on the placement of *Satyria*. Am J Bot 89:327–336

Krueger W (2004) Effects of future crop tree flagging and skid trail planning on conventional diameter-limit logging in a Bolivian tropical forest. For Ecol Manage 188:381–393

Kubota Y, Murata H, Kikuzawa K (2004) Effects of topographic heterogeneity on tree species richness and stand dynamics in a subtropical forest in Okinawa island, southern Japan. J Ecol 92:230–240

Kulli B, Gysi M, Fluhler H (2003) Visualizing soil compaction based on flow pattern analysis. Soil Till Res 70(1):29–40

Küper W, Kreft H, Nieder J, Köster N, Barthlott W (2004) Large-scale diversity patterns of vascular epiphytes in neotropical montane rain forests. J Biogeogr 31:1477–1487

Kupfer A, Langel R, Scheu S, Himstedt W, Maraun M (2006) Trophic ecology of a tropical aquatic and terrestrial food web: insights from stable isotopes (^{15}N). J Trop Ecol 22: 469–476

Küppers M, Giersch C, Schneider H, Kirschbaum MUF (1997) Leaf gas exchange in light- and sunflecks: response patterns and simulations. In: Rennenberg H, Eschrich W, Ziegler H (eds) Trees – contribution to modern tree physiology. Klein, Leiden, pp 77–96

Küppers M, Heiland I, Schneider H, Neugebauer PJ (1999) Lightflecks cause non-uniform stomatal opening – Studies with special emphasis on *Fagus sylvatica* L. Trees 14:130–144

Kürschner H, Parolly G (2004a) Ecosociological studies in Ecuadorian bryophyte communities. I. Syntaxonomy, life strategies and ecomorphology of the oreal epiphytic vegetation of S Ecuador. Nova Hedwigia 78:1–43

Kürschner H, Parolly G (2004b) Phytomass and water storing capacity of epiphytic bryophyte communities in Andean rain forests. Ecosociological studies in Ecuadorian bryophyte communities IV. Bot Jahrb Syst 125:489–504

Kürschner H, Parolly G (2008) Bryophyta (Musci). In: Liede-Schumann S, Breckle SW (eds) Provisional checklists of fauna and flora of the San Francisco valley and its surroundings (Reserva San Francisco/Prov. Zamore–Chinchipe, southern Ecuador). Ecotrop Monogr 4 (in press)

Kuzyakov Y (2002) Separating microbial respiration of exudates from root respiration in non-sterile soils: a comparison of four methods. Soil Biol Biochem 34:1621–1631

Lajtha K, Michener RH (1994) Stable isotopes in ecology and environmental science. Blackwell, London

Lamas G, Nielsen ES, Robbins RK, Häuser CL, Jong R de (2000) Developping and sharing data globally: the 'global butterfly information system' – GloBIS. In: Gazzoni DL (ed) XXI international congress of entomology, Foz do Iguassu 2000, Brazil: abstracts, vol 1. Embrapa Soja, Londrina, p. 196

Lamb D (1998) Large-scale ecological restoration of degraded tropical lands: the potential role of timber plantations. Restor Ecol 6:271–279

Lamoureux JF, Morrison JC, Ricketts TH, Olson DM, Dinerstein E, McKnight MW, Shugart HH (2006) Global tests of biodiversity concordance and the importance of endemism. Nature 440:212–214

Larcher W (1975) Pflanzenökologische Beobachtungen in der Páramostufe der venezolanischen Anden. Anz Math Naturwiss Klasse Oesterr Akad Wiss 11:194–213

Larcher W (2003) Physiological plant ecology, 4th edn. Springer, Berlin Heidelberg New York

Larrea M (1997) Respuesta de las epífitas vasculares a differentes formas de manejo del bosque nublado, Bosque Protegido Sierrazul, zona de amortiguamento de la Reserva Ecológica Cayambe-Coca, Napo, Ecuador. In: Mena PA, Soldi A, Alarcon R, Chiriboga C, Suarez L (eds) Estudios biológicos para la conservación. EcoCiencia, Quito, pp 321–346

Laubacher G, Megard F (1985) The Hercynian basement: a review. In: Pitcher WS, Atherton MP, Cobbing EJ, Beckinsale RD (eds) Magmatism at a plate edge – the Peruvian Andes. Wiley, London, pp 29–35

Lauer W (1976) Zur hygrischen Höhenstufung tropischer Gebirge. In: Schmithüsen J (ed) Neotropische Ökosysteme. Biogeographica 7:169–182

Lauer W (1978) Timberline studies in central Mexico. Arctic Alp Res 10:383–396

Lauer W, Erlenbach W (1987) Die tropischen Anden. Geoökologische Raumgliederung und ihre Bedeutung für den Menschen. Geogr Runds 39:86–95

Lauer W, Rafiqpoor MD (1986) Geoökologische Studien in Ecuador. Erdkunde 40:68–72

Lavigne MB, Boutin R, Foster RJ, Goodine G, Bernier PY, Robitaille G (2003) Soil respiration responses to temperature are controlled more by roots than by decomposition in balsam fir ecosystems. Can J Soil Sci 33:1744–1753

Lawton JH, Bignell DE, Bolton B, Bloemers GF, Eggleton P, Hammond PM, Hodda M, Holt RD, Larsen TB, Mawdsley NA, Stork NE, Srivastava DE, Watt AD (1998) Biodiversity inventories, indicator taxa and effects of habitat modification in tropical forest. Nature 391: 72–76

Le Brocque AF, Buckney RT (1997) Multivariate relationships between floristic composition and stand structure in vegetation of Ku-ring-gai Chase National Park, New South Wales. Aust J Bot 45:1033–1044

Lee M, Nakane K, Nakatsubo T, Koizumi H (2003) Seasonal changes in the contribution of root respiration to total soil respiration in a cool-temperate deciduous forest. Plant and Soil 255:311–318

Legendre P, Gallagher EC (2001) Ecologically meaningful transformations for ordination of species data. Oecologia 129:271–280

Legendre P, Legendre L (1998) Numerical ecology, 2nd edn. Elsevier, Amsterdam

Lehnert M, Kessler M, Salazar LI, Navarrete H, Werner FA, Gradstein SR (2008) Pteridophyta. In: Liede-Schumann S, Breckle SW (eds) Provisional checklists of fauna and flora of the San Francisco valley and its surroundings (Reserva San Francisco/Prov. Zamore–Chinchipe, southern Ecuador). Ecotrop Monogr 4 (in press)

Leischner B (2005) Phänologie, Saatgutproduktion, Keimung und Anzucht einheimischer Baumarten des tropischen Bergregenwaldes Südecuadors. PhD thesis, Technical University of Munich, Munich

Leischner B, Bussmann RW (2002) Mercado y uso de madera en el sur de Ecuador. In: Bussmann RW, Lange S (eds) Conservación de la biodiversidad en los Andes y la Amazonía – conservation of biodiversity in the Andes and the Amazon. Ceres, Lima, pp 651–660

Leischner B, Stimm B, Weber M (2004) Reforestation with native species – possibilities for alternative uses and conservation of biodiversity at the example of southern Ecuador (in German). Bundesamt Nat Treffpunkt Biol Vielfalt 4:71–76

León Yánez S, Gradstein SR, Wegner C (2006) Catálogo de hepáticas (Marchantiophyta) y antoceros (Anthocerophyta) del Ecuador. Herbario QCA, Quito

Leopold AC, Andrus R, Finkeldey A, Knowles D (2001) Attempting restoration of wet tropical forests in Costa Rica. For Ecol Manage 142:243–249

Leopoldo LFW (1993) Water balance and hydrologic characteristics of a rain forest catchment in the central Amazonian Basin. Water Resour Res 29:759–773

Leopoldo PR, Franken W, Villa-Nova N (1995) Real evaporation and transpiration through a tropical rainforest as estimated by the water balance method. For Ecol Manage 73:185–195

Lettau HH (1976) Dynamic and energetic factors which cause aridity along South America's Pacific coast. World Surv Climatol 12:188–192

Leuschner C, Hertel D, Coners H, Büttner V (2001) Root competition between beech and oak: a hypothesis. Oecologia 126:276–284

Levin SA, Paine RT (1974) Disturbance, patch formation, and community structure. Proc Natl Acad Sci USA 71:2744–2747

Li Y, Xu M, Sun OJ, Cui W (2004) Effects of root and litter exclusion on soil CO_2 efflux and microbial biomass in wet tropical forests. Soil Biol Biochem 36:2111–2114

Liao J (2004) Woodland development and soil carbon and nitrogen dynamics and storage in a subtropical savanna ecosystem. PhD thesis, Texas A&M University, Austin

Lieberman D, Lieberman M, Hartshorn G, Peralta R (1985) Growth rates and age-size relationships of tropical wet forest trees in Costa Rica. J Trop Ecol 1:97–109

Lieberman D, Lieberman M, Peralta R, Hartshorn GS (1996) Tropical forest structure and composition on a large-scale altitudinal gradient in Costa Rica. J Ecol 84:137–152

Likens GE, Bormann FH (1995) Biogeochemistry of a forested ecosystem, 2nd edn. Springer, Berlin Heidelberg New York

Lin Y-L (2003) The dynamics of orographic precipitation. (Preprints of the Harold D Orville symposium, 26 April 2003) Institute of Atmospheric Sciences, Rapid City, pp 68–89

Litherland M, Aspen JA, Jemielita RA (1994) The metamorphic belts of Ecuador. Overseas Mem Br Geol Surv 11:1–147

Lloyd CR, Marques AdO (1988) Spatial variability of throughfall and stemflow measurements in Amazonian rainforests. Agric For Meteorol 42:63–73

Loder N, Gaston KJ, Warren PH, Arnold HR (1998) Body size and feeding specificity: macrolepidoptera in Britain. Biol J Linn Soc 63:121–139

Lodhiyal N, Lodhiyal LS (2003) Biomass and net primary productivity of Bharbar Shisham forests in central Himalaya, India. For Ecol Manage 176:217–235

Loker WM (1997) Pasture performance and sustainability in the Peruvian Amazon: results of long-term on-farm research. Agric Syst 55:385–408

Lomolino MV (2001) Elevation gradients of species-density: historical and prospective views. Global Ecol Biogeogr 10:3–13

Londo G (1976) The decimal scale for releves of permanent quadrats. Vegetatio 33:61–64

Lopez R (2003) The policy roots of socioeconomic stagnation and environmental implosion: Latin America 1950–2000. World Dev 31:259–280

Lorence DH, Sussman RW (1986) Exotic species invasion into Mauritius wet forest remnants. J Trop Ecol 2:147–162

Lösch R (2001) Wasserhaushalt der Pflanzen. Quelle & Meyer, Wiebelsheim

Lowday JE (1986) A comparison of the effects of cutting with those of the herbicide asulam on the control of bracken. In: Smith RT, Taylor JA (eds) Bracken: ecology, land use and control technology. Parthenon, London pp 359–367

Lozano PE (2002) Los tipos de bosques en el sur de Ecuador. In: Aguirre Z, Madsen JE, Cotton E, Balslev H (eds) Botánica Austroecuatoriana. Estudios sobre los recursos vegetales en las provincias de El Oro, Loja y Zamora–Chinchipe. Abya-Yala, Quito. pp 29–49

lsenbeer H, Newton BE, Dunne T, Moraes JMD (1999) Soil hydraulic conductivities of latosols under pasture, forest and teak in Rondonia, Brazil. Hydrol Process 13:1417–1422

Lucky A, Erwin TL, Witman JD (2002) Temporal and spatial diversity and distribution of arboreal Carabidae (Coleoptera) in a western Amazonian rain forest. Biotropica 34:376–386

Lugo AE (1997) The apparent paradox of reestablishing species richness on degraded lands with tree monocultures. For Ecol Manage 99:9–19

Lundgren L (1978) Studies of soil and vegetation development on fresh landslide scars in the Mgeta valley, western Uluguru mountains, Tanzania. Geogr Ann 60:91–127

Lundström US (1993) The role of organic acids in the soil solution chemistry of a podzolized soil. J Soil Sci 44:121–133

Luxton M (1975) Studies on the oribatid mites of a Danish beech wood soil. II. Biomass, calirometry, and respirometry. Pedobiologia 12:434–463

Luxton M (1981) Studies on the oribatid mites of a Danish beech wood soil III. Introduction to the field population. Pedobiologia 21:301–311

Mackay WP, Lopez-Castro C, Fernandez F (2002) A new, high altitude Colombian species of the ant genus *Camponotus* with dimorphic males and females (Hymenoptera: Formicidae). Sociobiology 40:421–430

Mackensen J (1996) Nutrient transfer to the atmosphere by burning of debris in eastern Amazonia. For Ecol Manage 86:121–128

Madsen JE, Øllgaard B (1993) Inventario preliminar de las especies vegetales en el Parque Nacional Podocarpus. (Ciencias agrícolas 22–23) UN Loja, Loja, pp 66–87

Madsen JE, Øllgaard B (1994) Floristic composition, structure, and dynamics of an upper montane rain forest in southern Ecuador. Nord J Bot 14:403–423

MAG, INEC, SICA (2002) III censo nacional agropecuario. Ministerio de Agricultura y Ganaderia, Instituto Nacional de Estadística y Censos, Proyecto Servicio de Información y Censo Agropecuario, Quito

Mahecha MD, Schmidtlein A (2007) Revealing biogeographical patterns by nonlinear ordinations and derived anisotropic spatial filters. Global Ecol Biogeogr (in press)

Mahecha MD, Martinez A, Lischeid G, Beck E (2007) Nonlinear dimensionality reduction: Alternative ordination approaches for extracting and visualizing biodiversity patterns in tropical montane forest vegetation data. Ecol Inform, 2:138–149

Mamilov AS, Dilly OM (2002) Soil microbial eco-physiology as affected by short-term variations in environmental conditions. Soil Biol Biochem 34:1283–1290

Maraun M, Scheu S (1996) Seasonal changes in micorbial biomass and activity in leaf litter layers of beech (*Fagus sylvatica*) forests on a basalt-limestone gradient. Pedobiologia 40:21–31

Maraun M, Scheu S (2000) The structure of oribatid mite communities (Acari, Oribatida): patterns, mechanisms and implications for future research. Ecography 23:374–383

Maraun M, Alphei J, Beste P, Bonkowski M, Buryn R, Peter M, Migge S, Schaefer M, Scheu S (2001) Indirect effects of carbon and nutrient amendments on the soil meso- and microfauna of a beechwood. Biol Fertil Soils 34:222–229

Maraun M, Martens H, Migge M, Theenhaus A, Scheu S (2003a) Adding to the 'enigma of soil animal diversity': fungal feeders and saprophagous soil invertebrates prefer similar food substrates. Eur J Soil Biol 39:85–95

Maraun M, Salamon JA, Schneider K, Schaefer M, Scheu S (2003b) Oribatid mite and collembolan diversity, density and community structure in a moder beech forest (*Fagus sylvatica*): effects of mechanical disturbances. Soil Biol Biochem 35:1387–1394

Marin CT, Bouten IW, Dekker S (2000) Forest floor water dynamics and root water uptake in four forest ecosystems in northwest Amazonia. J Hydrol 237:169–183

Marrs RH, Proctor J, Heaney A, Mountford MD (1988) Changes in soil nitrogen-mineralization and nitrification along an altitudinal transect in tropical rain foret in Costa Rica. J Ecol 76:466–482

Marrs RH, Johnson SW, LeDuc MG (1998) Control of bracken and restoration of heathland. 7. The response of bracken rhizomes to 18 years of continued bracken control or 6 years of control followed by recovery. J Appl Ecol 35:748–757

Marrs RH, Johnson SW, Le Duc MG (2000) Control of bracken and restoration of heathland. VI. The response of bracken fronds to 18 years of continued bracken control or 6 years of control followed by recovery. J Appl Ecol 35:479–490

Martens R (1995) Current methods for measuring microbial biomass C in soil: potentials and limitations. Biol Fertil Soils 19:87–99

Martinez LJ, Zinck JA (2004) Temporal variation of soil compaction and deterioration of soil quality in pasture areas of Colombian Amazonia. Soil Tillage Res 75(1):3–17

Martínez Jerves JA (2007) Sekundäre Bergregenwälder in Südecuador: der Einfluß der Art der Störung auf das Spektrum der Pflanzenarten und die Waldstruktur, eine vegetationskundliche Analyse. PhD thesis, University of Bayreuth, Bayreuth

Marufu L, Dentener F, Lelieveld J, Andreae MO, Helas G (2000) Photochemistry of the African troposphere: influence of biomass-burning emissions. J Geophys Res 105:14513–14530

Marwan N, Trauth MH, Vuille M, Kurths J (2003) Comparing modern and Pleistocene ENSO-like influences in NW Argentina using nonlinear time series analysis methods. Clim Dynam 21:317–326

The MathWorks (2005a) Matlab ver 7.1. Available at: http://www.mathworks.com

The MathWorks (2005b) Image processing toolbox ver 5.1. Available at: http://www.mathworks.com/products/image/

Matt F (2001) Pflanzenbesuchende Fledermäuse im tropischen Bergregenwald: Diversität, Einnischung und Gildenstruktur – Eine Untersuchung der Fledermausgemeinschaften in drei Höhenstufen der Andenostabdachung des Podocarpus Nationalparks in Südecuador. PhD thesis, University of Erlangen–Nürnberg, Erlangen

Matzner E (2004) Biogeochemistry of a forested catchment in a changing environment. A German case study. (Ecological studies 172) Springer, Berlin Heidelberg New York

MBG (2006) Current specimen list for *Cedrela montana* and current specimen list for *Tabebuia chrysantha*. (w3 specimen data base) Missouri Botanical Garden. Available at: mobot.mobot.org/cgi-bin/search_vast

McCoy ED (1990) The distribution of insects along elevational gradients. Oikos 58:313–322

McDonald MA, Hofny-Collins A, Healy JR, Goodland TCR (2003) Evaluation of trees indigenous to the montane forest of the Blue Mountains, Jamaica for reforestation and agroforestry. For Ecol Manage 175:379–401

McDowell N, Barnard H, Bond BJ, Hinckley T, Hubbard RM, Ishii H, Köstner BMM, Magnani F, Marshall JD, Meinzer FC, Phillips N, Ryan MG, Whitehead D (2001) The relationship between tree height and leaf area: sapwood area ratio. Oecologia 132:12–20

McDowell WH (1998) Internal nutrient fluxes in a Puerto Rican rain forest. J Trop Ecol 14:521–536

McDowell WH, Asbury CE (1994) Export of carbon, nitrogen, and major ions from three tropical montane watersheds. Limnol Oceanogr 39:111–125

McGeachie WJ (1989) The effects of moonlight illuminance, temperature and wind speed on lighttrap catches of moths. Bull Entomol Res 79:185–192

McGroddy ME, Daufresne T, Hedin LO (2004) Scaling of C:N:P stoichiometry in forests worldwide: implications of terrestrial redfield-type ratios. Ecology 85:2390–2401

McNaughton KG, Jarvis PG (1983) Predicting effects of vegetation changes on transpiration and evaporation. In: Kozlowski TT (ed): Water deficits and plant growth, vol 7. Springer, Berlin Heidelberg New York, pp 1–47

McQueen DJ, Johannes MRS, Post JR, Stewart TJ, Lean DRS (1989) Bottom-up and top-down impacts on freshwater pelagic community structure. Ecol Monogr 59:289–309

Mecham J (2001) Causes and consequences of deforestation in Ecuador. CIBT. Available at: www.rainforestinfo.org.au/projects/jefferson.htm

Meijaard E, Sheil D, Nasi R, Augeri D, Rosenbaum B, Iskandar D, Setywati T, Lammertink M Rachmatika I, Wong A, Soehartono T, Stanley S, O'Brien T (2005) Life after logging: reconciling wildlife conservation and production forestry in Indonesian Borneo. Center for International Forestry Research, Bogor

Meinzer FC, Goldstein G, Andrade JL (2001) Regulation of water flux through tropical forest canopy trees. Do universal rules apply? Tree Physiol 21:19–26

Meiri S, Dayan T (2003) On the validity of Bergmann's rule. J Biogeogr 30:331–351

Merilä P, Strömmer R, Fritze H (2002) Soil microbial activity and community structure along a primary succession transect on the land-uplift coast in western Finland. Soil Biol Biochem 34:1647–1654

Merkl N (2000) Propagation of native tree species of South Ecuador. BSc thesis, Ruprecht-Karls University Heidelberg, Heidelberg

Messier C, Nikinmaa E (2000) Effects of light availability and sapling size on the growth, biomass allocation, and crown morphology of understory sugar maple, yellow birch, and beech. Ecoscience 7:345–356

Metcalfe D, Grubb P, Turner I (1998) The ecology of very small-seeded shade-tolerant trees and shrubs in lowland rain forest in Singapore. Plant Ecol 134:131–149

Meyer CR, Paulay G (2005) DNA barcoding: error rates based on comprehensive sampling. PLoS Biol 3:2229–2238

Migge S (2001) The effect of earthworm invasion on nutrient turnover, microorganisms and microarthropods in Canadian aspen forest soil. PhD thesis, University of Göttingen, Göttingen

Migge S, Maraun M, Scheu S, Schaefer M (1998) The oribatid mite community (Acarina) on pure and mixed stands of beech (*Fagus sylvatica*) and spruce (*Picea abies*) at different age. Appl Soil Ecol 9:119–126

Minet J, Scoble MJ (1999) The drepanoid/geometroid assemblage. In: Kristensen NP (ed) Lepidoptera: moths and butterflies. (Handbook of zoology, vol 1) Gruyter, Berlin, pp 301–320

Ministerio del Ambiente (2003) Ponencias del Ministerio del Ambiente para el fortalecimiento y consolidación del sistema nacional de áreas protegidas. (Primer congreso nacional de áreas naturales protegidas) Ministerio del Ambiente, Quito

Ministerio del Ambiente (2004) Norma para el manejo forestal sustentable para aprovechamiento de madera en bosque andino. Ministerio del Ambiente, Quito

Mitchell J (1973) Mobilization of phosphorus by *Pteridium aquilinum*. Plant Soil 38:489–491

Mitchell JM (1976) An overview of climatic variability and its causal mechanisms. Quat Res 6:481–493

Mittermeier RA, Robles Gil P, Mittermeier CG (2004) Megadiversity: Earth's biologically wealthiest nations, 1st English edn. Graphic Arts Center, Portland

Molina R, Trappe JM (1982) Lack of mycorrhizal specificity by ericaceous hosts *Arbutus menziesii* and *Arctostaphylos uva-ursi*. New Phytol 90:495–509

Montagnini F (2001) Strategies for the recovery of degraded ecosystems, experiences from Latin America. Interciencia 26:498–503

Montagnini F, González E, Rheingans R, Porras C (1995) Mixed and pure forest plantation in the humid neotropics: a comparison of early growth, pest damage and establishment costs. Commonw For Rev 74:306–314

Montagnini F, Uglade L, Navarro C (2003) Growth characteristics of some native tree species used in silvopastoral systems in the humid lowlands of Costa Rica. Agrofor Syst 59:163–170

Monteith JL (1965) Evaporation and environment. Symp Soc Exp Biol 19:206–234

Monteith JL, Unsworth MH (1990) Principles of environmental physics, 2nd edn. Arnold, London, 291 pp

Moore JV, Lysinger M (1997) The birds of Cabañas San Isidro, Ecuador. Moore Nature Recordings, San Jose

Moraes JL, Cerri CC, Melillo JM, Kicklighter D, Neill C, Skole DL, Steudler PA (1995) Soil carbon stocks of the Brazilian Amazon Basin. Soil Sci Soc Am J 59:244–247

Morales MA, Dodge GJ, Inouye DW (2005) A phenological mid-domain effect in flowering diversity. Oecologia 142:83–89

Moran MD, Scheidler AR (2002) Effects of nutrients and predators on an old-field food chain: interactions of top-down and bottom-up processes. Oikos 98:116–124

Mori SA, Cremers G, Gracie C, Granville JJ de, Hoff M, Mitchell JD (1997) Guide to the vascular plants of central French Guiana, part 1. New York Botanical Garden, New York

Mori SA, Cremers G, Gracie CA, Granville JJ de, Heald SV, Hoff M, Mitchell JD (2002) Guide to the vascular plants of central French Guiana, part 2. New York Botanical Garden, New York

Motzer T (2003) Bestandesklima, Energiehaushalt und Evapotranspiration eines neotropischen Bergregenwaldes. Forstmeteorologische und ökophysiologische Untersuchungen in den Anden Süd-Ecuadors. PhD thesis, University of Mannheim, Mannheim

Motzer T, Munz N, Küppers M, Schmitt D, Anhuf D (2005) Stomatal conductance, transpiration and sap flow of tropical montane rainforest trees in the southern Ecuadorian Andes. Tree Physiol 25:1283–1293

Motzer T, Anhuf D, Küppers M (2007) Transpiration and microclimate of a tropical montane forest in southern Ecuador. In: Juvik JO, Bruijnzeel LA, Scatena FN, Bubb P (eds) Mountains in the mist: science for conserving and managing tropical montane cloud forests. University of Hawaii, Honolulu (in press)

Mousseau TA (1997) Ectotherms follow the converse to Bergmanns's rule. Evolution 51:630–632

Moutinho P, Nepstad DC, Davidson EA (2005) Influence of leaf-cutting ant nests on secondary forest growth and soil properties in Amazonia. Ecology 84:1265–1276

Moyersoen B (1993) Ectomicorrizas y micorrizas vesículo-arbusculares en Caatinga Amazonica del Sur de Venezuela. Sci Guaianae 3:1–83

Muchhala N, Patricio Mena V, Luis Albuja V (2005) A new species of *Anoura* (Chiroptera: Phyllostomidae) from the Ecuadorian Andes. J Mammal 86:457–461

Muirhead-Thomson RC (1991) Trap responses of flying insects. Academic, London

Mulholland PJ, Wilson GV, Jardine PM (1990) Hydrogeochemical response of a forested watershed to storms: effects of preferential flow along shallow and deep pathways. Water Resour Res 26:3021–3036

Müller U, Frahm JP (1998) Diversität epiphytischer Moose eines westandinen Bergregenwaldes in Ecuador. Trop Bryol 15:29–43

Müller-Böker U (1999) The Chitawan Tharus in Southern Nepal. An ethnoecological approach. Nepal Research Centre Publications, 21 Stuttgart

Müller-Hohenstein K, Paulsch A, Paulsch D, Schneider R (2004) Vegetations- und Agrarlandscha ftsstrukturen in den Bergwäldern Südecuadors. Geogr Runds 56:48–55

Muñoz-Arango J (1990) Diversidad y hábitos alimenticios de murciélagos en transectos altitudinales a través de la Cordillera Central de los Andes en Colombia. Stud Neotrop Fauna Environ 25:1–17

Munroe EG, Solis MA (1999) The Pyraloidea. In: Kristensen NP (ed) Lepidoptera: moths and butterflies. (Handbook of zoology, vol 1) Gruyter, Berlin, pp 233–256

Munroe EG, Becker VO, Shaffer JC, Shaffer M, Solis MA (1995) Pyraloidea. In: Heppner, JB (ed) Atlas of neotropical Lepidoptera, vol 3, checklist: part 2 (Hyblaeoidea – Pyraloidea – Tortricoidea). Scientific, Gainesville, pp 34–105

Münzel M (1977) Jívaro-Indianer in Südamerika. Roter Faden zur Ausstellung, 4. Museum für Völkerkunde, Frankfurt

Münzel M (1987) Kulturökologie, Ethnoökologie und Ethnodesarrollo im Amazonasgebiet. Entwicklungsperspektiven 29

Münzel M (1989) Bemerkungen zum indianischen Umweltbewußtsein im Amazonasgebiet. Geogr Runds 41:431–435

Münzenberger B, Kottke I, Oberwinkler F (1992) Ultrastructural investigations of *Arbutus unedo–Laccaria amethystea* mycorrhizas sythesized in vitro. Trees 7:40–47

Murcia C (1995) Edge effects in fragmented forests: implications for conservation. Trends Ecol Evol 10:58–62

Myers N (1994) Tropical deforestation: rates and patterns. In: Brown K, Pearce DW (eds) The causes of tropical deforestation. UCL Press, London

Myers N, Mittermaier RA, Mittermaier CG, Fonseca GAB da, Kent J (2000) Biodiversity hotspots for conservation priorities. Nature 403:853–858

Nadelhoffer KJ, Fry B (1988) Controls on natural nitrogen-15 and carbon-13 abundance in forest soil organic matter. Soil Sci Soc Am J 52:1633–1640

Nadelhoffer K, Shaver G, Fry B, Giblin A, Johnson L, McKane R (1996) ^{15}N natural abundances and N use by tundra plants. Oecologia 107:386–394

Nadkarni NM (1984) Epiphyte biomass and nutrient capital of a neotropical elfin forest. Biotropica 16:249–256

Nazarea VD (1999) Ethnoecology, situated knowledge/located lives. University of Arizona Press, Tucson

Neal C, Reynolds B, Stevens P, Hornung M (1989) Hydrogeochemical controls for inorganic aluminium in acidic stream and soil waters at two upland catchments in Wales. J Hydrol 106:155–175

Nebel M, Kreier HP, Preußing M, Weiß M, Kottke I (2004) Symbiotic fungal associations of liverworts are the possible ancestors of mycorrhizae. In: Agerer R, Piepenbring M, Blanz P (eds) Frontiers in basidiomycote mycology. IHW, Eching, pp 339–360

Neill C, Piccolo MC, Steudler PA, Melillo JM, Feigl BJ, Cerri CC (1995) Nitrogen dynamics in soils of forests and active pastures in the western Brazilian Amazon Basin. Soil Biol Biochem 27:1167–1175

Neill C, Melillo JM, Steudler PA, Cerri CC, Moraes JFL, Piccolo MC, Brito M (1997) Soil carbon and nitrogen stocks following forest clearing for pasture in the southwestern Brazilian Amazon. Ecol Appl 7:1216–1225

Neill C, Cerri CC, Melillo JM, Feigl BJ, Steudler PA, Moraes JFL, Piccolo MC (1998) Stocks and dynamics of soil carbon following deforestation for pasture in Rondônia. In: Lal R, Kimble JM, Follet RF, Stewart BA (eds) Soil processes and the carbon cycle. CRC, Boca Raton, pp 9–28

Neill D (2005) Cordillera del Cóndor. Botanical treasures between the Andes and the Amazon. Plant Talk 41:17–21

Newstrom LE, Frankie GW, Baker HG (1994) A new classification for plant phenology based on flowering patterns in lowland tropical forest trees at La Selva, Costa Rica. Biotropica 26:141–159

Niedbala W, Illig J (2007) New species and new records of ptyctimous mites (Acari, Oribatida) from Ecuador. Trop Zool 20:107–122

Nkongmeneck BA, Lowman MD, Atwood JT (2002) Epiphyte diversity in primary and fragmented forests of Cameroon, Central Africa: a preliminary survey. Selbyana 23:121–130

Nöske NM (2005): Effekte anthropogener Störung auf die Diversität kryptogamischer Epiphyten (Flechten, Moose) in einem Bergregenwald in Südecuador. PhD thesis, University of Göttingen, Göttingen. Available at: webdoc.sub.gwdg.de/diss/2005/noeske/noeske.pdf

Nöske NM, Gradstein SR, Kürschner H, Parolly G, Torrachi S (2003) Cryptogams of the Reserva Biológica San Francisco (Province Zamora–Chinchipe, Southern Ecuador). I. Bryophytes. Cryptogamie, Bryologie 24:15–32

Nöske NM, Hilt N, Werner FA, Brehm G, Fiedler K, Sipman HJ, Gradstein SR (2007) Disturbance effects on diversity in montane forest in Ecuador: sessile epiphytes vs mobile moths. Basic Appl Ecol (in press)

Nöske NM, Mandl N, Sipman JM (2008) Lichens. In: Liede-Schumann S, Breckle SW (eds) Provisional checklists of fauna and flora of the San Francisco valley and its surroundings (Reserva San Francisco/Prov. Zamore–Chinchipe, southern Ecuador). Ecotrop Monogr 4 (in press)

Novotny V, Basset Y (2000) Rare species in communities of tropical insect herbivores: pondering the mystery of singletons. Oikos 89:564–572

Novotny V, Basset Y, Miller SE, Weiblen GD, Bremer B, Cizek L, Drozd P (2002) Low host specificity of herbivorous insects in a tropical forest. Nature 416:841–844

Novotny V, Miller SE, Basset Y, Cizek L, Darrow K, Kaupa B, Kua J, Weiblen GD (2005) An altitudinal comparison of caterpillar (Lepidoptera) assemblages on *Ficus* trees in Papua New Guinea. J Biogeogr 32:1303–1314

Novotny V, Drozd P, Miller SE, Kulfan M, Janda M, Basset Y, Weiblen GD (2006) Why are there so many species of herbivorous insects in tropical rainforests? Science 313:1115–1118

Núnez-Farfan J, Dirzo R (1988) Within-gap spatial heterogeneity and seedling performance in a Mexican tropical forest. Oikos 51:274–284

Oesker M, Homeier J, Dalitz H (2007) Spatial heterogeneity of canopy throughfall quantity and quality in a tropical mountain Forest in south Ecuador. In: Bruijnzeel LA, et al (eds): Mountains in the mist: science for conserving and managing tropical montane cloud forests. University of Hawaii, Honolulu

Ohl C, Bussmann RW (2004) Recolonisation of natural landslides in tropical mountain forests of Southern Ecuador. Feddes Repert Z Bot Taxon Geobot 115:248–264

Okuda T, Suzuki M, Adachi N, Quah ES, Hussein NA, Manokaran N (2003) Effect of selective logging on canopy and stand structure and tree species composition in a lowland dipterocarp forest in peninsular Malaysia. For Ecol Manage 175:297–320

Oosterhoorn M, Kappelle M (2000) Vegetation structure and composition along an interior–edge–exterior gradient in a Coast Rica montane cloud forest. For Ecol Manage 126:291–307

Opler PA, Frankie GW, Baker HG (1980) Comparative phenological studies of treelet and shrub species in tropical wet and dry forests in the lowlands of Costa Rica. J Ecol 68:167–188

Oren R, Zimmermann R, Terborgh J (1996) Transpiration in upper Amazonia floodplain and upland forests in response to drought-breaking rains. Ecology 77:986–973

Oren R, Phillips N, Katul G, Ewers BE, Pataki DE (1998) Scaling xylem sap flux and soil water balance and calculating variance: a method for partitioning water flux in forests. Ann Sci For 55:191–216

Orshan G (1986) Plant form as describing vegetation and expressing adaptation to environment. Ann Bot Roma 44:7–38

Osher LJ (2003) Effect of landuse change on soil carbon in Hawaii. Biogeochemistry 65:213–232

Ostertag R (2001) Effects of nitrogen and phosphorus availability on fine-root dynamics in Hawaiian montane forests. Ecology 82:485–499

Palacios WA (1996) Cuenca del Río Nangaritza. Rev Geogr 36:93–119

Paoletti MG, Taylor RAJ, Stinner BR, Stinner DH, Benzing DH (1991) Diversity of soil fauna in the canopy and forest floor of Venezuelan cloud forest. J Trop Ecol 7:373–383

Park A, Justiniano MJ, Fredericksen TS (2005) Natural regeneration and environmental relationships of tree species in logging gaps in a tropical forest. For Ecol Manage 217:147–157

Parker GG (1983) Throughfall and stemflow in the forest nutrient cycle. Adv Ecol Res 13:57–133

Parolly G, Kürschner H (2004a) Ecological studies in Ecuadorian bryophyte communities I. Nova Hedwigia Z Kryptogamenkd 78:1–43

Parolly G, Kürschner H (2004b) Ecological studies in Ecuadorian bryophyte communities II. Syntaxonomy of the submontane and montane epiphytic vegetation of S Ecuador. Nova Hedwigia Z Kryptogamenkd 79:377–424

Parolly G, Kürschner H (2005) Ecological studies in Ecuadorian bryophyte communities V. Syntaxonomy, life forms and life strategies of the bryophyte vegetation on decaying wood and tree bases in S Ecuador. Nova Hedwigia Z Kryptogamenkd 81:1–36

Parolly G, Kürschner H, Schäfer-Verwimp A, Gradstein SR (2004) Cryptogams of the Reserva Biológica San Francisco (Province Zamora–Chinchipe, southern Ecuador) III. Bryophytes – additions and new species. Cryptog, Bryol 25:271–289

Parrotta JA (1992) The role of plantation forests in rehabilitating degraded tropical ecosystems. Agric Ecosyst Environ 41:115–133

Parrotta JA, Turnbull JW, Jones N (1997) Catalyzing native forest regeneration on degraded tropical lands. For Ecol Manage 99:1–7

Parsons DJ (1975) Vegetation structure in the Mediterranean scrub communities of California and Chile. J Ecol 64:435–447

Patterson BD, Stotz DF, Solari S, Fitzpatrick JW, Pacheco V (1998) Contrasting patterns of elevational zonation for birds and mammals in the Andes of southeastern Peru. J Biogeogr 25:593–607

Patzelt E (1996) Flora del Ecuador, 2nd edn. Banco Central del Ecuador, Quito

Paulsch A (2002) Development and application of a classification system for undisturbed and disturbed tropical montane forest based on vegetation structure. PhD thesis, University of Bayreuth, Bayreuth. Available at: opus.ub.uni-bayreuth.de/volltexte/2002/1

Paulsch A, Czimczik C (2001) Classification of tropical montane shrub vegetation – a structural approach. Erde 132:25–39

Paulsch A, Schneider R, Hartig K (2001) Land-use induced vegetation structure in a montane region of southern Ecuador. Erde 132:93–102

Paulsch D (2007) Aves. In: Liede-Schumann S, Breckle SW (eds) Provisional checklists of fauna and flora of the San Francisco valley and its surroundings (Reserva San Francisco/Prov. Zamore–Chinchipe, southern Ecuador). Ecotrop Monogr 4 (in press)

Pazmiño L (2006) Estudio preliminar de hongos que forman micorrizas en helechos de las familias Grammitidaceae y Polypodiacea en el bosque de la ECSF y El Tiro, Loja, Ecuador. BSc thesis, Universidad Técnica Particular de Loja, Loja

Peacock AD, Mullen MD, Ringelberg DB, Tyler DD, Hedrick DB, Gale PM, White DC (2001) Soil microbial community responses to dairy manure or ammonium nitrate applications. Soil Biol Biochem 33:1011–1019

Pearson DL (1994) Selecting indicator taxa for the quantitative assessment of biodiversity. Philos Trans R Soc Lond B 345:75–79

Pearson TRH, Burslem DFRP, Goeriz RE, Dalling JW (2003) Interactions of gap size and herbivory on establishment, growth and survival of three species of neotropical pioneer trees. J Ecol 91:785–796

Pebesma EJ (2004) Multivariable geostatistics in S: the gstat package. Comput Geosci 30:683–691

PerssonT, Baath E, Clarholm M, Lundkvist H, Söderström BE, Sohlenius B (1980) Trophic structure, biomass dynamics and carbon metabolism of soil organisms in a Scots pine forest. In: Persson T (ed) Structure and function of northern coniferous forests – an ecosystem study. Ecol Bull 32:419–459

Pettitt AN, McBratney AB (1993) Sampling designs for estimating spatial variance components. Appl Stat 42:185–209

Phillips N, Oren R, Zimmermann R, Wright SJ (1999) Temporal patterns of water flux in trees and lianas in a Panamanian moist forest. Trees 14:116–123

Phillips N, Bond BJ, Ryan MG (2001) Gas exchange and hydraulic properties in the crowns of two tree species in a Panamanian moist forest. Trees 15:123–130

Phillips OL, Vargas PN, Monteagudo AL, Cruz AP, Zans M-EC, Sánchez WG, Yli-Halla M, Rose S (2003) Habitat association among Amazonian tree species: a landscape-scale approach. J Ecol 91:757–775

Piccolo MC, Neill C, Melillo JM, Cerri CC, Steudler PA (1996) 15N natural abundance in forest and pasture soils of the Brazilian Amazon Basin. Plant Soil 182:249–258

Pickett STA, White PS (1986) The ecology of natural disturbance and patch dynamics. Academic, Orlando

Piechowski D (2003) Vegetationsstrukturanalyse unterschiedlich beeinflusster submontaner Regenwälder in Ecuador. PhD thesis, Free University of Berlin, Berlin

Pinard MA, Putz FE, Rumíz D, Guzmán R, Jardim A (1999) Ecological characterization of tree species for guiding forest management decisions in seasonally dry forests in Lomerío, Bolivia. For Ecol Manage 113:201–213

Piotrowski JS, Denich T, Klironomos JN, Graham JM, Rillig MC (2004) The effects of arbuscular mycorrhizas on soil aggregation depend on the interaction between plant and fungal species. New Phytologist 164:365–373

Pitkin LM (2002) Neotropical ennomine moths: a review of the genera (Lepidoptera: Geometridae). Zool J Linn Soc 135:121–401

Pitkin LM (2005) Moths of the Neotropical genera Ischnopteris, Stegotheca and Rucana (Lepidoptera: Geometridae, Ennominae). Syst Biodivers 3:13–96

Pitman J (1989) Rainfall interception by bracken in open habitat – relations between leaf area, canopy storage and drainage rate. J Hydrol 105:317–334

Plowman KP (1981) Distribution of Cryptostigmata and Mesostigmata (Acari) within the litter and soil layers of two subtropical forests. Aust J Ecol 6:365–374

Pohle P (2004) Erhaltung von Biodiversität in den Anden Südecuadors. Geogr Runds 56:14–21

Pohle P, Reinhardt S (2004) Indigenous knowledge of plants and their utilization among the Shuar of the lower tropical mountain forest in southern Ecuador. Lyonia 7:133–149

Pokorny B (1997) Stand der Sekundärwaldforschung. Forstarchiv 68:228–237

Pollmann W, Hildebrand R (2005) Structure and the composition of species in timberline ecotones of the Southern Andes. In: Broll G, Keplin B (eds) Mountain ecosystems – studies in treeline ecology, Springer, Berlin Heidelberg New York, pp 117–151

Ponder F Jr, Tadros M (2002) Phospholipid fatty acids in forest soil four years after organic matter removal and soil compaction. Appl Soil Ecol 19:173–182

Poorter L, Arets EJMM (2003) Light environment and tree strategies in a Bolivian tropical moist forest: an evaluation of the light partitioning hypothesis. Plant Ecol 166:295–306

Posey DA (1985) Indigenous management of tropical forest ecosystems: the case of the Kayapó indians of the Brazialian Amazon. Agrofor Syst 3:139–158

Posey DA, Balée W (1989) Resource management in Amazonia: indigenous and folk strategies. Adv Econ Bot 7

Post DM (2002) Using stable isotopes to estimate trophic position: models, methods, and assumptions. Ecology 83:703–718

Powell A, Kron K (2003) Molecular systematics of the andean blueberries (Vaccinieae, Vaccinioideae, Ericaceae). Int J Plant Sci 164:987–995

Powell JA, Mitter C, Farrell B (1999) Evolution of larval food preferences. In: Kristensen NP (ed) Lepidoptera: moths and butterflies. (Handbook of zoology, vol 1) Gruyter, Berlin, pp 403–442

Powers JS, Veldkamp E (2005) Regional variation in soil carbon and ^{13}C in paried forests and pastures of northeastern Costa Rica. Biogeochemistry 72:315–336

Powers JS, Treseder KK, Lerdau MT (2005) Fine roots, arbuscular mycorrhizal hyphae and soil nutrients in four neotropical rain forests: pattern across large geographic distances. New Phytol 165:913–921

Predesur (2004) Megaproyecto de repoblación forestal de 300.000 ha en la provincia de Loja y parte alta de la provincia de El Oro. Predesur, Loja

Priess JA, Fölster H (2001) Microbial properties and soil respiration in submontane forests of Venezuelian Guyana: characteristics and response to fertilizer treatments. Soil Biol Biochem 33:503–509

Proctor J (1987) Nutrient cycling in primary and old secondary rainforests. Appl Geogr 7:135–152

Proctor J, Lee YF, Langely AM, Munro WR, Nelson T (1988) Ecological studies on Gunung Silam, a small ultrabasic mountain in Sabah, Malaysia: environment, forest structure and floristics. J Ecol 76:320–340

Quested HM, Callaghan TV, Cornelissen JHC, Press MC (2005) The impact of hemiparasitic plant litter on decomposition: direct, seasonal and litter mixing effects. J Ecol 93:87–98

Quizhpe W, Aguirre ZM, Cabrera O, Delgado TE (2002) Los páramos del Parque Nacional Podocarpus. In: Aguirre ZM, Madsen JE, Cotton E, Balslev H (eds) Parque Nacional Podocarpus. Abya-Yala, Quito pp 79–90

R Development Core Team (2005) R: a language and environment for statistical computing. R Foundation for Statistical Computing, Vienna, Austria. Available at: http://www.R-project.org

Rahbek C (2005) The role of spatial scale and the perception of large-scale species-richness patterns. Ecol Lett 8:224–239

Rahbek C, Bloch H, Poulsen MK, Rasmussen JF (1995) The avifauna of Podocarpus National Park – the 'Andean jewel in the crown' of Ecuador's protected areas. Ornithol Neotrop 6:113–120

Raich JW (2003) Interannual variability in global soil respiration on a 0.5 degree grid cell basis (1980–1994). Available at: cdiac.ornl.gov/epubs/ndp/ndp081/map_yrmn.html

Raich JW, Tufekcioglu A (2000) Vegetation and soil respiration: correlations and controls. Biogeochemistry 48:71–90

Raich JW, Russell AE, Vitousek PM (1997) Primary productivity and ecosystem development along an elevational gradient on Mauna Loa, Hawai'i. Ecology 78:707–721

Raich JW, Potter CS, Bhagawati D (2002) Interannual variability in global soil respiration, 1980–94. Global Change Biol 8:800–812

Rasmussen JF, Rahbek C (1994) Aves del Parque Nacional Podocarpus: una lista anotada. CECIA, Quito

Rasmussen JF, Rahbek C, Horstman E, Poulsen MK, Bloch H (1994) Aves del Parque Nacional Podocarpus, una lista anotada. CECIA, Quito

Read DJ, Perez-Moreno J (2003) Mycorrhizas and nutrient cycling in ecosystems – a journey towards relevance? New Phytologist 157:475–492

Reagan DP, Camilo GR, Waide RB (1996) The community food web: major properties and patterns of organization. In: Reagan DP, Waide RB (eds) The food web of a tropical rain forest. University of Chicago Press, Chicago, pp 461–510

Reichenberger S, Amelung W, Laabs V, Pinto A, Totsche KU, Zech W (2002) Pesticide displacement along preferential flow pathways in a Brazilian Oxisol. Geoderma 110(1–2):63–86

Reiners WA, Bouwman AF, Parsons WFJ, Keller M (1994) Tropical rainforest conversion to pasture: changes in vegetation and soil properties. Ecol Appl 4:363–377

Reisigl H, Keller R (1999) Lebensraum Bergwald. Spektrum, Heidelberg, 147 pp

Remsen JV (2003) Furnariidae. In: Hoyo J del, Elliott A, Christie D (eds) Broadbills to Tapaculos. (Handbook of the birds of the world, vol 8) Lynx, Barcelona, pp 162–357

Renck A, Lehmann J (2004) Rapid water flow and transport of inorganic and organic nitrogen in a highly aggregated tropical soil. Soil Sci 169(5):330–341

Restrepo C, Vitousek P (2001) Landslides, alien species, and the diversity of a Hawaiian montane mesic ecosystem. Ecotropica 33:409–420

Rey A, Pegorano E, Tedeschi V, Parri I de, Jarvis PG, Valentini R (2002) Annual variation in soil respiration and its components in a coppice oak forest in Central Italy. Global Change Biol 8:851–866

Rheenen HMPJB van, Boot RGA, Werger MJA, Ulloa Ulloa M (2004) Regeneration of timber trees in a logged tropical forest in north Bolivia. For Ecol Manage 200:39–48

Rhoades CC, Coleman DC (1999) Nitrogen mineralization and nitrification following land conversion in montane Ecuador. Soil Biol Biochem 31:1347–1354

Rhoades CC, Eckert GE, Coleman DC (2000): Soil carbon differences among forest, agriculture, and secondary vegetation in lower montane Ecuador. Ecol Appl 10:497–505

Ribeiro JELS, Hopkins MJG, Vicentini A, Sothers CA, Costa MAS, Brito JM, Souza MAD, Martins LHP, Lohmann LG, Assunção PACL, Pereira EC, Silva CF, Mesquita MR, Procópio LC (1999) Flora da Reserva Ducke. INPA, Manaus

Richards PW (1952) The tropical rain forest: an ecological study. Cambridge University Press, Cambridge

Richards PW, Tansley AG, Watt AS (1940) The recording of structure, life form and flora of tropical forest communities as a basis for their classification. J Ecol 28:224–239

Richter M (2001) Vegetationszonen der Erde. Klett Perthes, Gotha

Richter M (2003) Using epiphytes and soil temperature for eco-climatic interpretations in south Ecuador. Erdkunde 57:161–181

Richter M, Moreira-Muñoz A (2005) Heterogeneidad climática y diversidad vegetacional en el sur de Ecuador: un método de fitoindicación. In: Weigend M, Rodriguez E, Arana C (eds). Bosques relictos del NO de Perú y SO de Ecuador. Rev Peru Biol 12:217–238

Richter M, Schmidt D (2002) Cordillera de la Atacama – das trockenste Hochgebirge der Welt. Petermanns Geogr Mitt 146:24–33

Ricketts TH, Daily GC, Ehrlich PR, Fay JP (2001) Countryside biogeography of moths in a fragmented landscape: biodiversity in native and agricultural habitats. Conserv Biol 15:378–388

Ridgely RS, Greenfield PJ (2001a) The birds of Ecuador. Helm, London

Ridgely RS, Greenfield PJ (2001b) The birds of Ecuador, vol 2. Field Guide. Cornell University Press, Ithaca

Roberts J, Cabral OMR, Fisch G, Molion LCB, Moore CJ, Shuttleworth WJ (1993) Transpiration from an Amazonian rainforest calculated from stomatal conductance measurements. Agric Forest Meteorol 65:175–196

Roberts JM, Cabral OMR, Costa da JP, McWilliam ALC, A Sá de TD (1996) An over-view of the leaf area index and physiological measurements during Abracos. In: Victoria RL (ed) Amazonian deforestation and climate. Wiley, Chichester, pp 87–381

Röderstein M, Hertel D, Leuschner C (2005) Above- and below-ground litter production in three tropical montane forests in southern Ecuador. J Trop Ecol 21:483–492

Roelofs G-J, Lelieveld J, Smit HG, Klug D (1997) Ozone production and transport in the tropical Atlantic region during the biomass burning season. J Geophys Res 102:10637–10651

Rollenbeck R, Bendix J (2006) Experimental calibration of a cost-effective X-band weather radar for climate-ecological studies in southern Ecuador. Atmos Res 79:296–316

Rollenbeck R, Fabian P, Bendix J (2005) Precipitation dynamics and chemical properties in tropical mountain forests of Ecuador. Adv Geosci 6:1–4

Rollenbeck R, Bendix J, Fabian P, Boy J, Dalitz H, Emck P, Oesker M, Wilcke W (2007) Comparison of different techniques for the measurement of precipitation in tropical montane rain forest regions. J Atmos Ocean Technol 24:156–168

Rosenberger T, Williams K (1999) Response of vascular epiphytes to branch-fall gap formation in *Clusia* trees in a montane rain forest. Selbyana 20:49–58

Ruan HH, Zou XM, Scatena FN, Zimmermann JK (2004) Asynchronous fluctuations of soil microbial biomass and plant litterfall in a tropical wet forest. Plant Soil 260:147–154

Rudel TK, Horowitz W (1993) Tropical deforestation, small farmers and land clearing in the Ecuadorian Amazon. Columbia University Press, New York

Rudel TK, Bates D, Machinguiashi, R (2002) Ecologically noble Amerindians? Cattle ranching and cash cropping among the Shuar and colonists in Ecuador. Lat Am Res Rev 37:144–159

Rustad L (2001) Matter of time on the prairie. Nature 413:578–579

Rutter AJ, Morton AJ, Robins PC (1975) A predictive model of rainfall interception in forests 2. Generalization of the model an comparison with observations in some coniferous and hardwood stands. J Appl Ecol 12:367–380

Rydell J, Lancaster WC (2000) Flight and thermoregulation in moths were shaped by predation from bats. Oikos 88:13–18

Sabogal C (2005) Site-level rehabilitation strategies for degraded forest lands. In: ITTO/IUCN (eds) Restoring forest landscapes. ITTO Tech Ser 23:101–108

Sakai S (2001) Phenological diversity in tropical forests. Pop Ecol 43:77–86

Sakai S, Momose K, Yumoto T, Nagarnitsu T, Nagamasu H, Hamid AA, Nakashizuka T (1999) Plant reproductive phenology over four years including an episode of general flowering in a lowland dipterocarp forest, Sarawak, Malaysia. Am J Bot 86:1414–1436

Salamanca EF, Raubuch M, Joergensen RG (2002) Relationship between soil microbial indices in secondary tropical forest soils. Appl Soil Ecol 21:211–219

Salamon JA, Alphei J, Ruf A, Schaefer M, Scheu S, Schneider K, Sührig A, Maraun M (2006) Transitory dynamic effects in the soil invertebrate community in a temperate deciduous forest: effects of resource quality. Soil Biol Biochem 38:209–221

Saldarriaga JG, West DC, Tharp ML, Uhl C (1988) Longterm chronosequence of forest succession in the upper Rio Negro of Colombia and Venezuela. J Ecol 76:938–958

Sampaio E (2001) Effects of the forest fragmentation on the diversity and abundance patterns of central Amazonian bats. Logos, Berlin

Sanchez-Cordero V (2001) Elevation gradients of diversity for rodents and bats in Oaxaca, Mexico. Global Ecol Biogeogr 10:63–76

Sanders HL (1968) Marine-benthic diversity: a comparative study. Am Nat 102:243–282

Sanders IR (2004) Intraspecific genetic variation in arbuscular mycorrhizal fungi and its consequences for molecular biology, ecology, and development of inoculum. Can J Bot 82:1057–1062

Santiago LS (2000) Use of coarse woody debris by the plant community of a Hawaiian montane cloud forest. Biotropica 32:633–641

Santiago LS, Goldstein G, Meinzer FC, Fownes JH, Müller-Dombois D (2000) Transpiration and forest structure in relation to soil waterlogging in a Hawaiian montane cloud forest. Tree Physiol 20:673–681

Sauer W (1971) Geologie von Ecuador. Beiträge zur regionalen Geologie der Erde, vol 11. Bornträger, Berlin

Sautter R (2003) Waldgesellschaften in Bayern. Ecomed, Landsberg

Schaap MG, Leij FJ, Van Genuchten MT (2001) ROSETTA: a computer program for estimating soil hydraulic parameters with hierarchical pedotransfer functions. J Hydrol 251(3–4):163–176

Schaefer DA, Reiners WA (1990) Throughfall chemistry and canopy processing mechanisms. In: Lindberg SE, Page AL, Norton SA (eds) Acidic precipitation, vol 3. Sources, depositions, and canopy interactions. (Advances in environmental science) Springer, Berlin Heidelberg New York, pp 241–284

Schellekens J, Bruijnzeel LA, Scatena FN, Bink NJ, Holwerda F (2000) Evaporation from a tropical rain forest, Luquillo Experimental Forest, eastern Puerto Rico. Water Resour Res 36:2183–2196

Schemenauer RS, Cereceda P (1994) A proposed standard fog collector for use in high elevation regions. J Appl Meteorol 33:1313–1322

Scheu S, Falca M (2000) The soil food web of two beech forests (*Fagus sylvatica*) of contrasting humus type: Stable isotope analysis of a macro- and mesofauna-dominated community. Oecologia 123:285–296

Scheu S, Parkinson D (1994) Successional changes in microbial biomass, respiration and nutrient status during litter decomposition in an aspen and pine forest. Biol Fertil Soils 19:327–332

Scheu S, Schaefer M (1998) Bottom-up control of the soil macrofauna community in a beechwood on limestone: manipulation of food resources. Ecology 79:1573–1585

Scheuner ET, Makeschin F (2005) Impact of atmospheric nitrogen deposition on carbon dynamics in two Scots pine forest soils in northern Germany. Plant Soil 275:43–54

Schläpfer F, Schmid B (1999) Ecosystem effects of biodiversity: a classification of hypotheses and exploration of empirical results. Ecol Appl 9:893–912

Schläpfer F, Schmid B, Seidl I (1999) Expert estimates about effects of biodiversity on ecosystem processes and services. Oikos 84:346–352

Schlather M (2005) SoPhy: some soil physics tools for R. Available at: http://www.r-project.org/, contributed extension package

Schlather M, Huwe B (2005) A risk index for characterising flow pattern in soils using dye tracer distributions. J Contam Hydrol 79(1–2):25–44

Schlichting E, Blume HP, Stahr K (1995) Bodenkundliches Praktikum, 2nd ed. Enke, Stuttgart

Schmid E, Oberwinkler F, Gomez LD (1995) Light and electron microscopy of the host–fungus interaction in the roots of some epiphytic ferns from Costa Rica. Can J Bot 73: 991–996

Schmidt O, Curry JP, Dyckmans J, Rota E, Scrimgeour CM (2004) Dual stable siotope analysis ($\delta 13C$ and $\delta 15N$) of soil invertebrates and their food sources. Pedobiologia 48:171–180

Schneider K, Migge S, Norton RA, Scheu S, Langel R, Reineking A, Maraun M (2004) Trophic niche differentiation in oribatid mites (Oribatida, Acari): evidence from stable isotope ratios ($^{15}N/^{14}N$). Soil Biol Biochem 36:1769–1774

Schnitzer SA, Carson WP (2001) Treefall gaps and maintenance of species diversity in a tropical forest. Ecology 82:913–919

Schöngart J, Piedade MTF, Ludwigshausen F, Horna V, Worbes M (2002) Phenology and stem-growth periodicity of tree species in Amazonian floodplain forests. J Trop Ecol 18:581–597

Schrumpf M, Guggenberger G, Schubert C Valarezo C, Zech W (2001) Tropical montane rain forest soils – development and nutrient status along an altitudinal gradient in the south Ecuadorian Andes. Erde 132:43–59

Schuettpelz E, Trapnell DW (2006) Exceptional epiphyte diversity on a single tree in Costa Rica. Selbyana 27:65–71

Schulin R, Vangenuchten MT, Fluhler H, Ferlin P (1987a) An experimental-study of solute transport in a stony field soil. Water Resour Res 23(9):1785–1794

Schulin R, Wierenga PJ, Flühler H, Leuenberger J (1987b) Solute transport through a stony soil. Soil Sci Soc Am J 51(1):36–42

Schulze CH (2000) Auswirkungen anthropogener Störungen auf die Diversität von Herbivoren – Analysen von Nachtfalterzönosen entlang von Habitatgradienten in Ost-Malaysia. PhD thesis, University of Bayreuth, Bayreuth

Schulze CH, Fiedler K (2003) Vertical and temporal diversity of a species-rich moth taxon in Borneo. In: Basset Y, Novotny V, Miller S, Kitching RL (eds) Arthropods of tropical forests: spatio-temporal dynamics and resource use in the canopy. Cambridge University Press, Cambridge, pp 69–85

Schulze CH, Linsenmair KE, Fiedler K (2001) Understorey versus canopy – patterns of vertical stratification and diversity among Lepidoptera in a Bornean rainforest. Plant Ecol 153:133–152

Schulze CH, Waltert M, Kessler PJA, Pitopang R, Shahabuddin S, Veddeler D, Mühlenberg M, Gradstein SR, Leuschner C, Steffen-Dewenter I, Tscharntke T (2004) Biodiversity indicator groups of tropical land-use systems: comparing plants, birds, and insects. Ecol Appl 14:1321–1333

Schulze ED (1989) Ökosystemforschung – die Entwicklung einer jungen Wissenschaft. In: Gerwin R (ed) Wie die Zukunft Wurzeln schlug. Springer, Berlin Heidelberg New York, pp 55–64

Schulze ED, Mooney HA (1993) Ecosystem function of biodiversity: a summary. In: Schulze E-D, Mooney HA (eds) Biodiversity and ecosystem function. (Ecological studies, vol 99) Springer, Berlin Heidelberg New York, pp 497–510

Schulze ED, Beck E, Müller-Hohenstein K (2005) Plant ecology. Springer, Berlin Heidelberg New York, 702 pp

Schuster RL, Nieto AS, Orourke TD, Crespo E, Plazanieto G (1996): Mass wasting triggered by the 5 March 1987 Ecuador earthquakes. Eng Geol 42:1–23

Schuur EAG (2001) The effect of water on decomposition dynamics in mesic to wet Hawaiian montane forests. Ecosystems 4:259–273

Schuur EAG, Chadwick OA, Matson PA (2001) Carbon cycling and soil carbon storage in mesic to wet hawaiian montane forests. Ecology 82:3182–3196

Schwartz RC, McInnes KJ, Juo ASR, Cervantes CE (1999) The vertical distribution of a dye tracer in a layered soil. Soil Sci 164(8):561–573

Schwendenmann L, Veldkamp E, Brenes T, O'Brien JJ, Mackensen J (2003) Spatial and temporal variation in soil CO_2 efflux in an old-growth neotropical rain forest, La Selva, Costa Rica. Biogeochemistry 64:111–128

Scoble MJ (1992) The Lepidoptera: form, function, and diversity. Oxford University Press, Oxford

Scoble MJ (1999) Geometrid moths of the world – a catalogue (Lepidoptera: Geometridae). CSIRO, Collingwood

Selles F, Karamanos RE, Kachanoski RG (1986) The spacial variability of nitrogen 15 and its relation to the variability of other soil properties. Soil Sci Soc Am J 50:105–109

Seng HW, Ratnam W, Noor SM, Clyde MM (2004) The effects of the timing and method of logging on forest structure in Peninsular Malaysia. For Ecol Manage 203:209–228

Serra YL, A'Hearn P, Freitag HP, McPhaden MJ (2001) ATLAS self-siphoning rain gauge error estimates. J Atmos Ocean Technol 18:1989–2002

Setaro S, Weiß M, Oberwinkler F, Kottke I (2006a) Sebacinales form ectendomycorrhizas with *Cavendishia nobilis*, a member of the Andean clade of Ericaceae, in the mountain rain forest of southern Ecuador. New Phytol 169:355–365

Setaro S, Kottke I, Oberwinkler F (2006b) Anatomy and ultrastructure of mycorrhizal associations of neotropical Ericaceae. Mycol Progr 5:243–254

Shmida A, Wilson W (1985) Biological determinants of species diversity. J Biogeogr 12:1–20

Sierra R, Stallings J (1998) The dynamics and social organization of tropical deforestation in Northwest Ecuador, 1983–1995. Hum Ecol 26:135–146

Silva Matos DM, Bovi MLA (2002) Understanding the threats to biological diversity in southeastern Brazil. Biodivers Conserv 11:1747–1758

Silver WL, Scatena FN, Johnson AH, Siccima TG, Sanchez MJ (1994) Nutrient availability in a montane wet tropical forest: spatial patterns and methodological considerations. Plant Soil 164:129–145

Silver WL, Thompson AW, McGroddy M, Varner RK, Dias JD, Silva H, Crill PM, Keller M (2005) Fine root dynamics and trace gas fluxes in two lowland tropical forest soils. Global Change Biol 11:290–306

Simberloff D (1998) Flagships, umbrellas, and keystones: is single-species management passé in the landscape era? Biol Conserv 83:247–257

Simunek J, Sejna M, van Genuchten MT (1999) The HYDRUS-2D software package for simulating two-dimensional movement of water, heat, and multiple solutes in variably saturated media. ver. 2.0. IGWMC-TPS-53. International Ground Water Modeling Center, Colorado School of Mines, Golden, Colorado, 251 pp.

Smith K, Gholz HL, Oliveira FA (1998) Litterfall and nitrogen-use efficiency of plantations and primary forest in the eastern Brazilian Amazon. For Ecol Manage 109:209–220

Smith N, Mori SA, Henderson A, Stevenson DW, Heald SV (2004) Flowering plants of the neotropics. Princeton University Press, Princeton

Smith RGB, Nichols JD (2005) Patterns of basal area increment, mortality and recruitment were related to logging intensity in subtropical rainforest in Australia over 35 years. For Ecol Manage 218:319–328

Smith VC, Bradford MA (2003) Do non-additive effects on decomposition in litter-mix experiments result from differences in resource quality between litters? Oikos 102:235–242

Soethe N, Lehmann J, Engels C (2006a) The vertical pattern of rooting and nutrient uptake at different altitudes of a south Ecuadorian montane forest. Plant Soil 286:287–299

Soethe N, Lehmann J, Engels C (2006b) Root morphology and anchorage of six native tree species from a tropical montane forest and an elfin forest in Ecuador. Plant Soil 279:173–185

Soethe N, Lehmann J, Engels C (2007) Nutritional status of plants growing at different altitudes of a tropical montane forest in Ecuador. J Trop Ecol (in press)

Soler R, Bezemer TM, Putten WH van der, Vet1 LEM, Harvey JA (2005) Root herbivore effects on aboveground herbivore, parasitoid and hyperparasitoid performances via changes in plant quality. J Anim Ecol 74:1121–1130

Solis MA, Pogue MG (1999) Lepidopteran biodiversity: patterns and estimators. Am Entomol 45:206–212

Sørensen T (1948) A method of establishing groups of equal amplitude in plant sociology based on similarity of species content. K Danske Vidensk Selsk Biol Skr 5:1–34

Southgate D, Sierra R, Brown L (1991) The causes of tropical deforestation in Ecuador: a statistica analysis. World Dev 19:1145–1151

Spichtinger N, Wenig M, James P, Wagner T, Platt U, Stohl A (2001) Satellite detection of a continental-scale plume of nitrogen oxides from boreal forest fires. Geophys Res Lett 28:4579–4582

Srutek M, Dolezal J, Hara T (2002) Spatial structure and associations in a *Pinus canariensis* population at the treeline, Pico del Teide, Tenerife, Canary Islands. Arct Antarct Alp Res 34:201–210

Ståhl B (2004) Cyrillaceae. In: Smith N, Mori SA, Henderson A, Stevenson DW, Heald SV (eds) Flowering plants of the neotropics. Princeton University Press, Princeton

Steel D (1999) Trade goods and Jívaro warfare: the Shuar 1850–1957, and the Achuar 1940–1978. Ethnohistory 46:745–776

Steenwerth KL, Jackson LE, Calderón FJ, Stromberg MR, Scow KM (2002) Soil microbial community composition and land use history in cultivated and grassland ecosystems of coastal California. Soil Biol Biochem 34:1599–1611

Steinhardt U (1979) Untersuchungen über den Wasser- und Nährstoffhaushalt eines andinen Wolkenwaldes in Venezuela. Göttinger Bodenkd Ber 56:1–146

Stephens GL (2005) Cloud feedbacks in the climate system: a critical review. J Clim 18:237–273

Sterck FJ, Bongers F (2001) Crown development in tropical rain forest trees: patterns with tree height and light availability. J Ecol 89:1–13

Stern MJ (1995) Vegetation recovery on earthquake-triggered landslide sites in the Ecuadorian Andes. In: Churchill SP, et al (eds) Biodiversity and conservation of neotropical montane forests. New York Botanical Garden, New York, pp 207–222

Stevens GC (1989) The latitudinal gradient in geographic range: how so many species coexist in the tropics. Am Nat 133:240–256

Stevens GC (1992) The elevational gradient in altitudinal range: an extension of Rapoport's latitudinal rule to altitude. Am Nat 140:893–911

References

Stohl A, Forster C, Eckhardt S, Spichtinger N, Huntrieser H, Heland J, Schlager H, Wilhelm S, Arnold F, Cooper O (2003) A backward modeling study of intercontinental pollution transport using aricraft measurements. J Geophys Res 108:D12, 4370. DOI 10.1029/2002JD002862

Stoyan R (2000) Aktivität, Ursachen und Klassifikation der Rutschungen in San Francisco/ Südecuador. PhD thesis, Friedrich-Alexander University of Erlangen–Nürnberg, Erlangen

Stuart GH (1988) The influence of canopy cover on understory development in forests of the western Cascade Range, Oregon, USA. Vegetatio 76:79–88

Suárez JP, Weiß M, Abele A, Garnica S, Oberwinkler F, Kottke I (2006) Diverse tulasnelloid fungi form mycorrhizas with epiphytic orchids in an Andean cloud forest. Mycol Res 110:1257–1270

Suárez L (2002) Introducción. In: Granizo T, Pacheco C, Ribadeneira MB, Guerrero M, Suárez L (eds) Libro rojo de las aves del Ecuador. (Serie libros rojos del Ecuador, vol 2) SIMBIOE/ Conservacíon Internacional/EcoCiencia/Ministerio del Ambiente/UICN, Quito

Sulzman EW, Brant JB, Bowden RD, Lajhta K (2005) Contribution of aboveground litter, belowground litter and rhizosphere respiration to total soil CO_2 efflux in an old growth coniferous forest. Biogeochemistry 73:231–256

Süßenbach D (2003) Diversität von Nachtfaltergemeinschaften entlang eines Höhengradienten in Südecuador (Lepidoptera: Pyraloidea, Arctiidae). PhD thesis, University of Bayreuth, Bayreuth. Available at: opus.ub.uni-bayreuth.de/volltexte/2003/33

Svenning JC (1998) The effect of land-use on the local distribution of palm species in an Andean rain forest fragment in northwestern Ecuador. Biodiv Conserv 7:1529–1537

Tabarelli M, Mantovani W (2000) Gap-phase regeneration in a tropical montane forest: the effects of gap structure and bamboo species. Plant Ecol 148:149–155

Takyu M, Aiba SI, Kitayama K (2002) Effects of topography on tropical lower montane forests under different geological conditions on Mount Kinabalu, Borneo. Plant Ecol 159:35–49

Tanimoto H, Kaji Y, Hirokawa J, Akimoto H, Minko NP (2000) The atmospheric impact of boreal forest fires in far eastern Siberia on the seasonal variation of carbon monoxide: observations at Rishiri, a northern remote island in Japan. Geophys Res Lett 27:4073–4076

Tanner EVJ (1977) Four montane rain forests of Jamaica: a quantitative characterization of the floristics, the soils and the foliar mineral levels, and a discussion of interrelations. J Ecol 65:883–918

Tanner EVJ (1980) Studies on the biomass and productivity in a series of montane rain forests in Jamaica. J Ecol 68:573–588

Tanner EVJ, Kapos V, Freskos S, Healey JR, Theobald AM (1990) Nitrogen and phosphorus fertilization of Jamaican montane forest trees. J Trop Ecol 6:231–238

Tanner EVJ, Kapos V, Franco W (1992) Nitrogen and phosphorus fertilization effects on Venezuelan montane forest trunk growth and litterfall. Ecology 73:78–86

Tanner EVJ, Vitousek PM Cuevas E (1998) Experimental investigation of nutrient limitations of forest growth on wet tropical mountains. Ecology 79:10–22

Temme M (1972) Wirtschaft und Bevölkerung in Südecuador. Eine sozio-ökonomische Analyse des Wirtschaftsraumes Loja. Kölner Forsch Wirtsch Sozialgeogr 19, Wiesbaden

Tenenbaum YB, Silva V de, Langford JC (2000) A global geometric framework for nonlinear dimensionality reduction. Science 290:2319–2323

Thies W (1998) Resource and habitat use in two frugivorous bat species (Phyllostomidae: *Carollia perspicillata* and *C. castanea*) in Panama: mechanisms of coexistence. PhD thesis, University of Tübingen, Tübingen

Thompson J, Proctor J, Scott DA, Fraser PJ, Marrs RH, Miller RP, Viana V (1998) Rain forest on Macará Island, Roraima, Brazil: artificial gaps and plant response to them. For Ecol Manage 102:305–321

Tian G, Brussaard L, Kang B T (1995) Plant residue decomposition in the humid tropics. Influence of chemical composition and soil fauna. In: Reddy M (ed) Soil organisms and litter decomposition in the tropics. Westview, Boulder, pp 203–224

Tibaldi A, Ferrari L, Pasquare G (1995) Landslides triggered by earthquakes and their relations with faults and mountain slope geometry – an example from Ecuador. Geomorphology 11:215–226

Timm H-C, Küppers M, Stegemann J (2004) Non-destructive analysis of architectural expansion and assimilate allocation in different tropical tree saplings: consequences of using steady-state and dynamic photosynthesis models. Ecotropica 10:101–121

Toledo VM, Ortiz B, Medellín-Morales S (1994) Biodiversity islands in a sea of pasturelands: indigenous resource management in the humid tropics of Mexico. Etnoecologica 2:37–50

Townsend AR, Vitousek PM, Trumbore SE (1995) Soil organic matter dynamics along gradients in temperature and land use on the island of Hawaii. Ecology 76:721–733

Treseder K, Vitousek PM (2001) Effects of soil nutrient availability on investment in acquisition of N and P in Hawaiian rain forests. Ecology 82:946–954

Trueblood DD, Gallagher ED, Gould DM (1994) Three stages of seasonal succession on the Savin Hill Cove mudflat, Boston Harbor. Limnol Oceanogr 39:1440–1454

Tryon RM, Tryon AF (1982) Ferns and allied plants, with special reference to tropical America. Springer, Berlin Heidelberg New York

Turner IM (1996) Species loss in fragments of tropical rain forest: a review of the evidence. J Appl Ecol 33:200–209

Uhl C, Verissimo A, Mattos MM, Brandino Z, Vieira ICG (1991) Social economic, and social consequences of selective logging in an Amazon frontier: the case of Tailandia. For Ecol Manage 46:243–273

Ulloa C, Jørgensen PM (1995) Árboles y arbustos de los Andes del Ecuador. AAU Rep 30

Ulrich B (1983) Interactions of forest canopies with atmospheric constituents: SO_2, alkali and earth alkali cations and chloride. In: Ulrich B, Pankrath J (eds) Effects of accumulation of air pollutants in forest ecosystems. Reidel, Dordrecht, pp 33–45

Ulrich B (1984) Ion cycle and forest ecosystem stability. In: Ågren GI (ed) State and change of forest ecosystems – indicators in current research. (Research report 13, Department of ecology and environmental research) Swedish University of Agricultural Sciences, Gothenberg, pp 207–233

Ulrich B (1987) Stability, elasticity, and resilience of terrestrial ecosystems with respect to matter balance. In: Schulze ED, Zwölfer H (eds) Potentials and limitations of ecosystem analysis. (Ecological studies vol 61) Springer, Berlin Heidelberg New York, pp 11–49

UNEP (2002) Forest programme – Ecuador. Available at: www.unep-wcmc.org/forest/data/country/ecu.htm

UNESCO (1973) International classification and mapping of Vegetation. UNESCO, Paris

UNESCO (1984) Action plan for biosphere reserves. Nat Resour 20:11–22

Urgiles N (2003) Evaluacion del potencial de micorrizas en la propagacion de tres especies nativas forestales. BSc thesis. Universidad Nacional de Loja, Loja

US Library of Congress (2006) Whites and mestizos. Available at: countrystudies.us/ecuador/29.htm

Utrio P (1995) On flight temperatures and foraging strategies of nocturnal moths (in Finnish, with English summary). Baptria 20:113–121

Valarezo C (1996) Procesos erosivos en los sistemas de riego de la Zona Andina de la Región Sur del Ecuador. [Sistema Andino de Postgrado Agropecuario (SAPOA). Segundo curso intensivo] OUI/IICA/Universidad de McGill/UNL, Loja

Valencia R, Cerón C, Palacios W, Sierra R (1999) Las formaciones naturales de la sierra del Ecuador. In: Sierra R (ed) Propuesta preliminar de un sistema de clasificación de vegetación para el Ecuador continental. Proyecto INEFAN/GEF-BIRF y EcoCiencia, Quito, pp 79–108

Valencia R, Pitman N, Leó-Yánez S, Jørgensen PM (2000) Libro rojo de las plantas endémicas del Ecuador 2000. Pontificia Universidad Católica del Ecuador, Quito

Vance ED, Nadkarni NM (1992) Root mass distribution in a moist tropical montane forest. Plant Soil 142:31–39

Vance ED, Brookes PC, Jenkinson DS (1987) An extraction method for measuring soil microbial biomass C. Soil Biol Biochem 19:703–707

Vanderklift MA, Ponsard S (2003) Sources of variation in consumer-diet $\delta15N$ enrichment: a meta-analysis. Oecologia 136:169–182

Varnes DJ (1978) Slope movements: type and processes. In: Eckel EB (ed) Landslides – analysis and control. Transp Res Board Spec Rep 176:11–33

Vasconcelos HL, Vilhena JMS, Caliri GJA (2000) Responses of ants to selective logging of a central Amazonian forest. J Appl Ecol 37:508–514

Vasquez-Yanes C, Orozco-Segovia A (1993) Patterns of seed longevity and germination in the tropical rainforest. Annu Rev Ecol Syst 24:69–87

Veblen TT (1978) Forest preservation in the western highlands of Guatemala. Geogr Rev 68:417–434

Veldkamp E (1994) Organic carbon turnover in three tropical soils under pasture after deforestation. Soil Sci Soc Am J 58:175–180

Veneklaas EJ (1990) Nutrient fluxes in bulk precipitation and throughfall in two montane tropical rain forests, Colombia. J Ecol 78:974–992

Veríssimo A, Barreto P, Mattos M, Tarifa R, Uhl C (1992) Logging impacts and prospects for sustainable forest management in the Brazilian Amazon. For Ecol Manage 55:169–199

Vieira BC, Fernandes NF (2004) Landslides in Rio de Janeiro: the role played by variations in soil hydraulic conductivity. Hydrol Proc 18:791–805

Vitorello VA, Cerri CC, Andreux F, Feller C, Victoria RL (1989) Organic matter and natural carbon-13 distribution in forested and cultivated Oxisols. Soil Sci Soc Am J 53:773–778

Vitousek PM (1984) Litter fall, nutrient cycling, and nutrient limitation in tropical forests. Ecology 65:285–298

Vitousek PM, Farrington H (1997) Nutrient limitation and soil development: experimental test of a biogeochemical theory. Biogeochemistry 37:63–75

Vitousek PM, Aplet G, Turner D, Lockwood JJ (1992) The Mauna Loa environmental matrix: foliar and soil nutrients. Oecologia 89:372–382

Vitousek PM, Walker LR, Whiteaker LD, Matson PA (1993) Nutrient limitations to plant growth during primary succession in Hawaii Volcanoes National Park. Biogeochemistry 23:197–215

Vitousek PM, Turner DR, Kitayama K (1995) Foliar nutrients during long-term soil development in Hawaiian montane rain forests. Ecology 76:712–720

Vitt LJ, Avila-Pries TCS, Caldwell JP, Oliveira VRL (1998) The impact of individual tree harvesting on thermal environments of lizards in amazonian rain forest. Conserv Biol 12:654–664

Vogt KA, Grier CC, Vogt DJ (1986) Production, turnover, and nutrient dynamics of above- and belowground detritus of world forests. Adv Ecol Res 15:303–377

Voss RS, Emmons LH (1996) Mammalian diversity in neotropical lowland rainforests: a preliminary assessment. Bull Am Mus Nat Hist 230:1–115

Vuille M, Bradley RS, Keimig F (2000) Climate variability in the Andes of Ecuador and its relation to tropical Pacific and Atlantic sea surface temperature anomalies. J Clim 13:2520–2535

Walker LR, Zarin DJ, Fetcher N, Myster RW, Johnson AH (1996) Ecosystem development and plant succesion on landslides in the Caribbean. Biotropica 28:566–576

Walter DE (1985) The effect of litter type and elevation on colonization of mixed coniferous litterbags by oribatid mites. Pedobiologia 28:383–387

Waltert M, Mardiastuti A, Mühlenberg M (2004) Effects of land use on bird species richness in Sulawesi, Indonesia. Conserv Biol 18:1339–1346

Wardle DA, Ghani A (1995) A critique of the microbial metabolic quotient (qCO2) as a bioindicator of disturbance and ecosystem development. Soil Biol Biochem 27:1601–1610

Wardle DA, Nilsson MC, Zackrisson O, Gallet C (2003) Determinants of litter mixing effects in a Swedish boreal forest. Soil Biol Biochem 35:827–835

Warren DM, Slikkerveer LJ, Brokensha G (1995) The cultural dimension of development: indigenous knowledge systems. London

Watson JW, Eyzaguirre PB (2002) Home gardens and in situ conservation of plant genetic resources in farming systems. (Proceedings of the second international home gardens workshop, 17–19 July 2001, Witzenhausen, Germany) International Plant Genetic Resources Institute, Rome

Watt AS (1940) Contributions to the ecology of bracken (*Pteridium aqulinum*). I. The rhizom. New Phytol 39:401–422

Wattenberg I (1996) Struktur, Diversität und Verjüngungsdynamik eines prämontanen Regenwaldes in der Cordillera de Tilarán in Costa Rica unter besonderer Berücksichtigung der

Rolle der Gaps für die Aufrechterhaltung der Baumartendiversität. PhD thesis, University of Bielefeld, Bielefeld

Weaver PL, Murphy PG (1990) Forest structure and productivity in Puerto-Rico Luquillo Mountains. Biotropica 22:69–82

Weaver PL, Medina E, Pool D, Dugger K, Gonzales-Liboy J, Cuevas E (1986) Ecological observations in the dwarf cloud forest of the Luquillo mountains of Puerto Rico. Biotropica 18:79–85

Webb LJ, Tracey JG, Williams WT, Lance GN (1970) A comparison of the properties of floristic and physiognomic–structural data. J Ecol 58:203–232

Weber H (1958) Die Páramos von Costa Rica. Abh Math Naturwiss Klasse Akad Wissensch Lit Mainz 3:1–78

Webster BD, Steeves TA (1958) Morphogenesis in *Pteridium aquilinum* (L.) Kuhn. General morphology and growth habit. Phytomorphology 8:30–41

Webster GL, Rhode RM (2001) Plant diversity of an Andean cloud forest. Checklist of the vascular flora of Maquipucuna, Ecuador. (University of California publications in botany, vol 82) University of California Press, Berkeley

Wegner C, Wunderlich M, Kessler M, Schawe M (2003) Foliar C:N ratio of ferns along an Andean elevational gradient. Biotropica 35:486–490

Weigend M (2002) Observations on the biogeography of the Amotape–Huancabamba zone in northern Peru. Bot Rev 68:38–54

Weigend M (2004) Additional observations on the biogeography of the Amotape–Huancabamba zone in northern Peru: defining the southeastern limits. Rev Peru Biol 11:127–134

Weiler M, Flühler H (2004) Inferring flow types from dye patterns in macroporous soils. Geoderma 120(1–2):137–153

Weiß M, Selosse M-A, Rexer K-H, Urban A, Oberwinkler F (2004) Sebacinales: a hitherto overlooked cosm of heterobasidiomycetes with a broad mycorrhizal potential. Mycol Res 108:1003–1010

Weller SJ, Jacobson NL, Conner WE (1999) The evolution of chemical defences and mating systems in tiger moths (Lepidoptera: Arctiidae). Biol J Linn Soc 68:576–578

Weninger B, Jöris O, Danzeglocke U (2004) Calpal – the Cologne radiocarbon CALibration and PALaeoclimate research package. Available at: www.calpal.de

Werger MJA, Sprangers JTC (1982) Composition of floristic and structural classification of vegetation. Vegetatio 50:175–183

Werner FA, Homeier J, Gradstein SR (2005) Diversity of vascular epiphytes on isolated remnant trees in the montane forest belt of southern Ecuador. Ecotropica 11:21–40

Whitmore TC (1989) Canopy gaps and the two major groups of forest trees. Ecology 70:536–538

Whitmore TC (1996) A review of some aspects of tropical rain forest seedling ecology with suggestions for further enquiry. In: Swaine MD (ed) The ecology of tropical forest tree seedlings. (Man and the biosphere series, vol 17) UNESCO, Paris, pp 3–39

Whitmore TC (1998) An introduction to tropical rain forests. Oxford University Press, Oxford

Wiens JA, Stenseth NC, Horne B van, Ims RA (1993) Ecological mechanisms and landscape ecology. Oikos 66:369–380

Wilcke W, Yasin S, Valerezo C, Zech W (2001a) Change in water quality during passage through a tropical montane rain forest in Ecuador. Biogeochemistry 55:45–72

Wilcke W, Yasin S, Valerezo C, Zech W (2001b) Nutrient budget of three microcatchments under tropical montane rain forest in Ecuador – preliminary results. Erde 132:61–74

Wilcke W, Lilienfein J (2004) Soil carbon-13 natural abundance under native and managed vegetation in Brazil. Soil Sci Soc Am J 68:827–832

Wilcke W, Yasin S, Abramowski U, Valerezo C, Zech W (2002) Nutrient storage and turnover in organic layers under tropical montane rain forest in Ecuador. Eur J Soil Sci 53:15–27

Wilcke W, Valladarez C, Stoyan R, Yasin S, Valarezo C, Zech W (2003) Soil properties on a chronosequence of landslides in montane rain forest, Ecuador. Catena 53:79–95

Wilcke W, Hess T, Bengel C, Homeier J, Valarezo C, Zech W (2005) Coarse woody debris in a montane forest in Ecuador: mass, C and nutrient stock, and turnover. For Ecol Manage 205:139–147

Wilcke W, Schmidt A, Homeier J, Valarezo C, Zech W (2007) Soil properties and tree growth along an altitudinal transect in Ecuadorian tropical montane forest. J Plant Nutr Soil Sci (in press)

Will KW, Mishler BD, Wheeler QD (2005) The perils of DNA barcoding and the need for integrative taxonomy. Syst Biol 54:844–851

Williamson, MH (1978) The ordination of incidence data. J Ecol 66:911–920

Willig MR, Camilo GR, Noble S (1993) Dietary overlap in frugivorous and insectivorous bats from edaphic cerrado habitats of Brazil. J Mammal 74:117–128

Willott SJ (1999) The effects of selective logging on the distribution of moths in a Bornean rainforest. Philos Trans R Soc Lond B 354:1783–1790

Wirooks L (2005) Die ökologische Aussagekraft des Lichtfangs – eine Studie zur Habitatbindung und kleinräumigen Verteilung von Nachtfaltern und ihren Raupen. Wolf & Kreuels, Havixbeck–Hohenholte

Wolf JHD (1993) Diversity patterns and biomass of epiphytic bryophytes and lichenas along an altitudinal gradient in the northern Andes. Ann Missouri Bot Gard 80:928–960

Wolff D (2006) Nectar sugar composition and volumes of 47 species of Gentianales from a southern Ecuadorian montane forest. Ann Bot 97:767–777

Wolff D, Braun M, Liede S (2003) Nocturnal versus diurnal pollination success in *Isertia laevis* (Rubiaceae): a sphingophilous plant visited by hummingbirds. Plant Biol 5:71–78

Woodward CL (1996) Soil compaction and topsoil removal effects on soil properties and seedling growth in Amazonian Ecuador. For Ecol Manage 82:197–209

Wright SG, Carrasco C, Calderón, O, Paton S (1999) The El Niño southern oscillation, variable fruit production, and famine in a tropical forest. Ecology 80:1632–1647

Wright SJ, Schaik CP van (1994) Light and the phenology of tropical trees. Am Nat 143:192–199

Wubet T, Kottke I, Teketay D, Oberwinkler F (2003) Mycorrhizal status of indigenous trees in dry Afromontane forests of Ethiopia. For Ecol Manage 179:387–399

Wunder S (1996a) Deforestation and the uses of wood in the Ecuadorian Andes. Mount Res Dev 16:367–382

Wunder S (1996b) Los caminos de la madera. DDA Intercorporation/UICN, Bern

Wunder S (2000) The economics of deforestation: the example of Ecuador. St. Martin, New York

Yamakura T, Hagihara A, Sukardjo S, Ogawa H (1986) Aboveground biomass of tropical rain-forest stands in Indonesian Borneo. Vegetatio 68:71–82

Yanai RD, Majdi H, Park BB (2003) Measured and modelled differences in nutrient concentrations between rhizosphere and bulk soil in a Norway spruce stand. Plant Soil 257:133–142

Yao H, He Z, Wilson MJ, Campbell CD (2000) Microbial biomass and community structure in a sequence of soils with increasing fertility and changing land use. Microb Ecol 40:223–237

Yasin S (2001) Water and nutrient dynamics in microcatchments under montane forest in the south Ecuadorian Andes. (Bayreuther Bodenkundliche Berichte 73) University of Bayreuth, Bayreuth

Young A, Boshier D, Boyle T (2000) Forest conservation genetics. Principles and practice. CSIRO, Canberra

Young KR, Ulloa Ulloa C, Luteyn JL, Knapp S (2002) Plant evolution and endemism in Andean South America: an introduction. Bot Rev 68:4–21

Zagt RJ, Boot RGA (1997) The response of tropical trees to logging: a cautious application of matrix models. In: Zagt RJ (eds) Tree demography in the tropical rain forest of Guyana. (Tropenbos–Guyana ser 3) Tropenbos Guyana Program, Georgetown, pp 167–215

Zapata FA, Gaston KJ, Chown SL (2003) Mid-domain models of species richness gradients: assumptions, methods and evidence. J Anim Ecol 72:677–679

Zarin DJ, Johnson AH (1995a) Nutrient accumulation during primary succession in a montane tropical forest, Puerto Rico. Soil Sci Soc Am J 59:1444–1452

Zarin DJ, Johnson AH (1995b) Base saturation, nutrient cation, and organic matter increases during early pedogenesis on landslide scars in the Luquillo Experimental Forest, Puerto Rico. Geoderma 65:317–330

Zelles L (1999) Fatty acid patterns of phospholipids and lipopolysaccharides in the characterisation of microbial communities in soil: a review. Biol Fertil Soils 29:111–129

Ziegler AD, Giambelluca TW, Tran LT, Vana TT, Nullet MA, Fox J, Vien TD, Pinthong J, Maxwell JF, Evett S (2004) Hydrological consequences of landscape fragmentation in mountainous northern Vietnam: evidence of accelerated overland flow generation. J Hydrol 287(1–4):124–146

Zimmermann M (2005) Reaktion von Nachtfalter-Gemeinschaften im ecuadorianischen Bergregenwald auf einen experimentellen forstlichen Eingriff. PhD thesis, University of Bayreuth, Bayreuth

Zimmermann R, Soplin H, Börner A, Mette T (2002) Tree growth history, stand structure, and biomass of premontane forest types at the Cerro Tambo, Alto Mayo, Northern Peru. In: Bussmann RW, Lange S (eds) Conservación de la biodiversidad en los Andes y la Amazonía – conservation of biodiversity in the Andes and the Amazon. Ceres, Lima, pp 327–335

Zotz G (1998) Demography of the epiphytic orchid, *Dimerandra emarginata*. J Trop Ecol 14:725–741

Zotz G (2004) Long-term observation of the population dynamics of vascular epiphytes. In: Breckle SW, Schweizer B, Fangmeier A (eds) Results of worldwide ecological studies. (Second symposium of the A.F.W.) Schimperstiftung, Hohenheim, pp 97–117

Zotz G, Tyree MT, Patiño S, Carlton MR (1998) Hydraulic architecture and water use of selected species from a lower montane forest in Panama. Trees 12:302–309

Zou XM, Ruan HH, Fu Y, Yang XD, Sha LQ (2005) Estimating soil labile organic carbon and potential turnover rates using a sequential fumigation–incubation procedure. Soil Biol Biochem 37:1923–1928

Subject Index

A
accumulation zone 326
acidity 79
acrisol 6
afforestation 13, 37
AGB 239
age 412
Agoyan unit 4
agrobiodiversity 335
agroforestry 462
agropastoral 39
A horizon 79
AHZ 275
air
   humidity 313
   temperature 7, 64, 313
alfisol 6
allocation 229
allometric studies 246
alpha diversity 19, 21, 178, 412, 449, 454
altitudinal belt 123
altitudinal gradient 7, 19, 50, 63, 75, 87, 101, 113, 149, 157, 167, 181, 217, 229, 295, 304, 327, 452, 456
altitudinal transects 157
aluminum 197, 213, 241, 269, 398
Alzateetalia verticillatae 413
Amazon rain forest 10
ammonia 305, 452
Amotape–Huancabamba depression 453
Andean forests 99
Andes-occurring System 9
animal-husbandry systems 12
ANPP 259
anthropogenic activity 215
anthropogenic emission 205
appressoria 146
arbuscular mycorrhiza 137, 453
Atlantic circulation 451

B
basal area 231, 251, 272
   tree 260
baseflow 213
base metal 80, 212
   saturation 84
   status 397
bat frugivore 158
Bergmann's rule 51
beta diversity 100, 177, 412
BGB 239
B horizon 79
bio-macropores 393
biodiversity 15, 19, 41, 449, 452
   conservation 42, 347
   feedback 458
   feedback on ecosystem 460
   hotspot 10, 15, 19, 22, 41, 52, 279, 455, 458
   indication 52, 177
   indicator 19
   inventory 15
   loss of 52
   management 331
   regeneration 460
bioelement
   losses 462
biogeography 10
biological activity 81
biomass 436
   aboveground 229, 239
   allocation 229
   belowground 229, 239
   burning 309
   microbial 187, 217, 223
   plant 51, 58
biosphere reserve 343
bird community 149
black carbon 408
bog core 103

bosque siempreverde montano 12
bottom-up 188
boundary
   aerodynamic 254
bracken fern
   control 373
   problem 461
bryophyte
   abundance 364
buffer capacity 461
   rate 205
   system 81
bulk density 80, 376
burning 309, 371
bush stage 367

C

^{13}C
   abundance 399
C/N ratio 76, 187, 223, 457
calcium 78, 205, 260, 269, 462
   exchangeable 326
Caledonian orogenic event 4
cambial activity 291
Cambisol 6, 76
canopy 159, 194, 205
   budget model 204
   closed 314
   gaps 311
   guild 159
   height 59, 313
   openings 313
   openness 270
   structure 267, 270
carbon 76, 189, 217, 218, 222, 270, 276, 307, 415
   allocation 240, 241
   black 408
   microbial 223
   stocks 457
carrying capacity 372
Catamayo Valley 9
catchment 75, 204, 355
   budget 213
cation-exchange capacity 84
cattle 12, 342
CBD 41
CEC_{eff} 75
CEC_{pot} 75
Ceja Andina 10
ceja de montaña 89, 98
charcoal 103, 461

Chiguinda unit 4
chloride 204, 452
chronosequence 325
circulation 285
CITES 41
classification
   landscape level 116
   plot level 117
classification system
   structure based 114
clay
   concentration 79
   destruction 80
clearing 436
climate 7, 63, 105, 451
   change 203
   diagram 55
   history 453
   variability 281
climax 373
climber 16, 271
cloud
   frequency 7, 64, 284
   water 63, 70, 199, 303
cloudiness 282
Cmic/Nmic ratio 224
Cmic/SOC ratio 225
CNESS index 444
CO_2-fixation 51
coarse root 232
coarse woody debris 208
cohorts 133
cold air surges 289
cold events 282
colonization 39
   de novo 371
colonizers
   first 362
colonos 12, 31, 371, 462
competition 440
   crown 297
composition
   floristic 87, 88, 125
   taxonomic 171
conductance
   canopy 254
   leaf 246
conductivity
   electrical 355
conservation 23, 347, 461
   dynamic 418
   ex situ 418
constraints
   thermal 51

Subject Index

Convention
   of Biological Diversity 41
conversion 38, 52, 88, 374, 402
Cordillera del Condor 4
Cordillera Occidental 4
Cordillera Real 1
Costa 4
coumaryl 76
coupling
   atmospheric 254
cover abundances 411
crop tree
   potential 349
crown competition 297
crown projection 251
cryptogamic stage 370
cultivation
   shifting 462
cutting cycles 348
cycling 209

D
database 44
dbh 59, 117, 231, 249, 268, 434
dead wood 204
deciduous forest 10
decomposer 181, 457
decomposition 187
   constant 210
deep leaching 213
deforestation 37, 358, 428, 463
   Ecuador 38
   reasons 39
   South America 38
degradation 61
dendrometer 291
dendrometry 453
densiometers 349
density 182
   stem 231
deposition 207, 303, 452
   bulk 207
   dry 205
   rainfall 207
   throughfall 207
depth distribution 81
diameter growth 351
dioecy 427
dispersal 157
   capacity 446
distance 432
disturbance 291, 352, 361, 446
   anthropogenic 52, 375

areas 164
   linear 361
   natural 375
diurnal cycle 286
diversity 51, 87, 182, 320, 443
   alpha 19, 178, 412, 449, 454
   altitudinal pattern 454
   animal 353
   beta 100, 177, 412
   cryptogam 275
   functional 462
   host plant 176
   index 368
   moth 176, 446
   plant 11, 87, 125, 275
   protection 343
   species 271
   structural 462
   structure 164
   tree 349
   tree species 59
DNA barcoding 15
domestication 358
driving factor 460
dry forests 88
dye coverage function 388
dye tracer experiment 387
dynamics
   spatio-temporal 456
Dystric Cambisols 325

E
earthquake 320
easterlies 452
EC 267
ecological experiment 348
ecosystem
   analysis 49
   buffer capacity 459
   categories of components 458
   components 455
   development 203
   dynamic composition 460
   flux 211
   functioning 330
   interactions 460
   man-made 52
   resilience 52, 460
   services 460
   stability 52, 458
ecotone 123, 125, 429, 453
ecotourism 344

ECSF 1
ectomycorrhiza 138
effect
　dilution/concentration 207
　geophysical 50
efflux
　CO_2 358
element
　budget 203
　concentration 205
　flux 204
　redistribution 85
　total concentrations 84
　turnover times 81
Elfin forest 10, 89, 98, 244
El Libano 56
El Niño 281
El Tiro Pass 5, 102, 127, 158, 276, 284
emergency situation 458
emission 132, 205
endemism 10, 87
energy balance 253
enrichment planting 439, 463
ENSO 207, 281, 451
epiphyte coverage 194
epiphyte 16, 87, 348, 457
　understory 352
　vascular 21, 91
erogosterol 189
erosion 319, 326, 365
Estacion Científica San Francisco 1
ethnicity 32, 332
ethnology 331
evaporation 193
　annual 56
evapotranspiration 194, 253
excrement
　bat 163
　bird 163
　trap 162
exotic tree species 463
experimentation 49
exploitation 39
extinction 428

F
factors
　limiting 189
fauna 12
feeding guild 151
felling 463
ferrolysis 80
fertilization 462

fine litter 208
fine root 233, 259
fire 13, 101, 307, 371, 461
　recurrent 373
floristic analysis 413
floristic change 328
flow
　lateral 197
　patterns 387
　preferential 387
　velocities 394
flow-weighted mean 204
flowering 289, 291, 419
　seasonality 420, 459
fog 130
　water 70, 303
foliar analysis 262
food 158
food web 181, 457
　complexity 49
forest
　clearing 371, 464
　cover 39
　deciduous 10
　dynamics 313
　Elfin 10, 89, 98, 244
　fires 307
　gap dynamics 311
　garden 332
　lower montane 89, 91, 94, 99, 244
　management 40
　margins 461
　natural 1, 40
　perhumid montane broad-leaved 10
　premontane rainforest 89
　primary 117, 347, 387, 413
　products 420
　Purdiaea nutans 12, 275
　ravine 117
　recovery of secondary forest 460
　regeneration 411
　ridge 96, 118
　secondary 39, 367, 411, 460
　semi-deciduous 10, 90
　stand parameter 348
　submontane rainforest 89
　type classification 452
　types 91, 150, 153, 243, 313
　upper montane 89, 90, 94, 96
fragmentation
　population 456
freezing 132
friajes 282
fructification
　seasonality 423

fruiting 289, 291, 419
functional group 256
functional relationship 59, 114, 120, 149, 394, 461
FWM 204

G

gaging station 204
Gallagher's CNESS index 444
gap 127, 319, 349
    dynamics 100, 311, 456
    size 311
gelifluction 320
gene-pool 418
gene flow 418
general flowering 429
genetic drift 456
genetic resources 418
genetic variation 420
geology 4
geomorphology 4
geostatistics 384
germination 417
    protocols 426
gleysols 6
glomalin 362
glucose 189
glyphosate 436
gorge 267
gradient
    altitudinal 7, 19, 50, 63, 75, 87, 101, 113, 149, 157, 167, 181, 217, 229, 295, 304, 327, 452, 456
    analysis 49, 55
    climatic 452
    disturbance 52, 59, 446, 460
    environmental 49
    horizontal 55, 272
    humidity 451
    land use 52, 59, 114, 461
    latitudinal 21
    local 304
    meteorological 63
    natural 55, 451
    regeneration 53, 61, 381, 443, 462
    reverse disturbance 53
    succession 443, 445
    temperature 64
    topographical 87
    vertical 57, 70
grass-Páramo 456
groundwater 193
    vulnerability 388

growth
    dynamics 291
    heterogeneities 292, 295
    performance 439
    rate 272, 291, 315, 351
Guelph parameter 197
guild 150, 158
gully forest 244
Guyana shield 4

H

habitat 418
    fragmentation 157
heat shock 371
Hellmann gages 194
hemi-epiphytes 127
hemispherical images 267
herbal stage 368
herbarium 45
herbicide application 463
herbivorous 59
herbs 16
heterogeneities 303
heterogeneity
    environmental 100
high-flow condition 213
histosol 6, 76
Holocene 103, 456
home garden 12, 335, 353, 462
horizon 399
    A 79
    B 79
    O 78
    organic 80
horizontal gradient 55, 272
host-plant 177
hotspot 10, 15, 19, 22, 41, 52, 279, 455, 458
Huancabamba depression 1
Huancabamba Zone 275, 456
huertas 335
human activity 105
    impact 38, 320
humidity 130, 270, 313
    gradient 451
    relative 132
Hurlbert
    rarefied species numbers 158
Hurlbert's
    rarefaction method 444
hydraulic conductivity 375
hydrologic conditions 85
hydromorphic properties 75
Hydrus-2D 387

I

identification 42, 47
   taxonomic 43
illite 212
image processing techniques 387
images
   digital 44
imaging 44
immobilization 81
improvement felling 348, 463
Inca 110
Indians 25, 332
indicator
   integrative 51
initial stages 362
inoculation 426
input
   atmospheric 303
inputs 203
insect herbivorous 51, 59, 167
integrative indicators 51
Inter-Andean basin 4
Inter-Andean region 1
interaction
   biotic 457
interandean forests 100
interception 258
   loss 193
interflow 193
intermediate disturbance hypothesis 53, 354
inventory 21, 42
iron 75
   oxides 76
irradiance 64
irradiation 132
isolation 456
ISOMAP ordination technique 410
isotope ratios 181

J

Jaccard indices 410

K

key species 460
keystone species 475
K_L value 210
KO_i value 210
KO_L value 210

L

Lagunas de los Compadres 5
LAI 194, 232, 243, 270
landforms 320
landscape level 85
landslide 6, 90, 127, 162, 311, 319, 361, 375, 387, 444
   activity 324
   anthropogenic 330
   hazards 323
   natural 330
   succession 319, 329
land use 12, 116, 379, 387
   concept 12, 462
   gradient 52, 59, 114, 463
   intensity 60
   non-sustainable 462
   pattern 11
   practices 463
   sustainable 52, 331, 338, 462
La Niña 281, 323, 451
latitudinal gradient 21
leaching of nutrients 80, 396
leaf shedding 291
leptosol 6
life forms 127, 258, 413
light trap 444
lignin 78
   VSC lignin 76
liquefaction 323
litter
   decomposition 58
   leachate 210
   quality 187, 457, 459
litterfall 208
liverwort
   thalloid 364
LMMC 41
logging
   intensity 349
   reduced impact 348
   selective 353
lower montane forest 89, 91, 94, 244
luvisols 6
lysimeter 204

M

macronutrient 78, 260
magnesium 78, 205, 260, 269, 462
Malacatos 10, 12
man-made ecosystem 52
management 52
   biodiversity 331
   natural forest 347, 463
   options 462
   pasture 462
   sustainable 40, 347

manganese 269
mass flowering 429
mass movement 319
matter
   atmospheric 451
   deposition 303
   input 304
megadiversity 41
   challenges 42
mesh grid collector 63
Mestizo-Colonos 462
mestizos 25, 31
meta-siltstones 398
meta data 46
metal 76
meteorological station 63
mica 212
microbial activity 462
microcatchment 213
microclimate 314
microhabitat 190
mid-domain concept 51
migration 25, 428
   species 456
   trends 32
mineralization 79, 215, 326, 457
   rate 203
mineralogy 398
mineral soil 76, 211, 265
minimum area 410
mist-net 149
moderate disturbance theory 460
montane forest 99, 311
Monte Olivo 4
morphology 85
morphospecies 43
mortality 313
mosaic climax 456
mottling 80
mudflow 6, 322
museum 43
mycorrhiza 16, 137, 426, 453, 457
   arbuscular 137, 457
   cavendishioid 144
mycorrhiza-like interactions 363
mycorrhizal fungi 16
mycorrhizal partner 365
mycorrhization 362

N

^{15}N
   abundance 399
National Parks 428
native tree species 463

natural forest 1, 40
   management 463
naturalness 60
natural vegetation 88
niche partitioning 313
nitrate 305, 452, 472
nitrogen 76, 189, 205, 217, 260
   microbial 223
   total soil 399
   uptake activity 265
nunatak landscape 6
nurse effect 439
nutrient 205
   acquisition 259, 264
   allocation 241
   availability 78
   budget 213
   cycling 348, 355
   deposition 51
   export 212
   flux 203
   input 208
   plant-available 210
   release 210
   status 203, 262
   stocks 326
   supply 79, 260
nutritional status 259

O

$\delta^{18}O$ values 198
Oa horizon 78
Oe horizon 78
Oi horizon 78
O isotopic signature 198
orchard 12
organic horizon 80
   layer 76, 209, 260, 326
   material accumulation at high
      elevations 459
   matter 76, 203, 457
   turnover 210
Oriente 1, 4
overexploitation 342
oxidation potential 396

P

Pacific circulation 451
Palta 111
páramo 89, 101, 127, 453
   grass-rich 10
   shrubby 10
   vegetation 6

páramo arbustivo 12
parent rock 212
Pareto distribution 388
parthenogenesis 183
pasture 1, 61, 163, 335, 387, 397, 431
   abandoned 373, 462
   grasses 371
   management 12, 462
   systems 12
pattern
   detection 49
   distribution 51
PCT 349
pedogenetic process 79
pedotransfer functions 386
penetration resistance 376
people 25
percolation 193
perhumid climate 6
perhumid montane broad-leaved forest 10
periodicity 423
pH 6, 79, 190, 205, 231, 268, 304, 371, 399
phenology 295, 419
   reproductive 417
phenotypic plasticity 428
phosphorus 78, 189, 205, 260
   available 397
phyllite 212, 398
phytodiversity 459
pioneer 366
   species 292, 311
   tree 251
   vegetation 364
planosol 6, 76
plant
   biomass 51, 58
   community 114
   diversity 11, 87
   ethnospecific knowledge 333
   formations 88
   life forms 51
   species richness 100
   use 333
   vascular 16
plantation 40, 417, 435
planting 463
Pleistocene 103, 456
PLFA 225, 455
Podocarpus National Park 10
podzolic feature 7
point count 149

pollen 107
   analysis 461
   record 101
   percentage diagram 103
pollinators 157, 418
polyphenol 79
population 25, 418, 456
   dynamics 353
potassium 78, 205, 260, 269, 462
precipitation 63, 130, 195, 285, 303, 452
   annual 56, 255
   horizontal 200
   ionic concentrations 451
preferential pathways 390
preservation through use 338
pressure head 394
primary forest 117, 347, 387, 413
primary production 50
process 49
   biogeochemical 203
   chemical 395
   chemical reduction 212
   degradation 61
   geomorphologic 320
   nocturnal circulation 451
   redoximorphic 7
productivity 462
propagation technique 429
properties
   hydraulic 256
protection by use 343
provenance 418
Purdiaea nutans
   forest 12

Q

quartzites 398

R

R 389
radiation 130, 270
radiation balance 255
radiocarbon 102
rain 304
rainfall 8, 70, 130, 194, 205
   anomalies 282
   incident 207
rain rate 71
rainstorm 197, 213

rainy season 451
range
　size 279
　spatial 158
Rapoport's rule 52, 456
Rapoport effect 51
rarefaction 444
rarefied species number 158, 271
ravine forest 117
RBSF 1
RC 218
re-trapping 159
recovery 375, 381
recruitment 313
redoximorphic process 7
Red Queen Hypothesis 184
reductive dissolution 80
redundancy of species 458
reforestation
　abandoned pastures 463
　native species 463
reforestation 52, 374, 417, 431
regeneration 61, 315, 319, 327, 361,
　　366, 409, 443, 460
　biodiversity 460
　natural 432
　patterns 328
regosol 6
rehabilitation 61, 418
remote areas 303
reproductive material 418
Reserva Biologíca San Francisco 1
residence time 210
resource balance hypothesis 241
resource limitation 51, 188, 457
resources
　natural 331
respiration 355
　root 220
　soil 218, 355, 371
rhizoids 147, 365
rhizome 371
ridge 267, 276
　forest 118, 317
　top 83
ridge-top forest 277
Rio San Francisco 1
risk
　from exotic species 435
　index 387
RLD 264
road construction 361

rock
　paleozoic metamorphic 4
root 232
　contribution 218
　length density 264
　respiration 220
root/shoot ratio 51, 58, 229, 237
runoff 212

S

Sabanilla unit 4
sandstones 398
sap flow 243
　density 247
sap wood area 243
Saraguro 29, 462
Saraguros 12, 25, 348, 478
savanna 10
scar area 325
scatterometer 304
sea salt 309
season 64, 285, 453
seasonality 291, 419, 457
　growth 453
secondary forest 39, 367, 409
　regeneration stages 384
　types 409
sediment archive 101
seed
　centers 429
　collection 425
　dispersal 159, 162
　disperser 157, 420
　ecology 417
　harvesting calendar 429
　procurement 417
　reservoir 370
　sources 417, 435
seedling 426
semi-deciduous forest 10
Shannon-Wiener index 271
sheet erosion 6
shrub páramo 90
shrubs 16
Shuar 12, 25, 26, 462
Sierra 4
silvicultural systems 348
silvicultural treatment 349
silviculture 462
similarity analysis 125
simulation of water flow 389

sinapyl alcohol 76
skidding 350
slash and burn 371, 462
slash burning 399
slates 398
slope 76
slope forest 244, 276
smoke 307
SOC 218
social status 297
sodium 452
Soerensen indices 411
soil 6, 58, 193, 375
   chemical properties 462
   CO_2 efflux 218
   core 105, 107, 109
   creeping 127
   fertility 78, 271, 319, 455
   food web 183
   heterogeneity 76, 387
   hydrology 375
   layers 381
   microarthropods 12
   microorganisms 454
   mineral 76, 211, 265
   moisture 132, 394
   nitrogen 222
   nutrient hypothesis 457
   organic carbon 222, 399
   organic matter 457
   organisms 454
   physical properties 375
   profile 389
   properties 75, 260, 325
   reaction 6, 397
   stony forest 387
   structure 387
   temperature 135, 313, 455
   texture 376, 387
   water 455
soil respiration
   heterotrophic 220
   total 218
SoPhy 389
south Pacific anticyclone 7
space-for-time substitution 368
Spanish conquests 111
spatial logistic regression 323
spatial structure 377
species
   composition 91, 126, 154, 177, 187, 317, 369, 383, 447, 453, 459
   description 16, 43
   distribution 149

   diversity 59, 320
   exotic 419, 435
   forest types 153
   formation 157
   identification 15
   indigenous 419
   introduced 417
   native 417
   shared' index 444
   turnover 19, 174, 448, 456
species richness 18, 41, 50, 87, 100, 128, 328
   estimation 42
   patterns 19
spore 107
stand
   dynamics 291
   evapotranspiration 256
   structure 229, 235, 243
   transpiration 243
statistics
   extreme value 388
steady state cycle 323
steepness 6
stem density 231
stemflow 195, 209
stem wood density 232
stomata 247
storage 78
   capacity 194
storm 212
strato-volcanoes 4
stream water 212
stress 203
strontium 269
structural analysis 413
stunted forests 251
Sub-Andean zone 4
sub-páramo 98, 105, 456
subpopulation 427
subsoil 212
succession 164, 319, 328, 364, 397, 409, 460
   habitats 53
   sequence 373
   stages 366
   vegetation 409
sulfate 305
sulfur 260, 452
surface flow 194
survival
   of trees 436
sustainability 348
symbiosis 137

synecological knowledge 157
synoptical circulation 72
systematics 15

T
tail distribution 388
taxonomy 15, 44
temperature 55, 64, 130, 132, 174, 270, 313
   adiabatic 58
   air 7
   anomalies 282
texture 79
TF 267
thermal physiology 51
throughfall 195, 209, 267
tierra caliente 7
tierra fria 7
tierra helada 7
tierra nevada 7
tierra templada 7
timberline 123, 453
timberline ecotone 10
TN 218, 399
TOC 399
top-down 188
TOPMODEL 194
topographic position 76
topography 2, 267
topsoil 462
total soil N 399
trajectories 308
transects
   permanent 313
translation slides 323
transpiration 51, 193, 243
   annual 200, 255
   daily 254
   leaf 246
   plot 244
   stand 253
   total tree 251
   understorey 253
transport
   of nutrients 395
   of pollutants 395
   of solutes 395
trapping 159
treatment
   silvicultural 349
tree
   age 246
   fall 90, 313

   growth 260
   nitrogen-fixing 436
   potential crop 349
   regeneration 313
   ring chronology 453
tree line 10, 453
   ecotone 59
   upper 51
tree species
   exotic 463
   multipurpose 462
   native 463
trenching 218
tributary slides 324
tropical trade wind regime 7
turnover time 7, 203, 325
type material 44
Typic Dystropepts 325

U
ultrastructure 142
understorey 159
up-scaling 253
upper mountain forest 275
UV-B 132

V
valley-breeze system 452
valley bottom 83
valley forest 317
vapour pressure deficit 67, 251
variability
   intraspecific 424
   spatial 84
variogram 383
vascular pioneers 366
vascular plants 41, 453, 455
   abundance 364
vegetation 10
   cover 327
   development 107, 363
   embankment 362
   islands 327
   past 101
   potential 87
   potential natural 10
   regeneration 319
   structure 113, 149
   types 10, 88, 89, 280, 328
Velamen radicum 366
Veranillo del Niño 133, 285
vertical dimension 456

Vilcabamba 12, 56
vines 246
Visual Plants 46
volcanism 309
volume-weighted mean 204
VPD 67, 314
VSC 76
VWM 204

W
water
  budget 196, 198
  conductivity 197
  flow 387
  intake 71
  level 200
  relations 193
watershed 193
weathering 212
weed 13
weir 194
wildfires 127

wind 128, 133
  direction 68
  velocity 67

X
xylem hydro-activity 253

Y
Yalca 10

Z
Zamora 55
Zamora batholith 4
zonal type 59
zonation 452
zonation models 120
zones
  gene-ecological 418
  vertical 72
zonobiome 456

Taxonomic Index

A

Ageratina dendroides 370, 373
Alnus acuminata 300, 436, 463
Alzatea verticillata 139
Amerioppia 184
amphibians 18
Andropogon bicornis 366
Aneuraceae 139
ants 22
Araceae 22
Arachnida 17, 189
Archaea 16
Arctiidae 167, 353, 443, 460
arctiid moths 455
Arecaceae 22
arthropod 17, 353
Artibeus phaeotis 159
Ascomycota 138
Asplenium 91, 352
Astigmata 189

B

Baccharis genistelloides 368, 370
Baccharis latifolia 370, 373
bacteria 16
bamboo 127
Basidiomycota 137
bat 12, 16, 59, 157, 455
beans 371
Bejaria aestuans 368
bird 12, 16, 22, 59, 149, 157, 454, 461
Blechnum 109
Blechnum cordatum 366
Blechnum schomburgkii 279
Brachyotum 370
bracken fern 371, 436, 461
bryophyte 16, 21, 87, 91, 275, 453
bush cricket 16

C

Calypogeia 365
Campyloneurum amphostenon 146
Carollia brevicauda 159
Cecropia montana 247
Cedrela 342, 348, 465
Cedrela montana 292, 420, 438
Cephalozia 365
Ceradenia farinosa 146
chiroptera 157
Chusquea 460
Cladonia cf. cervicornis ssp. *vertilicillata* 362
Clethra 107
Clethra revoluta 292, 420, 433
Clusia 95, 107
Clusia cf ducuoides 295, 349
Cochlidium serrulatum 146
Coleoptera 17
Colocasia esculenta 335
Cortaderia cf. jubata 362
Critoniopsis floribunda 139
cryptogams 16, 275, 453, 457
Cupressus 417, 435
Cyathea conjugate 107
Cyathea peladensis 279

D

Dictyonema glabratum 362
diplopods 181
diptera 17

E

earthworms 181
Elaphoglossum 91, 352
Elleanthus aurantiacus 366
Eois 45

E

Epidendrum carpophorum 366
Epidendrum lacustre 366
ericads 137
Escallonia paniculata 366
Eucalyptus 13, 417, 437
Eucalyptus saligna 436

F

Fabaceae 22
fern 16, 91, 137, 275
Ficus 91, 292
Ficus subandina 297, 349
Formicidae 189
fungi 16, 137
Fuscocephaloziopsis subintegra 280

G

Gamasina 184
Gaultheria erecta 366
Geometridae 22, 44, 167, 443, 454, 460
geometrid moths 16, 18, 22, 455
Gleicheniaceae 327
Glomeromycota 15, 16, 137, 362
Graffenrieda emarginata 90, 142, 296, 404, 423, 433
Grammitidaceae 139
Grammitis paramicola 146
Gynoxis 10, 453

H

Hedyosmum 103
Heliocarpus americanus 91, 292, 315, 420, 436
Huperzia 103
Hyeronima asperifolia 91, 298, 349
Hyeronima moritziana 298, 349
Hymenomycetes 137
Hymenoptera 17
Hymenoscyphus ericae 144

I

Ilex 97, 105
Inga 91, 292, 342, 420
Inga acreana 137, 298, 349
insect 15, 43, 176, 457
Isertia laevis 292, 420
isopods 181

J

Jensenia spinosa 146
Juglans neotropica 436

L

Lauraceae 21
Lellingeria subsessilis 146
Lepidoptera 17, 45, 176, 443, 459
liana 127, 457
lichen 16, 275
liverwort 137, 279, 362
Lophocolea 365
Lycopodiaceae 328
Lycopodiella glaucescens 328, 362
Lycopodium 105
Lycopodium clavatum 107
Lycopodium magellanicum 362

M

Macrocarpia 459
Macromitrium perreflexum 280
maize 371
mammal 12
Manihot esculenta 335
Marcgraviaceae 459
Marchantia 365
Marchantia chenopoda 362
Melinis minutiflora 362, 371
Melpomene assurgens 146
Melpomene pseudonutans 146
microarthropods 12, 182
Micropholis guyanensis 139
Micropolypodium sp 146
mites 181
molluscs 17
Monochaetum lineatum 368, 373
mosses 279, 455
moth 12, 16, 45, 167, 348, 443, 454
Musa sp. 335
Myrica pubescens 292, 368, 420
Myrsine 105

N

Nectandra acutifolia 139
Nectandra membranacea 91, 298, 349

Nematoda 188
Niphidium carinatum 146
Nyctaginaceae 141

O
orchid 21, 100, 137
oribatid mites 16, 59, 181, 454
Orthoptera 17

P
Pallaviciniaceae 139
Paracromastigum bifidum 365
Pergalumna sura 187
Phyllostomidae 157
Pinus 417, 435
Pinus patula 13, 436
Pinus radiata 435
Piper 21, 91, 100
Piptocoma discolor 91, 292, 315, 342, 420, 453, 456
Platyrrhinus infuscus 159
Pleopeltis percussa 146
Pleurozia 97
Pleurozia heterophylla 280
Pleurozia paradoxa 280
Podocarpaeceae 425
Podocarpus 105
Podocarpus oleifolius 90, 139, 295, 348, 439
Pogonatum tortile 362
Polylepis 10, 123, 453
Polypodiaceae 139
Prostigmata 189
Prumnopitys montana 300, 424, 439, 454
Pselaphidae 184
Psydium guayava 342
Pteridium aquilinum 342, 404
Pteridium arachnoideum 371, 461
pteridophytes 278
Pterozonium brevifrons 279
Purdiaea forest 276
Purdiaea nutans 90, 99, 107, 275, 295, 314, 420, 453
Pyraloidea 167
pyraloid moth 16, 22, 168, 454

R
Riccardia amazonica 362

S
Scelochilus 456
Scheloribates 187
Sebacinales 139
Setaria sphacelata 371, 404
spermatophytes 88
Staphylinidae 184
Stelis 100
Stereocaulon sp. 362
Sticherus rubiginosus 366
Sticherus spp 362
Sturnira bidens 159
Sturnira erythromos 159
Sturnira ludovici 159
Symphyogyna 365
Symphyogyna brongniartii 365
Symplocos 107

T
Tabebuia chrysantha 56, 91, 139, 292, 342, 348, 420, 436, 463
Teagueia 456
Tectona grandis 417
Testacea 16, 188
Tettigoniidae 16
Thelephora 141
Tibouchina laxa 373
Tibouchina lepidota 366
Tomentella 141
Truncozetes sturmi 187
Tulasnella 139

U
Uropodina 184

V
Viburnum sp. 420
Vismia 459
Vismia tomentosa 292, 420, 456
Vittaria 352

W
Weinmannia 94, 103, 127
Weinmannia loxensis 238

Z
Zea mais 107

Printing: Krips bv, Meppel, The Netherlands
Binding: Stürtz, Würzburg, Germany